This book addresses the molecular bases of some of the most important biochemical rhythms known at the cellular level. Clarifying the mechanism of these oscillatory phenomena is of key importance for understanding the origin as well as the physiological function of these rhythms, and the conditions in which simple periodic behaviour transforms into complex oscillations, including bursting and chaos.

The approach rests on the analysis of theoretical models closely related to experimental observations. Among the main rhythms considered are glycolytic oscillations observed in yeast and muscle, oscillations of cyclic AMP in *Dictyostelium* amoebae, intracellular calcium oscillations observed in a variety of cell types, the mitotic oscillator that drives the cell division cycle in eukaryotes, and circadian oscillations of the *period* protein (PER) in *Drosophila*. For each of these phenomena, experimental facts are reviewed and mathematical models presented. These models throw light on the mechanism of periodic behaviour at the molecular and cellular levels and show how enzyme regulation or receptor desensitization can give rise to oscillations. The theoretical models are also used to analyse the transition between simple periodic oscillations and more complex oscillatory phenomena such as birhythmicity (i.e. the coexistence between two stable rhythms), bursting oscillations and chaos. The model for cyclic AMP oscillations in *Dictyostelium* is further used to discuss the function of pulsatile signalling in intercellular communication. The frequency encoding of cyclic AMP signals is related to the fact that many hormones are secreted in pulses whose frequency governs their physiological effect.

This book, which contains more than 1200 references, provides a wide survey of work on biochemical oscillations and cellular rhythms. The author has made numerous contributions to the subject since the early developments in the field.

The book will be of interest to life scientists such as biochemists, cell biologists, chronobiologists, medical scientists and pharmacologists. In addition, it will appeal to scientists studying nonlinear phenomena, including oscillations and chaos, in chemistry, physics, mathematics and theoretical biology.

# BIOCHEMICAL OSCILLATIONS
# AND CELLULAR RHYTHMS

# BIOCHEMICAL OSCILLATIONS AND CELLULAR RHYTHMS

## The molecular bases of periodic and chaotic behaviour

ALBERT GOLDBETER

*Faculté des Sciences, Université Libre de Bruxelles*

CAMBRIDGE
UNIVERSITY PRESS

PUBLISHED BY THE PRESS SYNDICATE OF THE UNIVERSITY OF CAMBRIDGE
The Pitt Building, Trumpington Street, Cambridge CB2 1RP, United Kingdom

CAMBRIDGE UNIVERSITY PRESS
The Edinburgh Building, Cambridge CB2 2RU, United Kingdom
40 West 20th Street, New York, NY 10011-4211, USA
10 Stamford Road, Oakleigh, Melbourne 3166, Australia

First published in French in 1990 by Masson, Paris,
as *Rythmes et Chaos dans les Systèmes Biochimiques
et Cellulaires* and © Masson 1990

This revised and enlarged edition published in 1996 by
Cambridge University Press as *Biochemical
Oscillations and Cellular Rhythms*
First paperback edition 1997

Typeset in Times 10/13 pt

*A catalogue record for this book is available from the British Library*

*Library of Congress Cataloguing in Publication data*

Goldbeter, A.
[Rythmes et chaos dans les systemes biochimiques et cellulaires.
English]
Biochemical oscillations and cellular rhythms : the molecular
bases of periodic and chaotic behaviour / Albert Goldbeter.
p.   cm.
Includes bibliographical references and index.
ISBN 0 521 40307 3 (hc)
1. Biological rhythms.   2. Oscillating chemical reactions.
3. Cell Physiology.   4. Chaotic behaviour in systems   I. Title.
QP84.6.G6513   1996
574.1′882 – dc20   95–10166 CIP

ISBN 0  521 40307 3 hardback
ISBN 0  521 59946 6 paperback

Transferred to digital printing 2004

WS

Avez-vous visité les hauts laboratoires,
Où l'on poursuit, de calcul en calcul,
De chaînon en chaînon, de recul en recul,
A travers l'infini, la vie oscillatoire?

Emile Verhaeren
*Les Forces Tumultueuses*, 1902

# Contents

## Contents

# Foreword to the English edition

All life is rhythmical. Just to feel the human pulse is to marvel at how the heart beats out its regular rhythm. It is a sobering thought that, for the whole of our lives, a small group of pacemaker cells dutifully send out a signal every second or so to drive each heart beat. For someone living to 70 years with a heart rate of 60 beats/minute, we can calculate that this oscillator will have faithfully transmitted well over two billion beats! Such rhythmicity occurs in many other cell types especially when they are being stimulated. For example, a liver cell generates regular pulses of calcium when it responds to the hormones that signal the breakdown of glycogen. Another striking example is the burst of spikes that occur in mammalian eggs following fertilization. In humans, as in other mammals, the egg remains quiescent for many years until it is coaxed into life by the train of spikes triggered by the arrival and attachment of the sperm. Exactly the same mechanisms are employed to generate the calcium spikes observed in liver cells and eggs. On a separate time scale, we are all aware of our circadian clock through its ability to drive our diurnal sleep–wake cycle. The strength of this circadian oscillator is all too obvious to jet travellers as they cross time zones. A small population of neurons located within a tiny region of the brain function as a 'clock' that becomes active at exactly the same time each day. One of the outstanding unsolved problems in biology concerns the biochemical nature of this clock. How is it that cells can measure a time interval of approximately 24 hours with such accuracy?

All these examples illustrate the enormous variety of different kinds of biological rhythms that have provided such an irresistible fascination to Albert Goldbeter. He revels in their complexity, which is further compounded by the enormous plasticity of these oscillatory mechanisms. This plasticity is clearly evident for some of the examples

xv

described above. We are all familiar with the heart speeding up when we become excited. The frequency of the calcium spikes in liver cells can vary depending on hormone concentration. And we also know that we recover from jet lag as our circadian clock gradually resets to the new time zone. This flexibility of oscillator frequency further highlights the complexity of these cellular chronometers.

The only way to cope with all the parameters that determine such oscillatory activity and its complex modulation is to formulate mathematical models, which is the subject of this book. At this stage, I must confess an earlier scepticism concerning the value of such modelling. In 1970 I began to record oscillations in the membrane potential of an insect salivary gland. I was immediately captivated by the way the cell spontaneously emitted very regular pulses and I set about formulating a working model that I was very proud of because it seemed to incorporate everything we knew then about how calcium signalling regulated the activity of these salivary gland cells. My pride was somewhat dented when I showed my ideas to the mathematician Paul Rapp, who soon pointed out that my model mitochondria would explode within seconds. At that point I began to appreciate the value of good modelling, especially now that we have learnt so much more about the basic biochemical details regulating cellular activity. For each system now being investigated, there are so many variables that it becomes impossible to use our intuition to assess how each parameter influences the oscillatory cycle. The only way to understand these biological rhythms, therefore, is to become more quantitative and to develop rigorous mathematical models. Albert Goldbeter is at the very forefront of this new approach.

Having been converted to the importance of modelling I was most enthusiastic about accepting an offer to join Albert Goldbeter in trying to model calcium spiking and waves. Everyone who knows Albert will be familiar with his infectious enthusiasm, which pervades all his work and soon captivates all his colleagues and collaborators. There is nothing he enjoys more than a vigorous discussion on oscillatory mechanisms. At times such discussions often turn into intense interrogation sessions as Albert searches and probes for bits of information.

What is particularly impressive about his work is that his modelling is always firmly based on a very detailed understanding of the basic biology underlying each oscillatory system. This deep biological understanding extends across the very wide range of subjects covered in this book. He delves into the intricacies of glycolysis, the aggregation of slime moulds, the generation of calcium waves, the cell cycle oscillator and circadian rhythms. In analysing these diverse systems he seeks to uncover the basic molecular mechanisms responsible for all these rhythmical processes. Anyone reading this book not only will learn about the mathematics of oscillatory systems but also will come away with an up-to-date synopsis of all these biological systems that are currently at the very forefront of modern biology.

*Michael J. Berridge*

The Laboratory of Molecular Signalling,
The Babraham Institute, Department of Zoology, Cambridge

# Foreword to the French edition

It is a privilege to introduce the book of my friend and colleague Albert Goldbeter.

Our century has witnessed basic changes in direction in science, and foremost among them I would place the 'rediscovery of time'. Only a few decades ago, most of the research efforts in physical chemistry were devoted to equilibrium states. Not only did the notion of equilibrium dominate chemistry, but it was also considered to be the fundamental paradigm for the extension of scientific ideas to other fields, including economics and social sciences. The arrow of time was considered to be an artefact reflecting human ignorance. Today, the situation has radically changed. Everywhere, from astrophysics to biology, the essential role of nonequilibrium is well recognized.

It is indeed far from equilibrium that we encounter self-organization processes that lead to the breaking of temporal and spatial symmetries as well as to the appearance of coherent behaviour on a supramolecular scale. Chemical oscillations provide a particularly striking example. I remember my astonishment when I was shown for the first time a chemical oscillatory reaction, more than 20 years ago. I still think that this astonishment was justified, as the existence of chemical oscillations illustrates a quite unexpected behaviour. We are used to thinking of molecules as travelling in a disordered way through space and colliding with each other according to the laws of chance. It is clear, however, that this molecular disorder may not give rise by itself to supramolecular coherent phenomena in which millions and millions of molecules are correlated over macroscopic dimensions. These supramolecular correlations permit the molecules to change their identity as a result of chemical reactions in a synchronized fashion. It is this synchronization that breaks temporal symmetry. Near equilibrium, two instants are equivalent; this is no longer the case when the evolution of the system becomes periodic.

By now we know a large number of oscillatory reactions, not only in chemistry – as exemplified by the famous Belousov–Zhabotinsky reaction – but also in biochemical and cellular systems. It is interesting to observe that in oscillations in inorganic chemistry the molecules are simple but the mechanisms are highly complex, whereas in biochemistry the molecules responsible for rhythmic phenomena possess a complex structure (enzymes or receptors ...) whereas the mechanisms often are simple.

Albert Goldbeter was associated from the beginning with our studies on periodic chemical processes. He focused, particularly, on two examples that remain today among the most popular: glycolytic oscillations in yeast, and the periodic synthesis of cyclic AMP in cellular slime moulds. Indeed, when at the end of the sixties we took up the study of dissipative structures far from equilibrium, we immediately thought that an interesting application would be the study of biochemical oscillations. One of the first works devoted to dissipative structures was an article published with R. Lefever, A. Goldbeter and M. Kaufman in *Nature* in 1969 (vol. **223**, pp. 913–16). But what a spectacular development there has been over these last two decades! I just stressed the fact that in biochemistry the mechanism of oscillatory reactions is often simple. This is well illustrated in this book by Albert Goldbeter, as he reaches truly remarkable conclusions about a huge variety of rhythmic phenomena, using models involving only two independent variables. The simplicity of these models allows him to go very far in analysing the molecular basis of biochemical rhythms and to investigate numerous questions such as the role of enzyme cooperativity.

Within this restrictive framework of two-variable models, Albert Goldbeter derives fascinating original results such as birhythmicity, which allows a system to choose between two simultaneously stable oscillatory regimes. With the number of variables, the repertoire of dynamic phenomena increases rapidly. Now, besides simple periodic behaviour we can also predict and observe complex oscillations of the bursting type, the coexistence between more than two rhythms, or the evolution toward chaos. As the author shows, small variations in the values of some control parameters permit the switch from one mode of behaviour to the other. The essential elements, in all cases, are the feedback mechanisms of biochemical reactions and the fact that these reactions occur far from equilibrium.

This book is an excellent introduction to some fundamental problems of biochemistry and biology. The work of Goldbeter will certainly play a major role in the developments that we may expect to occur in the near future. First, there is the temporal aspect of intercellular communications. I find it fascinating that there exist optimal periodicities in these communication processes, and I would like to emphasize the importance of the model developed in chapter 8, which relates the existence of these optimal periodicities to the kinetics of desensitization and resensitization of receptors. Such models make us grasp the importance of rhythms at all levels of functioning of living systems, from metabolism and hormonal regulation to the nervous system. The clinical implications of these results are also considered by the author, who suggests that the pulsatile administration of certain drugs at the appropriate frequency could represent a new approach to chronopharmacology.

Another essential question is the ontogenesis of biological rhythms. Besides the problem of the emergence of rhythms in the course of development, there are the questions of the origin of the sensitivity of the organism to pulsatile signals, and of the mutations that, on a molecular scale, have permitted the selection of optimal frequencies. Conversely, studying the ontogenesis of rhythms may contribute to elucidating some of the mechanisms of evolution.

The classical picture of a physical world at equilibrium was well illustrated by the phase diagrams that are found in textbooks. There we can discern the regions of existence of the various states of matter: liquid, solid, vapour. But what additional richness is there when we examine nonequilibrium phase diagrams such as those appearing here! One distinguishes regions of simple oscillations, regions of bursting or birhythmicity, and routes to chaos. The diversity of dissipative phenomena contrasts with the universality of equilibrium. When we lower the temperature of a substance, the latter necessarily passes through the gaseous, liquid and solid states. But when we subject a nonlinear system to nonequilibrium constraints, we obtain limit cycles, steady states, chaotic behaviour, or complex oscillations, without any emerging universal scheme. It is precisely in this fashion that nonequilibrium phenomena reflect the diversity of the world in which we live.

I cited one of the first collaborative papers of Albert Goldbeter, published in 1969. Since then he has produced an imposing body of work, which reflects creative imagination together with a constant concern for the dialogue between theoretical concepts and experimental research. It is an exemplary work in that it emphasizes, in an important domain of biology, the link which exists between molecular phenomena on the one hand and global behaviour on the other. This volume summarizes efforts pursued over many years, but it also opens up new research themes and ever more fascinating conclusions. This book will be of interest to a wide audience, from chemists and physicists to biologists and medical scientists. I wish it the great success it deserves.

*Ilya Prigogine*
Nobel laureate in Chemistry

International Solvay Institutes
for Physics and Chemistry, Brussels

# Preface

This book grew out of a passion for rhythmic phenomena in biological systems, which has impelled my work for 25 years. I had the chance of being introduced to that field when I joined the group of Ilya Prigogine at the University of Brussels, for what I thought would be a one-year experience at the end of my studies in chemistry. I was rapidly hooked, however, and still have the pleasure of pursuing my research there. I would like to seize this opportunity to express my gratitude to Ilya Prigogine, who kindly agreed to write a preface (reproduced here as a foreword) for the French edition of this book; I have been marked by the enthusiasm that drives his incessant questioning of the world. The atmosphere in Brussels has always been ideal for developing my work; in addition to Ilya Prigogine and Paul Glansdorff, I am indebted to Grégoire Nicolis and René Lefever for making this friendly intellectual climate so fruitful.

This book owes much to the graduate students and co-workers with whom I had the pleasure of collaborating over the years: Jean-Louis Martiel, Olivier Decroly, Federico Morán, Yue-Xian Li, Geneviève Dupont, and José Halloy. They made decisive contributions to the various topics considered in this book, by actually carrying out much of the research described. Special thanks are due to Thomas Erneux, with whom I am always happy to pursue new ventures.

Collaborations outside Brussels have brought me some of my fondest experiences: Benno Hess and Arnold Boiteux in Dortmund, Roy Caplan and Lee Segel at the Weizmann Institute in Rehovot, and Dan Koshland in Berkeley – the latter two of whom, to my great enjoyment, I continue to work with – have enriched my life experience and influenced my views in many ways. The same is true for all the fruitful contacts I had over the years with numerous scientists, whom I shall not list here, many of whom became friends. More recently I was lucky to interact with Michael Berridge, from Cambridge, at an early stage of the wave of investigations on intracellular calcium oscillations. I am grateful to Michael for writing the Foreword to this volume.

Five years have passed since the publication of the French version of this book. The text has been largely expanded in this English version, which contains three new chapters – on calcium oscillations, the mitotic oscillator, and circadian rhythms – as well as a more comprehensive list of references. All other chapters have been revised and updated.

My work over the years has been supported by a number of organizations among which the Belgian National Fund for Scientific Research (FNRS), the European Molecular Biology Organization (EMBO), the Belgian Fund for Medical Scientific Research (FRSM), the Scientific Policy Unit of the Prime Minister's Office (SPPS), and the Division of Scientific Research, Ministry of Science and Education, French Community of Belgium, in the framework of the programme 'Actions de Recherche Concertée' (ARC 94-99/180).

I would like to thank all authors who kindly gave me permission to reproduce figures from their work. My thanks go also to Simon Capelin, who has been my contact editor at Cambridge University Press, and to Sandi Irvine, thanks to whom the style of this book is far better than it would have been without her patient efforts. Both of them, at CUP, provided the support necessary to carry this project to its end.

The painting reproduced on the jacket of the book is due to Gottfried Wiegand, whom I wish to thank for allowing me to use it for the cover. The painting is entitled *Doppel-Handler*, which means, literally, the man who does two things at the same time with his hands. Besides having a direct link to the phenomena and curves described in the book, this painting also illustrates the deep underlying movement that I have attempted to follow, between experiment and theory. In a way I have tried to be like the man on the painting, on the one hand analysing experimental data and, on the other, constructing theoretical models, in a never-ending feedback linking the two complementary approaches to investigating the molecular bases of biological rhythms.

*Albert Goldbeter*

Université Libre de Bruxelles

# 1

# Introduction

## 1.1 Chemical oscillations and biological rhythms

Rhythms are among the most conspicuous properties of living systems. They occur at all levels of biological organization, from unicellular to multicellular organisms, with periods ranging from fractions of a second to years (Table 1.1). In humans, the cardiac and respiratory functions and the circadian rhythm of sleep and wakefulness point to the key role of periodic processes in the maintenance of life. In spite of their close association with physiology, however, periodic phenomena are by no means restricted to living systems. Since the discovery of the oscillatory chemical reaction of Belousov and Zhabotinsky (Belousov, 1959; Zhabotinsky, 1964; Tyson, 1976), it has become clear over the last three decades (Nicolis & Portnow, 1973; Noyes & Field, 1974; Pacault *et al.*, 1976; Epstein, 1983, 1984; Field & Burger, 1985) that numerous chemical reactions studied in the laboratory display periodic behaviour in the course of time. Biological rhythms nevertheless remain more common than oscillations in purely chemical systems.

However, oscillatory behaviour does not always possess a simple periodic nature. Thus, both in chemistry and biology, oscillations sometimes present complex patterns of bursting, in which successive trains of high-frequency spikes are separated at regular intervals by phases of quiescence. Yet another mode of complex oscillatory behaviour, which has received increased attention over the last decade, is characterized by its aperiodic nature and sensitivity to initial conditions. Such chaotic oscillations have been observed in chemical reactions (Scott, 1991; Field & Györgi, 1993) and in a variety of biological contexts (Olsen & Degn, 1985; Holden, 1986; Glass & Mackey, 1988).

What is the link between chemical oscillations and periodic phenomena observed in living systems? Is it possible to uncover the molecular

1

Table 1.1. *A list of the main biological rhythms, classified according to increasing period. Rhythms marked by an asterisk occur at the cellular level (sometimes they may also arise from interactions between cells, e.g. in neural networks)*

| Rhythm | Period |
|---|---|
| Neural rhythms* | 0.01 to 10 s (and more?) |
| Cardiac rhythm* | 1 s |
| Calcium oscillations* | 1 s to several minutes |
| Biochemical oscillations* | 1 min to 20 min |
| Mitotic cycle* | 10 min to 24 h (or more) |
| Hormonal rhythms* | 10 min to several hours (also 24 h) |
| Circadian rhythms* | 24 h |
| Ovarian cycle | 28 days (human) |
| Annual rhythms | 1 year |
| Epidemiology and ecological oscillations | years |

mechanisms and the physicochemical bases of rhythmic processes in biology? As explained in detail later in this chapter, the main goal of this book is to address the molecular bases of periodic and chaotic behaviour by considering theoretical models for some of the best-known examples of cellular oscillations of a biochemical nature. Elucidating the molecular bases of biochemical and biological rhythms first requires the identification of the types of molecule involved and of their regulatory interactions. An isolated enzyme or receptor will never produce any oscillations. It is the network of molecular species, controlled by positive or negative feedback loops and driven by a flow of matter, that gives rise to periodic or chaotic behaviour. The oscillatory mechanism is only clarified when a theoretical model based on these observations successfully accounts for the observed periodic or chaotic phenomenon. Such a model permits a better comprehension of the molecular mechanism as it allows us to determine in a detailed manner the effect of each parameter on oscillatory behaviour.

Dealt with in turn are: (i) glycolytic oscillations that occur in yeast and muscle cells, with a period of several minutes; (ii) oscillations of cyclic AMP (cAMP) that govern with a similar period the transition from the unicellular to the multicellular stage in the life cycle of *Dictyostelium* amoebae; (iii) intracellular $Ca^{2+}$ oscillations, which occur with a period ranging from seconds to minutes in many cell types, either spontaneously or after stimulation by a hormone or a neurotransmitter;

(iv) the mitotic oscillator that drives the cell division cycle in amphibian embryos; and (v) the biochemical oscillator that governs circadian rhythms in *Drosophila*.

Extensions of the model for glycolytic oscillations are moreover used to analyse the transition from simple to complex oscillatory behaviour, including chaos, bursting and the coexistence between two simultaneously stable rhythms (birhythmicity). The model for cAMP signalling in *Dictyostelium* also serves to address this question as well as the ontogenesis of cAMP oscillations in the course of development and the function of pulsatile signalling in intercellular communication. The latter topic is of physiological and clinical interest as it bears on the role of the pulsatile secretion observed for a large number of hormones and on the search for optimal patterns of drug delivery. In addition to temporal oscillations, which constitute the main theme of this book, the origin of closely related spatiotemporal patterns in the form of cAMP or $Ca^{2+}$ waves is also considered.

Although they represent an important class of rhythmic behaviour at the cellular level, oscillations of the membrane potential that originate in nerve and muscle cells from the interplay between several voltage-dependent membrane conductances are not addressed. This book focuses on oscillations that occur within cells as a result of: feedback processes involving various modes of enzyme regulation (e.g. glycolytic oscillations and the biochemical cascade controlling the periodic onset of mitosis), receptor–enzyme interactions (as in the case of pulsatile cAMP signalling in *Dictyostelium*), transport between the cell and its environment or between various intracellular compartments (as for $Ca^{2+}$ oscillations), or the control of transcription (as in the case of circadian oscillations of the *period* protein in *Drosophila*).

Based on experimental observations, the theoretical models considered throw light on the origin of simple periodic phenomena and complex oscillations, including chaos, in biochemical systems. Besides illustrating the variety of molecular mechanisms producing oscillations at the cellular level, the models allow us to delineate in a qualitative and quantitative manner the conditions in which sustained oscillations occur. The theoretical approach thereby contributes to a better understanding of the physiological roles of such rhythms. Furthermore, in such an approach the relative frequency of occurrence of simple versus complex patterns of oscillations becomes amenable to quantitative assessment. Finally, theoretical models for biochemical and cellular rhythms allow us to relate these phenomena to the different modes of simple or complex oscillatory behaviour observed in neurobiology and in the nonequilibrium dynamics of chemical systems.

## 1.2 Rhythms as limit cycles and temporal dissipative structures

Until the 1950s, the rare periodic phenomena known in chemistry, such as the reaction of Bray (1921), represented laboratory curiosities and elicited little but incredulity. Some oscillatory reactions were also known in electrochemistry (see Wojtowicz, 1973; Koper, 1992). The link between these phenomena and the cardiac rhythm had intrigued some, while others, such as van der Pol & van der Markt (1928), stressed the relationship between that rhythm and electrical oscillators. The first kinetic model for oscillatory reactions was analysed by Lotka (1920, 1925), while similar equations were proposed soon after by Volterra (1926) to account for oscillations of predators and preys in ecology. The next important advance came from neurobiology, with the experimental and theoretical studies of Hodgkin & Huxley (1952), which clarified the physicochemical bases of the action potential in electrically excitable cells. The theory that they developed was later applied (Huxley, 1959) to account for sustained oscillations of the membrane potential in these cells. Remarkably, the classic study by Hodgkin & Huxley (1952) appeared in the very same year as Turing's pioneering analysis of spatial patterns in chemical systems.

The approach of periodic phenomena in physicochemical terms made further progress when Prigogine & Balescu (1956) showed that sustained oscillations can occur far from thermodynamic equilibrium in open chemical systems governed by appropriate, nonlinear kinetic laws. These results were extended by the analysis of abstract models of oscillatory reactions, such as the **Brusselator**, which name was given by Tyson (1973) to the theoretical model studied in detail in Brussels by Lefever, Nicolis & Prigogine (1967). As shown by these studies, chemical oscillations can occur at a critical distance from equilibrium, around a steady state that has become unstable owing to the presence of autocatalytic steps in the reaction kinetics (Lefever *et al.*, 1967; Prigogine, 1967, 1969; Lefever & Nicolis, 1971; Nicolis & Prigogine, 1977; Lefever, Nicolis & Borckmans, 1988).

In the phase space formed by the concentrations of the chemical variables involved in the reaction, sustained oscillations correspond to the evolution towards a closed curve called a **limit cycle** by Poincaré, because of its uniqueness and independence from initial conditions (Minorsky, 1962). The time taken to travel along the closed curve represents the period of the oscillations. When a single limit cycle exists, the system always evolves towards the same closed curve characterized by a fixed amplitude and period, for a given set of parameter values, regardless of the initial state of the system. It is in this sense that oscillations of the limit cycle type differ from Lotka–Volterra oscillations, for which an infinity of closed curves, corresponding to oscillations of different periods and amplitudes, surround the steady state in the phase space. In such cases the choice of any one of the closed trajectories depends on the initial conditions (Minorsky, 1962; Nicolis & Prigogine, 1977).

The fact that the limit cycle surrounds a singular point has important consequences for the response of the oscillating system to external perturbations. Pulses can be used to reset the phase of the oscillations. Winfree (1970, 1974, 1980) showed that, if the dynamics is governed by a limit cycle, there exists a critical perturbation applied at the appropriate time, with the appropriate duration, that brings the system back to the steady state. If the latter is unstable, as is usually the case, the system will return more or less rapidly to the limit cycle, but with an indefinite phase. Such an approach was applied by Winfree first to the circadian clock controlling pupal eclosion in *Drosophila* (Winfree, 1970) and later to glycolytic oscillations in a suspension of yeast cells (Winfree, 1972b). In both cases the three-dimensional graph obtained by plotting the new phase of the rhythm as a function of time and duration of the stimulus had the appearance of a helicoidal surface spiralling around a vertical axis. For perturbations located on that axis, phase resetting becomes erratic.

While other studies were devoted to the thermodynamic properties of biological membranes in the linear domain (Katchalsky & Curran, 1965), the developments of the Thermodynamics of Irreversible Processes in the nonlinear domain permitted location of periodic phenomena in the field of nonequilibrium processes of self-organization (Prigogine, 1967, 1968; Glansdorff & Prigogine, 1970, 1971). Much as spatial structures arise in chemical systems beyond a critical point of instability with respect to diffusion (Turing, 1952), rhythms correspond to a temporal organization that appears beyond a critical point of instability of a nonequilibrium steady state. These two types of nonequilibrium self-organization represent **dissipative structures** (Prigogine, 1968, 1969) that can be maintained only by the energy dissipation associated with the exchange of matter between the chemical system and its environment. Sustained oscillations of the limit cycle type can thus be viewed as **temporal dissipative structures** (Prigogine, 1968, 1969; Glansdorff & Prigogine, 1970, 1971; Nicolis & Prigogine, 1971, 1977; Prigogine & Stengers, 1979). These structures are the indubitable sign, probably the clearest, that a chemical or biological system functions beyond a nonequilibrium instability: endogenous rhythms, produced by a system and not by its environment, are indeed the signature of an instability.

From a mathematical point of view, the appearance of sustained oscillations generally corresponds to the passage through a **Hopf bifurcation point** (Guckenheimer & Holmes, 1983): for a critical value of a control parameter, the steady state becomes unstable as a focus. Before the bifurcation point, the system displays damped oscillations and eventually reaches the steady state that is a stable focus. Beyond the bifurcation point, a stable solution arises in the form of a small-amplitude limit cycle surrounding the unstable steady state (Minorsky, 1962; Nicolis & Prigogine, 1977). In the case of a subcritical bifurcation, a large-amplitude, stable limit cycle can coexist with a stable steady state; this form of bistability gives rise to the phenomenon of **hard excitation** (Minorsky, 1962): starting from the steady state, the system evolves towards the limit cycle only after a suprathreshold perturbation, in contrast with the spontaneous evolution to the cycle, observed when the system starts from an unstable focus.

By reason of their stability or regularity, most biological rhythms correspond to oscillations of the limit cycle type rather than to Lotka–Volterra oscillations. Such is the case for the periodic phenomena in biochemical and cellular systems discussed in the following chapters. Similarly, the phase space analysis of two-variable models indicates that the oscillatory dynamics of neurons corresponds to the evolution towards a limit cycle (Fitzhugh, 1961; Kokoz & Krinskii, 1973; Krinskii & Kokoz, 1973; Hassard, 1978; Rinzel, 1985; Av-Ron, Parnas & Segel, 1991). A similar evolution is predicted (May, 1972) by models for predator–prey interactions in ecology.

The seventies saw a true explosion of theoretical and experimental studies devoted to oscillating reactions. Nowadays, this domain continues to expand as more and more complex phenomena are observed in the experiments or predicted theoretically. The initial impetus for the study of oscillations owes much to the concomitance of several factors. The discovery of temporal and spatiotemporal organization in the

Belousov–Zhabotinsky reaction (Belousov, 1959; Zhabotinsky, 1964; Zaikin & Zhabotinsky, 1970), which has remained the most important example of a chemical reaction giving rise to oscillations and waves, and the elucidation of its reaction mechanism (Field, Köros & Noyes, 1972) occurred at a time when advances in thermodynamic studies were establishing the theoretical bases of temporal and spatial self-organization in chemical systems under nonequilibrium conditions (Prigogine, 1967, 1969; Prigogine *et al.*, 1969; Glansdorff & Prigogine, 1970, 1971; Nicolis & Prigogine, 1971, 1977).

At the same time as the Belousov–Zhabotinsky reaction provided a chemical prototype for oscillatory behaviour, the first experimental studies on the reaction catalysed by peroxidase (Yamazaki, Yokota & Nakajima, 1965; Nakamura, Yokota & Yamazaki, 1969) and on the gly-colytic system in yeast (Chance, Schoener & Elsaesser, 1964; Ghosh & Chance, 1964; Pye, 1969; Chance *et al.*, 1973) indicated the possibility of observing biochemical oscillations *in vitro*. These advances opened the way to the study of the molecular bases of oscillations in biological systems.

## 1.3 Biochemical and cellular rhythms

Although rhythmic phenomena in biology have been known for a very long time, their mechanism often remains unclear. This is particularly true for the nearly ubiquitous circadian rhythms. These endogenous rhythms, whose period is close to 24 h, are observed at the unicellular level (Vanden Driessche, 1980; Schweiger, Hartwig & Schweiger, 1986; Edmunds, 1988) as well as in more evolved organisms, including humans (Moore-Ede, Sulzman & Fuller, 1982). Their mechanism has not yet been elucidated, although good progress has been made in several organisms such as *Drosophila* (see chapter 11). It is possible that various mechanisms produce circadian rhythmicity in different organisms, but protein and mRNA synthesis repeatedly appear to be involved, suggesting that a unified mechanism might be at work (Takahashi, 1992). In mammals, these rhythms originate in the suprachiasmatic nuclei in the hypothalamus (Moore-Ede *et al.*, 1982), by a still unknown mechanism; membrane processes seem to play a prominent role in unicellular organisms such as *Acetabularia* (Vanden Driessche, 1980; Schweiger *et al.*, 1986). Recent evidence (Michel *et al.*, 1993) indicates that in multicellular organisms there are circadian oscillations even at the level of single neurons. Current advances in molecular biology (Takahashi, 1992, 1993; Stehle *et al.*, 1993; Takahashi *et al.*, 1993; Young, 1993) bear much promise for rapid progress in understanding the molecular bases of circadian rhythmicity.

Early theoretical studies interpreted circadian rhythms in terms of limit cycle behaviour (Pavlidis & Kauzmann, 1969; Pavlidis, 1973; Winfree, 1970, 1980). The recent advances in the mechanism of circadian rhythmicity pave the way for the construction of realistic models that will address this issue in more precise biochemical terms. One such model is proposed in chapter 11 of this book.

Among biological rhythms whose molecular origin has largely been clarified, an important category is that of high-frequency neuronal rhythms. The oscillatory properties of nerve cells have been known for a long time (Carpenter, 1982; Jacklet, 1989a) and have become a major theme of theoretical studies in neurobiology (Koch & Segev, 1989a). These properties play fundamental roles in the activity of the central nervous system (Llinas, 1988; Steriade, Jones & Llinas, 1990; Basar & Bullock, 1992; Buzsaki *et al.*, 1994). In 1936 the French neurophysiologist Alfred Fessard devoted a book to them entitled *Propriétés rythmiques de la matière vivante*. As is shown in this book, that title is much more general than Fessard envisaged, since rhythmic properties characterize nonexcitable as well as electrically excitable cells.

Thanks to the studies of Hodgkin & Huxley, which culminated in 1952 with the publication of a series of articles, of which the last was of theoretical nature, the physicochemical bases of neuronal excitability giving rise to the action potential were elucidated. Soon after, Huxley (1959) showed how a nerve cell can generate a train of action potentials in a periodic manner (see also Connor, Walter & McKown, 1977; Aihara & Matsumoto, 1982; Rinzel & Ermentrout, 1989). Even if the properties of the ionic channels involved have not yet been fully elucidated, cardiac oscillations originate in a similar manner from the pacemaker properties of the specialized, electrically excitable tissues of the heart (Noble, 1979, 1984; Noble & Powell, 1987; Noble, DiFrancesco & Denyer, 1989; DiFrancesco, 1993). These examples remained the only biological rhythms whose molecular mechanism was known to some extent, until the discovery of biochemical oscillations.

The oscillations observed *in vitro* in the glycolytic system of muscle (Frenkel, 1968; Tornheim & Lowenstein, 1974, 1975) and yeast cells (Pye & Chance, 1966; Hess & Boiteux, 1968a,b, 1971; Hess, Boiteux & Krüger, 1969; Pye, 1969, 1971) are still the prototype for biochemical oscillations resulting from the regulation of enzyme activity. These periodicities provided the opportunity to address, in an experimental and theoretical manner, the question of the molecular bases of a biological rhythm whose origin differs from that of electrical oscillations in neurons. Another biochemical rhythm, discovered a decade later (Gerisch & Hess, 1974; Gerisch & Wick, 1975), provided a further example of cellular oscillations of nonelectrical nature. That rhythm underlies the periodic generation of cAMP signals that control the aggregation of *Dictyostelium* amoebae after starvation. As with glycolytic oscillations, the period of cAMP oscillations is of the order of 5–10 min. The particular interest of periodic signals of cAMP stems from the fact that they allow us to assess the function of biochemical and hormonal rhythms in intercellular communication. The analysis of the molecular bases of these two periodic processes by means of theoretical models based on experimental observations is at the core of this book.

Other examples of biochemical oscillations are known, but are not considered here in further detail. Among these are oscillations in the above-mentioned peroxidase reaction (Yamazaki *et al.*, 1965; Nakamura *et al.*, 1969; Olsen & Degn, 1978; Degn, Olsen & Perram, 1979), mitochondrial oscillations (Chance & Yoshioka, 1966), and pH oscillations in an artificial membrane containing papain, a pH-dependent, acid-producing enzyme (Naparstek, Thomas & Caplan, 1973). These oscillations have been reviewed elsewhere (Hess & Boiteux, 1971; Goldbeter & Caplan, 1976; Berridge & Rapp, 1979). Oscillations in membranes, known for long (see Kachalsky & Spangler, 1968), have been reviewed by Larter (1990). Also noteworthy are the oscillations that occur with a period of several minutes in microtubule polymerization (Carlier *et al.*, 1987; Melki, Carlier & Pantaloni, 1988; Mandelkow *et al.*, 1988, 1989). The latter phenomenon is of particular interest in view of the fact that these proteins are key components of the cytoskeleton in eukaryotic cells.

In sketching initial developments in the field of biochemical and cellular rhythms, besides early theoretical studies of biochemical oscillations (Spangler & Snell, 1961; Higgins, 1964, 1967), special mention should be made of the efforts devoted in the sixties and seventies to the question of whether oscillations can occur in the synthesis of enzymes as a result of genetic regulation. Following the classic studies of Jacob & Monod (1961), which established the molecular mechanisms governing gene expression, Goodwin (1963, 1965) was the first to investigate theoretically the conditions in which negative feedback in the form of repression can produce oscillations in protein synthesis. Some of the conclusions reached in Goodwin's study by analogue computer simulations had, however, to be modified in view of analytical results later obtained by Griffith (1968). The most thoroughly studied example of periodic protein synthesis, both from an experimental and theoretical point of view, is that of the tryptophan operon of *Escherichia coli*. Bliss, Painter & Marr showed (1982) that oscillations due to cooperative repression by tryptophan do not occur in the wild type. In a mutant with reduced feedback inhibition of tryptophan on anthranilate synthetase, however, oscillations of the operon accompanied by the periodic variation of tryptophan are predicted theoretically and observed in the experiments. This example has been addressed further in more recent publications (Painter & Tyson, 1984; Sen & Liu, 1990). Interestingly, mechanisms based on periodic repression have recently been proposed to account for the circadian rhythm in the expression of the *per* gene in *Drosophila* (Hardin, Hall & Rosbach, 1990, 1992). Along these lines, a model based on the control of transcription of the *period* (*per*) gene by its protein product is proposed in chapter 11.

Goodwin's studies were influential in triggering a wave of theoretical studies devoted to the related problem of how oscillations develop in metabolic chains controlled by end-product inhibition. Such enzyme inhibition and genetic repression indeed give rise to formally similar kinetic equations. The studies of enzyme end-product inhibition were aimed (Morales & McKay, 1967; Walter, 1970; Hunding, 1974; Rapp, 1975, 1980; Mees & Rapp, 1978; Tyson & Othmer, 1978) at finding the minimum degree of feedback cooperativity required for sustained oscillatory behaviour, as a function of the number of enzymic steps in the pathway. Conditions for oscillations in regulated futile cycles were also obtained (Ricard & Soulié, 1982). The theoretical prediction of metabolic oscillations due to such negative feedback processes has, however, not yet been corroborated by experimental observations. That negative feedback is nevertheless involved in some important biochemical and cellular rhythms is shown by some of the models considered in this book.

Glycolytic oscillations and cAMP oscillations were, respectively, discovered around 1965 and 1975. Might there be a rough periodicity of some 10 years in progress on biochemical and cellular rhythms? The field of biochemical oscillations has indeed changed drastically due to the discovery in 1985 of intracellular $Ca^{2+}$ oscillations that occur in a variety of cells, either spontaneously or as a result of stimulation by an external signal such as a hormone or a neurotransmitter. Since their direct observation in oocytes and hepatocytes (Cuthbertson & Cobbold, 1985; Woods, Cuthbertson & Cobbold, 1987), which followed their indirect measurement (Rapp & Berridge, 1981) and some earlier theoretical predictions (Rapp & Berridge, 1977; Kuba & Takeshita, 1981), the number of experimental studies of $Ca^{2+}$ oscillations and the variety of cells in which they occur have increased so much that by now these oscillations represent the most widespread example of oscillatory behaviour at the cellular level, besides membrane potential oscillations in electrically excitable cells.

Several reviews (Berridge & Galione, 1988; Cuthbertson, 1989; Berridge, 1990; Cobbold & Cuthbertson, 1990; Jacob, 1990a; Petersen & Wakui, 1990; Tsien & Tsien, 1990; Meyer & Stryer, 1991; Tsunoda, 1991; Dupont & Goldbeter, 1992b; Amundson & Clapham, 1993; Fewtrell, 1993; Berridge & Dupont, 1994) have been devoted to $Ca^{2+}$ oscillations and the associated propagation of intracellular $Ca^{2+}$ waves. These phenomena are being characterized in a still increasing variety of cells. As a next step, the physiological significance of $Ca^{2+}$ oscillations and waves will most probably become a subject of active research in coming years.

A second, no less important cellular rhythm whose mechanism has largely been uncovered, thanks to recent breakthroughs in genetic and biochemical studies, is that underlying the onset of mitosis in eukaryotic cells. In support of previous conjectures (Kauffman & Wille, 1975; Winfree, 1980, 1984), experimental progress in this area indicates that a continuous biochemical oscillator is involved in the timing of cell division in amphibian embryos (Hara, Tydeman & Kirschner, 1980; Murray & Kirschner, 1989a,b; Murray & Hunt, 1993). The oscillator relies on a cascade of phosphorylation–dephosphorylation cycles governing the activity of an enzyme known as cdc2 kinase (Murray & Kirschner, 1989a,b; Nurse, 1990). It is the periodic activation of cdc2 kinase, driven by cyclin accumulation, that triggers the onset of mitosis. The cell cycle in yeast and somatic cells appears to be more complex but it probably involves similar oscillatory mechanisms subjected to tight control by protein inhibitors.

The molecular bases of $Ca^{2+}$ oscillations, as well as those of the mitotic oscillator, are addressed at the end of this book, where minimal models closely related to recent experimental observations are analysed for the two phenomena and shown to admit limit cycle oscillations.

## 1.4 Complex oscillations: bursting and chaos

Simple periodic behaviour is far from being the only mode of oscillation observed in chemical and, even more, biological systems. For many nerve cells, indeed, particularly in molluscs, oscillations take the form of bursts of action potentials, recurring at regular intervals representing a phase of quiescence. The best-characterized example of this mode of oscillatory behaviour known as **bursting** is provided by the R15 neuron of *Aplysia* (Alving, 1968; Adams & Benson, 1985). Neurons of the central nervous system of mammals (Johnston & Brown, 1984) also present this type of oscillations. In addition, complex oscillations have been observed and modelled in chemical systems (see, for example, Janz, Vanacek & Field, 1980; Rinzel & Troy, 1982, 1983; Petrov, Scott & Showalter, 1992).

Another type of oscillation, of aperiodic nature, discovered theoretically in a hydrodynamic model by Lorenz (1963), has been the subject of an increasing number of experimental and theoretical studies over more than a decade. These complex oscillations, called **chaos**, were first observed, in a biological context, in simple ecological models based on one-dimensional return maps (May, 1976; May & Oster, 1976). The analysis of these equations permitted the demonstration of a universal route for the appearance of the phenomenon: a cascade of period-doubling bifurcations often characterizes the oscillations before the latter become aperiodic. The sequence of values of the control parameter that correspond to the successive bifurcations preceding chaos is then characterized by a universal constant (Feigenbaum, 1978). Following these theoretical studies and the analysis of simple models (Rössler, 1976; Sparrow, 1982), a great number of experimental studies have been devoted to the search for chaos and the associated **strange attractors** in chemistry (Bergé, Pomeau & Vidal, 1984; Scott, 1991) as well as in various fields of biology ranging from neurobiology, cardiac physiology and biochemistry (Olsen & Degn, 1985; Holden, 1986; Glass & Mackey, 1988; Field & Györgyi, 1993) to the dynamics of biological populations (May, 1987) and epidemiology (Olsen & Schaffer, 1990).

Chaotic behaviour was thus observed experimentally in the Belousov–Zhabotinsky reaction (Schmitz, Graziani & Hudson, 1977; Hudson & Mankin, 1981; Roux, 1983, 1993; Roux, Simoyi & Swinney, 1983) before being studied in a theoretical manner (Tomita & Tsuda, 1979; Györgyi & Field, 1993; Zhang, Györgyi & Peltier, 1993). The phenomenon has also been studied in electrochemical systems (Koper & Gaspard, 1992) and combustion (Johnson, Griffith & Scott, 1991). In biochemical systems, an example of autonomous chaos was obtained *in vitro* for the peroxidase reaction (Olsen & Degn, 1977; Degn *et al.*, 1979; Olsen, 1983; Aguda & Larter, 1991; Larter *et al.*, 1993), which had previously been investigated for simple periodic behaviour. Also in that reaction the phenomenon appears through a sequence of period-doubling bifurcations (Geest *et al.*, 1992; Steinmetz, Geest & Larter, 1993). Another example of chaotic dynamics is provided by pancreatic β-cells (Lebrun & Atwater, 1985) whose membrane potential oscillations have been studied theoretically by Chay & Rinzel (1985; see also Chay, 1993). At the single-cell level, putative chaotic dynamics has been discussed in the locomotor behaviour of *Halobacterium*, which spontaneously reverses its swimming direction every 10–15 s (Schimz & Hildebrand, 1992). Finally, examples of chaotic dynamics in physiology were analysed, such as the irregular production of blood cells in the course of time (Mackey, 1978). Following that study the idea has emerged that many clinical troubles could be interpreted as resulting from passage through a bifurcation point. Such a transition in dynamic behaviour could lead, for example, from periodic to chaotic oscillations. The associated physiological troubles have been referred to as **dynamic diseases** (Mackey & Glass, 1977; Glass & Mackey, 1979, 1988).

Since it was realized that chaos may occur in a relatively easy manner in systems subjected to periodic forcing (see, e.g. Tomita & Daido, 1980), an increasing number of experimental and theoretical studies have been devoted to the effect of a periodic perturbation, generally of sinusoidal nature, on the response of oscillating systems. Such an analysis has been performed, for example, for glycolysis in yeast extracts (Markus, Kuschmitz & Hess, 1984, 1985), and for cardiac cells (Guevara, Glass & Shrier, 1981; Glass *et al.*, 1983, 1984; Glass & Mackey, 1988; Lewis & Guevara, 1990) and certain neurons (Matsumoto *et al.*, 1984; Holden, 1986).

Oscillations of the bursting type have been modelled mainly in neurobiology, which is the field where they are most frequently observed. These models consist in modifications of the Hodgkin & Huxley equations, which generally take into account the effect of $Ca^{2+}$ and the slow, cyclical variation of some ionic conductances (Plant & Kim, 1976; Plant, 1978; Carpenter, 1979; Hindmarsh & Rose, 1984; Chay & Rinzel, 1985; Ermentrout & Koppell, 1986; Rinzel & Lee, 1986; Canavier *et al.*, 1991; Smolen & Keizer, 1992; Destexhe, Babloyantz & Sejnowski, 1993; Bertram, 1994).

A great unity therefore appears in the different modes of

nonequilibrium, dynamic behaviour observed in chemical and biological systems. Nothing fundamentally distinguishes one from another, apart from the particular nature of the molecules involved and the kinetics that gives rise to simple or complex oscillations in each case.

## 1.5 Objectives and summary of the book

This book is devoted to the study of the molecular bases of simple and complex patterns of temporal organization in biochemical and cellular systems. The approach followed relies on the analysis of models whose temporal behaviour is governed by a set of chemical kinetic equations. In most examples considered, the latter have the form of nonlinear, ordinary differential equations. Except for a few cases, the spatial aspects of biological organization are not included in the analysis. Indeed, the experiments to which the models relate often allow us to resort to the approximation of spatial homogeneity. This approximation permits us to focus on the origins of temporal organization that form the principal subject of this book. It should be stressed that in the absence of spatial homogenization, spatial or spatiotemporal organization phenomena may readily occur in the chemical or biological systems whose reaction kinetics gives rise to simple or complex oscillations in conditions of continuous stirring (Turing, 1952; Prigogine, 1969; Prigogine *et al.*, 1969; Zaikin & Zhabotinsky, 1970; Nicolis, 1971; Winfree, 1972a, 1991b; Tyson, 1976; Nicolis & Prigogine, 1977; Tyson & Keener, 1988; De Kepper *et al.*, 1991; Swinney & Krinsky, 1991; Lee, K.J., *et al.*, 1994). The relationship of such spatial pattern formation to biological morphogenesis has been explored (see Meinhardt, 1982; Segel, 1984; Murray, J.D., 1989). Here, the close relationship between temporal and spatial organization is illustrated in the chapters devoted to cAMP and intracellular $Ca^{2+}$ oscillations.

Other approaches to the dynamics of biological systems have been proposed, among which is logical analysis (Glass & Kauffman, 1973; Thomas, 1978, 1979, 1991; Thomas & d'Ari, 1990) based on Boolean algebra, which permits us to address in a simplified manner the temporal evolution of systems characterized by a large number of variables and feedback loops. This approach, initially developed for genetically regulated systems, is currently being extended (Thomas & d'Ari, 1990) to other types of cellular regulation, e.g. in immunological systems (Kauffman, Urbain & Thomas, 1985). The approach in terms of ordinary differential equations nevertheless remains the natural framework for the study of the kinetics of biochemical reactions, even if the analysis of these equations may sometimes prove more arduous than that of their Boolean counterpart. It appears that the description in terms of ordinary differential equations allows a better assessment of the effect of a control parameter on the dynamics of the system, particularly when a high degree of resolution is needed, as in the case of complex phenomena such as bursting or chaos.

Finally, neither the effect of external noise, which affects nonequilibrium transitions in chemical and biological systems (Lefever, 1981; Horsthemke & Lefever, 1984; Lefever & Turner, 1986), nor the stochastic aspects of these transitions (Nicolis, Baras & Malek-Mansour, 1984) are considered – with the exception of the glycolytic system (chapter 2). Such a simplification, justified in the first approximation by the absence of systematic noise in the biological systems considered, permits us to avoid complicating from the outset the analysis of systems whose kinetics is already complex.

The first goal of this book is to analyse precise molecular models based on experimental observations for some of the best-known examples of biochemical rhythms. Among these are glycolytic oscillations in yeast and muscle, and oscillations of cAMP in the slime mould *Dictyostelium discoideum*. Besides pacemaker behaviour in nerve and cardiac cells, these oscillations are among the biological rhythms whose mechanisms have been identified thoroughly at the molecular level. In addition to these two prototypical examples of biochemical oscillations, subsequent chapters in the book are devoted to the modelling of two other oscillatory phenomena recently characterized at the cellular level. One pertains to the occurrence of intracellular $Ca^{2+}$ oscillations in a variety of cells stimulated by an external signal such as a hormone or neurotransmitter, the other to the relatively simple biochemical oscillator driving the cell division cycle in amphibian embryos. Finally, a chapter is devoted to the analysis of a model for circadian oscillations of the *period* protein in *Drosophila*; the model is based on the recent experimental advances made in characterizing the biochemical nature of the circadian clock in that organism.

Part I of the book is devoted to glycolytic oscillations. A two-variable allosteric model is analysed in chapter 2 for the phosphofructokinase reaction, which is responsible for the oscillations. The autocatalytic regulation of this reaction, which results from the cooperative activation of the multisubunit enzyme by one of its products, is at the core of the mechanism that produces the nonequilibrium instability beyond which

the glycolytic system undergoes simple periodic oscillations of the limit
cycle type. The model, developed in the framework of the concerted
transition theory proposed by Monod, Wyman & Changeux (1965) for
allosteric enzymes, allows determination of the influence exerted on the
period and amplitude by the main parameters amenable to experimental control, i.e. the enzyme concentration and the substrate injection
rate. Phase plane analysis of the two-variable model accounts for the
main experimental observation: namely, the existence of a domain of
sustained oscillations that is bounded by two critical values of the substrate input.

Besides the positive feedback exerted by the product, enzyme cooperativity is the second component of the mechanism of instability
that leads to glycolytic oscillations. The role of enzyme cooperativity in
the mechanism of instability is brought to light by analysing, for various
numbers of enzyme subunits, the Hill coefficient at the unstable steady
state as well as in the course of oscillations (the Hill coefficient provides
a widely used measure of cooperativity in enzyme kinetics). An extension of the model to the case of a Michaelian (i.e. saturable) sink of
product shows that oscillations of very large magnitude can nevertheless occur in the absence of cooperativity, when the enzyme consists of
a single subunit. While all these results are established for the case of a
constant substrate input, the last part of the chapter is devoted to the
experimental and theoretical study of the entrainment of oscillations by
a periodic source of substrate. The chapter ends with a discussion of the
possible function of glycolytic oscillations. Interest in the latter question
has been renewed by recent observations, which indicate that glycolytic
oscillations also occur in insulin-secreting pancreatic β-cells (Chou,
Berman & Ipp, 1992), and heart cells (O'Rourke, Ramza & Marban
1994).

Building on these results, chapters 3 and 4 present extensions of the
two-variable model for glycolytic oscillations. These somewhat abstract
extensions, not directly based on experimental observations, permit a
detailed analysis of the transition from simple to complex oscillatory
phenomena, which forms the subject of part II.

The model analysed in chapter 3 is again that of an enzyme reaction
regulated by positive feedback. To this reaction, which forms the core
of the mechanism for glycolytic oscillations, is added a nonlinear recycling of product into substrate. The advantage of this extension is to
keep only two variables while increasing the repertoire of dynamic
behaviour. In particular, the model allows the verification of a conjec-

ture, based on phase plane analysis, regarding the conditions that permit the coexistence between two simultaneously stable limit cycles, for a given set of parameter values. This phenomenon, previously observed in the three-variable model analysed in chapter 4, is called **birhythmicity** (Decroly & Goldbeter, 1982). The model shows how the system can switch reversibly from one limit cycle to the other after chemical perturbation delivered at the right phase, with the appropriate magnitude. The two oscillatory regimes possess different sensitivities towards such perturbations, owing to the different sizes of their basins of attraction in the phase plane.

The interest of the theoretical study of birhythmicity stems from the fact that the phenomenon has not been firmly demonstrated in biological systems. Some studies, however, do suggest its occurrence in the heart, as well as in a neuronal system (Hounsgaard *et al.*, 1988). Birhythmicity has been observed in a number of chemical oscillatory systems (Alamgir & Epstein, 1983; Roux, 1983; Lamba & Hudson, 1985; Citri & Epstein, 1988).

In the model, the appearance of birhythmicity is closely linked to the existence of multiple oscillatory domains as a function of the substrate injection rate, which is taken as the control parameter. In these conditions, the increase or decrease of this parameter from a value corresponding to a stable steady state gives rise to either one of two stable rhythms which markedly differ in period and amplitude. In a very different context, a similar property characterizes neurons of the thalamus. Thalamic neurons are indeed capable of oscillating with a frequency of 6 Hz or 10 Hz when the membrane, initially in a stable resting state, is slightly depolarized or hyperpolarized (Jahnsen & Llinas, 1984a,b; Llinas, 1988). The phase plane analysis of the biochemical model provides a clue for this behaviour and for the existence of multiple excitability thresholds, which are also observed in these neurons.

The models analysed in chapters 2 and 3 contain a single positive feedback loop. The question arises as to what happens when two instability-generating mechanisms interact within a given system. The model considered in chapter 4 was constructed to investigate such a situation. It consists in the coupling in series of two enzyme reactions controlled by positive feedback, and comprises three variables. Here, the repertoire of dynamic behaviour is much richer than in the models based on a single instability-generating mechanism.

Besides simple periodic oscillations and the coexistence between a stable steady state and a stable limit cycle, the multiply regulated biochemical model possesses at least two domains of birhythmicity in parameter space. When these domains partially overlap, we observe a phenomenon of trirhythmicity: three stable limit cycles then coexist for the same set of parameter values. The evolution towards a particular cycle depends on initial conditions; each cycle possesses its own basin of attraction. A similar multiplicity of oscillatory regimes has recently been obtained in a neuronal model (Canavier *et al.*, 1993). The boundaries between multiple periodic attractors in the multiply regulated, biochemical model may have a simple or complex structure. In particular, when two stable rhythms are separated by a regime of unstable chaos, the system acquires the property of **final state sensitivity** (Grebogi *et al.*, 1983): the continuous variation of the initial value of one of the variables results in the alternating evolution to one or the other of the two stable cycles. This unpredictable alternation is characterized by self-similarity: the same irregular pattern of evolution towards one or the other of the two cycles is recovered when the phenomenon is studied on a finer scale.

The particular interest of the multiply regulated model is that it provides a three-variable prototype for analysing the transition from simple periodic behaviour to complex periodic oscillations of the 'bursting' type, and chaos. The latter phenomenon, associated with the evolution towards a strange attractor in the three-variable phase space, occurs through a cascade of period-doubling bifurcations that can originate from each of the different limit cycles identified in the model. Bursting oscillations are observed when the system emerges from the domain of chaos. The model allows a detailed study of the transition between simple periodic behaviour and bursting, as well as the transition between a bursting pattern with $n$ peaks per period and a pattern containing $(n + 1)$ peaks. Two methods are utilized to analyse these transitions. The first, proposed by Rinzel (1987), consists in the reduction of the number of variables from three to two, by considering that one of the variables behaves as a slowly changing parameter that governs the evolution of a two-variable, 'fast' subsystem. The second method rests on the construction, by numerical simulations, of a one-dimensional return map characterizing complex periodic behaviour and chaos.

A piecewise linear map constructed on the basis of these results permits us to account for most patterns of bursting observed by numerical integration of the three differential equations. In particular, this approach explains the alternation between relatively simple and complex patterns of bursting. This alternation is also characterized by a property of self-similarity. A slight modification of the map that makes it nonlinear permits us to account for chaos in addition to simple and complex patterns of bursting.

Part III is devoted to cAMP oscillations in the slime mould *Dictyostelium discoideum*. The mechanism of these oscillations is addressed in chapter 5, which marks the return to models based directly on experimental observations. The wavelike aggregation of *D. discoideum* cells represents a prototype of spatiotemporal organization in biological systems. After reviewing the role of cAMP oscillations in the development and aggregation of the amoebae after starvation, an allosteric model based on the self-amplifying properties of the cAMP signalling system is proposed. This three-variable model rests on the activation of adenylate cyclase (the enzyme that synthesizes cAMP) by extracellular cAMP, via the binding of the latter to a membrane receptor. As in the phosphofructokinase reaction responsible for glycolytic oscillations, a self-amplification loop thus governs the synthesis of cAMP in *D. discoideum*. The analysis of the model indicates that this process can lead to a nonequilibrium instability beyond which sustained cAMP oscillations occur.

The phase plane analysis of a reduced version of the model containing two variables indicates the necessary link that exists between oscillations and excitability. The latter behaviour accounts for the relay of suprathreshold cAMP pulses, a phenomenon observed in cell suspensions as well as in the course of slime mould aggregation on agar. The coupling between oscillations and relay is responsible for the wavelike nature of aggregation: the amoebae respond chemotactically to the cAMP signals emitted by the aggregation centres, with a periodicity close to 5 min, and relay these signals towards the periphery of the aggregation field.

Models must be amended when they cannot account for the results of new experiments. Thus, several observations indicate the necessity of modifying certain assumptions of the allosteric model for cAMP signalling. This is done in the second half of chapter 5, which is devoted to the analysis of a model based on desensitization of the cAMP receptor (Martiel & Goldbeter, 1987a). In *D. discoideum*, the receptor is phosphorylated upon incubation with cAMP. This phosphorylation is associated with desensitization of the receptor and with the progressive decline in its capacity to activate adenylate cyclase (Devreotes &

Sherring, 1985; Klein, Lubs-Haukeness & Simons, 1985). The model based on these experimental observations contains seven variables, although this number can be reduced to only three.

For parameter values collected in the literature, this model yields qualitative and quantitative agreement with the experiments on oscillations of cAMP and on the relay of suprathreshold cAMP pulses. In particular, the model allows us to explain the role of receptor desensitization in the molecular mechanism of relay and oscillations. The phase plane analysis of a two-variable reduction of the model, which represents the true core of the relay and oscillatory mechanism, once again demonstrates a close link between the two modes of dynamic behaviour. The next section of chapter 5 is devoted to the response of the amoebae to constant stimulation by extracellular cAMP. Here also the model yields good agreement with the experimental observations, which indicate that, like many other sensory and hormonal systems, cAMP signalling adapts to constant stimuli. Yet a further extension of the model takes into account the recently established role of G-proteins in signal transduction from the receptor to adenylate cyclase. Finally, considered at the end of the chapter are the conditions in which concentric or spiral waves of cAMP similar to those observed in the course of aggregation on agar occur over a two-dimensional surface in the absence of stirring.

In a somewhat surprising manner, complex oscillatory phenomena occur in the model for cAMP signalling, in the form of birhythmicity, bursting and chaos. These phenomena are dealt with in chapter 6. Their appearance is analysed as in chapter 4, by means of a reduction in which a fast two-variable subsystem is governed by a third variable that behaves as a slowly changing parameter. Whereas in the model analysed in chapter 4 complex oscillatory phenomena resulted from the coupling in series of two instability-generating reactions, similar dynamic phenomena in cAMP synthesis originate from the interplay of two endogenous oscillatory mechanisms coupled in parallel. Here, the two mechanisms share the positive feedback loop exerted by extracellular cAMP on adenylate cyclase, and differ by the process limiting this self-amplification. These results provide a unique example of autonomous chaos in a realistic biochemical model based on experimental observations.

The model suggests that the aperiodic aggregation of the mutants *Fr17* and *HH201* of *D. discoideum* (Durston, 1974b; Coukell & Chan, 1980) might be interpreted in terms of chaos. The transition from periodic behaviour to autonomous chaos might result from a mutation affecting cAMP synthesis. The results of an experimental study of the light-scattering properties of the mutant *HH201* in cell suspensions indicate that this putatively chaotic mutant can display rather regular oscillations (Goldbeter & Wurster, 1988). This observation, however, does not contradict previous results on the irregular dynamics of the *HH201* and *Fr17* mutants in the course of aggregation on agar. Given the smallness of the domains of chaos in parameter space found in the models, the cells might easily have reached a domain where they function in a periodic manner. Alternatively, chaos might be lost in cell suspensions, in which intercellular coupling is much stronger than during aggregation on agar. Such a conjecture is corroborated by the study of the coupling of aperiodic and periodic oscillations of cAMP in cell suspensions. This study shows that the initial presence of a tiny fraction of periodic cells sometimes suffices to suppress chaos in such conditions. Such a result is extended to the suppression of chaotic behaviour through periodic forcing by a small-amplitude sinusoidal perturbation (Halloy *et al.*, 1990; Li *et al.*, 1992a,b). The latter approach provides a particular strategy for controlling chaos (Ott, Grebogi & Yorke, 1990; Shinbrot *et al.*, 1993).

Chapter 7 is devoted to the study of the appearance of cAMP oscillations during development of the amoebae *D. discoideum*. The theoretical analysis in the space formed by key parameters shows how the system can follow a **developmental path** along which it acquires successively the capability of relaying a signal of given amplitude, and then that of oscillating in an autonomous manner (Goldbeter & Segel, 1980). This developmental path corresponds to the continuous increase, observed after starvation, in the activity of enzymes such as adenylate cyclase and phosphodiesterase and in the quantity of cAMP receptor within the membrane. In the light of this discussion, the difference between cells behaving as centres and those behaving as relays in the course of aggregation would be of a dynamic nature. Thus, centres would be the cells that are the first to enter the domain of instability of the adenylate cyclase reaction, while cells following them on the developmental path would only be excitable and capable of relay. Apart from the case of *Dictyostelium*, the onset of cAMP oscillations during development provides a good model for studying in molecular and nonlinear dynamic terms the ontogenesis of biological rhythms.

Part IV is devoted to the function of pulsatile signalling in intercellular communication. The function of cAMP pulses in *Dictyostelium* is first addressed in chapter 8. Experiments have shown that cAMP signals delivered every 5 min elicit cell differentiation, whereas constant cAMP stimuli or pulses separated by 2 min intervals have no effect. The model based on receptor desensitization provides an explanation for these observations. It appears that a periodic signal of appropriate frequency allows target cells to generate repeatedly quasi-maximal responses, while continuous stimulation leads to desensitization. The model for cAMP synthesis in *D. discoideum* therefore suggests that periodic signalling represents an optimal mode of intercellular communication.

A link is established between periodic cAMP signalling and the pulsatile secretion observed for an increasing number of hormones. Thus, the gonadotropin-releasing hormone (GnRH) secreted by the hypothalamus induces the secretion of the gonadotropin hormones, i.e. luteinizing hormone (LH) and follicle-stimulating hormone (FSH), by the pituitary only when released into the circulation at the physiological frequency of one pulse per hour. Constant stimulation by GnRH or administration of this hormone with a frequency of two or more pulses per hour fail to produce the physiological effect (Knobil, 1980). It therefore appears that the temporal pattern of hormonal stimulation is as important as the level of hormone in the blood (Knobil, 1981). Similar results have recently been obtained for other hormones such as insulin, glucagon or growth hormone (GH). These observations have led to clinical applications in endocrinology, for example in the treatment of some reproductive disorders by means of pumps programmed to deliver GnRH at the physiological frequency.

The analysis of the role of periodic stimuli is pursued in chapter 8 in the framework of a general model for the response of a receptor subjected to desensitization. The analysis of the response of this model to periodic stimulation indicates that there exists optimum values of the duration of each pulse and of the interval between two successive stimuli, which maximize cellular responsiveness. The optimum pattern of periodic signalling is closely dependent on the kinetic parameters that govern desensitization and resensitization of the receptor (Li & Goldbeter, 1989b). Moreover, the efficiency of periodic stimuli is greater than that of chaotic or stochastic pulses (Li & Goldbeter, 1992). The results obtained in the general model based on receptor desensitization apply to intercellular communication by cAMP pulses in the amoebae *D. discoideum* as well as to the periodic secretion of hormones such as GnRH. In both cases the pulsatile signal is encoded in terms of its frequency. Frequency encoding of pulsatile signals and the existence of an optimum pattern of pulsatile stimulation bear on other examples of hormonal signalling and on the search for optimum patterns of drug delivery.

As indicated in section 1.3, cytosolic $Ca^{2+}$ oscillations, which occur in a variety of cell types as a result of stimulation by hormones or neurotransmitters, are among the most widespread of cellular rhythms, besides oscillations driven by periodic variations of the membrane potential in electrically excitable cells. These oscillations, whose period varies from seconds to minutes depending on the cell type, sometimes occur spontaneously. Part V is devoted to this phenomenon, which clearly represents the most significant addition to the field of biochemical oscillations over the last decade, in addition to the evidence that has accumulated to show that a continuous biochemical oscillator controls the eukaryotic cell cycle (see below). Experimental work on $Ca^{2+}$ oscillations has increased so much over the last years that it is by now the most studied biochemical rhythm.

After a brief review of the main experimental observations and of the different mechanisms proposed for the phenomenon, a model for intracellular $Ca^{2+}$ oscillations (Goldbeter, Dupont & Berridge, 1990) is examined in chapter 9, based on the process of $Ca^{2+}$-induced $Ca^{2+}$ release (CICR). The model is minimal as it contains only two variables, namely the concentration of cytosolic $Ca^{2+}$ and the level of $Ca^{2+}$ in the store sensitive to CICR. This model accounts for a large number of experimental observations and shows, in particular, how the oscillations can be controlled by the level of extracellular $Ca^{2+}$ or by the level of inositol 1,3,4-trisphosphate ($IP_3$), a second messenger that is synthesized in response to external stimulation (Berridge, 1993). A first version of the CICR model is based on two pools of $Ca^{2+}$, one sensitive to $IP_3$ and the other to $Ca^{2+}$. A second version considers a single $Ca^{2+}$ pool sensitive to both $Ca^{2+}$ and $IP_3$ behaving as co-agonists for $Ca^{2+}$ release. The predictions of the two versions of the model concerning $Ca^{2+}$ oscillations are compared. Experiments indicate that in a window of oscillatory behaviour the frequency of $Ca^{2+}$ spikes rises with the degree of stimulation. A plausible mechanism for the frequency encoding of $Ca^{2+}$ oscillations is examined, on the basis of protein phosphorylation by a $Ca^{2+}$-activated protein kinase.

The last part of chapter 9 is devoted to a study of intracellular $Ca^{2+}$ waves. Computer simulations show that the model based on CICR can account for the two types of wave seen in the experiments in different cell types (Dupont & Goldbeter, 1992b, 1994). When the period of the oscillations is of the order of 1 s, as in cardiac cells, the wave takes the form of a narrow band of $Ca^{2+}$ propagating at a relatively high rate along the cell. In contrast, when the period of the oscillations is of the order of 1 min, as in hepatocytes or endothelial cells, the wave looks more like a 'tide' (Tsien & Tsien, 1990) progressively rising along the cell and propagating at a slower rate. The model shows that the kinetics of $Ca^{2+}$ exchange between the cytosol and $Ca^{2+}$ pools governs the period of the oscillations and, hence, the appearance of the waves.

Part VI turns to what is perhaps the fundamental oscillator in cell biology as it controls the very process of cell division that plays a key role in development as well as in pathological states of unrestrained cellular proliferation. In line with some earlier suggestions, the idea that the cell cycle is controlled by a continuous biochemical oscillator of the limit cycle type was first tested in a detailed manner in a theoretical and experimental study of mitosis in the slime mould *Physarum* (Kauffman & Wille, 1975). Subsequent experiments in *Xenopus* oocytes lent further support to the existence of a cytoplasmic oscillator governing cell division (Hara *et al.*, 1980). Because of the lack of biochemical information, the precise chemical nature of the variables in these studies remained unspecified. Experimental advances made over recent years indicate that the periodic occurrence of mitosis in eukaryotic cells is brought about by a biochemical oscillator based on a cascade of phosphorylation–dephosphorylation cycles driven by proteins named cyclins. This cascade leads to the activation of a protein kinase, product of the gene *cdc2* in fission yeast and of its homologues in other eukaryotes (Murray & Kirschner, 1989a,b; Nurse, 1990; Murray & Hunt, 1993). The simplest form of the oscillator is encountered in early amphibian embryos, while the cell cycle in yeast and somatic cells appears to be subjected to additional controls or checkpoints (Cross, Roberts & Weintraub, 1989; Murray, 1992; Norbury & Nurse, 1992; Murray & Hunt, 1993).

A minimal cascade model is presented in chapter 10, for the embryonic cell cycle. There, a particular cyclin promotes the activation of cdc2 kinase, while the subsequent activation of a cyclin protease by cdc2 kinase leads to cyclin degradation. Such a regulation gives rise to a negative feedback loop. The three-variable model indicates (Goldbeter, 1991a, 1993a) how mitotic oscillations may readily arise from thresholds and time delays characterizing the activation by cyclin of cdc2 kinase through dephosphorylation by phosphatase cdc25, and the activation by cdc2 kinase of cyclin proteolysis. The thresholds arise in a natural manner from the kinetics of phosphorylation–dephosphorylation cycles, owing to the phenomenon of 'zero-order ultrasensitivity' (Goldbeter & Koshland, 1981). The minimal cascade model shows how the enzyme cascade that controls the onset of mitosis may function in a periodic manner corresponding to limit cycle oscillations.

The diagrams of stability established as a function of the main parameters allow us to discuss various ways in which the mitotic oscillator may be halted by inducing the exit from the oscillatory domain in parameter space and the concomitant entrance into a domain of stable steady states corresponding to cell cycle arrest. Such a question is of importance for the control of cell proliferation. In the last part of the chapter, the minimal cascade model is extended to take into account the control of phosphatase cdc25 by reversible phosphorylation. If cdc2 kinase were the kinase activating cdc25 (which phosphatase activates cdc2 kinase), this would create a positive feedback loop that could provide a second mechanism of instability capable of generating oscillations in cdc2 kinase activity. The effect of such a positive feedback, which is at the core of other models proposed for the mitotic oscillator (Norel & Agur, 1991; Tyson, 1991; Novak & Tyson, 1993a,b), is investigated in the extended cascade model. Finally the possibility of controlling the mitotic oscillator by rendering cyclin synthesis dependent on the presence of a growth factor is investigated theoretically. One goal of this study is to account for the effect of pulsatile signals of platelet-derived growth factor (PDGF) on lens development (Brewitt & Clarke, 1988).

Biological clocks are, more than often, synonymous with **circadian rhythms** (Chovnik, 1960). Part VII is devoted to these oscillations whose period is close to 24 h. Circadian rhythms are widespread in biological systems, and directly affect our human experience. The origin of circadian rhythms has, for a long time, remained an enigma. Remarkable advances on the molecular bases of circadian rhythmicity have, however, been made in recent years (Hall & Rosbash, 1988; Takahashi, 1992, 1993; Young, 1993). A multiplicity of organisms are being studied with regard to circadian rhythms, ranging from unicellular (Edmunds, 1988) to multicellular, including plants (Bünning, 1973), insects, rodents and humans (Moore-Ede, Sulzman & Fuller, 1982). Molluscs such as *Aplysia* and *Bulla* also provide highly useful models; a recent study of *Bulla* (Michel *et al.*, 1993) has shown that a circadian rhythm in membrane conductance already occurs at the level of a single neuron in that organism.

One of the organisms that has proved particularly fruitful for the study of the molecular bases of circadian rhythmicity is the fly *Drosophila*, which exhibits circadian rhythms in locomotor activity and eclosion. Mutants in which these rhythms are altered were obtained more than 20 years ago (Konopka & Benzer, 1971), and led to the identification of the *period* (*per*) gene as a leading actor in the mechanism of circadian rhythmicity in that organism. Much progress has since been made on the characterization of the structure and function of the *per* gene (see Young, 1993, for a presentation of recent advances, and chapter 11 for further references). Sequencing of the gene permitted the identification of mutations responsible for the loss of rhythmicity in the $per^0$ mutant and for the lengthened or shortened periods in the mutants $per^l$ and $per^s$, respectively (Baylies *et al.*, 1987; Yu *et al.*, 1987b).

Recently, evidence has accumulated that points to a role for the *per* gene product as a regulator of transcription (Huang *et al.*, 1993; see also Takahashi, 1992). It appears that the *per* gene product exerts a negative feedback on the transcription of the *per* gene (Hardin *et al.*, 1990, 1992). Similar results have recently been obtained for the *frq* gene that controls circadian rhythmicity in *Neurospora* (Aronson *et al.*, 1994). A model for circadian oscillations of the *period* protein (PER) in *Drosophila*, based on this regulatory loop and on the recently described phosphorylation of the protein (Edery *et al.*, 1994) is presented in chapter 11. The model shows how the regulation exerted by PER can give rise to sustained oscillations of the limit cycle type in *per* mRNA and PER protein. The model thus provides a molecular basis for the limit cycle nature of circadian rhythms, which has been under discussion for a long time (see, e.g. Winfree, 1980). Moreover, the results suggest a possible, biochemical explanation for the change in period observed in *per* mutants as compared to the wild type.

The results presented in preceding chapters are drawn together and discussed in chapter 12, with respect to other examples of nonlinear dynamics in chemistry and biology. The different models considered in the book allow us to clarify the molecular bases of some of the most important biochemical and cellular rhythms, ranging from glycolytic and cAMP oscillations to signal-induced $Ca^{2+}$ oscillations, the biochemical oscillator controlling the onset of mitosis in amphibian embryonic cells, and circadian rhythms in *Drosophila*. The analysis of the models also throws light on the transition between simple periodic behaviour and complex oscillations such as bursting or chaos. The theoretical results further indicate the possibility of a phenomenon not yet demonstrated in biology, namely the coexistence and reversible transition between distinct, simultaneously stable oscillations (bi- and trirhythmicity). Finally the models show how chaos can be controlled through forcing aperiodic oscillations by a small-amplitude sinusoidal perturbation.

From a physiological point of view, the theoretical study of cAMP oscillations allows us to discuss in molecular and dynamic terms the question of the ontogenesis of biological rhythms. Furthermore, by linking pulsatile cAMP signalling in *Dictyostelium* to the rhythmic secretion of a large number of hormones, the study of models based on receptor desensitization suggests that an important role of such biological rhythms is to provide optimal modes of intercellular communication. The existence of optimal patterns of pulsatile stimulation bears not only on clinical applications in endocrinology but also on the search for optimal patterns of drug delivery.

The study of biochemical oscillations thus extends in many directions, from simple and complex periodic behaviour to chaotic dynamics, and from cellular rhythms to chronopharmacology. It is the purpose of this book to explore the molecular bases of these oscillatory phenomena and the richness of their physiological implications.

# Part I

Glycolytic oscillations

# 2

# Oscillatory enzymes:
# simple periodic behaviour in an
# allosteric model for glycolytic oscillations

## 2.1 Experimental observations on glycolytic oscillations

Glycolytic oscillations are the prototype for periodic phenomena in bio-chemistry, even if the function of these oscillations, if there is any, has for long remained obscure. Their study began some 30 years ago, and still continues. The first observation indicating the existence of oscilla-tory behaviour in glycolysis is due to Duysens & Amesz (1957), who reported, by studying the fluorescence of some glycolytic intermediates in yeast, that one of these underwent damped oscillations in the course of time (fig. 2.1).

This observation was pursued from 1964 on by Chance and cowork-ers (Chance, Schoener & Elsaesser, 1964; Ghosh & Chance, 1964; Pye & Chance, 1966), who showed that glycolytic oscillations could be maintained in yeast suspensions for relatively long periods of time. Quickly thereafter observations were conducted on yeast extracts, with very similar results. Regular oscillations were recorded over an extend-ed period of time, provided that the glycolytic substrate was produced in a continuous manner (Chance, Hess & Betz, 1964; Hess, Brand & Pye, 1966; Pye & Chance, 1966; Pye, 1969, 1971). The latter condition was fulfilled when trehalose was used, this metabolite being trans-formed into a glycolytic substrate by the enzyme trehalase at a low, con-stant rate (fig. 2.2). The periodic phenomenon is also observed in a single yeast cell (Chance, Pye & Higgins, 1967). All glycolytic interme-diates oscillate with the same frequency (fig. 2.3), but with different phases (Betz & Chance, 1965). Recording the fluorescence of NADH allows the continuous measurement of the phenomenon, which can also be studied by chemically assaying the concentrations of glycolytic inter-mediates (Ghosh & Chance, 1964; Hommes, 1964; Betz & Chance,

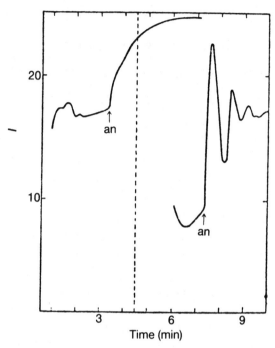

Fig. 2.1. Damped oscillations in the fluorescence of a glycolytic intermediate, NADH, following the injection of glucose (right) in a suspension of yeast cells. This observation was the first indication of the possibility of oscillatory behaviour in glycolysis. The curve on the left shows the addition of ethanol. an, anaerobic condition (Duysens & Amesz, 1957).

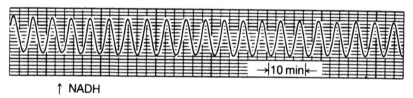

↑ NADH

Fig. 2.2. Sustained oscillations in an extract of the yeast *Saccharomyces carlsbergensis* utilizing trehalose as the glycolytic substrate. The slow degradation of this substrate gives rise to regular oscillations that can be maintained for more than 100 cycles (Pye, 1971). The oscillations are recorded by measuring the fluorescence of the glycolytic intermediate, NADH.

1965; Betz & Moore, 1967; Betz & Sel'kov, 1969; Hess *et al.*, 1969) or following changes in pH (Hess *et al.*, 1969; Hocker *et al.*, 1994). Yeast cells undergoing glycolytic oscillations rapidly synchronize in stirred suspensions (Ghosh, Chance & Pye, 1971). The synchronizing factor

Table 2.1. *Range of substrate injection rate for which glycolytic oscillations occur in yeast extracts*

| Input rate[a] mM/h | Period (min) | Amplitude in mM NADH | Damping | Waveform |
|---|---|---|---|---|
| <20 | — | Steady high level of NADH | — | — |
| 20 | 8.6 | 0.2–0.4 | — | Double periodicities, nonsinusoidal |
| 40 | 6.5 | 0.6 | — | Nonsinus–sinus |
| 60–80 | 5.0 | 0.3 | — | Sinus |
| 120 | 3.5 | 0.2 | — | Sinus |
| >160 | — | Steady low level of NADH | +++ | |

[a] Fructose or glucose serve as substrates. Cell-free extract of ~60 mg/ml. The data, obtained when using glucose or fructose as glycolytic substrate, indicate that oscillations occur for substrate injection rates between 20 and 160 mM/h.
Hess *et al.*, 1969.

could be acetaldehyde or pyruvate (Betz & Becker, 1975), but it was recently suggested (Aon *et al.*, 1992) that it might be external ethanol.

The essential property of glycolytic oscillations is illustrated by fig. 2.4a and b as well as table 2.1: sustained periodic behaviour is observed only in a precise range of substrate injection rates. This observation, carried out in yeast extracts (Hess & Boiteux, 1968b, 1973; Hess *et al.*, 1969), was confirmed (Von Klitzing & Betz, 1970) in suspensions of intact yeast cells (fig. 2.5). Below a critical value of the substrate injection rate, the system reaches a stable steady state. When this rate increases, oscillations occur, but they disappear when the substrate injection rate exceeds a second, higher, critical value. This disappearance is reversible, as shown by fig. 2.4b. The period of glycolytic oscillations is of the order of several minutes and diminishes as the substrate injection rate increases (Hess *et al.*, 1969; Hess & Boiteux, 1973; see table 2.1).

The control by the substrate injection rate of the period, amplitude and very existence of glycolytic oscillations makes these a particularly appropriate model for the study of biological rhythms. Moreover, in contrast with many of these rhythms for which the mechanism of oscillations remains obscure, the molecular basis of glycolytic oscillations is known. This allows the construction of realistic models, which in turn

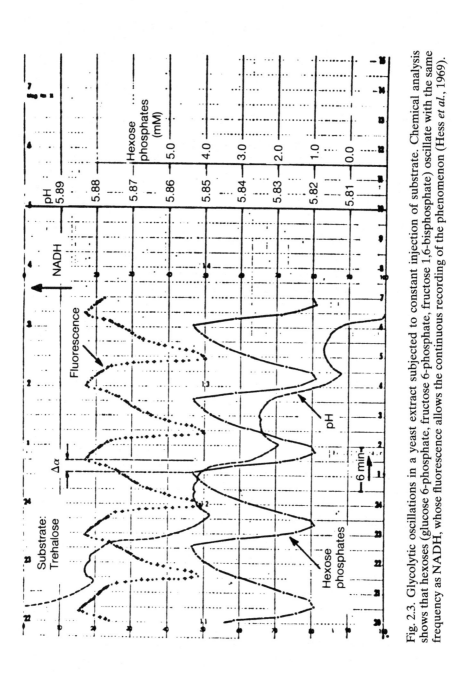

Fig. 2.3. Glycolytic oscillations in a yeast extract subjected to constant injection of substrate. Chemical analysis shows that hexoses (glucose 6-phosphate, fructose 6-phosphate, fructose 1,6-bisphosphate) oscillate with the same frequency as NADH, whose fluorescence allows the continuous recording of the phenomenon (Hess et al., 1969).

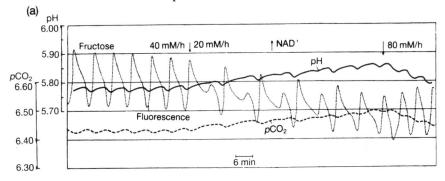

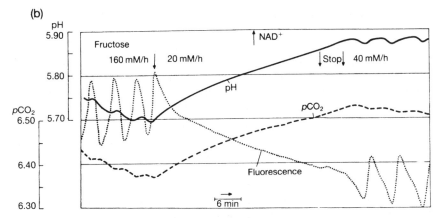

Fig. 2.4. Control of glycolytic oscillations in yeast extracts by the substrate injection rate. (a) The diminution of the rate of injection of fructose from 40 to 20 mM/h causes a lengthening of the period as well as a change in the waveform of oscillations; this change is reversible. (b) Decreasing the injection rate below 20 mM/h causes the reversible suppression of the oscillations (Hess & Boiteux, 1968b).

permit better understanding of the origin as well as the various properties of the periodic phenomenon.

Very early on, the source of oscillations within the glycolytic system was identified (Ghosh & Chance, 1964; Hess & Boiteux, 1968b, 1971). Glycolysis represents a chain of enzyme reactions which in yeast transforms a sugar such as glucose or fructose into ethanol and $CO_2$ (fig. 2.6). When a hexose such as glucose 6-phosphate or fructose 6-phosphate (F6P) is taken as the glycolytic substrate, periodic behaviour is observed. This observation indicates that the source of oscillations lies beyond the first two enzymes of the chain, hexokinase and glucose-

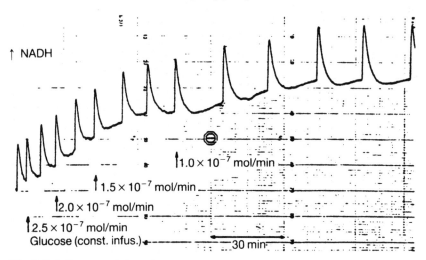

Fig. 2.5. Influence of the substrate injection rate on glycolytic oscillations in a suspension of intact yeast cells. The successive decrements in the glucose injection rate cause the progressive lengthening of the period (Von Klitzing & Betz, 1970).

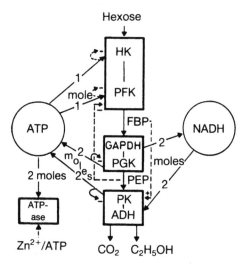

Fig. 2.6. Control of glycolysis in yeast. The main enzyme reactions are indicated vertically, from hexokinase (HK) to alcohol dehydrogenase (ADH) via phosphofructokinase (PFK), which is responsible for the oscillations. This chain of reactions constitutes the glycolytic pathway whose function is to produce ATP during the transformation of hexoses such as glucose or fructose into ethanol and $CO_2$. Also indicated are the reactions utilizing or producing ATP and NADH, as well as some of the main regulations. GAPDH, glyceraldehyde phosphate dehydrogenase; PGK, phosphoglycerokinase; PK, pyruvate kinase (Hess & Boiteux, 1968b).

phosphate isomerase. However, when the phosphofructokinase (PFK) step is bypassed by injecting fructose 1,6-bisphosphate (FBP) as glycolytic substrate, the oscillations disappear. Periodic behaviour therefore originates at the enzymic step catalysed by PFK.

The role of phosphofructokinase is corroborated by the fact that inhibitors or activators of the enzyme affect periodic behaviour. This is the case, for example, for ammonium ions that activate yeast PFK; their addition suppresses the oscillations (Hess & Boiteux, 1968b). In muscle, where glycolytic oscillations are also observed (Frenkel, 1968; Tornheim & Lowenstein, 1974, 1975; Tornheim, 1988; Tornheim, Andrés & Schulz, 1991; Marynissen, Sener & Malaisse, 1992), the molecular mechanism is identical: the enzyme responsible for the phenomenon is again PFK. Citrate ion, an inhibitor of PFK, suppresses the oscillations in muscle extracts (Frenkel, 1968). Moreover, as in yeast (Hess, 1968), the addition of purified PFK produces a phase shift of the oscillations and permits modulation of their period and amplitude; above a threshold, the addition of PFK suppresses the oscillations, but these reappear when the system is supplied with hexokinase (fig. 2.7). The latter observation, together with the data from figs. 2.2–2.5, indicates that periodic behaviour in glycolysis depends on a delicate

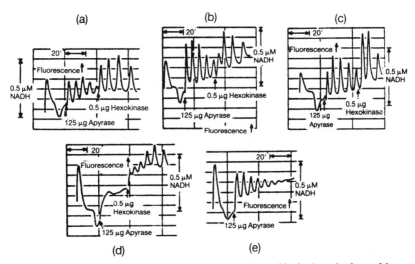

Fig. 2.7. Effect of adding increasing amounts of purified phosphofructokinase (PFK) on glycolytic oscillations in muscle extracts. The quantities of PFK added are: (a) 0; (b) 0.2 unit; (c) 0.4 unit; (d) 0.6 unit; (e) 0. The addition of apyrase each time induces oscillations, because of the transformation by this enzyme of ATP into the activator ADP (Frenkel, 1968).

balance between processes producing and transforming the substrate in the reaction catalysed by PFK.

Besides yeast and muscle, glycolytic oscillations have also been observed in: Ehrlich ascites tumour cells (Ibsen & Schiller, 1967); an insect, the blowfly *Phormia terraenovae*, for which age-dependent changes in the oscillations have been described (Collatz & Horning, 1990; Horning & Collatz, 1990); pancreatic β-cells (Chou *et al.*, 1992); and, very recently, heart cells (O'Rourke *et al.*, 1994).

*In vitro* experiments carried out in a reconstituted system containing PFK show that such a minimal system is capable of exhibiting transitions between multiple steady states or sustained oscillations (Eschrich, Schellenberger & Hofmann, 1980, 1983). So far, however, experiments on a reconstituted system have led primarily to the observation of multiple steady states (Hofmann, Eschrich & Schellenberger, 1985; Eschrich *et al.*, 1990).

Once the step responsible for the origin of periodic behaviour within the glycolytic system is identified, the question arises as to what are the properties that confer on PFK the capability of producing oscillations. The distinctive property of PFK, in this regard, was rapidly identified (Ghosh & Chance, 1964): the enzyme is activated by a reaction product, ADP, and by AMP, which is converted into ADP in the reaction catalysed by adenylate kinase. The control exerted by ADP is readily demonstrated by the immediate phase shift induced upon addition of this metabolite (fig. 2.8). The activation of an enzyme by its product is relatively rare in biochemistry. Regulation by negative feedback is much more common, as exemplified by the inhibition of an enzyme by the end product of a metabolic chain (Cohen, 1983). Glycolysis, however, does not provide the only example of activation of an enzyme by its reaction product in biochemistry: another case of autocatalytic regulation of enzyme activity, considered in chapter 5, is that of adenylate cyclase in *Dictyostelium* amoebae.

The reason why PFK is activated by its product can be clarified (Goldbeter, 1974) by resorting to the concept of adenylate energy charge proposed by Atkinson (1968, 1977). According to this author, metabolic regulation as a whole can be comprehended in terms of this charge, which increases from zero up to unity as adenylates progressively transform from AMP into ADP and, finally, into ATP. The adenylate energy charge, defined as the ratio of concentrations ([ATP] + 0.5 [ADP])/([ATP] + [ADP] + [AMP]), thus provides a measure of the energetic resources of the cell.

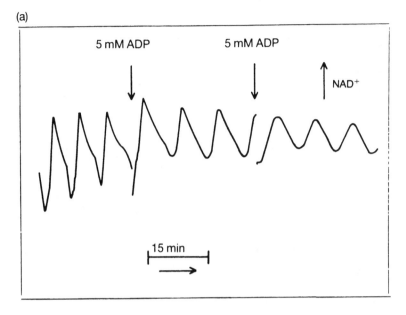

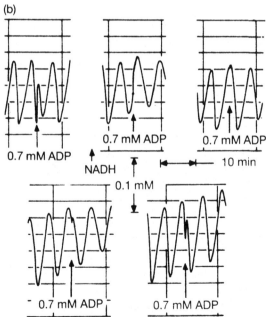

Fig. 2.8. Phase shift of glycolytic oscillations by addition of ADP. The maximum effect is observed when the perturbation occurs at the minimum of NADH (maximum of NAD); NADH oscillates in phase with ADP. (a) Addition of 5 mM ADP (Hess & Boiteux, 1968b); (b) addition of 0.7 mM ADP (Pye, 1969).

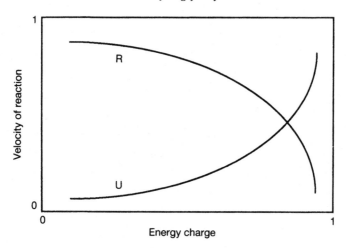

Fig. 2.9. Schematic representation of the control exerted by the adenylate energy charge on the enzymes utilizing (U) and regenerating (R) ATP (after Atkinson, 1968, 1977).

The enzymes belonging to biosynthetic pathways that consume ATP would be active at high values of the adenylate charge, i.e. would be activated by ATP and inhibited by an excess of ADP and AMP. In contrast, enzymes belonging to ATP-regenerating pathways would be inhibited by an excess of ATP and activated by AMP and ADP (fig. 2.9). What makes PFK unique is that the enzyme belongs to a pathway that globally produces ATP while at the local level, i.e. in the enzymic step that it catalyses, PFK utilizes ATP as substrate (see fig. 2.6). Because PFK is regulated as a function of the global role of glycolysis, which is to produce ATP, it follows that this enzyme is activated by ADP, its reaction product.

How the autocatalytic regulation of PFK leads to glycolytic oscillations is a question that benefits from being put in theoretical terms. The knowledge of the mechanism of oscillations and the availability of numerous experimental data early on prompted the construction of models for the PFK reaction. The first model for glycolytic oscillations, proposed by Higgins (1964), was based on the activation of the enzyme by its second product, FBP. This model, however, only admitted relatively unstable oscillations of the Lotka–Volterra type (see chapter 1, and Nicolis & Prigogine, 1977).

A second model, based on a similar autocatalytic regulation, is due to Sel'kov (1968a; see also Sel'kov, 1968b, 1972a,b). In this model, which leads to stable oscillations of the limit cycle type, the activation kinetics considered for the enzyme remains global and relies on a trimolecular step; the allosteric nature of the enzyme kinetics is not considered explicitly. As in the model due to Higgins, the kinetic equations can be put in polynomial form, under suitable conditions. Because of its simplicity, which represents its main virtue, the model proposed by Sel'kov does not account for one of the most conspicuous properties observed in the experiments, namely (see table 2.1 and fig. 2.4) the existence of two critical values of the substrate injection rate bounding a domain in which sustained oscillations occur. The closely related allosteric model developed below (Goldbeter & Lefever, 1972; Goldbeter, 1974; Goldbeter & Nicolis, 1976), which leads to a somewhat more complicated form of the rate function for PFK, yields better agreement with experimental data.

## 2.2 Allosteric model for glycolytic oscillations

### *Allosteric kinetics of phosphofructokinase*

Phosphofructokinase belongs to the class of regulatory proteins known as allosteric enzymes (Monod, Changeux & Jacob, 1963; Perutz, 1990). Such enzymes comprise multiple subunits that carry catalytic sites specific for the substrate, or regulatory sites on which activators or inhibitors bind and thereby modulate the enzyme activity. The essential property of allosteric enzymes is their cooperativity: the subunits forming the protein interact in such a manner that the binding of an effector or of the substrate to one of the sites facilitates (positive cooperativity) or impedes (negative cooperativity) the binding of the ligand to the remaining free sites. These interactions are responsible for the sigmoidal form of kinetic curves obtained for the majority of allosteric enzymes in the most common case, namely positive cooperativity.

In contrast with Michaelian enzymes, which have hyperbolic kinetics, allosteric enzymes, thanks to their sigmoidal kinetics, possess an enhanced sensitivity towards variations in the concentration of an effector or of the substrate. This is the reason why many enzymes that play an important role in the control of metabolism are of the allosteric type.

Several models have been proposed to account for the sigmoidal kinetics observed as a function of the concentration in substrate or effector. The first, developed by Hill (1910) for the binding of oxygen to haemoglobin, is phenomenological and supposes the simultaneous binding of $n$ molecules of ligand to the protein. Models based on more plausible molecular mechanisms were later proposed, successively, by Adair (1925), Monod *et al.* (1965), and Koshland, Némethy & Filmer (1966). In the latter two models, each subunit of the enzyme exists in two conformational states, which may differ in their affinity for the ligand. For Monod *et al.*, the transition of all subunits between these two conformations is concerted, like an umbrella whose elements would simultaneously (i.e. in a concerted manner) turn inside out (the conformational transition) in a strong wind, to use a metaphor due to Kirschner (1968). In contrast, the transition is sequential in the model of Koshland *et al.*, where hybrid states formed by a variable number of subunits in either one of the two conformations are considered.

Allosteric control is not the only mode of regulation that can give rise to sigmoidal kinetics associated with more or less abrupt thresholds. Another mode of enzyme regulation, as common as allosteric control, is that based on the covalent modification of proteins. Such regulation is exemplified by phosphorylation–dephosphorylation (Krebs & Beavo, 1979): an enzyme can be activated or inhibited through phosphorylation by a protein kinase, which can itself be activated by an intracellular signal such as cyclic AMP or calcium; the modified enzyme can be dephosphorylated in a second reaction catalysed by a protein phosphatase. The theoretical analysis of phosphorylation–dephosphorylation cycles has revealed the possibility of threshold phenomena in the covalent modification of proteins (Goldbeter & Koshland, 1981, 1982a, 1984, 1987; Koshland, Goldbeter & Stock, 1982). Thus, phosphorylation curves can possess a sigmoidal shape much steeper than that of allosteric enzymes when the enzymes catalysing the reversible modification are saturated by their substrate. Such threshold-generating mechanisms are considered in chapter 10 in relation to the model developed for the mitotic oscillator. That oscillations can occur in an enzyme system controlled by covalent modification, e.g. by phosphorylation–dephosphorylation, has also been been shown in a more abstract model of a product-activated reaction (Martiel & Goldbeter, 1981).

The regulation of PFK by a large number of positive or negative effectors, which reflects the key role of this enzyme in cellular metabolism (Krebs, 1972; Hofmann, 1978), is achieved mainly through allosteric control (Mansour, 1972; Laurent *et al.*, 1978; Laurent & Seydoux, 1977; Nissler *et al.*, 1977), although the protein can be phosphorylated (Kitajima, Sakakibara & Uyeda, 1983). As shown by the effect of pH, which modulates the allosteric properties of the enzyme as well as the number of cycles observed in the course of oscillations (Hess & Boiteux, 1968a,b), the allosteric nature of PFK plays an important role in the mechanism of glycolytic oscillations. This is the reason why a realistic model for this reaction must be based on such a mode of regulation.

### Hypotheses of the allosteric model for glycolytic oscillations

Phosphofructokinase possesses two substrates, ATP and F6P, which it transforms into ADP and FBP. A complete model for this reaction should therefore take into account the evolution of these four metabolites. However, studies carried out in yeast indicate that the couple ATP–ADP plays a more important role than the couple F6P–FBP in the control of oscillations. Indeed, the addition of ADP elicits an immediate phase shift of the oscillations (fig. 2.8) while the effect of FBP is much weaker (Hess & Boiteux, 1968b; Pye, 1969). The predominant regulation is thus exerted by ADP. In order to keep the model as simple as possible and to limit the number of variables to only two, which allows us to resort to the powerful tools of phase plane analysis, the situation in which an allosteric enzyme is activated by its unique reaction product is considered (fig. 2.10). This monosubstrate, product-activated,

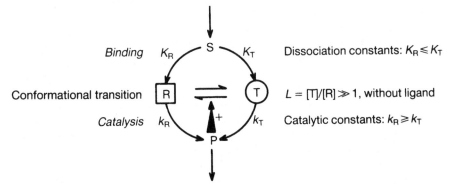

Fig. 2.10. Scheme of an autocatalytic enzyme reaction. The allosteric enzyme catalysing the reaction is formed by several subunits (not shown), which exist in the conformational states R and T. These states differ in their affinity for the substrate and/or in their catalytic activity. In the model by Monod *et al.* (1965), the transition between the two conformational states is concerted for all subunits. Here the product is a positive effector, i.e. an activator, as it binds in an exclusive or preferential manner to the most active, R state, of the enzyme. The system is open as the substrate S is injected at a constant rate while the product P disappears as a result of its utilization in a subsequent enzyme reaction.

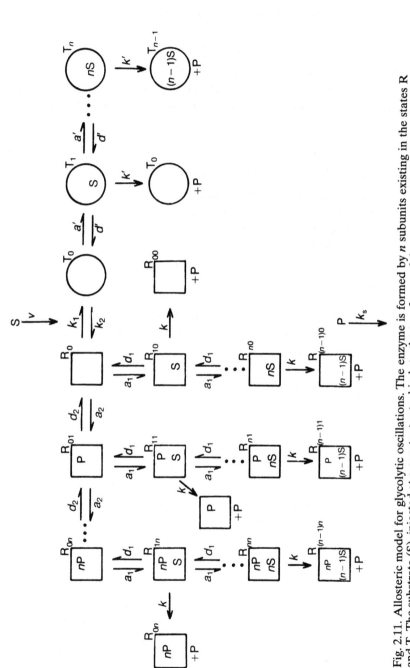

Fig. 2.11. Allosteric model for glycolytic oscillations. The enzyme is formed by $n$ subunits existing in the states R and T. The substrate (S), injected at a constant rate, binds to the two forms of the enzyme with different affinities. The complexes thus formed in the two states decompose with different rates to yield the product (P). The latter binds in an exclusive manner to the the most active, R, form of the enzyme, and disappears from the reaction medium in an apparent first-order reaction (Goldbeter & Lefever, 1972; Venieratos & Goldbeter, 1979; Goldbeter, 1980).

allosteric enzyme reaction can be viewed as the core of the oscillatory mechanism.

The model is represented in fig. 2.11 in the general case of an enzyme formed by $n$ subunits or protomers. The hypotheses of the model are as follows:

(1)  As in the experiments performed in yeast extracts, the substrate (S) is injected at a constant rate, $v_i$.
(2)  Each subunit, or protomer, can exist in two states, R and T, which differ by their affinity for the substrate ($K$ effect) and/or by their catalytic activity ($V$ effect). The protomers undergo a concerted transition between the states $R_0$ and $T_0$ free from ligands.
(3)  The reaction product binds exclusively to a regulatory site of the enzyme in the R state, which has the greatest affinity for the substrate and/or the highest catalytic activity.
(4)  The reaction product leaves the system at a rate proportional to its concentration; such a kinetic assumption corresponds to a step catalysed by a Michaelian enzyme not saturated by its substrate (the effect of nonlinear kinetics for the product sink will be considered in section 2.7 below).

The autocatalytic regulation of PFK is thus introduced in this model in a natural manner, by assuming that the product binds to a regulatory site that is accessible only when the enzyme is in the R state. Exclusive binding of the product to the regulatory site in that state therefore stabilizes the enzyme in its most active form and thereby contributes to the enhancement of the reaction rate.

In the following we shall restrict the analysis to a spatially homogeneous system; such an assumption corresponds to the experiments on glycolytic oscillations in continuously stirred yeast extracts. In the absence of such stirring, the effect of diffusion may lead to the formation of spatial or spatiotemporal dissipative structures in the form of propagating waves (Goldbeter & Lefever, 1972; Goldbeter, 1973; Goldbeter & Nicolis, 1976). The formation of these structures is not discussed in detail here (see, however, section 2.10), as this book focuses primarily on the mechanisms of temporal organization in biochemical systems.

### Kinetic equations of the allosteric model

For an enzyme with $n$ protomers, the time evolution of the metabolites and of the free or complexed enzyme forms is governed by the following ordinary differential equations:

$$\frac{dR_0}{dt} = -k_1 R_0 + k_2 T_0 - n a_2 P R_0 + d_2 R_{01} - n a_1 S R_0 + (d_1 + k) R_{10}$$

.
.
.

$$\frac{dR_{0n}}{dt} = a_2 P R_{0(n-1)} - n d_2 R_{0n} - n a_1 S R_{0n} + (d_1 + k) R_{1n}$$

.
.
.

$$\frac{dR_{nn}}{dt} = a_1 S R_{(n-1)n} - n (d_1 + k) R_{nn}$$

$$\frac{dT_0}{dt} = k_1 R_0 - k_2 T_0 - n a' S T_0 + (d' + k') T_1$$

.
.
.

$$\frac{dS}{dt} = v_i - n a_1 S \Sigma_0 - (n-1) a_1 S \Sigma_1 - \ldots - a_1 S \Sigma_{n-1} + d_1 \Sigma_1 + 2 d_1 \Sigma_2 + \ldots$$
$$+ n d_1 \Sigma_n - n a' S T_0 - (n-1) a' S T_1 - \ldots$$
$$- a' S T_{n-1} + d' T_1 + 2 d' T_2 + \ldots + n d' T_n$$

$$\frac{dP}{dt} = -n a_2 P R_0 - (n-1) a_2 P R_{01} - \ldots - a_2 P R_{0(n-1)} + d_2 R_{01} + 2 d_2 R_{02} + \ldots$$
$$+ n d_2 R_{0n} + k \Sigma_1 + 2 k \Sigma_2 + \ldots + n k \Sigma_n + k' T_1 + 2 k' T_2 + \ldots$$
$$+ n k' T_n - k_s P \tag{2.1}$$

together with the conservation relation:

$$R_0 + R_{ij} + T_0 + T_i = D_0 \; (i,j = 1, \ldots n) \tag{2.2}$$

In these equations, $S$ and $P$ denote the concentrations of substrate and product, while $R_{ij}$ represents the concentration of the enzyme form in the R state carrying $i$ molecules of substrate bound to the catalytic site and $j$ molecules of product bound to the regulatory site; $T_i$ denotes the concentration of the enzyme in the T state carrying $i$ molecules of $S$.

Moreover, the sum of R forms carrying $i$ molecules of substrate is denoted:

$$\Sigma_i = \sum_{j=0}^{n} R_{ij} \ (i = 0, \ldots n) \tag{2.3}$$

Kinetic constants are defined in fig. 2.11.

When the concentration of the enzyme is much smaller than those of the substrate and product, the enzyme forms evolve much more rapidly than the metabolites. A quasi-steady-state hypothesis can then be made for the enzyme (Heineken, Tsuchiya & Aris, 1967; Reich & Sel'kov, 1974; Segel, 1988). It is useful, at this point, to normalize the concentration of the substrate by dividing it by the dissociation constant for the enzyme in the R state; similarly, the concentration of the product is divided by its dissociation constant for the regulatory site of the enzyme in the R state. Thus normalized, the concentrations of substrate and product become dimensionless:

$$\alpha = S/K_R, \gamma = P/K_P \tag{2.4}$$

with:

$$K_R = d_1/a_1, K_P = d_2/a_2 \tag{2.5}$$

The algebraic equations $(dR_0/dt) = 0$, $(dR_{ij}/dt) = 0$, $(dT_i/dt) = 0$ give the following expressions, which link the enzyme forms to the substrate ($\alpha$) and product ($\gamma$) normalized concentrations:

$$\Sigma_0 = R_0 (1 + \gamma)^n, \Sigma_1 = n \ \alpha e \Sigma_0, \ldots, \Sigma_n = (\alpha e)^n \Sigma_0$$

$$T_0 = L R_0, \ldots, T_n = (\alpha c e')^n T_0$$

with:

$$R_0 = \frac{D_0}{L (1 + \alpha c e')^n + (1 + \alpha e)^n (1 + \gamma)^n} \tag{2.6}$$

Inserting these relations into the evolution equations for $S$ and $P$ finally leads to the two-variable system:

$$\frac{d\alpha}{dt} = v - \sigma\phi = f(\alpha, \gamma)$$

$$\frac{d\gamma}{dt} = q\sigma \ \phi - k_s \ \gamma = g(\alpha, \gamma) \tag{2.7}$$

where the rate function $\phi$ is given by relation:

$$\phi = \frac{\alpha e (1 + \alpha e)^{n-1} (1 + \gamma)^n + L\theta\alpha ce' (1 + \alpha ce')^{n-1}}{L (1 + \alpha ce')^n + (1 + \alpha e)^n (1 + \gamma)^n} \tag{2.8}$$

Parameters $v$ and $\sigma$ in eqns (2.7) represent the substrate injection rate and the maximum rate of the enzyme reaction, divided by constant $K_R$:

$$v = v_i/K_R, \sigma = n\, k\, D_0/K_R = V_M/K_R \tag{2.9}$$

Furthermore:

$$q = K_R/K_P, e = (1 + \epsilon)^{-1}, e' = (1 + \epsilon')^{-1} \text{ with } \epsilon = k/d \text{ and } \epsilon' = k'/d'$$

$$\theta = k'/k, L = k_1/k_2, c = K_R/K_T \quad \text{with } K_T = d'/a' \tag{2.10}$$

Parameters $c$ and $L$ are, respectively, the nonexclusive binding coefficient of the substrate and the allosteric constant of the enzyme (Monod *et al.*, 1965). The enzyme represents a **perfect $K$ system** when $\theta = 1$, i.e. when the R and T states possess the same catalytic activity and differ only by their affinity for the substrate ($c < 1$). When $c = 1$ and $\theta < 1$, the enzyme behaves as a **perfect $V$ system**. In the general case of a $K$–$V$ system, parameters $c$ and $\theta$ are both less than unity. Parameters $L$ and $c$ markedly influence the degree of cooperativity of the enzyme.

A simple form of eqns (2.7–2.8), which is particularly useful in analysing the dynamics of models based on allosteric regulation with positive feedback, is obtained in the limit case where the enzyme is a dimer and the substrate binds exclusively to the R state. When incorporating parameter $e$ into the normalization of the substrate concentration, one obtains the simplified expression (2.11) for the rate function $\phi$ appearing in eqns (2.7):

$$\phi = \frac{\alpha (1 + \alpha) (1 + \gamma)^2}{L + (1 + \alpha)^2 (1 + \gamma)^2} \tag{2.11}$$

A large part of the numerical simulations presented below relate to this particular case. Two enzyme subunits suffice to give rise to the phenomenon of positive cooperativity whose role in the origin of oscillations we wish to assess in this theoretical study. The effect of a larger number of subunits is considered further below.

### 2.3 Linear stability analysis and periodic behaviour

Equations (2.7) admit a unique steady-state solution obtained by solving the equations $d\alpha/dt = d\gamma/dt = 0$. The product concentration at steady state, $\gamma_0$, is given by eqn (2.12), while the corresponding concentration of substrate, $\alpha_0$, is obtained by solving eqn (2.13):

$$\gamma_0 = qv/k_s \qquad (2.12)$$

$$v = \sigma \phi (\alpha_0, \gamma_0) \qquad (2.13)$$

The latter equation admits at most a single real, positive root when $\sigma > v$. In the opposite case, the substrate injection rate exceeds the maximal capacity of the enzyme to transform the substrate into product, so that the former accumulates in the course of time.

#### *Linearization of the kinetic equations and condition of instability of the steady state*

The stability properties of the steady state can be analysed by slightly perturbing the concentrations of substrate and product around their steady-state values, and by determining the time evolution of the infinitesimal perturbations $\delta\alpha$, $\delta\gamma$ defined by relations (2.14) (see Nicolis & Prigogine, 1977):

$$\alpha = \alpha_0 + \delta\alpha, \ \gamma = \gamma_0 + \delta\gamma \qquad (2.14)$$

with:

$$|\delta\alpha/\alpha_0|, \ |\delta\gamma/\gamma_0| \ll 1.$$

Inserting these relations into eqns (2.7) leads to the linearized evolution equations for the perturbations:

$$\begin{aligned}
\frac{d(\delta\alpha)}{dt} &= (\partial f/\partial\alpha)_0 \, \delta\alpha + (\partial f/\partial\gamma)_0 \, \delta\gamma \\
&= -\sigma \, (\partial\phi/\partial\alpha)_0 \, \delta\alpha - \sigma \, (\partial\phi/\partial\gamma)_0 \, \delta\gamma \\
&= -AC \, \delta\alpha - BC \, \delta\gamma
\end{aligned}$$

$$\begin{aligned}
\frac{d(\delta\gamma)}{dt} &= (\partial g/\partial\alpha)_0 \, \delta\alpha + (\partial g/\partial\gamma)_0 \, \delta\gamma \\
&= q\sigma \, (\partial\phi/\partial\alpha)_0 \, \delta\alpha + [q\sigma \, (\partial\phi/\partial\gamma)_0 - k_s] \, \delta\gamma \\
&= qAC \, \delta\alpha + (qBC - k_s) \, \delta\gamma \qquad (2.15)
\end{aligned}$$

where the subscript zero indicates that the partial derivatives are evaluated at the steady state $(\alpha_0, \gamma_0)$. The quantities $A$, $B$ and $C$ are given by relations (2.16), where $\alpha$ and $\gamma$ denote the steady-state values:

$$
\begin{aligned}
A = {}& e(1+\gamma)^{2n}(1+\alpha e)^{2n-2} + L^2\theta ce'(1+\alpha ce')^{2n-2} \\
& + L(1+\gamma)^n(1+\alpha e)^{n-2}[\theta ce'(1+n\alpha ce') \\
& (1+\alpha e)^2 + e(1+n\alpha e)(1+\alpha ce')^2 \\
& - ncee'\alpha(1+\alpha e)(1+\alpha ce')(1+\theta)]
\end{aligned}
$$

$$
B = nL\alpha(1+\gamma)^{n-1}(1+\alpha e)^{n-1}(1+\alpha ce')^{n-1}[e(1+\alpha ce') - \theta ce'(1+\alpha e)]
$$

$$
C = \sigma/[(1+\alpha e)^n(1+\gamma)^n + L(1+\alpha ce')^n]^2 \tag{2.16}
$$

The linearized system (2.15) admits solutions of the form:

$$
\delta\alpha = ae^{\omega t}, \, \delta\gamma = be^{\omega t} \tag{2.17}
$$

Insertion of these relations into system (2.15) yields the algebraic system:

$$
(\omega + AC)\,\delta\alpha + BC\,\delta\gamma = 0
$$

$$
-qAC\,\delta\alpha + (\omega + k_s - qBC)\,\delta\gamma = 0 \tag{2.18}
$$

which admits nontrivial solutions only if its determinant is nil. Expressing this condition leads to the second-degree **characteristic equation**:

$$
\omega^2 + \omega[C(A - qB) + k_s] + k_s AC = 0 \tag{2.19}
$$

Perturbations will grow in the course of time if Re $\omega > 0$, and will regress in the opposite case. The condition of instability of the steady state thus takes the form of condition:

$$
C(A - qB) + k_s < 0, \text{ with } AC > 0 \tag{2.20}
$$

Quantity $C$ is always positive. In the absence of inhibition by the substrate ($\theta = 1$ or $c < 1$ with $\theta < 1$), quantity $A$ is also positive. The roots $\omega$ are thus imaginary when their real part becomes positive. The steady state therefore becomes unstable as a focus (**Hopf bifurcation**): the passage through the critical point of instability corresponds to the occurrence of sustained oscillations in the course of time.

Linear stability analysis permits the construction of stability diagrams as a function of the main parameters of the model. Such diagrams are represented in fig. 2.12 as a function of the nonexclusive binding coefficient of the substrate, $c$, and of the allosteric constant, $L$ for two values

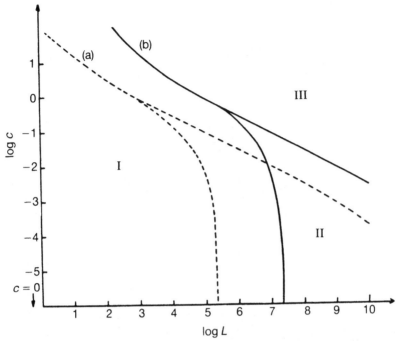

Fig. 2.12. Stability diagram for the allosteric model with positive feedback, as a function of the nonexclusive binding coefficient $c$ and of the allosteric constant $L$, for (a) $\epsilon = 10^{-3}$ and (b) $\epsilon = 10^{-1}$. System (2.7) admits a unique steady state. This state is stable in domain I and unstable in domain II; in the latter domain, the system evolves towards a limit cycle. For parameter values corresponding to domain III no steady state is reached as the substrate can only accumulate in the course of time (Goldbeter & Lefever, 1972).

of the ratio $\epsilon = k/d$. Three domains are visible in each stability diagram: in domain I, the steady state is stable; in the dashed domain II, the steady state is unstable and the system evolves toward a stable limit cycle (see below), while parameter values are such in domain III that the system does not admit any steady state as the substrate accumulates in the course of time (this occurs at large values of $c$ and $L$ in fig. 2.12).

### Sustained oscillations: evolution toward a limit cycle

In the phase plane $(\alpha, \gamma)$, the steady state corresponds to the point $(\alpha_0, \gamma_0)$. When this steady state becomes unstable, the system moves away from it and evolves towards a closed curve surrounding the steady state;

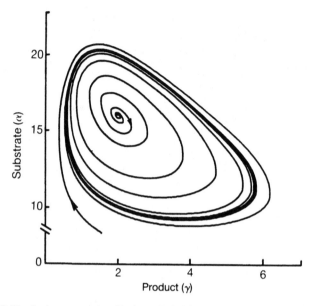

Fig. 2.13. Evolution towards a limit cycle in the model for the product-activated, allosteric enzyme. As shown by the figure, the stable limit cycle is reached in the phase plane $(\alpha, \gamma)$, starting from the unstable steady state $(\alpha = 16, \gamma = 2)$ or from an initial condition located outside the cycle. The curves are obtained by numerical integration of eqns (2.7) for $n = 2$, $v = 0.2 \text{ s}^{-1}$, $\sigma = 10^3 \text{ s}^{-1}$, $k_s = 0.1 \text{ s}^{-1}$, $q = 1$, $L = 7.5 \times 10^6$, $c = 0.01$, $\epsilon = 0.1$, $\epsilon' = \theta = 0$. These parameter values correspond to a point in domain II in fig. 2.12. The period of oscillations is equal to 145 s (Goldbeter & Lefever, 1972).

such a curve is known as a **limit cycle** (fig. 2.13). This closed trajectory is stable as it can be reached from any initial condition. Thus, in fig. 2.13, the limit cycle is approached from two such initial conditions, one close to the unstable steady state, within the cycle, and one located outside the asymptotic trajectory. The limit cycle corresponds to sustained oscillations of the substrate and product in the course of time (fig. 2.14); the period of the oscillations is precisely the time required for turning once around the cycle.

The existence of the limit cycle can be established (Erle, Mayer & Plesser, 1979) by means of the Poincaré–Bendixson theorem (see, for example, Boyce & Di Prima, 1969; Nicolis & Prigogine, 1977). The demonstration relies on the fact that it is possible to find finite values $\alpha^*$, $\gamma^*$ that permit us to define in the phase plane a domain $(0 < \alpha < \alpha^*;\ 0 < \gamma < \gamma^*)$ containing the steady state and no other singular point; all trajectories defined by eqns (2.7) move into this rectangular domain.

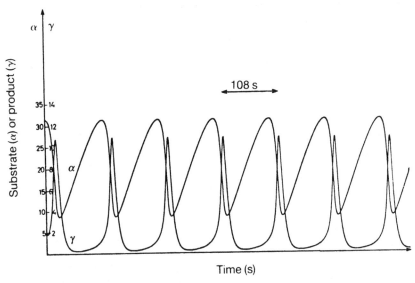

Fig. 2.14. Sustained oscillations of the substrate ($\alpha$) and product ($\gamma$) concentrations in the course of time. The curves correspond to the evolution to a limit cycle of the type shown in fig. 2.13 (Goldbeter & Lefever, 1972).

Since trajectories starting in the vicinity of the unstable steady state will necessarily tend to move away from it, the theorem ensures the existence of a least one closed trajectory in the domain considered. The numerical study of the model confirms the existence of a stable limit cycle and indicates, moreover, that this cycle is unique for a given set of parameter values. How a modification of the model permits the coexistence of several, simultaneously stable limit cycles is examined in the following chapter.

The oscillations of fig. 2.14 correspond to a periodic variation of the enzyme activity, which passes from less than 1% to some 75% of the maximum reaction rate (fig. 2.15). This range is comparable to that observed in the course of glycolytic oscillations in yeast for the periodic variation in PFK activity (Hess *et al.*, 1969). As indicated in table 2.2, the mean value and the maximum reached by the enzyme activity during oscillations depend on the value of the substrate injection rate.

### *Effect of the substrate injection rate and of the maximum enzyme activity*

The allosteric model for glycolytic oscillations thus shows how periodic behaviour originates from the peculiar regulation of PFK. The period

Table 2.2. *Variation of the enzyme activity in the oscillatory domain as a function of the normalized substrate injection rate, $v$*

| $v$ (s$^{-1}$) | 0.2 | 0.3 | 0.4 | 0.5 | 0.6 | 0.7 | 0.8 | 0.9 | 1 |
|---|---|---|---|---|---|---|---|---|---|
| $\langle v_r \rangle$ (% $V_{max}$) | 5 | 7.5 | 10 | 12.5 | 15 | 17.5 | 20 | 22.5 | 25 |
| $v_M$ (% $V_{max}$) | 38 | 52 | 61 | 66 | 70 | 73 | 75 | 75 | 72 |
| Activation factor | 118 | 155 | 146 | 124 | 100 | 77 | 56 | 36 | 18 |
| $T$ (s) | 420 | 315 | 255 | 225 | 200 | 180 | 170 | 160 | 145 |

Indicated are the mean value of the reaction rate, $\langle v_r \rangle$, the maximum value $v_M$ of that rate, and the activation factor equal to the ratio of the maximum and minimum values of the enzyme reaction rate in the course of one period, $T$, of the oscillations.
Goldbeter & Nicolis, 1976.

and amplitude of the phenomenon predicted by the model for physiologically acceptable values of the parameters are of the order of those observed in yeast. We should, however, try to go beyond such a general statement by comparing theoretical predictions with the large number of available experimental observations. A straightforward and most important comparison bears on the effect of the substrate injection rate on the oscillations. Numerical integration of eqns (2.7) as a function of parameter $v$ indicates (fig. 2.16) the existence of a domain of substrate injection rates within which sustained oscillations occur. Moreover, the period of the oscillations decreases, while the amplitude passes through a maximum, as the rate of substrate injection increases. These results are in agreement with experimental observations (Hess *et al.*, 1969; Hess & Boiteux, 1973). The brief passage through a maximum in the period at low values of $v$ has not been tested experimentally and therefore remains a theoretical prediction. While the data in fig. 2.16 have been obtained by using expression (2.11) for the rate function $\phi$ that holds for exclusive binding of the substrate to the R state, similar results are obtained when using the full expression (2.8) in the case where the substrate binds nonexclusively to the R and T states of the enzyme (Goldbeter & Lefever, 1972; Goldbeter & Nicolis, 1976).

The effect of the enzyme concentration on periodic behaviour (fig. 2.17) is similar to that exerted by the substrate input rate. There exists a precise domain of enzyme concentrations within which sustained oscillations occur. This domain is bounded by two critical values of the maximum enzyme activity. As the latter increases, the period is seen to diminish. These results account for the observations of Frenkel

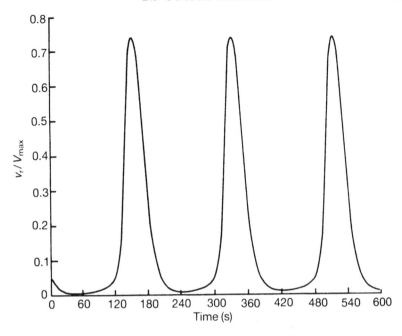

Fig. 2.15. Periodic variation of phosphofructokinase (PFK) activity in the allosteric model with positive feedback. The ratio $v_r/V_{max}$, equal to the rate function $\phi$, goes from less than 1% to nearly 80% in the course of oscillations (Goldbeter & Nicolis, 1976). The curve is obtained by numerical integration of eqns (2.7) for the following parameter values: $n = 2$, $v = 0.7\,s^{-1}$, $k_s = 0.1\,s^{-1}$, $\sigma = 4\,s^{-1}$, $L = 10^6$, $c = 10^{-5}$, $\epsilon = 0.1$, $\epsilon' = 0.1$, $\theta = 1$. For a dissociation constant $K_R = 5 \times 10^{-5}$ M, these values correspond to the following values of the parameters before normalization: substrate injection rate: $v_i = 126$ mM/h; PFK concentration: $D_0 = 5 \times 10^{-7}$ M; catalytic constant of the enzyme in the R state: $k = 200\,s^{-1}$; bimolecular association constant of the substrate to the enzyme in the R state: $a_1 = 4 \times 10^7\,M\,s^{-1}$. The order of magnitude of these parameters is consistent with the data available for yeast PFK (Hess *et al.*, 1969; Boiteux *et al.*, 1975); the values of $K_R$ and of the allosteric constant $L$ are taken from a study of PFK from *Escherichia coli* (Blangy, Buc & Monod, 1968).

(1968), who showed that the amplitude of glycolytic oscillations in muscle extracts passes through a maximum and then decreases as increasing amounts of PFK are added to the medium (see fig. 2.7); oscillations reappear when hexokinase is injected into the extracts. In the model, similarly, oscillations resume at high values of the enzyme activity $\sigma$ when parameter $v$, measuring the substrate input, increases sufficiently (this behaviour, however, occurs only in a certain range of $\sigma$ and $v$, as shown by stability diagrams established as a function of these parameters).

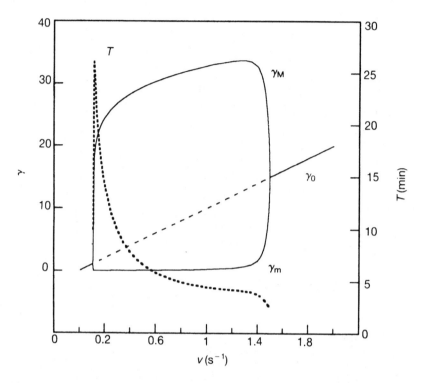

Fig. 2.16. Variation of the period ($T$) and amplitude of the oscillations as a function of the substrate injection rate. The bifurcation diagram shows the envelope of the product oscillations as a function of the substrate injection rate $v$; solid lines denote the stable steady state ($\gamma_0$) or the maximum ($\gamma_M$) and minimum ($\gamma_m$) of the product in the course of limit cycle oscillations; the dashed line denotes the unstable steady state. The curves have been established by numerical integration of eqns (2.7), function $\phi$ being given by the simple expression (2.11). Parameter values are $\sigma = 4 \text{ s}^{-1}$, $q = 1$, $L = 5 \times 10^6$, $k_s = 0.1 \text{ s}^{-1}$.

### Phase shift of the oscillations

Another comparison with the experiments bears on the phase shift of the oscillations that is elicited by the addition of reaction product at different moments over the period. As indicated by fig. 2.8b, the product ADP, which oscillates in phase with NADH, induces an immediate phase shift that is reflected by a delay of more than 1 min when 0.7 mM ADP is added near the minimum of the product oscillations. However, the same perturbation administered near the maximum of ADP leaves the phase unaltered. The model provides an explanation at the molecu-

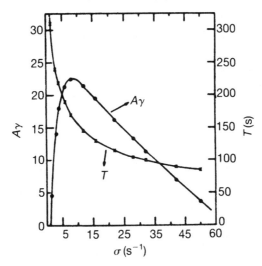

Fig. 2.17. Variation of the period ($T$) and amplitude ($A\gamma$) of the oscillations as a function of phosphofructokinase concentration, linked to parameter $\sigma$ (Goldbeter & Nicolis, 1976).

lar level for these observations, based on the allosteric nature of the oscillatory enzyme.

When 14 units of $\gamma$, corresponding to a 0.7 mM pulse, are added at various phases of an oscillation cycle (fig. 2.18), the maximum effect of the perturbation is to induce a delay of more than 1 min when the addition occurs near the minimum of $\gamma$, while the perturbation remains almost without effect near the maximum. This differential effect can be explained by the dynamic behaviour of the enzyme in the course of oscillations: when the product reaches its maximum, the enzyme is mainly in the R state, which is the most active, given that the reaction product is a positive effector and favours the transition from the T to the R state of the enzyme. Adding product at that phase of the oscillations has practically no effect, because the enzyme is already predominantly in its most active state.

At the minimum of the product, i.e. at low concentrations of the positive effector, the less active form of the enzyme predominates (oscillations indeed occur at large values of the allosteric constant $L$, which measures the ratio of enzyme in the T to R states in the absence of ligand; see fig. 2.12). It is precisely then that a pulse of activator elicits a significant transition from the T to the R state. Such a transition is sufficient for a nonnegligible quantity of substrate to be consumed, but the

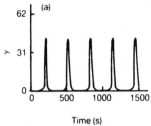

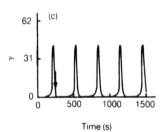

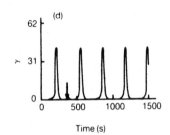

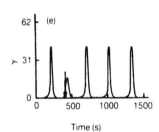

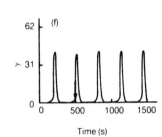

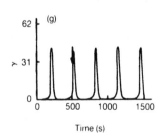

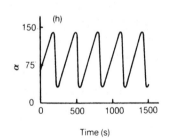

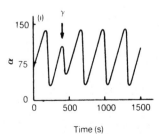

resulting peak in product remains reduced because the substrate level at that time is still below its maximum value (see fig. 2.18i). The aborted peak in product is associated with a delay of the oscillations because the substrate has to re-accumulate before the synthesis of the next normal peak of product can occur.

A systematic determination of the phase delay ($\Delta\Phi < 0$) or phase advance ($\Delta\Phi > 0$) as a function of the phase $\Phi$ of the oscillations yields the phase response curves of fig. 2.19. The two curves were established at different levels of the perturbation in product $\gamma$. In both cases, a discontinuity occurs as a delay abruptly transforms into a phase advance. In the terminology of Winfree (1980), such phase response curves are of **type 0**. If the magnitude of the product perturbation were sufficiently small, a very weak phase shift would occur at all phases $\Phi$ so that a plot of the new versus the old phase would have a slope close to unity; the corresponding phase response curve would then be of **type 1**. The phase response curves of fig. 2.19 are reminiscent of those observed for a large number of biological oscillations, including circadian rhythms (Winfree, 1980).

### *Molecular mechanism of oscillations*

The preceding results permit a detailed description of the sequence of the events whose spontaneous merging gives rise to sustained oscillations. Starting from a low value of the product concentration, the latter slowly builds up owing to its production by the small amount of enzyme present in the R conformation ($L \neq \infty$). This production increases as the substrate, injected at a constant rate, accumulates (see fig. 2.14). When the product reaches the level at which autocatalysis becomes effective – which level corresponds to $\gamma \approx 1$ – the enzyme begins to shift from the T state into the more active R conformation, as a result of the exclusive binding of product to the latter state. This enhances the synthesis of the product, and the resulting positive feedback loop results in the accelera-

Fig. 2.18.(b)–(g) Effect of an addition of 14 units of $\gamma$ (arrow), equivalent to 0.7 mM ADP, at various phases of the oscillations. The maximum effect on oscillations is observed in (e) where a delay of more than 1 min in the normal peak in the product occurs when the latter is added near its minimum (see text). (a) and (h) Evolution of the product ($\gamma$) and of the substrate ($\alpha$) in the absence of perturbation. (i) Substrate evolution corresponding to the situation described in (e) (Goldbeter & Nicolis, 1976).

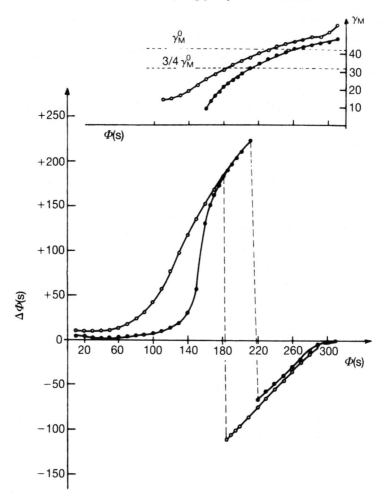

Fig. 2.19. Phase response curve for the phosphofructokinase allosteric model, indicating the phase shift caused by ADP addition. The phase shift $\Delta\Phi$ corresponding to a delay ($\Delta\Phi > 0$) or advance ($\Delta\Phi < 0$) of the oscillations is given as a function of the time of addition, $\Phi$, over the period $T = 312$ s. The phase $\Phi = 0$ corresponds to the maximum in ADP concentration in the course of unperturbed oscillations, i.e. $\gamma = \gamma_M^0 = 42.4$; the corresponding substrate concentration is $\alpha = 57.2$. The phase shift is considered as phase advance when the first maximum induced by ADP addition exceeds the value $(3/4)\gamma_M^0$. The two curves correspond, respectively, to the addition of a total of 14 units of $\gamma$ (0.7 mM ADP) within 2 s (open circles), and of 6 units of $\gamma$ (0.3 mM ADP) within 2 s (filled circles). Data are obtained by integration of eqns (2.7)–(2.8) for $v = 0.5$ s$^{-1}$, $k_s = 0.1$ s$^{-1}$, $\sigma = 8$ s$^{-1}$, $e = e' = 10^{-3}$, $c = 10^{-5}$, $\theta = 1$, $L = 5 \times 10^6$.

tion of the reaction rate, which leads to the abrupt synthesis of a peak of product (fig. 2.14). The enzyme is now predominantly in the R state.

The subsequent decrease in the level of product originates both from its increased rate of degradation and from the limitation of its production resulting from the drop in substrate consumption during the phase of product synthesis. The decrease in product signals the return of the enzyme towards the less active T state, which further accelerates the drop in product until the latter reaches its minimum. A new cycle of oscillations can then begin, with the slow accumulation of substrate and product until the latter reaches again the level at which autocatalysis becomes significant.

It is clear from fig. 2.14 that the phase of substrate re-accumulation occupies the largest part of one period of the oscillations. The duration of this phase is directly linked to the magnitude of the input of substrate. This explains why the period of the oscillations decreases as the rate of substrate injection increases, as observed in the experiments.

The mechanism of glycolytic oscillations can thus be closely related to the periodic alternation of the allosteric enzyme between its two conformational states. Such an alternation occurs in an autonomous manner, driven by the constant substrate input and by the autocatalytic regulation of the enzyme by its reaction product.

## 2.4 Phase plane analysis: explanation of the control of oscillations by the substrate injection rate

In the phase plane, two curves possess a special significance. These are the **nullclines** that correspond to the equations $d\alpha/dt = 0$ and $d\gamma/dt = 0$. The intersection of these two curves defines the steady state $(\alpha_0, \gamma_0)$. The substrate and product nullclines are given, respectively, by eqns (2.21a) and (2.21b) obtained from eqns (2.7):

$$v = \sigma \phi \tag{2.21a}$$

$$\gamma = \lambda \phi \tag{2.21b}$$

with $\lambda = q\sigma/k_s$.

### Instability condition in the phase plane

It is possible to relate the instability condition (2.20) obtained from linear stability analysis to the form of the two nullclines and, more specifically, to the position of the steady state on the product nullcline

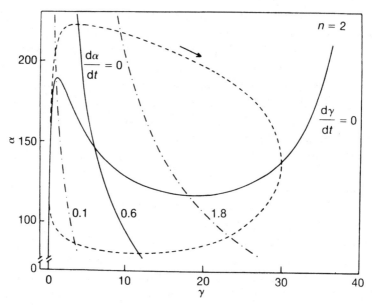

Fig. 2.20. Phase plane analysis explains the effect of the substrate injection rate on glycolytic oscillations. The substrate nullcline is represented for three values of the normalized injection rate $v$ (indicated on the curves labelled 0.1, 0.6, 1.8, in $s^{-1}$). The product nullcline is also represented; this curve does not depend on the value of $v$. When the steady state, located at the intersection of the two nullclines, lies in a region of sufficiently negative slope on the product nullcline, this state is unstable and the system evolves towards a limit cycle. Such a limit cycle is represented by a dashed line for the intermediate value of the substrate input $v$ (Venieratos & Goldbeter, 1979).

$d\gamma/dt = 0$. As seen below, a necessary condition for the occurrence of sustained oscillations is that the product nullcline possesses a region of negative slope $(d\alpha/d\gamma) < 0$ at the steady state. For appropriate parameter values, the product nullcline satisfies this condition and takes the form of an S-shaped or, rather, N-shaped sigmoid – depending on the choice of axes (fig. 2.20). In order to obtain an expression of the instability condition directly related to the location of the steady state on that nullcline, we differentiate the two sides of eqn (2.21b) with respect to $\gamma$. After dividing by $d\gamma$, this procedure yields:

$$1 = \lambda \left[ (\partial \phi / \partial \alpha)(d\alpha/d\gamma) + (\partial \phi / \partial \gamma) \right] \qquad (2.22)$$

from which we obtain:

$$\frac{d\alpha}{d\gamma} = - \frac{[\lambda (\partial \phi / \partial \gamma) - 1]}{\lambda(\partial \phi / \partial \alpha)} \qquad (2.23)$$

Relation (2.22) holds in all points of the product nullcline and, in particular, at the steady state that lies at the intersection of the nullcline $\gamma = \lambda\phi$ with the substrate nullcline.

Taking into account relations (2.15), we can rewrite the instability condition (2.20), obtained from linear stability analysis, in the form:

$$\sigma \, (\partial\phi/\partial\alpha)_0 + k_s \, [1 - \lambda \, (\partial\phi/\partial\gamma)_0] < 0 \qquad (2.24)$$

or:

$$k_s > \frac{\sigma \, (\partial\phi/\partial\alpha)_0}{[\lambda \, (\partial\phi/\partial\gamma)_0 - 1]} \qquad (2.25)$$

Given the definition $\lambda = q\sigma/k_s$, the comparison of relations (2.23) and (2.25) allows us to express the instability condition in the phase plane as a function of the slope $(d\alpha/d\gamma)$ at steady state, in the compact form of inequality (2.26):

$$(d\alpha/d\gamma)_0 < -\frac{1}{q} \qquad (2.26)$$

Thus the steady state is unstable if and only if the slope of the product nullcline is sufficiently negative in that point. For oscillations to occur it is therefore essential that the nullcline $(d\alpha/d\gamma) = 0$ possesses at least one region of negative slope. In the case of exclusive binding of substrate to the R state of a dimeric enzyme, function $\phi$ takes the form of eqn (2.11). We can show in that simple case that the curve defined by eqn (2.21b) admits three values of $\gamma$ for certain values of $\alpha$, i.e. the curve takes the form of an S-shaped sigmoid possessing one region of negative slope, as soon as $\lambda > 8$ (T. Erneux & A. Goldbeter, unpublished results; Segel & Goldbeter, 1994). In such a case, therefore, oscillations will never be observed as long as parameter values are such that $(q\sigma/k_s) < 8$.

### Existence of a domain of substrate injection rates producing oscillations

Phase space analysis provides an explanation for the existence of a domain of sustained oscillations bounded by two critical values of the substrate injection rate. As indicated by relation (2.12), the steady state, located at the intersection of the two nullclines, moves towards the right when the substrate injection rate increases, since $\gamma_0$ is proportional to $v$. Since parameter $v$ enters only into eqn (2.21a) of the substrate nullcline and not into eqn (2.21b), the product nullcline remains unchanged when parameter $v$ increases while the substrate nullcline is displaced to the right (fig. 2.20).

For sufficiently low values of the substrate injection rate, the steady state lies on the left limb of the sigmoid nullcline; since the slope $(d\alpha/d\gamma)$ is positive, the steady state is stable. As $v$ progressively increases, the steady state moves to the right and remains at first stable until it reaches the region of negative slope on the nullcline. As soon as condition (2.26) is satisfied, the steady state becomes unstable and oscillations occur. A further increase in $v$ results in the displacement of the steady state further to the right on the product nullcline, until a second critical value of the substrate injection rate is reached beyond which condition (2.26) ceases to be satisfied: the steady state recovers its stability and oscillations disappear.

Phase plane analysis thus readily accounts for the main experimental observation on the control of glycolytic oscillations by the substrate injection rate. Below the lower critical value of the substrate injection rate $v$, a stable steady state is established, corresponding to a low level of reaction product and to an enzyme predominantly in the inactive T state. Above the higher critical value of $v$, the stable steady state is associated with a higher level of product and with an enzyme predominantly in the active state R. Sustained oscillations, in the course of which the enzyme switches back and forth between the R and T states, occur in the range delimited by the two critical values of the substrate input.

A similar analysis can be carried out for the effect of the enzyme concentration on periodic behaviour. The existence of an oscillatory domain bounded by two critical values of the maximum enzyme activity, which is proportional to the enzyme concentration, is apparent from the data shown in fig. 2.17. That there exists a minimum value of $\sigma$ required for oscillations is due to the fact that the product nullcline admits a region of negative slope, necessary for instability of the steady state, only if $\lambda = (q\sigma/k_s)$ exceeds a critical value (equal to 8 in the case $n = 2$, $c = 0$). The existence of a second, higher critical value of $\sigma$ comes from the fact that the slope $(d\alpha/d\gamma)$ becomes less and less negative at high values of $\sigma$, so that condition (2.26) ceases to be satisfied.

## 2.5 Influence of the number of enzyme subunits on oscillations

The numerical integration of eqns (2.7) for different values of $n$ permits us to determine how oscillations depend on the number of subunits that form the allosteric enzyme. Figure 2.21 shows the limit cycle, together

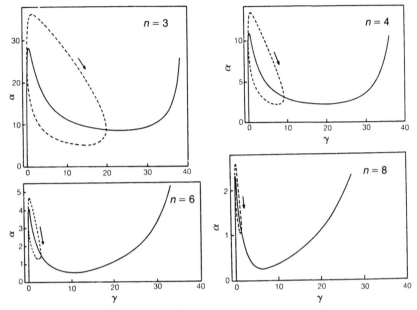

Fig. 2.21. Dependence of limit cycle oscillations and product nullcline on the number $n$ of subunits of the allosteric enzyme. The dashed curves are obtained by numerical integration of eqns (2.7) for the values $n = 3$, 4, 6 and 8. In each case, the dashed curve represents the limit cycle of maximum amplitude determined as a function of the substrate injection rate (Venieratos & Goldbeter, 1979).

with the product nullcline, for different values of parameter $n$. In each case, the trajectory shown by the dashed line represents the limit cycle of maximum amplitude obtained in the domain of oscillations defined by parameter $v$. The curves can be compared to the limit cycle obtained in similar conditions in the case of a dimeric enzyme (fig. 2.20).

A surprising result from this comparative study is that the amplitude of oscillations markedly diminishes for both substrate and product as the number of subunits progressively increases. This originates from the fact that the slope of the trajectory followed by the system when it leaves the left limb of the sigmoid to move towards the right becomes steeper and steeper when $n$ rises. Since the maximum product concentration in the course of oscillations corresponds to the point where the trajectory intersects the sigmoid nullcline, we can see how this maximum will decrease as $n$ increases. As regards the substrate, the reduced amplitude results from the shift of the product nullcline towards lower and lower levels of $\alpha$ as $n$ rises from 2 to 8.

The decrease in the product amplitude attenuates, however, when the

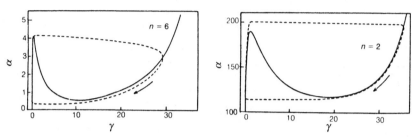

Fig. 2.22. Limit cycles obtained for $n = 2$ and 6, when the product varies much faster than the substrate. The existence of two time scales results from the large values $q = 100$, $k_s = 10 \, \text{s}^{-1}$, which do not modify the shape of the product null-cline. Other parameter values are as in fig. 2.21 (Venieratos & Goldbeter, 1979).

variation of the product becomes more rapid than that of the substrate. Such a situation occurs at high values of $q$ and $k_s$. The concomitant increase in these two parameters corresponds to the appearance of two time scales without affecting either the steady state or the form of the product nullcline $\gamma = \lambda \phi$. Because of the faster variation of the product, the transitions from the left to the right limb of the sigmoid nullcline are now quasi-horizontal, as indicated by fig. 2.22; the resulting relaxation oscillations should be compared with the curves of fig. 2.21 for corresponding values of $n$. A detailed analysis of the limit cycle in the conditions of relaxation oscillations has been performed recently (Segel & Goldbeter, 1994). In this analysis, the limit cycle is dissected into four regions; the choice of appropriate scaling in each region allows us to compare the relative rates of evolution on the different segments of the limit cycle and to obtain an approximate expression for the period of sustained oscillations.

The amplitude of the product oscillations in fig. 2.22 varies only slightly with the subunit number, in contrast to what happens when the product and the substrate evolve on comparable time scales (see also fig. 2.23). The influence of parameter $n$ on the other characteristics of the oscillatory phenomenon is detailed in table 2.3 both in the absence ($q = 1$) and in the presence ($q = 100$) of two distinct time scales.

Besides the effect of the number of subunits on the amplitude of the oscillations, the domain of values of parameter $v$ giving rise to oscillations increases with $n$, as shown by table 2.4, where the minimum and maximum values of $v$ producing periodic behaviour are indicated for $n$ varying from 2 to 8 (the range observed for the majority of allosteric enzymes). For a given number of enzyme subunits, the oscillatory domain widens when the product varies more rapidly than the sub-

Table 2.3. *Amplitude and period of the oscillations as a function of the number of enzyme subunits*

| | | $q = 1$ | | | | $q = 100$ | | | |
|---|---|---|---|---|---|---|---|---|---|
| | | | | Enzyme activity ($\% V_{max}$) | | | | Enzyme activity ($\% V_{max}$) | |
| | | | Period | | | | Period | | |
| $n$ | $v\,(\mathrm{s}^{-1})$ | Amplitude | (s) | Min | Max | Amplitude | (s) | Min | Max |
| 2 | 0.6 | 23.8 | 365 | 0.42 | 78.9 | 29.6 | 186 | 0.40 | 88.3 |
| 3 | 0.4 | 15.8 | 115 | 0.37 | 82.2 | 34.2 | 63 | $1.7 \times 10^{-2}$ | 95.7 |
| 4 | 0.2 | 7.1 | 85 | 0.29 | 61.3 | 33.7 | 53 | $2.3 \times 10^{-3}$ | 90.6 |
| 6 | 0.07 | 2.1 | 77 | 0.10 | 19.7 | 28.5 | 58 | $3.7 \times 10^{-4}$ | 76.3 |
| 8 | 0.05 | 0.9 | 60 | 0.16 | 7.4 | 23.1 | 48 | $1.7 \times 10^{-4}$ | 67.7 |

Venieratos & Goldbeter, 1979.

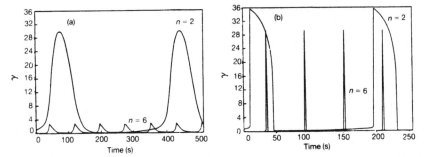

Fig. 2.23. Oscillations of the product in (a) the absence and (b) the presence of two time scales, for $n = 2$ and 6. Parameters $q$ and $k_s$ are respectively equal to 1 and 0.1 $\mathrm{s}^{-1}$ in (a), and to 100 and 10 $\mathrm{s}^{-1}$ in (b) (Venieratos & Goldbeter, 1979).

strate. The main reason for this is that the instability condition (2.26) is satisfied with more and more ease as parameter $q$ rises, in the region of negative slope on the product nullcline.

## 2.6 Role of enzyme cooperativity in the mechanism of oscillations

The instability mechanism that underlies the oscillations comprises two main components. The first is the positive feedback exerted by the reaction product on the enzyme; without it the system would remain stable whatever the values of the parameters. The nullcline $\gamma = \lambda\phi$ indeed admits no region of negative slope when the term $(1 + \gamma)^2$, which

Table 2.4. *Domain of values of the substrate injection rate producing oscillations as a function of the number of enzyme subunits, for one (q = 1) and two (q = 100) time scales*

| | $q = 1$ | | | $q = 100$ | | |
|---|---|---|---|---|---|---|
| $n$ | Lower bound | Upper bound | Width of oscillatory domain | Lower bound | Upper bound | Width of oscillatory domain |
| 2 | 0.13 | 1.51 | 11.61 | 0.12 | 1.87 | 15.50 |
| 3 | 0.058 | 0.66 | 11.38 | 0.053 | 2.27 | 42.90 |
| 4 | 0.041 | 0.32 | 7.80 | 0.034 | 1.89 | 54.94 |
| 6 | 0.030 | 0.14 | 4.60 | 0.021 | 0.99 | 48.77 |
| 8 | 0.027 | 0.08 | 2.96 | 0.015 | 0.58 | 39.69 |

For each value of $n$, the minimum and maximum values of the normalized substrate injection rate $v$ bounding the instability domain are indicated, as well as the extent of this domain measured by the ratio of these quantities. Venieratos & Goldbeter, 1979.

expresses the activation by the product, disappears from the numerator and from the denominator of function $\phi$ defined by eqn (2.11). The second component of the instability mechanism is enzyme cooperativity, which provides a source of nonlinearity that proves essential in amplifying the activation exerted by the product.

That enzyme cooperativity plays a primary role in the occurrence of periodic behaviour is already indicated by stability diagrams such as that shown in fig. 2.12. There, the domain of instability of the steady state corresponds to large values of the allosteric constant $L$, of the order of $10^5$ or more. Such values are associated with strong sigmoidicity of the kinetic curve characterizing the allosteric enzyme; this curve becomes hyperbolic as the allosteric constant approaches unity (Monod *et al.*, 1965; Rubin & Changeux, 1966). Cooperative effects require, indeed, that the enzyme be predominantly in the T state in the absence of substrate or effector. In the opposite case, which corresponds to low values of the allosteric constant, cooperative interactions that favour the transition from the T to the R conformation would have only minimal importance, since the enzyme would already be present in significant amounts of the latter state in the absence of ligand.

A phenomenological measure of positive or negative cooperativity commonly used in biochemistry for characterizing protein saturation by a ligand or velocity curves in enzyme kinetics is provided by the Hill coefficient, defined by eqn (2.27) with respect to some ligand of concentration $J$ (see, e.g. Cornish-Bowden & Koshland, 1975):

$$n_H = \frac{d \log M}{d \log J} \tag{2.27}$$

where $M = Y/(1 - Y)$ or $v_r/(V_M - v_r)$, according to whether experimental data are expressed as a function of the saturation function $Y$ (which varies from zero up to unity) or of the enzyme reaction rate $v_r$, whose maximum value is denoted by $V_M$. A value of the Hill coefficient above unity denotes positive cooperativity, and is associated with sigmoidicity of the experimental curve; a value equal to unity corresponds to hyperbolic curves characteristic of Michaelian enzymes, while a value of $n_H$ smaller than unity denotes a phenomenon of negative cooperativity (Levitzki & Koshland, 1969).

For the allosteric model based on a concerted transition between the two conformational states, as considered here in the model for glycolytic oscillations, application of definition (2.27) in the case where the two states of the enzyme differ only by the affinity toward the substrate ($\theta = 1$) leads to expression (2.28) for the Hill coefficient related to saturation of the enzyme by the substrate (Goldbeter, 1976, 1977):

$$n_H = 1 + \frac{L'\alpha W}{Q} \tag{2.28}$$

with:

$$W = (1 + \alpha)^{n-2} (1 + \alpha c)^{n-2} (1 + L'cn) (1 - c)^2 (n - 1)$$

$$Q = [(1 + \alpha)^{n-1} + L'c (1 + \alpha c)^{n-1}] (1 + L'cn) [(1 + \alpha)^{n-1} + L' (1 + \alpha c)^{n-1}]$$

where the apparent allosteric constant $L'$ is defined as:

$$L' = \frac{L}{(1 + \gamma)^n} \tag{2.29}$$

In these expressions, $\alpha$ is defined as the ratio $S/(K_R/e)$, while $c$ is equal to the ratio $(K_R e'/K_T e)$, where $e'$ and $e$ are defined by relations (2.10).

The analytical expression (2.28) allows us to establish an explicit link between enzyme cooperativity and instability of the steady state. This expression also permits us to determine the periodic variation of enzyme cooperativity in the course of oscillations.

Figure 2.24 shows how the Hill coefficient at steady state varies as a function of the allosteric constant $L$ for different values of the number

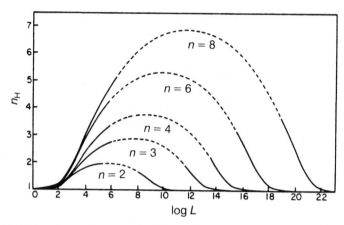

Fig. 2.24. Role of enzyme cooperativity in the mechanism of oscillations. The curves yield the Hill coefficient, $n_H$, at steady state (eqn (2.28)) as a function of the allosteric constant $L$, for increasing values of the number of enzyme subunits ($n = 2$ to $8$). On each curve, the dashed domain denotes the instability of the steady state (Goldbeter & Venieratos, 1980).

of enzyme subunits. The curves are established for a finite value of the nonexclusive binding coefficient $c$. As shown by Rubin & Changeux (1966), the curves $n_H$ vs $L$ possess a bell shape in such a case, given that the Hill coefficient tends towards unity, and cooperativity vanishes, at both low and large values of $L$. Indeed, the enzyme then behaves as a protein existing nearly completely in either the R or the T state; in these two extreme cases, the enzyme displays Michaelian behaviour. At intermediate values of the allosteric constant, the Hill coefficient approaches its maximum value equal to the number of subunits constituting the enzyme.

On each curve of fig. 2.24, the domain of instability of the steady state, determined according to condition (2.20), is represented by a dashed line. Two conclusions can be drawn from such an analysis. First, the domain of $L$ values corresponding to instability and oscillations broadens as the number of subunits increases. Second, the degree of positive cooperativity required for oscillations rises with $n$; the domain of values of $n_H$ associated with instability extends on both sides of the maximum of each curve, i.e. it is centred around the maximum cooperativity that can be reached for a given number of subunits. Moreover, the Hill coefficient remains close to its maximum value in the course of oscillations (fig. 2.25), at least when the substrate and product evolve on similar time scales.

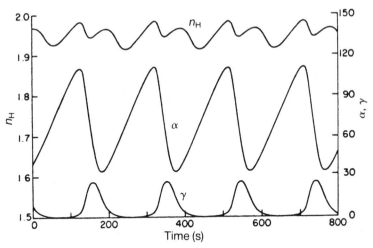

Fig. 2.25. Periodic variation of the Hill coefficient characteristic of substrate binding to the enzyme, in the case $n = 2$. The corresponding oscillations in substrate ($\alpha$) and product ($\gamma$) are also represented (Goldbeter, 1977).

The results of fig. 2.24 support the conclusion that for sustained oscillations to occur, the degree of positive cooperativity must exceed a threshold value. This conclusion suggests an explanation in terms of enzyme cooperativity for the inhibition of glycolytic oscillations by positive or negative effectors of PFK (Goldbeter & Venieratos, 1980). How does an activator of the enzyme such as the ammonium ion (Hess & Boiteux, 1968b), or an inhibitor such as citrate (Frenkel, 1968), suppress the oscillations? In the concerted allosteric model, the presence of an activator has the effect of decreasing the apparent value of the allosteric constant, as indicated by eqn (2.29), whereas the opposite holds for an inhibitor (Monod *et al.*, 1965). Consider a situation where the system oscillates and is therefore inside the instability domain of one of the bell-shaped curves of fig. 2.24. The addition of an activator such as the ammonium ion will result in a lowering of the allosteric constant; the point representative of the steady state of the system will thus be displaced towards the left. For a sufficiently large addition of positive effector, the steady state will become stable if the degree of cooperativity drops below the threshold required for instability. Likewise, the addition of an inhibitor such as citrate will result in a displacement of the steady state towards the right. If this displacement is sufficiently important, cooperativity should again fall below the threshold required for instability. Thus, for both types of effector, oscillations will disappear as

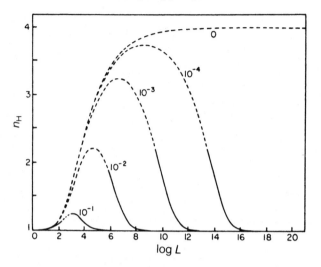

Fig. 2.26. The Hill coefficient at steady state and domain of instability (dashed line) as a function of the allosteric constant $L$ and of the nonexclusive binding coefficient of the substrate, $c$, for a tetrameric enzyme. The value of $c$ is indicated for each curve (Goldbeter & Venieratos, 1980).

soon as the positive cooperativity of the enzyme decreases below a critical value.

This analysis predicts that the addition of a negative effector should not inhibit the oscillations in the case of an exclusive binding of the substrate to the R state of the enzyme. As indicated in fig. 2.26, indeed, the curve $n_H$ vs $L$ loses its bell shape when coefficient $c \to 0$; the degree of cooperativity remains maximal at large values of the allosteric constant so that the steady state remains unstable regardless of the amount of inhibitor added to the system. The fact that an inhibitor of PFK, such as citrate, suppresses glycolytic oscillations would therefore suggest that the substrate binds in a nonexclusive manner to the two conformations of the enzyme.

The existence of two time scales within the system lowers the threshold of cooperativity required for oscillations. As indicated in fig. 2.27, oscillations can indeed occur for values of the Hill coefficient close to unity when the product concentration varies more rapidly than that of the substrate. The importance of enzyme cooperativity in the mechanism of oscillations appears to be reduced in these conditions.

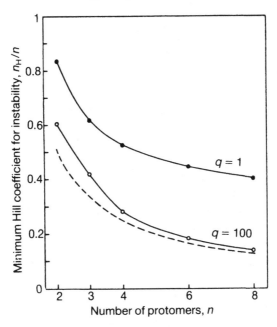

Fig. 2.27. Minimum Hill coefficient required for instability of the steady state in the presence ($q = 100$) and absence ($q = 1$) of two time scales. The curves are established as a function of the number of subunits constituting the enzyme ($n = 2$ to $8$). The value of $n_H$ has been normalized by dividing by $n$. The dashed curve denotes the value $n_H = 1$ (Goldbeter & Venieratos, 1980).

## 2.7 Effect of a saturable sink of the reaction product

In the model studied so far, the only source of nonlinearity arises from the allosteric nature of PFK kinetics and from the regulation of the enzyme by positive feedback. The sink of product was supposed to remain linear. This assumption, which greatly simplifies the calculations, corresponds to the fact that ADP is used as substrate by several glycolytic enzymes and therefore does not accumulate within the system. This would not occur if the sink reaction(s) were saturated by ADP. It is nevertheless instructive to consider the case of a saturable sink, of the Michaelian type, in order to better understand the role of nonlinearities in the mechanism of oscillations. Maybe the degree of cooperativity of the allosteric enzyme required for oscillations might diminish once an additional nonlinearity is present within the system?

When the removal of product is catalysed by a Michaelian enzyme, eqns (2.7) take the form of:

$$\frac{d\alpha}{dt} = v - \sigma \phi$$

$$\frac{d\gamma}{dt} = q\sigma \phi - \frac{r_s \gamma}{\mu + \gamma} \qquad (2.30)$$

where $r_s$ denotes the maximum rate of the sink reaction divided by the dissociation constant $K_P$ of the product for the regulatory site of the allosteric enzyme, and $\mu$ denotes the Michaelis constant of the enzyme catalysing the transformation of the product, also divided by $K_P$. The rate function $\phi$ remains defined by eqn (2.8) or (2.11).

The linear stability analysis of eqns (2.30) shows that the domain of instability of the unique steady state is larger than in the case of a linear sink of product. The main effect of the Michaelian sink of product is, however, to allow the occurrence of sustained oscillations in the absence of enzyme cooperativity – but not of autocatalytic regulation (Goldbeter & Dupont, 1990).

In the phase plane, the product nullcline is now defined by:

$$\gamma = q\sigma \phi \frac{\mu}{r_s - q\sigma \phi} \qquad (2.31)$$

This nullcline retains an S-shape for $n > 1$. In the case of an enzyme comprising a single subunit, for which the phenomenon of cooperativity is necessarily absent, the nullcline $\gamma = \lambda \phi$, obtained for a linear sink, did not possess any region of negative slope. This is no more the case for eqn (2.31), which can admit up to two distinct values of $\gamma$ for a given value of $\alpha$ (fig. 2.28).

When the steady state lies in a region of sufficiently negative slope on the product nullcline given by eqn (2.31) for $n = 1$, it becomes unstable and sustained oscillations occur. These oscillations are characterized by an extremely large amplitude of the product, due to the absence of an ascending right limb of the product nullcline at high values of $\gamma$ (fig. 2.28): the maximum of the product concentration indeed occurs only when the trajectory encounters the limb of the nullcline close to the axis of the abscissas.

The above results show that cooperativity is not required absolutely for oscillations, and that the strong nonlinearity of the regulated step can be replaced by **diffuse nonlinearities** distributed over several steps of the system. These milder nonlinearities, of the Michaelian type, combine to raise the overall degree of nonlinearity up to the level required for sustained oscillatory behaviour. In biochemical systems controlled

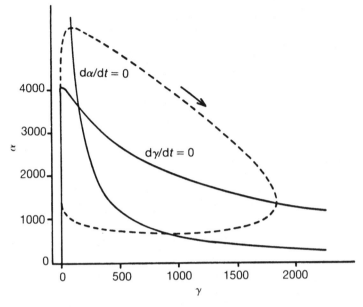

Fig. 2.28. Product nullcline and limit cycle in the absence of cooperativity ($n = 1$) in the allosteric model with positive feedback when the product sink is of Michaelian nature (Goldbeter & Dupont, 1990).

by end-product inhibition, the effect of a saturable rather than linear sink for the end product is, similarly, to reduce the degree of feedback cooperativity required for sustained oscillations (Thron, 1991).

If, as shown by the analysis of the situation of a Michaelian sink of reaction product, enzyme cooperativity is not a necessary prerequisite for sustained glycolytic oscillations, the analysis of the model nevertheless indicates that the nonlinearity associated with enzyme cooperativity favours periodic behaviour and, in most circumstances, remains essential for its occurrence.

## 2.8 Effect of a periodic source of substrate: entrainment and chaos

In what precedes, the input of substrate in the model remains constant in the course of time. This hypothesis fits the conditions adopted for most experiments on glycolytic oscillations. In order to compare experimental and theoretical results in different conditions, the analysis of the model was extended to the case of a periodic or stochastic source of substrate and the corresponding experiments were performed in yeast extracts (Boiteux, Goldbeter & Hess, 1975; Goldbeter & Nicolis, 1976).

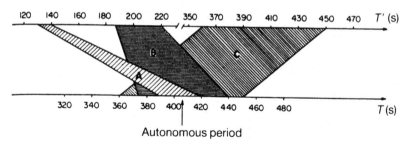

Fig. 2.29. Entrainment of the oscillations by a periodic source of substrate in the allosteric model for glycolytic oscillations. The period of the sinusoidal source and the resulting period of the oscillatory enzyme are denoted by $T'$ and $T$, respectively. Domains C, B and A correspond to entrainment by the fundamental frequency, and by the 1/2 and 1/3 harmonics of the forcing input (Boiteux *et al.*, 1975).

The sinusoidal source of period $T'$ used in the simulations is given by the relation:

$$v = 0.5 + 0.25 \sin (2 \pi t/T') \qquad (2.32)$$

Numerical integration of eqns (2.7) in which source $v$ is given by relation (2.32) permits the construction of a diagram (fig. 2.29) where the main domains of entrainment of the oscillatory enzyme by the external periodicity are indicated. For a constant source $v = 0.5 \text{ s}^{-1}$ equal to the mean value chosen for the periodic input, the autonomous period of the system is equal to 406 s. Domains C, B and A indicate, respectively, entrainment of the oscillatory enzyme to the fundamental frequency of the external source $(T = T')$, and entrainment by the harmonics $\frac{1}{2} (T = 2T')$ and $\frac{1}{3} (T = 3T')$. In all these cases, the period and the phase of the oscillations with respect to the source remain constant after entrainment.

As indicated in fig. 2.29, subharmonic entrainment at a fraction of the fundamental frequency, which is characteristic of nonlinear systems (Hayashi, 1964), occurs in smaller and smaller domains of the external period as the value of that fraction diminishes. A similar phenomenon is observed in the entrainment of circadian rhythms by light–dark cycles (Pittendrigh, 1960, 1965); the latter property underlies the adaptation of most living organisms to periodic variations in their environment.

Experiments corresponding to the simulations of fig. 2.29 were conducted in yeast extracts (Boiteux *et al.*, 1975), by means of a pump capable of injecting the substrate at 10 different rates. The periodic source utilized had a piecewise, square-wave shape, represented by the dark trace at the bottom of each part of fig. 2.30. Parts b and c in that figure

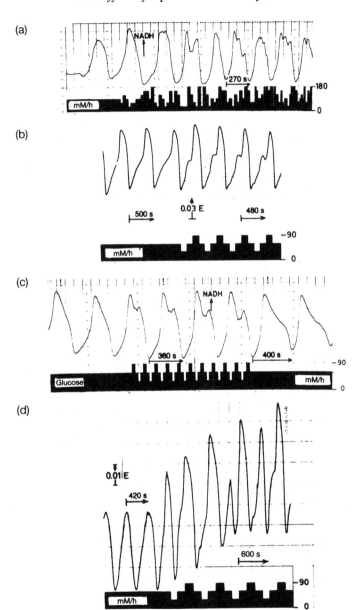

Fig. 2.30. Experimental study of the effect of a (a) stochastic or (b–d) periodic variation of the substrate source on glycolytic oscillations in yeast extracts. The variation of the substrate input is indicated in black at the bottom of each panel. The entrained system oscillates with a period equal to the driving period in (b) and with a period equal to the triple of the external period in (c). In (d), the enzyme fails to be entrained by the external source and oscillates in an irregular (chaotic?) manner (Boiteux *et al.*, 1975).

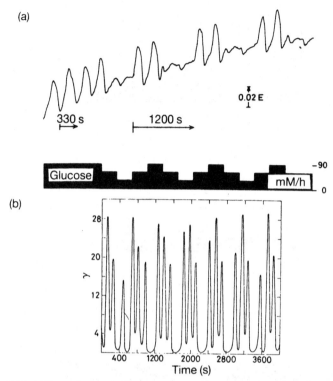

Fig. 2.31. Effect of a periodic source of substrate whose period exceeds that of glycolytic oscillations. (a) The experiments on yeast extracts and (b) the simulations of the allosteric model indicate the occurrence of oscillations characterized by the superposition of internal and external frequencies (Boiteux *et al.*, 1975).

illustrate, respectively, entrainment of period $T$ to the value $T'$ of the periodic input, and subharmonic entrainment to the value $3T'$. In each case, the entrained system recovers its autonomous period when the source of substrate regains its constancy. The agreement between model and experiment extends to the case where the system is entrained by longer periods. As shown in fig. 2.31a and b, the external and internal frequencies can both be distinguished in the behaviour of the entrained system when the external periodicity is much longer than the autonomous period.

The effect of a stochastic source of substrate in the model is represented in fig. 2.32. Although parameter $v$ changes in a random manner over the entire domain of values, producing sustained oscillations with a period that may differ by a factor of up to 2, the interval between suc-

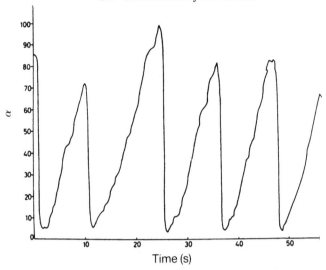

Fig. 2.32. Effect of a stochastic source of substrate on the oscillatory behaviour of the allosteric model with positive feedback (Goldbeter & Nicolis, 1976).

cessive peaks does not present such a large variation and remains rather regular. This stability of the oscillations with respect to external noise is also seen in the experiments, as shown in fig. 2.30a. This study of the effect of a stochastic source, of course, is not a thorough one, but it gives general indications of the filtering of such perturbations. In addition, external noise is known to influence nonequilibrium transitions by altering the values at which control parameters elicit bifurcations (Horsthemke & Lefever, 1984).

Markus *et al.* (1984) have pursued the experimental analysis of the effect of an external periodicity on glycolytic oscillations. Their study confirmed the results obtained for entrainment and showed, moreover, the possibility of chaotic behaviour (see chapter 4). The same authors (Markus & Hess, 1984; Markus *et al.*, 1985) also conducted a parallel study of a four-variable model for glycolytic oscillations, centred around the allosteric kinetics of PFK; in the presence of a periodic source of substrate, that model predicted the appearance of aperiodic oscillations and of multiple rhythms in the forced glycolytic system (see also Tomita & Daido, 1980).

## 2.9 Extensions of the model for glycolytic oscillations

Although the allosteric model for glycolytic oscillations is relatively simple, it provides a qualitative and, in significant measure, a

Table 2.5. *Interaction of the glycolytic oscillator with a periodic source of substrate*

| Relation between $T'$ and $T_0$ | Interaction |
|---|---|
| $T'/T_0 \approx 1/n \, (n = 2, 3 \ldots)$ | Entrainment by the $1/n$ subharmonic of the input frequency (fig. 2.30c) |
| $0.7 \leq T'/T_0 \leq 1.2$ | Entrainment by the fundamental frequency of the input (fig. 2.30b) |
| $1.2 < T'/T_0 < 1.6$ $T'/T_0 > 3$ | No entrainment (fig. 2.30d) Double periodicity: separation of autonomous and input frequencies (fig. 2.31a) |

The table summarizes experimental observations in yeast extracts; $T'$ and $T_0$ denote, respectively, the external period and the autonomous period of the oscillatory enzyme.
Boiteux *et al.*, 1975.

quantitative agreement with a large number of experimental observations. This agreement pertains to the situation of a periodic source of substrate (see table 2.5) as well as to that of a constant source (table 2.6). The fact that the model contains only two variables allows us to understand, by means of phase plane analysis, the effect of the substrate injection rate, which is the most important parameter controlling the oscillations. This relative simplicity of the model permits us to study, in a more abstract manner, in the two following chapters, how suitable modifications can lead biochemical systems from simple periodic oscillations to more and more complex modes of oscillatory behaviour.

More complete models for glycolytic oscillations, centred around the PFK autocatalytic reaction, have been developed; these models incorporate a larger number of enzymes of the glycolytic chain (Sel'kov, 1968b; Schellenberger, Eschrich & Hofmann, 1978; Demongeot & Seydoux, 1979; Tornheim, 1979; Reich & Sel'kov, 1981; Termonia & Ross, 1981; Markus & Hess, 1984; Hocker *et al.*, 1994). A model specifically based on the kinetic data obtained for skeletal muscle PFK has been presented by Smolen (1995). These developments permit the refinement of the quantitative agreement with experimental observations by taking into account the detailed kinetic properties of PFK and of the other glycolytic enzymes. However, this gain is often counterbalanced by the increased complexity of the system of kinetic equations.

Table 2.6. *Comparison of oscillatory behaviour in the allosteric model with positive feedback and in the experiments on glycolytic oscillations in yeast extracts, in the case of a constant source of substrate*

| Sustained oscillations | Model | Experiment |
|---|---|---|
| Oscillatory range of substrate injection rate ($v_i$) | 19–246 mM/h | 20–160 mM/h |
| Period | Of the order of min; decreases by a factor $\leq 10$ as $v_i$ increases | Of the order of min; decreases by a factor $\leq 10$ as $v_i$ increases |
| Amplitude | In the range $10^{-5}$–$10^{-3}$ M; passes through a maximum as $v_i$ increases | In the range $10^{-5}$–$10^{-3}$ M, passes through a maximum as $v_i$ increases |
| Periodic change in PFK activity (in per cent $V_{max}$) | Minimum 0.95; maximum 73; mean 17.5; activation factor 77 | Minimum 1; maximum 80; mean 16; activation factor 80 |
| Phase-shift by ADP | Delay of 1–2 min upon addition of 0.7 mM ADP (14 units of $\gamma$) around the minimum of ADP oscillations of 5 min period; small phase advance when the addition precedes ADP maximum | Delay of 1.5 min upon addition of 0.7 mM ADP at the minimum of ADP oscillations of 5 min period; small phase advance when the addition precedes ADP maximum |

The comparison bears on the following aspects of oscillations: range of substrate injection rate producing the phenomenon, period and amplitude, periodic variation of phosphofructokinase (PFK), and phase shift induced by the addition of ADP.
Boiteux *et al.*, 1975.

As indicated by the experiments on the reconstitution of a minimum system oscillating *in vitro* (Eschrich *et al.*, 1983), the allosteric kinetics of PFK and its autocatalytic nature represent the core of the mechanism of glycolytic oscillations; these elements are central to the two-variable model presented above.

The question arises as to whether this simple model is not invalidated by the fact that it does not take into account the existence of important effectors of PFK such as fructose 2,6-bisphosphate (Hers, 1984). The role of the latter positive effector can be incorporated in the two-variable model in an implicit manner, if we consider that the effective value of the allosteric constant $L$ diminishes when the concentration of such an activator increases. The stability diagrams of figs. 2.12 and 2.24 indicate that oscillations should disappear when the effective value of the allosteric constant decreases below a critical level. Such a result agrees with the observation in yeast (Boiteux & Müller, 1983) and muscle (Tornheim, 1988) that the addition of sufficient amounts of fructose 2,6-bisphosphate abolishes glycolytic oscillations.

### 2.10  Spatial and spatiotemporal patterns in the allosteric model for the phosphofructokinase reaction

The analysis of the allosteric model can be extended to the spatially inhomogeneous case; eqns (2.7) then take the form of partial differential equations. Besides instabilities that lead to temporal organization in the form of sustained oscillations, instabilities with respect to diffusion can also occur in such conditions. The theoretical analysis of the model indicates (Goldbeter & Lefever, 1972; Goldbeter & Nicolis, 1976) that, as for the Sel'kov model (Prigogine *et al.*, 1969), dissipative structures of the Turing (1952) type corresponding to a spatially inhomogeneous distribution of substrate and product can in principle be established. The relative values of the diffusion coefficients of the two metabolites required for the establishment of such stationary spatial structures are, however, too much apart from each other to make such a situation plausible. When the constraint set by taking a realistic ratio of diffusion coefficients for the product and substrate is bypassed, such spatial patterns can readily be observed, as shown recently (Hasslacher, Kapral & Lawniczak, 1993) for the Sel'kov model.

Other types of spatiotemporal organization may nevertheless be observed for realistic values of the diffusion coefficients in the allosteric enzyme model. Thus, in the case of boundary conditions corresponding to fixed concentrations of substrate and product, the coupling between reaction and diffusion leads to the propagation of concentration waves when the system possesses an appropriate size (Goldbeter, 1973; Goldbeter & Nicolis, 1976). Similar results have been obtained in two-dimensional simulations of the model (Marmillot, Hervagault & Welch, 1992).

As indicated above, the spatial aspects were disregarded in this chapter, which was devoted primarily to the analysis of temporal oscillatory behaviour. The neglect of diffusion is justified by the fact that the experiments on glycolytic oscillations were performed in continuously stirred extracts of yeast.

## 2.11 What function for glycolytic oscillations?

Since the observation of glycolytic oscillations in yeast and muscle, the question of their possible function has remained a puzzle. Chance *et al.* (1964), noticing that the period of the oscillations can increase to more than 1 h when the temperature decreases (a similar phenomenon is observed upon diluting the glycolytic enzymes in yeast extracts; see Das & Busse, 1985), suggested at first that such high-frequency biochemical oscillations could underlie circadian rhythms. The latter, whose period is close to 24 h, are observed in nearly all living organisms. However, circadian rhythms are characterized by the stability of their period with respect to temperature (Pittendrigh, 1954; Bünning, 1973). Pavlidis & Kauzmann (1969) have nevertheless proposed mechanisms displaying temperature compensation, based on oscillatory enzyme reactions such as that present in glycolysis.

Although the model for PFK can in principle generate circadian periodicities when the concentration of the enzymes of the system and the source of substrate are sufficiently low, it is doubtful whether glycolytic oscillations are responsible for any circadian rhythm: these rhythms appear to originate from regulatory mechanisms involving protein synthesis and its modulation by transcriptional control (Hardin *et al.*, 1990, 1992; Takahashi, 1993; see also chapter 11).

Another possible role for glycolytic oscillations has been proposed in terms of the adenylate energy charge (Goldbeter, 1974). According to Atkinson (1968, 1977), enzymes belonging to biosynthetic pathways utilizing ATP are activated at values of the charge higher than 0.8, while enzymes belonging to ATP-regenerating pathways would rather be inhibited at such values (see fig. 2.9). The model for glycolytic oscillations allows us to compare the values of the energy charge at steady state and during oscillations. In the system governed by eqns (2.7), when the dissociation constants for ATP and ADP are equal ($q = 1$), the adenylate charge is given by the ratio $[\alpha + (\gamma/2)]/[\alpha + \gamma]$, given that AMP is not included in the two-variable model. That AMP is not considered should not matter too much, since the concentration of this metabolite is negligible with respect to that of ADP and ATP in the course of glycolytic oscillations (Betz & Moore, 1967).

Figure 2.33 shows the value of the energy charge at the unstable

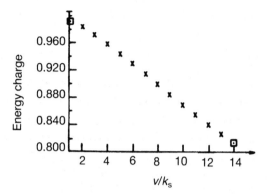

Fig. 2.33. Adenylate energy charge at steady state in the two-variable, allosteric model for glycolytic oscillations (Goldbeter, 1974).

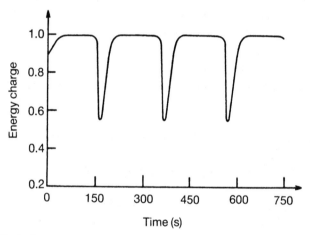

Fig. 2.34. Periodic variation of the adenylate charge in the course of sustained oscillations in the allosteric model with positive feedback. At the unstable steady state, the charge is equal to 0.945 (Goldbeter, 1974).

steady state in the range of substrate injection rates giving rise to oscillations. In this domain, the energy charge is larger than 0.8. The periodic variation of the charge is represented in fig. 2.34 for a value of $v$ located in the middle of the instability domain. While at steady state the elevated value of the charge should favour the operation of biosynthetic reactions at the expense of ATP-regenerating pathways, a periodic variation of the energy charge between values above and below 0.8 would permit, in contrast, a regular alternation of biosyntheses and ATP regeneration. It is conceivable that the cell could acquire in such a manner a versatility that would permit better adaptation to environmental variations. The possible role of glycolytic oscillations in permitting an alternation between opposite biochemical pathways and thereby avoiding futile cycles in metabolism has been further stressed by Sel'kov and coworkers (Boiteux, Hess & Sel'kov, 1980; Reich & Sel'kov, 1981).

Other workers have also speculated about the possible function of glycolytic oscillations. Thus, Tornheim (1979) suggested that these oscillations in muscle enhance the mean ATP/ADP ratio; this suggestion is supported by recent experimental observations on glycolytic oscillations in skeletal muscle extracts (Tornheim, 1988; Andrés, Schultz & Tornheim, 1990; Tornheim *et al.*, 1991). A related thermodynamic argument was put forward by Richter & Ross (1980), who proposed that PFK acts as a periodic force driving the lower part of glycolysis, where a second enzyme, pyruvate kinase, would behave as a damped oscillator; a resonance phenomenon would result in increasing the efficiency of the glycolytic system by 5–10%. There exists, however, no data on glycolytic oscillations *in vivo* that would permit us to corroborate or disprove the latter conjecture. The possibility of a second instability-generating mechanism in the lower part of glycolysis has been considered by Dynnik & Sel'kov (1973; see also Dynnik, Sel'kov & Ovtchinnikov, 1977).

An alternative, plausible view is that glycolytic oscillations have no particular function in cellular metabolism. Although yeast cells can rapidly synchronize their oscillatory behaviour (Ghosh *et al.*, 1971), it is doubtful whether they use it as a means of intercellular communication, as do *Dictyostelium* amoebae (see chapter 5). Glycolytic oscillations, therefore, may appear as a metabolic accident that would necessarily result from the autocatalytic regulatory properties of PFK; these properties in turn originate from the role played by this ATP-utilizing enzyme within a metabolic pathway whose function is to generate ATP.

The fact that the autocatalytic regulation of PFK was not eliminated in the course of evolution indicates, in any case, that glycolytic oscillations are by no means unfavourable to the cell. If they do represent nothing more than an accident, it is a fortunate one since it provides us with a precious prototype for studying the mechanism of biological rhythms at the molecular level. This function of glycolytic oscillations could well turn out to be the most significant, and suffices to make them a fascinating subject for investigation. Because the phenomenon is observed in cell-free extracts, it also provides one of the best biochemical systems for pursuing experiments on the occurrence of more complex oscillatory phenomena such as multiple oscillations and chaotic dynamics (see chapters 3 and 4).

Recent observations have nevertheless revived interest in the physiological significance of glycolytic oscillations. These highly interesting but still preliminary observations have led to the suggestion, which remains to be corroborated by further experiments, that glycolytic oscillations drive the pulsatile secretion of insulin in pancreatic β-cells. The suggestion was first made in a study by Lipkin, Teller & de Haën (1983), who showed that glycolytic oscillations occur with a period close to 14 min in perifused rat fat cells synchronized by insulin or hydrogen peroxide. On the basis of experimental results obtained in permeabilized insulinoma cells mixed with skeletal muscle extracts, Corkey *et al.* (1988) elaborated this conjecture in further detail by showing that oscillations in the ATP/ADP ratio produced by the glycolysing muscle extracts causes concomitant oscillations of free $Ca^{2+}$ in these cells.

Corkey *et al.* propose a mechanism, schematized in fig. 2.35, by which glycolytic oscillations in pancreatic β-cells would drive the pulsatile secretion of insulin: oscillations in the ATP/ADP ratio, caused by the periodic evolution of glycolysis, would periodically change the state of an ATP-sensitive $K^+$ channel that is known to control glucose-induced electrical activity in β-cells (Henquin, 1988). Closure of this channel when the ATP/ADP ratio rises would cause cell depolarization and the subsequent entry of $Ca^{2+}$ through voltage-sensitive channels; the high ATP/ADP ratio would also stimulate $Ca^{2+}$ pumping into cellular reservoirs while the increase in free cytosolic $Ca^{2+}$ would lead to insulin secretion (Corkey *et al.*, 1988; Tornheim *et al.*, 1991).

An attractive aspect of this proposal is that it provides a direct link between glucose metabolism – which is needed for insulin secretion – and the secretory process; an increase in the glucose level would indeed transform into an increase in the frequency of glycolytic oscillations and would cause the more frequent depolarization of pancreatic β-cells, thus leading to enhanced secretion of insulin. In support of this scheme, preliminary evidence has been obtained for the occurrence of glycolytic oscillations in the pancreatic islets of Langerhans, in which oscillations in lactate production have been observed, with a period close to that of insulin secretion (Chou *et al.*, 1990, 1992). In line with the suggested mechanism of fig. 2.35, experiments in which NADH fluorescence has been measured in single islet β-cells indicate that the $Ca^{2+}$ influx through voltage-sensitive channels follows the metabolic response to nutrient stimuli. These experiments pave the way for a detailed study of

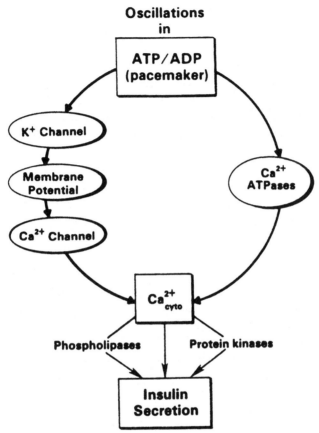

Fig. 2.35. Scheme showing the putative mechanism by which glycolytic oscillations might drive the pulsatile secretion of insulin in pancreatic β-cells (Corkey *et al.*, 1988).

glycolytic oscillations in pancreatic β-cells and their relation to pulsatile insulin secretion.

Although the attractive possibility that glycolytic oscillations in β-cells drive the pulsatile secretion of insulin remains to be established, additional questions remain to be addressed. Thus, there exists a hierarchy of rhythms in the electrical activity of these cells; besides the putative long-period rhythm associated with glycolytic oscillations, bursting of the membrane potential induced by glucose has been extensively studied, with a period of the order of 10 s (Perez-Armendariz, Atwater & Rojas, 1985). The question arises as to the relevance of this phenomenon to pulsatile insulin secretion, which takes place on a much longer time scale. In addition, glucose-induced oscillations in intracellular

$Ca^{2+}$ of a period of the order of 2 min have been reported (Valdeolmillos *et al.*, 1989); their link with insulin secretion remains to be established.

While oscillations in cytosolic free $Ca^{2+}$ with a period close to 5 min have also been demonstrated in conjunction with oscillations in oxygen consumption and insulin secretion in glucose-stimulated rat pancreatic cells (Longo *et al.*, 1991), the question arises as to the origin of the periodic phenomenon driving these processes. It will be important to determine whether these oscillations in $Ca^{2+}$ are brought about by glycolytic oscillations or whether they originate, as in many other cell types (see chapter 9), from a mechanism based on the regulation of $Ca^{2+}$ transport between the cytosol and intracellular pools. The possibility exists of an interplay between such a $Ca^{2+}$ oscillator and the one based putatively on glycolytic oscillations.

Glycolytic oscillations have recently been demonstrated in yet another type of cell, belonging to the cardiac tissue (O'Rourke *et al.*, 1994). Here, the physiological function of glycolytic oscillations is unknown, but it is likely that they may give rise to certain forms of arrhythmia, since they can modulate the membrane potential in these cells. In a similar manner, oscillations of intracellular $Ca^{2+}$ (see chapter 9) in heart cells provide another source of cardiac arrhythmia.

# Part II

From simple to complex oscillatory
behaviour

# 3

# Birhythmicity: coexistence between two stable rhythms

## 3.1 A two-variable biochemical model for birhythmicity

### Birhythmicity: coexistence of two stable limit cycles

The model governed by eqns (2.7) or (2.30) admits at most a single limit cycle, i.e. only one type of periodic behaviour for a given set of parameter values. The analysis of a three-variable model considered in chapter 4 has led to the fortuitous observation of a phenomenon of coexistence between two types of oscillation corresponding to two simultaneously stable limit cycles (Decroly & Goldbeter, 1982). This phenomenon represents a particular type of **bistability**, much less common than the one involving the coexistence between two simultaneously stable steady states, of which several examples are known in chemistry (Pacault *et al.*, 1976; Epstein, 1984) as well as biochemistry (Degn, 1968; Naparstek *et al.*, 1974; Eschrich *et al.*, 1980, 1990), or that involving the coexistence between a stable steady state and a stable limit cycle; the latter phenomenon is known as **hard excitation** (Minorsky, 1962). To differentiate it from these two types of bistability, Decroly & Goldbeter (1982) coined the term **birhythmicity** to denote the coexistence between two simultaneously stable limit cycles.

Birhythmicity has not yet been clearly demonstrated in biological systems, although some observations suggest that this type of dynamic behaviour might occur in cardiac tissue (Mines, 1913; Gilmour *et al.*, 1983) and in a neuronal preparation (Hounsgaard *et al.*, 1988). Following the theoretical predictions of the phenomenon by the model analysed in chapter 4, an experimental study of a chemical system involving two coupled oscillatory reactions permitted the demonstration of a coexistence between two simultaneously stable periodic

91

regimes (Alamgir & Epstein, 1983; Citri & Epstein, 1988). Other examples of chemical birhythmicity have been observed (Roux, 1983; Lamba
& Hudson, 1985; Johnson, Griffith & Scott, 1991). As regards biochemical systems, besides the models discussed in this and following chapters,
birhythmicity has also been found in models of the peroxidase reaction
(Larter *et al.*, 1993), of an enzyme reaction regulated by negative feedback (Palsson & Groshans, 1988), and of calcium metabolism (Tracqui
*et al.*, 1987; Tracqui, 1993), while preliminary evidence (Markus & Hess,
1990) suggests its occurrence in yeast glycolysis. Because of its potential
interest for biological systems, it is desirable to comprehend the origin
and the dynamic characteristics of this new type of bistability.

The analysis of the three-variable biochemical model (Decroly &
Goldbeter, 1982) developed further below (see chapter 4) suggests a
scenario for the origin of birhythmicity, which leads to a hypothesis for
the occurrence of the phenomenon. Testing this conjecture in a two-
variable system has the advantage of allowing a direct resort to phase
plane analysis. To this end, Moran & Goldbeter (1984) have modified
the allosteric model for glycolytic oscillations considered in chapter 2 by
supplementing it with a reaction in which the product is recycled into
substrate (fig. 3.1). The reason underlying this modification is that such
a reaction, without increasing the number of variables, allows the creation of a second region of negative slope on the product nullcline; as
shown below, the existence of two regions of negative slope on the sigmoid nullcline plays an essential role in the origin of birhythmicity.
That multiple stable limit cycles may arise in models for product-
activated enzymes was also suggested by Erle (1981). His mathematical

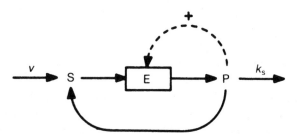

Fig. 3.1. Model of an enzyme reaction with positive feedback and recycling of
product (P) into substrate (S). This model serves as a two-variable prototype
for the study of birhythmicity and multiple domains of oscillations (Moran &
Goldbeter, 1984).

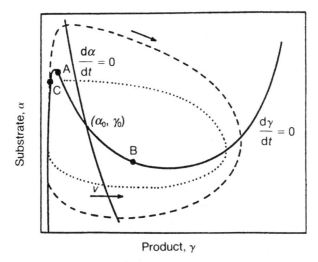

Product, $\gamma$

Fig. 3.2. Phase plane portrait in the absence of product recycling into substrate. The substrate ($\alpha$) and product ($\gamma$) nullclines are represented in a situation corresponding to an unstable steady state ($\alpha_0$, $\gamma_0$) surrounded by a stable limit cycle (dashed line). The instability domain extends, schematically, from A to B in the region of negative slope on the product nullcline. The arrow across the substrate nullcline indicates the direction of its displacement when the substrate injection rate $v$ increases. When the two nullclines intersect at C, on the left of the instability domain, the steady state is stable but excitable: an increase in product is then amplified in a pulsatory manner (dotted line) when the amplitude of the perturbation exceeds a threshold (Goldbeter & Moran, 1987).

arguments were also based on the form of nullclines, but this form was assumed *a priori*, rather than following from a biochemical mechanism.

### *Properties of the model in the absence of recycling: excitability with a single threshold, oscillations and bistability*

The main properties of the two-variable allosteric model for glycolytic oscillations, in the absence of product recycling into substrate, are summarized in fig. 3.2. In the phase plane ($\alpha$, $\gamma$), the system governed by eqns (2.7) evolves toward a limit cycle (dashed line) when the steady state, located at the intersection of the nullclines $(d\alpha/dt) = 0$, $(d\gamma/dt) = 0$, lies in a region of sufficiently negative slope $(d\alpha/d\gamma)$ on the latter nullcline. In fig. 3.2, this region extends, schematically, from A to B.

Two other phenomena follow from the structure of the nullclines as represented in fig. 3.2. First, when the steady state lies in the region of stability located to the left of point A, for example in C, the system becomes excitable: in a pulsatory manner it amplifies a perturbation once the latter exceeds a threshold. Indeed, when the perturbation takes the form of an instantaneous addition of product, the initial condition corresponds to a horizontal displacement to the right of the steady state; when this displacement passes the median limb of the sigmoid nullcline, the system undergoes a large excursion in the phase plane (dotted line in fig. 3.2) towards the right limb of the sigmoid before returning to the stable steady state.

The median branch of the product nullcline thus defines an abrupt threshold for excitability: when the initial condition corresponding to the addition of a pulse of product lies to the left of the threshold, the perturbation regresses and the system immediately returns to the steady state; the perturbation, on the contrary, is amplified when it exceeds the threshold. The role of the nullcline in this regard results from the fact that this curve, which obeys the equation $(d\gamma/dt) = 0$, separates the regions of the phase plane in which $\gamma$ will tend to grow or decay in the course of time.

On the other hand, the S (or rather the N) shape of the product nullcline implies the existence of a phenomenon of bistability when the substrate concentration is held constant in the course of time. Three steady states can indeed be obtained when the horizontal corresponding to the fixed value of $\alpha$ intersects the product nullcline in the region where the latter possesses a region of negative slope. The numerical integration of eqns (2.7) in these conditions reveals the evolution to either one of two simultaneously stable steady states; the choice of the final state depends on the initial product concentration, $\gamma_i$ (fig. 3.3). Each of the two stable states possesses its own basin of attraction, which is the set of all initial conditions in the phase plane from which the system evolves to reach eventually that particular state.

### Model with recycling

The two-variable model for birhythmicity is built on the basis of eqns (2.7) by incorporating into them a term related to the transformation of product into substrate, in a reaction catalysed by an enzyme whose cooperative kinetics is described by a Hill equation, characterized by a degree of cooperativity $n$. The kinetic equations of the model thus take the form of eqns (3.1) where the various parameters remain defined as for eqns (2.7) and (2.11):

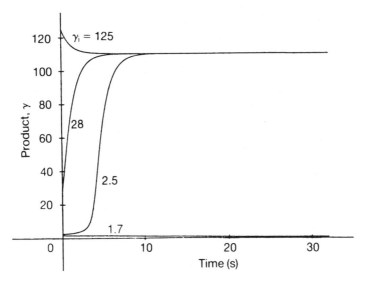

Fig. 3.3. Bistability. Two stable steady states corresponding to two distinct values of product $\gamma$ coexist in the model without recycling when the substrate concentration is held constant in the domain of existence of a region of negative slope on the product nullcline in fig. 3.2 (Goldbeter & Moran, 1987).

$$\frac{d\alpha}{dt} = v + \frac{\sigma_i \, \gamma^n}{K^n + \gamma^n} - \sigma \, \phi$$

$$\frac{d\gamma}{dt} = q\sigma \, \phi - k_s \, \gamma - \frac{q\sigma_i \, \gamma^n}{K^n + \gamma^n} \tag{3.1}$$

Parameter $\sigma_i$ denotes the maximum rate of recycling, divided by the Michaelis constant of the substrate for the autocatalytically regulated allosteric enzyme; constant $K$ is equal to the product concentration for which the recycling rate reaches its half-maximum value. The rate function of the product-activated allosteric enzyme is identical with that defined for eqns (2.7) with $c = 0$. It is given by eqn (3.2), identical with relation (2.11), which corresponds to a dimeric enzyme that binds the substrate only in the R state:

$$\phi = \frac{\alpha \, (1 + \alpha) \, (1 + \gamma)^2}{L + (1 + \alpha)^2 \, (1 + \gamma)^2} \tag{3.2}$$

The linear stability analysis of these equations, coupled to phase plane analysis, again leads to the instability condition (2.26) obtained for the model in the absence of product recycling.

To understand how the term for recycling influences the form of the product nullcline, it is first necessary to grasp intuitively the reason why the curve $(d\gamma/dt) = 0$ in fig 3.2 possesses a region of negative slope. The existence of such a region is due to the autocatalytic regulation of the enzyme. In the absence of autocatalysis, indeed, the curve $\gamma = \lambda\phi$, which is the locus of steady states as a function of $v$, admits at most a single intersection with a horizontal corresponding to a given value of $\alpha$; this implies the absence of any region of negative slope on the nullcline and, consequently, the stability of the steady state.

The position of the steady state on the nullcline is given by relation (2.12) as a function of parameter $v$. As the substrate injection rate increases, the steady state progressively moves towards the right. When starting with a low value of the rate, the system is initially on the left limb of the sigmoid nullcline. A rise in $v$ results in an increase in the level of substrate and product at steady state. In the presence of auto-catalysis, the enzyme reaction begins to be activated by the reaction product as soon as concentration $\gamma$ approaches unity; indeed, the auto-catalytic terms appear in function $\phi$ in the form $(1 + \gamma)^2$, and remain negligible as long as $\gamma$ is less than unity.

When the activation of the enzyme by the product ceases to be negligible, autocatalysis becomes stronger and stronger as $\gamma$ rises with $v$. The concentration of substrate at steady state then starts to decrease as $v$ increases, because of the enhanced consumption of substrate in the autocatalytically activated enzyme reaction. The region of negative slope thus created on the nullcline extends until the value of $v$ is such that the enzyme at steady state is predominantly in the active R state, owing to the high value of the product concentration: the rate of the enzyme reaction cannot augment any more, and the further increase in substrate input again leads to a rise in the steady-state level of sub-strate. The latter phase corresponds to the upward move of $\alpha$ on the right limb of the sigmoid nullcline.

### A conjecture for the origin of birhythmicity, based on phase plane analysis

The conjecture for the origin of birhythmicity relies on the creation of a domain of stability within a domain of instability corresponding to the existence of a large-amplitude periodic solution (see section 3.2). The recycling of product into substrate provides a mechanism for the cre-ation of such a zone of stability in the core of an oscillatory domain, without introducing any additional variable. The values of parameters $K$ and $\sigma_i$ that characterize the nonlinear recycling process play an essential role in this phenomenon, as shown by figs. 3.4 and 3.5 where the product nullcline, given by eqn (3.3), is represented for different values of these two parameters:

$$\gamma = q\sigma\,\phi - \frac{q\sigma_i\,\gamma^n}{K^n + \gamma^n} \qquad (3.3)$$

For the values of the two parameters that nullify the recycling term, i.e. for $\sigma_i$ and $K$ tending, respectively, toward zero and infinity, the curve keeps its characteristic aspect with a single region of negative slope (the ascending branch of the nullcline on the right occurs at higher values of $\gamma$, outside the range seen in figs. 3.4 and 3.5). In a narrow domain of $\sigma_i$ and $K$ values, the nullcline acquires a second region of negative slope. This occurs because the recycling of product into substrate acts to counterbalance the decrease in substrate resulting from autocatalysis: the enhanced consumption of the substrate in the enzyme reaction is indeed counteracted by the increase in its level through recycling of the product when the concentration of the latter rises.

Thereby, the slope locally becomes less and less negative for values of $\gamma$ close to $K$; this leads to the appearance of a hump separating a

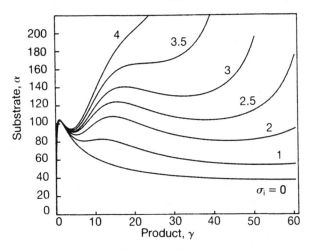

Fig. 3.4. Deformation of the product nullcline in the presence of nonlinear recycling of product into substrate. The curves are established for increasing values of the maximum rate of recycling, $\sigma_i$, ranging from 0 to 4 (in s$^{-1}$). Two regions of negative slope are observed at intermediate values of $\sigma_i$. The curves are obtained according to eqn (3.3) for the following parameter values: $q = 20$, $\sigma = 5.8\ \mathrm{s}^{-1}$, $n = 4$, $K = 12$, $L = 5 \times 10^6$, $k_s = 1\ \mathrm{s}^{-1}$ (Goldbeter & Moran, 1988).

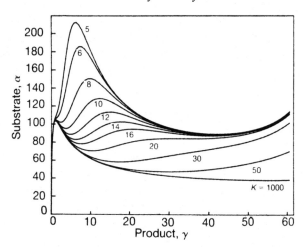

Fig. 3.5. Deformation of the product nullcline as a function of constant $K$, which characterizes the threshold of nonlinear recycling of product into substrate. Two regions of negative slope are apparent at intermediate values of this parameter, for example from $K = 10$ to $14$. The other parameters have the same values as in fig. 3.4, with $\sigma_i = 2.2 \text{ s}^{-1}$ (Goldbeter & Moran, 1988).

domain of negative slope into two parts. This phenomenon occurs when the values of the two parameters are such that recycling reaches 50% of its maximum in the middle of the pre-existing region of negative slope; it is therefore necessary that the magnitude of constant $K$ correspond to a value of $\gamma$ in that domain (fig. 3.5). However, the maximum rate of recycling, given by $\sigma_i$, should be neither too large nor too small (fig. 3.4); if not, the hump would be so large as to suppress the second region of negative slope or too weak to create it.

### 3.2 Birhythmicity: coexistence and transition between two simultaneously stable periodic regimes

Let us now see how the nullcline deformation due to recycling gives rise to birhythmicity. Shown in fig. 3.6 are the bifurcation diagrams obtained as a function of parameter $v$ for eight increasing values of the maximum rate of recycling, $\sigma_i$, for a fixed value of constant $K$. In each part, the steady-state value of the substrate $\alpha_0$ is indicated, as well as the maximum value $\alpha_M$ reached by the substrate during oscillations. Solid lines denote stable steady-state or periodic solutions; both types of solution are indicated by dashed lines when unstable. The stability properties of the steady state were determined by linear stability analysis of eqns

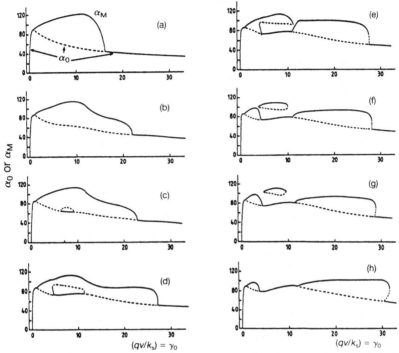

Fig. 3.6. Effect of the recycling reaction on birhythmicity. A series of bifurcation diagrams are represented for increasing values of the maximum rate of product recycling, $\sigma_i$ (in s$^{-1}$): (a) 0; (b) 0.5; (c) 0.6; (d) 1.2; (e) 1.3; (f) 1.4; (g) 1.5; (h) 2. Each diagram shows the steady-state concentration of substrate, $\alpha_0$, and the maximum concentration of substrate in the course of oscillations, $\alpha_M$, as a function of parameter $(qv/k_s)$ equal to the steady-state concentration of product. The curve yielding the steady-state level of substrate is therefore identical with the product nullcline. The solid and dashed lines denote, respectively, stable and unstable branches of periodic or steady-state solutions. Parameter values are $\sigma = 10$ s$^{-1}$, $L = 5 \times 10^6$, $K = 10$, $n = 4$, $q = 1$, $k_s = 0.06$ s$^{-1}$. Periodic regimes were obtained by numerical integration of eqns (3.1). The stability properties of the steady state were determined by linear stability analysis. Birhythmicity is apparent in (d)–(f), while in (h) two distinct instability domains appear as a function of parameter $v$ (Moran & Goldbeter, 1984).

(3.1), while stable or unstable periodic regimes were obtained by numerical integration of these equations (Moran & Goldbeter, 1984). The curve yielding $\alpha_0$ as a function of $(qv/k_s)$ is nothing less than the product nullcline, i.e. the locus of steady states, since the ratio $(qv/k_s)$ is equal to $\gamma_0$.

Figure 3.6a recalls the behaviour of the system in the absence of re-cycling (see also chapter 2). When the substrate injection rate exceeds a critical value, sustained oscillations appear beyond a Hopf bifurcation. As $v$ progressively increases, the amplitude of the oscillations passes through a maximum, and the periodic phenomenon disappears when the injection rate exceeds a second critical value.

When recycling is weak, the hump is at first barely perceptible and is not sufficient for creating a domain of stability of the steady state inside the instability domain (fig. 3.6b). The instability domain, however, expands because it covers a wider range of $v$ values; at the same time, the curve giving the amplitude is slightly deformed. An increase in $\sigma_i$ permits the creation of a zone of stability of the steady state as soon as the instability condition $(d\alpha/d\gamma)_0 < -(1/q)$ ceases to be satisfied, because of the increase in the slope, due to recycling. This phenomenon is local, however, and does not affect the existence of the large-amplitude limit cycle. As the system contains only two variables and admits a single steady state, the latter, stable state must necessarily be separated from the stable limit cycle by an unstable limit cycle. Such a situation is represented in fig. 3.6c.

Further increase in the rate of product recycling gives rise to a supercritical Hopf bifurcation corresponding to the creation of a stable, small-amplitude limit cycle. This phenomenon occurs at the border of the stability domain induced by recycling, inside the domain of existence of the large-amplitude limit cycle (fig. 3.6d). Birhythmicity arises from the coexistence of these two stable limit cycles.

At larger values of $\sigma_i$, the branch of the large-amplitude periodic solution breaks by merging, to the left, with the unstable limit cycle and, to the right, with a new stable limit cycle that originates from a second supercritical Hopf bifurcation, creating another region of birhythmicity (fig. 3.6e). Figure 3.6f and g shows how an **isola** forms when the rate of recycling increases; the birhythmicity present in part f has disappeared in part g. The final part, h, illustrates the situation obtained at higher values of $\sigma_i$ for which the system admits two distinct oscillatory domains as a function of the control parameter. The consequences of such a situation, distinct from birhythmicity, are considered further below.

A typical example of birhythmicity in the phase plane is illustrated in fig. 3.7. The large-amplitude limit cycle encloses a limit cycle of more reduced amplitude; these two stable cycles are separated by an unstable limit cycle. Birhythmicity is of interest because it allows the existence of two distinct oscillations in the same conditions, i.e. for a given set of parameter values. Moreover, the passage from one stable rhythm to the other can be effected by means of the same type of perturbation. Thus,

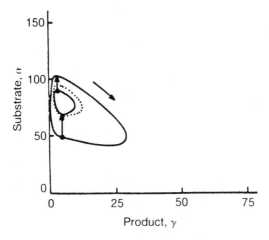

Fig. 3.7. Birhythmicity. In the phase plane, two stable limit cycles (solid lines) are separated by an unstable limit cycle (dotted line). The situation is that of the bifurcation diagram of fig. 3.6e, with $v = 0.255$ s$^{-1}$, i.e. $(qv/k_s) = 4.25$. Vertical arrows indicate how to switch from one stable cycle to the other, by adding substrate (Moran & Goldbeter, 1984).

the vertical arrows in fig. 3.7 indicate how the addition of an appropriate quantity of substrate near the oscillation minimum on the large cycle permits passage to the small cycle; the addition of substrate at the maximum of the latter permits a return to the large-amplitude oscillations. The numerical experiments corresponding to these predictions are represented in fig. 3.8.

The addition of an appropriate amount of substrate at the right phase thus allows the system to switch reversibly from one periodic regime to the other. The same transitions can of course occur in response to an addition of product, as suggested by the phase plane diagram of fig. 3.7.

Phase plane analysis indicates that the two limit cycles possess different sensitivities toward perturbations. It is much easier to pass from the small limit cycle to the large one than to achieve the reverse transition. This differential sensitivity results from the relative sizes of the attraction basins of the two cycles. Moreover, to pass from the large cycle to the small one, the quantity of substrate must be sufficient to cross the border defined by the unstable trajectory, but not so large so as to avoid bringing the system across the basin of the small cycle, into the other side of the attraction basin of the large cycle; in such a case, the perturbation would only cause a phase shift of the large oscillations. In

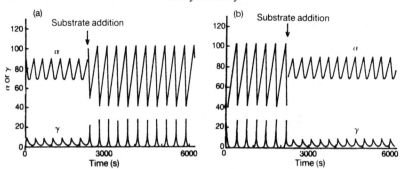

Fig. 3.8. Reversible transition between two simultaneously stable periodic regimes. The transitions are elicited by the addition of adequate amounts of substrate, at the appropriate phase of each oscillation, as indicated in fig. 3.7 established for the same parameter values. The curves are obtained by numerical integration of eqns (3.1). In (a), the transition occurs upon increasing $\alpha$ up to the value 100 in $t = 2300$ s (initial conditions: $\alpha = 83.9$; $\gamma = 3.2$), whereas the inverse transition occurs in (b) when $\alpha$ is raised up to the value 70 in $t = 2220$ s, starting from the initial conditions $\alpha = 102.9$ and $\gamma = 2.6$ (Moran & Goldbeter, 1984).

contrast, there exists only a lower bound on the perturbation that must be exceeded for inducing the passage from the small to the large cycle.

### 3.3 Excitability with multiple thresholds, and tristability

As a result of the particular structure of its nullclines in the phase plane, the two-variable biochemical model with recycling possesses a number of novel dynamic properties with respect to the model studied for glycolytic oscillations (Moran & Goldbeter, 1985; Goldbeter & Moran, 1987). In addition to the coexistence between two stable periodic regimes, another novel property concerns excitability. This phenomenon generally consists in the pulsatory amplification of perturbations whose amplitude exceeds a threshold. Accordingly, the dose–response curve associated with excitability generally possesses a sigmoidal shape characterized by a more or less abrupt, unique threshold (see, for example, in chapter 5, the curve of fig. 5.19 obtained for the relay of cyclic AMP signals in *Dictyostelium* amoebae). Such curves characterize neuronal excitability (Fitzhugh, 1961), as well as excitable behaviour in chemical systems such as the Belousov–Zhabotinsky (Winfree, 1972a) or Briggs–Rauscher reactions (De Kepper, 1976). In the model governed by eqns (3.1), in contrast, excitability corresponds to the existence of *two* distinct thresholds. A perturbation of low amplitude is

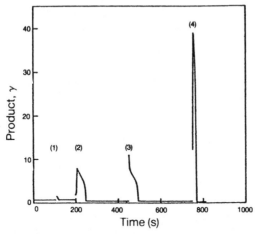

Fig. 3.9. Excitability with multiple thresholds. The response of the system to four increasing stimuli is shown; the initial product concentration is $\gamma_i = 1.6, 1.8,$ 11 and 12, respectively. Amplification of an instantaneous increase in product occurs beyond two distinct thresholds. The curves are established by numerical integration of eqns (3.1) for $v = 0.04$ s$^{-1}$, $\sigma = 4$ s$^{-1}$, $L = 5 \times 10^6$, $K = 8.7$, $n = 4$, $q = 50$, $k_s = 3$ s$^{-1}$, $\sigma_i = 1$ s$^{-1}$ (Moran & Goldbeter, 1985).

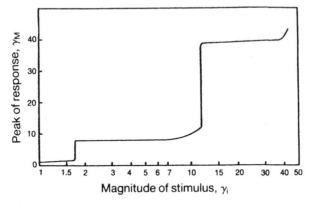

Fig. 3.10. Dose–response curve for excitability in the conditions of fig. 3.9. Two abrupt thresholds are apparent, corresponding to distinct initial product concentrations, $\gamma_i$ (Moran & Goldbeter, 1985).

amplified beyond a first threshold (curves 1 and 2 in fig. 3.9); larger perturbations are at first not amplified, until they exceed a second threshold (curves 3 and 4). The dose–response curve associated with such behaviour (fig. 3.10) takes a staircase shape characterized by two abrupt thresholds.

The origin of multiple excitability thresholds can readily be explained by the existence of two regions of negative slope on the product null-cline. When the steady state lies in a region of stability on the left limb of that nullcline, the two fragments of negative slope on that curve each define a threshold beyond which a perturbation in the form of an instantaneous addition of product will be amplified (fig. 3.11). The analysis predicts (fig. 3.12) that the second excitability threshold will disappear when the steady state is located below the second region of negative slope, or when the second local maximum on the nullcline is shifted downward, below the horizontal issued from the steady state, for lower values of the rate of recycling.

The same analysis indicates the existence of three stable steady-state values of the product when the substrate concentration is held constant in the range where the sigmoid nullcline admits two regions of negative slope. Such a tristability is illustrated in fig. 3.13. As indicated by comparison with fig. 3.3, recycling of product into substrate induces a new steady state in conditions where the enzyme system previously admitted only two steady states at most. Examples of tristability have been reported for chemical systems (Epstein, 1984), but the phenomenon has not yet been observed in biological systems.

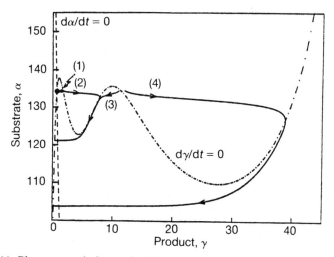

Fig. 3.11. Phase portrait for excitability with multiple thresholds. The trajectories corresponding to the four stimuli of fig. 3.9 are indicated. The existence of two regions of negative slope on the product nullcline give rise to the appearance of two distinct thresholds for the amplification of perturbations that move the system away from the stable steady state (filled circle) (Moran & Goldbeter, 1985).

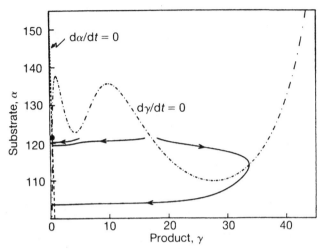

Fig. 3.12. Excitability with a single threshold. Despite the presence of two regions of negative slope on the product nullcline, a unique threshold is obtained for the excitable response, as a result of the position of the steady state with respect to the first of these regions (Moran & Goldbeter, 1985).

## 3.4 Coexistence between a stable steady state and a stable limit cycle

The bifurcation diagrams of fig. 3.6 indicate, for certain values of parameter $v$, the possibility of a coexistence between a steady state and a periodic regime, both of these being stable. This type of behaviour, illustrated by the phase portrait of fig. 3.14a, is encountered in this model only in the presence of recycling. Starting from the steady state located in the second region of positive slope on the product nullcline, the perturbation, in the form of a pulse of product, will induce the transition toward the limit cycle only if the threshold defined by the second region of negative slope is surpassed. The existence of such a threshold is illustrated by fig. 3.14b, where the time evolution of the system is represented following two perturbations, of which only the second leads to oscillations.

The phenomenon of **hard excitation** (Minorsky, 1962) required in these conditions for the evolution to the limit cycle should be contrasted with the **soft excitation** that characterizes the spontaneous evolution towards such a cycle, starting from an unstable steady state (see, for example, fig. 2.13). The interest of reversible transitions between a stable steady state and an oscillatory regime stems from the fact that such behaviour is observed in some nerve cells (Best, 1979; Guttman, Lewis & Rinzel, 1980).

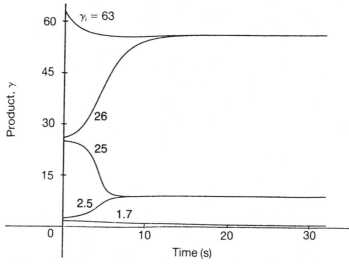

Fig. 3.13. Tristability in the model with nonlinear product recycling into substrate. When the substrate concentration is maintained constant in the domain
of existence of two regions of negative slope on the product nullcline, three stable steady states coexist. These states correspond to distinct values of $\gamma$, which
can be reached starting from appropriate initial conditions (Goldbeter &
Moran, 1987).

## 3.5 Consequences of a multiplicity of oscillatory domains: link with multiple modes of neuronal oscillations

### *Multiple modes of oscillations as a function of a control parameter*

A last type of dynamic phenomenon introduced by the recycling reaction is that of a multiplicity of oscillatory domains as a function of a
control parameter. This phenomenon is apparent in the bifurcation diagram of fig. 3.6h. Here again, the interest of the phenomenon stems
from its relationship to the behaviour of certain neurons; the model
provides a straightforward explanation for the neuronal behaviour in
terms of phase plane analysis.

The existence of two distinct oscillatory domains not only depends on
the shape of the product nullcline, but also on the time scale structure
of the system. Thus, fig. 3.15 shows three bifurcation diagrams obtained
for increasing values of parameters $q$ and $k_s$; the ratio $(q/k_s)$ is held constant so that the product nullcline remains unchanged. As indicated by
eqns (3.1), the time evolution of the product becomes more and more

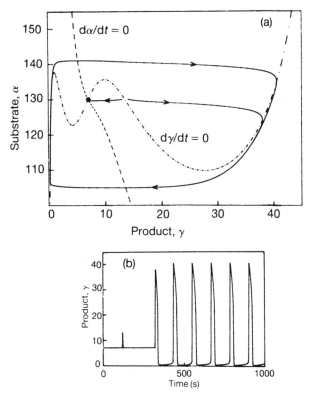

Fig. 3.14. (a) Phase portrait and (b) associated temporal dynamics illustrating the transition between a stable steady state and a stable limit cycle, following suprathreshold perturbation (Moran & Goldbeter, 1985).

rapid as parameters $q$ and $k_s$ increase. Whereas two distinct domains of periodic behaviour are observed for $q = 1$ in fig. 3.15, only one domain of oscillations persists at higher values of the parameter, even though the system still possesses two distinct domains of instability of the steady state! A domain of birhythmicity remains present at the three values of $q$ considered. The existence of two time scales within system (3.1) also influences the shape of the product oscillations, which can reach a higher amplitude when the product varies more rapidly than the substrate.

The existence of two distinct modes of oscillations for two different values of the control parameter $v$ is illustrated by fig. 3.16. In part a, the phase portrait is shown for $v = 0.15$ s$^{-1}$ and $v = 1.5$ s$^{-1}$. The periodic variation of the product in these two situations is indicated in part b. Phase plane analysis shows that the two limit cycles obtained in different con-

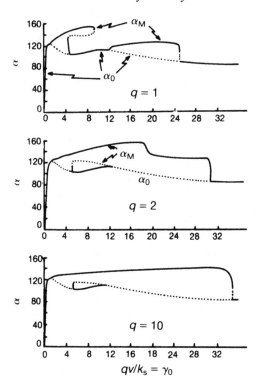

Fig. 3.15. Effect of multiple time scales on the oscillatory behaviour of the model. The bifurcation diagram, similar to those of fig. 3.6, is established as a function of parameter $(qv/k_s)$ for three increasing values of parameter $q$. The ratio $(q/k_s)$ is fixed in such a manner that the product nullcline, locus of the steady state, remains unchanged. The existence of two distinct domains of oscillations disappears when the product varies more rapidly than the substrate, at elevated values of $q$ (Goldbeter & Moran, 1988).

ditions – which distinguishes the phenomenon from that of birhythmicity – share a section of trajectory between points A and B in fig. 3.16a; the direction of movement on this common portion differs, however, for each of the two cycles.

For the two values of $v$ considered in fig. 3.16, the substrate nullcline crosses the product nullcline in either of the regions of negative slope where the condition of instability of the steady state is satisfied. For some intermediary values of parameter $v$, the intersection of the nullclines will lie in the region of positive slope separating the two instability domains. Starting from a value of $v$ corresponding to a stable steady

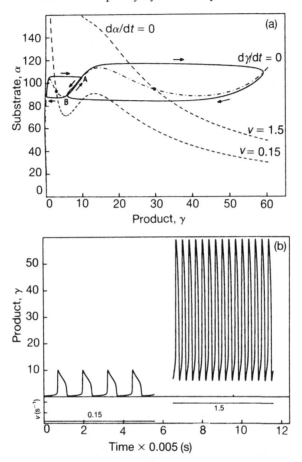

Fig. 3.16. (a) Phase portrait and (b) associated temporal evolution illustrating the existence of two distinct types of oscillation for two different values of parameter $v$ (Goldbeter & Moran, 1988).

state in that region, a decrease or increase in the control parameter will induce the appearance of two types of periodic behaviour that will differ markedly in both amplitude and period (fig. 3.17).

### Link with two types of oscillation in thalamic neurons

This behaviour of the biochemical model bears a striking resemblance to the observations of Jahnsen & Llinas (1984a,b) on neurons of the thalamus (see also Llinas, 1988; Steriade *et al.*, 1990). Under appropriate conditions, these neurons are in a resting, nonoscillatory state corresponding to a stable value of the membrane potential. A slight

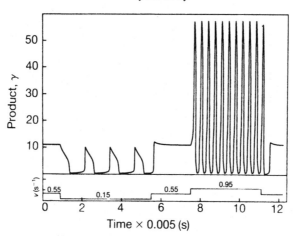

Fig. 3.17. Effect of a decrease and an increase in parameter $v$, starting from a value corresponding to a stable steady state. The two types of variation of the parameter reveal the existence of two distinct periodic regimes (Goldbeter & Moran, 1988).

depolarization of the neurons by a few millivolts gives rise to oscillations characterized by a frequency of 10 Hz; in contrast, a slight hyperpolarization of the initial state produces a rhythm of 6 Hz frequency (fig. 3.18).

The above analysis of the biochemical model of fig. 3.1 suggests that the phenomenon of multiple oscillations observed in thalamic neurons results from a similar nullcline structure in the phase plane, giving rise to a multiplicity of oscillatory domains as a function of a control parameter. The fact that the variables are of biochemical nature in the enzymic model but represent current and voltage in the neuronal system in no way detracts from the fact that a common dynamics in the phase plane is probably at work in each of the two systems (Goldbeter & Moran, 1988; Llinas, 1988; Steriade *et al.*, 1990). In particular, the existence of a nullcline possessing two regions of negative slope could explain a number of observations on the dynamic behaviour of thalamic neurons.

This conjecture is corroborated by the fact that the enzymic model possesses other properties shared by neurons of the thalamus. Reminiscent of the phenomenon described in section 3.3, two distinct thresholds of excitability, linked to two voltage-dependent $Ca^{2+}$ conductances, have been demonstrated in these neurons by Jahnsen & Llinas (1984a). Moreover, the evolution towards a stable steady state takes a different form, depending on the initial state of the system. Thus, in

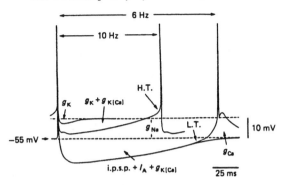

Fig. 3.18. Existence of two types of periodic behaviour in thalamic neurons. Starting from a stable membrane potential, a slight hyperpolarization gives rise to oscillations characterized by a frequency of 6 Hz, whereas depolarization produces oscillations of 10 Hz frequency. The occurrence of these two types of oscillations is linked to the existence of a high threshold (H.T.) and a low threshold (L.T.) characterizing the excitability of these neurons. i.p.s.p., inhibitory postsynaptic potential (Jahnsen & Llinas, 1984b).

fig. 3.19, three successive increments of the substrate injection rate in the model lead first to a steady state reached after an overshoot, then to a steady state attained without overshoot, and eventually to sustained oscillations. Similar curves are obtained for thalamic neurons in response to successive depolarizations (fig. 3.20).

Finally, suppressing the oscillations by changing the substrate injection rate in the model can lead, depending on the final value of this parameter, to a phenomenon of rebound (fig. 3.21) that is also observed in neurons of the thalamus. According to Llinas (1984), the rebound property could play an important role in neuronal excitability involved in the origin of muscular tremor.

The qualitative agreement between the biochemical system and the behaviour of thalamic cells suggests that birhythmicity, predicted by the model, could also occur in these neurons. To demonstrate such a phenomenon, each type of oscillation should be perturbed by depolarization or hyperpolarization, in order to determine whether the perturbation can elicit the switch from one rhythm to the other. Alternatively, an increase followed by a decrease in the applied current could reveal the existence of hysteresis; two different rhythms would then be observed at a given value of the membrane potential. Such a procedure, suggestive of birhythmicity, has been followed in another neuronal system (Hounsgaard *et al.*, 1988).

The idea that the two types of rhythmic behaviour in thalamic neu-

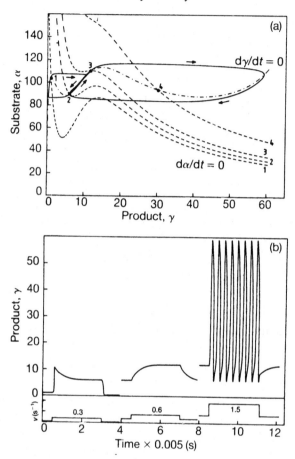

Fig. 3.19. Three types of temporal response (b) resulting from increments of increasing magnitude in parameter $v$, starting from a value corresponding to a stable steady state. The associated phase portrait is represented in (a). Numbers in this diagram relate to the following values of $v$, in $s^{-1}$: (1) 0.05; (2) 0.3; (3) 0.6; (4) 1.5 (Goldbeter & Moran, 1988).

rons are associated with a nullcline possessing two regions of negative slope is also at the basis of a model proposed by Rose & Hindmarsh (1985). In their model, this nullcline obeys a polynomial equation of the fifth degree chosen so as to give two regions of negative slope. The present model has the relative advantage of relying on a precise biophysical mechanism that allows us to modify the shape of the nullcline by modulating the characteristics of a single process, namely the recycling of product into substrate (see figs. 3.4 and 3.5). In the Rose–Hindmarsh

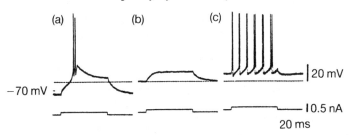

Fig. 3.20. (a)–(c) Three types of response of thalamic neurons to depolarizations of increasing magnitude (Jahnsen & Llinas, 1984b).

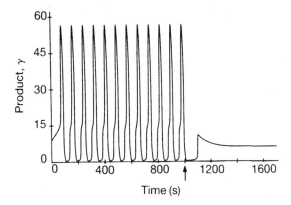

Fig. 3.21. Rebound excitation in the biochemical model with positive feedback and product recycling into substrate. The variation of parameter $v$ (arrow) brings an end to the oscillations while eliciting, concomitantly, a peak in product before the latter reaches a stable steady state (Goldbeter & Moran, 1988).

model, the existence of two regions of negative slope is attributed to two distinct ionic currents, corresponding to the same number of conductances controlled in a self-amplified manner at different values of the membrane potential. In subsequent work, Rose & Hindmarsh (1989a–c; see also Hindmarsh & Rose, 1989) have obtained similar results when describing the ionic currents in thalamic cells more realistically, in terms of Hodgkin–Huxley equations. A recent account of theoretical models for thalamic oscillations can be found in a review by Destexhe & Sejnowski (1995).

The preceding results could apply to neurons other than those of the thalamus. Thus, two distinct excitability thresholds are observed in neurons of the inferior olive (Llinas & Yarom, 1981a,b), which also possess the property of oscillating in an autonomous manner (Llinas & Yarom,

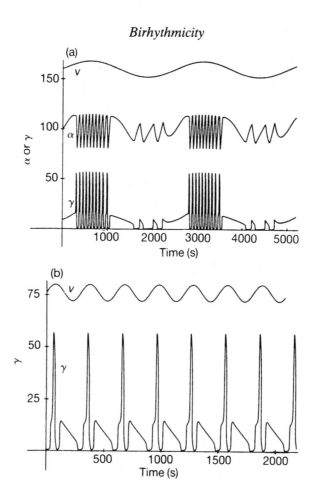

Fig. 3.22. Forcing of the biochemical system by a periodic variation of the source of substrate across the two domains of oscillations. The resulting periodic behaviour presents the characteristics of the two types of endogenous oscillations; the number of peaks of each of the latter is determined by the external periodicity. This number can be important for a slow variation of parameter $v$ (a), but decreases until a single peak of each of the oscillations is retained when the variation of $v$ is sufficiently rapid (b) (Goldbeter & Moran, 1988).

1986). The autocatalytic biochemical model with recycling of product suggests that birhythmicity could occur in neurons of both the inferior olive and the thalamus. A detailed appraisal of the role played by the oscillatory dynamics of the olivary neurons in the coordination of motor control has recently been given by Welsh *et al.* (1995).

**Effect of a periodic variation of the control parameter across the two oscillatory domains**

As the model of fig. 3.1 contains only two variables, it cannot give rise to complex periodic oscillations of the 'bursting' type (these oscillations are characterized by the regular alternation between active phases of spiking and phases of relative quiescence). Such oscillations, which are studied in further detail in the model analysed in the following chapter, are often observed in neurobiology. Bursting occurs in the present model when a control parameter such as the substrate injection rate is varied periodically. When two domains of oscillations exist as a function of that rate, a periodic variation of parameter $v$ across the two instability domains produces oscillations of the bursting type that are even more complex than those obtained usually (see chapter 4). Both types of oscillation indeed alternate in the active phases of bursting, the number of peaks being dictated by the rate of variation of $v$ (compare fig. 3.22a and b).

The effect of a periodic variation in the substrate input over a narrower range of values of $v$ has also been studied. While the periodic forcing of a system admitting a single limit cycle can generally give rise to chaos associated with a single strange attractor in the phase space, two such coexisting strange attractors can be obtained in these conditions when the periodic input spans a domain in which birhythmicity occurs (Moran & Goldbeter, 1987).

### 3.6 Oscillatory isozymes: another two-variable model for birhythmicity

Another two-variable model has been constructed to test the conjecture presented above as to the origin of birhythmicity. This model, schematized in fig. 3.23, relies on the coupling of two enzyme reactions catalysing the transformation of the same substrate, with different kinetic properties (Li & Goldbeter, 1989a). The two isozymes are regulated in an autocatalytic manner, since both are activated by their common reaction product. As in the model of fig. 3.1, birhythmicity originates from the creation of a small-amplitude limit cycle within the domain of existence of a limit cycle of larger amplitude (fig. 3.24). The phenomenon is accompanied by the breaking apart of a domain of instability into two parts, as in the sequence of bifurcation diagrams of fig. 3.6.

In the model of oscillatory isozymes, the product nullcline also

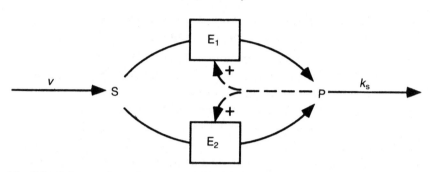

Fig. 3.23. Scheme of another two-variable biochemical model admitting birhythmicity. Substrate S, injected at a constant rate, is transformed into product P by two isozymes that differ in their allosteric properties and in their catalytic activity. This model represents one of the simplest examples of coupling between two biochemical oscillators (Li & Goldbeter, 1989a).

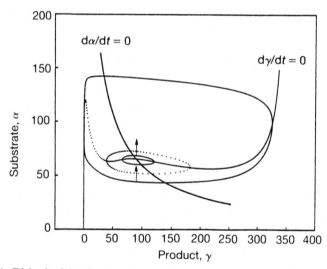

Fig. 3.24. Birhythmicity in the phase plane of the two-variable biochemical model represented in fig. 3.23. The arrows show how to pass from one stable limit cycle to the other (Li & Goldbeter, 1989a).

possesses two regions of negative slope. Consequently, this model also displays the property of excitability with multiple thresholds (see figs. 3.9 and 3.10) and the capability of oscillating in two markedly different ways for parameter values separated by a range corresponding to the evolution toward a stable steady state (fig. 3.25).

The system of two oscillatory isozymes can be viewed as the simplest model of coupled biochemical oscillators. Indeed, the number of vari-

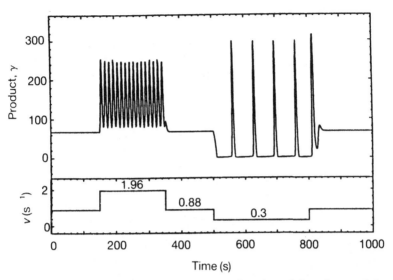

Fig. 3.25. Two distinct modes of oscillations as a function of the substrate injection rate, $v$, in the model for the two oscillatory isozymes of fig. 3.23 (Li & Goldbeter, 1989a).

ables remains limited to only two, the substrate and the product being common to the two enzyme reactions, which are both capable of producing oscillations on their own because of their autocatalytic regulation. The coupling between these two oscillatory reactions is here of chemical nature. Any model in which two oscillators are coupled by a transport process such as diffusion would contain at least four variables; the occurrence of birhythmicity in such a model of two, coupled, identical oscillators has been studied by Volkov & Stolyarov (1991).

The model based on oscillatory isozymes, which can be extended to a larger number of oscillatory units coupled in parallel, is closer to the neuronal situation than the model of fig. 3.1. Indeed, the situation of two isozymes activated by their common product is somewhat analogous to that of two ion channels carrying the same ionic species and activated in a self-amplified manner by the membrane potential.

The scenario described above for the origin of birhythmicity is not the only one conceivable. Different routes to birhythmicity exist, as shown by the theoretical study of other models containing two or three variables (Schulmeister & Sel'kov, 1978; Tracqui *et al.*, 1987; Tracqui, 1993). This question is discussed further in chapter 4, in relation to models based on the coupling between two instability-generating mechanisms.

# 4

# From simple periodic behaviour to complex oscillations, including bursting and chaos

## 4.1 A biochemical model with two instability mechanisms

In the two-variable models studied for glycolytic oscillations and birhythmicity, periodic behaviour originates from a unique instability mechanism based on the autocatalytic regulation of an allosteric enzyme by its reaction product. The question arises as to what happens when two instability-generating mechanisms are present and coupled within the same system: can new modes of dynamic behaviour arise from such an interaction?

An example of such a situation was considered at the end of the preceding chapter: the system with two oscillatory isozymes (fig. 3.23) contains two instability mechanisms coupled in parallel. Compared with the model based on a single product-activated enzyme, new behavioural modes may be observed, such as birhythmicity, hard excitation and multiple oscillatory domains as a function of a control parameter. The modes of dynamic behaviour in that model remain, however, limited, because it contains only two variables. For complex oscillations such as bursting or chaos to occur, it is necessary that the system contain at least three variables.

The coupling in series of two enzyme reactions with autocatalytic regulation (fig. 4.1) permits the construction of a three-variable biochemical prototype containing two instability-generating mechanisms (Decroly, 1987a,b; Decroly & Goldbeter, 1982). As in the model for glycolytic oscillations, the substrate S of the first enzyme is introduced at a constant rate into the system; this substrate is transformed by enzyme $E_1$ into product $P_1$, which serves as substrate for a second enzyme $E_2$ that transforms $P_1$ into $P_2$. The two allosteric enzymes are both activated by their reaction product; $P_1$ and $P_2$ are thus positive effectors of enzymes $E_1$ and $E_2$, respectively.

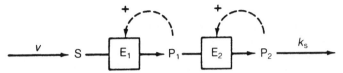

Fig. 4.1. Three-variable biochemical model for the coupling in series of two enzyme reactions with autocatalytic regulation (Decroly & Goldbeter, 1982, 1987).

Each of the two enzymes thus behaves as phosphofructokinase in the model considered for glycolytic oscillations (chapter 2). To limit the study to temporal organization phenomena, the system is considered here as spatially homogeneous, as in the case of experiments on glycolytic oscillations (Hess *et al.*, 1969). In the case where the kinetics of the two enzymes obeys the concerted allosteric model (Monod *et al.*, 1965), the time evolution of the model is governed by the kinetic equations (4.1), which take the form of three nonlinear, ordinary differential equations:

$$\frac{d\alpha}{dt} = v - \sigma_1 \, \phi \, (\alpha, \beta)$$

$$\frac{d\beta}{dt} = q_1 \, \sigma_1 \, \phi \, (\alpha, \beta) - \sigma_2 \, \eta \, (\beta, \gamma)$$

$$\frac{d\gamma}{dt} = q_2 \, \sigma_2 \, \eta \, (\beta, \gamma) - k_s \, \gamma \tag{4.1}$$

where the rate functions $\phi$ and $\eta$ of the allosteric enzymes $E_1$ and $E_2$ are given, respectively, by expressions (4.2) and (4.3). These equations, as well as system (4.1), are obtained as indicated in chapter 2 for the glycolytic model based on a single product-activated enzyme:

$$\phi \, (\alpha, \beta) = \frac{\alpha \, (1 + \alpha) \, (1 + \beta)^2}{L_1 + (1 + \alpha)^2 \, (1 + \beta)^2} \tag{4.2}$$

$$\eta \, (\beta, \gamma) = \frac{\beta \, (1 + \gamma)^2}{L_2 + (1 + \gamma)^2} \tag{4.3}$$

For simplicity, we consider that the rate of enzyme $E_2$ depends in a linear manner on the concentration $\beta$ of its substrate, which is the same as assuming that the enzyme is never saturated by it. Equations (4.1)–(4.3) were obtained by means of a quasi-steady-state hypothesis for the free and complexed forms of the two enzymes, and by further assuming that the latter are dimers that bind their substrate exclusively in the R state.

In the above equations, $\alpha$, $\beta$ and $\gamma$ denote the (dimensionless) normalized concentrations of S, $P_1$ and $P_2$, respectively; $L_1$ and $L_2$ are the allosteric constants of enzymes $E_1$ and $E_2$, while $\sigma_1$ and $\sigma_2$ are the normalized maximum rates (in $s^{-1}$) of these two enzymes; $v$ and $k_s$ (both in $s^{-1}$) denote the substrate injection rate and the apparent first-order rate constant for the removal of the final product in a reaction catalysed by a Michaelian enzyme far from saturation by its substrate.

## 4.2 Repertoire of the different modes of dynamic behaviour

To bring to light the various types of behaviour of which the multiply regulated enzyme system is capable, it is useful to focus on a few control parameters and to determine how their variation affects the dynamic behaviour of the model. The two parameters to be considered are the substrate injection rate, $v$, and the rate constant $k_s$ characterizing the sink of the final product $P_2$.

Figure 4.2 is the bifurcation diagram obtained when $k_s$ varies over some five orders of magnitude. Equations (4.1) admit a single steady state; the value of the substrate concentration in this state, $\alpha_0$, is indicated as a function of $k_s$, as well as the maximum amplitude $\alpha_M$ reached by the substrate in the course of oscillations. The steady state is unstable (dashed line) for most values of $k_s$, except those ranging from 0.792 to 1.584 $s^{-1}$ (see also table 4.1). In this domain, the stable steady state coexists with a stable limit cycle, denoted LC1, which appears at low values of $k_s$. When $k_s$ exceeds the value 1.584 $s^{-1}$, a Hopf bifurcation occurs, beyond which a new stable limit cycle, LC2, arises. As limit cycle LC1 has not yet disappeared, a phenomenon of birhythmicity takes place (see chapter 3) as a result of the coexistence of two stable rhythms for the same set of parameter values.

The structure of behavioural modes becomes more and more complex in the interval of $k_s$ values between 1.6 and 2.2 $s^{-1}$; two successive enlargements of this interval in the bifurcation diagram are therefore presented in fig. 4.2. The salient feature is the presence of a hysteresis loop through which the cycle LC2 gives rise to the stable limit cycle LC3, and coexists with it over a brief interval; this phenomenon corre-

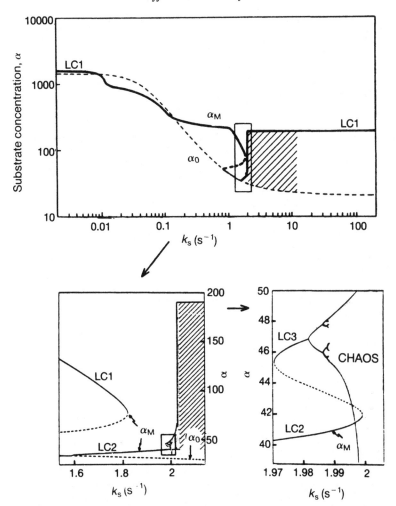

Fig. 4.2. Bifurcation diagram as a function of parameter $k_s$. The steady-state concentration of substrate, $\alpha_0$, as well as the maximum concentration of substrate in the course of oscillations, $\alpha_M$, is indicated. Solid and dashed lines denote, respectively, stable and unstable (steady or periodic) solutions. The diagram is obtained by numerical integration of eqns (4.1), and by linear stability analysis of the unique steady-state solution that these equations admit. LC1, LC2 and LC3 denote stable limit cycles. The dashed area represents a domain of complex periodic oscillations of the 'bursting' type. To clarify the details of dynamic transitions to chaos and birhythmicity, two successive enlargements of the domain of $k_s$ values between 1.6 and 2 s$^{-1}$ are shown. Parameter values are $v = 0.45$ s$^{-1}$, $\sigma_1 = \sigma_2 = 10$ s$^{-1}$, $q_1 = 50$, $q_2 = 0.02$, $L_1 = 5 \times 10^8$ and $L_2 = 100$ (Decroly & Goldbeter, 1982).

Table 4.1. *Dynamic behaviour of the multiply regulated enzymic system as a function of parameter* $k_s$ *(see fig. 4.2)*

| Parameter domain $(s^{-1})$ | Observed behaviour |
| --- | --- |
| $k_s \leq 0.792$ | A single limit cycle (LC1) |
| $0.792 \leq k_s \leq 1.584$ | One stable limit cycle (LC1) and one stable steady state (**hard excitation**) |
| $1.584 \leq k_s \leq 1.820$ | Two stable limit cycle, LC1 and LC2 (**birhythmicity**) |
| $1.820 \leq k_s \leq 1.974$ | A single limit cycle (LC2) |
| $1.974 \leq k_s \leq 1.990$ | Two limit cycles, LC2 and LC3, the second of which undergoes a cascade of period-doubling bifurcations leading to chaos |
| $1.990 \leq k_s \leq 2.034$ | Aperiodic oscillations (**chaos**) |
| $2.034 \leq k_s \leq 12.80$ | Complex periodic oscillations (**bursting**) |
| $k_s > 12.80$ | A single limit cycle (LC1) |

Decroly & Goldbeter, 1982.

sponds to a new mode of birhythmicity. The limit cycle LC3 then undergoes a series of bifurcations through which the oscillations successively acquire 2, 4, 8, 16, ... $2^n$ distinct maxima per period; this cascade of period-doubling bifurcations leads to aperiodic oscillations, i.e. **chaos**. Higher values of $k_s$ elicit the transition from chaos to complex periodic oscillations, before the system returns to the simple periodic behaviour prolonging the branch of limit cycle LC1.

A few of the behavioural modes revealed by the bifurcation diagram of fig. 4.2 are illustrated by fig. 4.3 for four increasing values of $k_s$. In fig. 4.3a, the system displays simple periodic behaviour, as in the monoenzyme model studied for glycolytic oscillations. Figure 4.3b illustrates the coexistence between a stable steady state and a limit cycle that the system reaches only after a suprathreshold perturbation (hard excitation). The aperiodic oscillations of fig. 4.3c represent chaotic behaviour, while the complex periodic oscillations shown in fig. 4.3d correspond to the phenomenon of bursting that is associated with series of spikes in product $P_1$, alternating with phases of quiescence. These various modes of dynamic behaviour, as well additional ones identified by the analysis of the model, are considered in more detail below.

### 4.3 Multiple periodic attractors: birhythmicity and trirhythmicity

It was before establishing the bifurcation diagram as a function of $k_s$ that the first indication of a coexistence between two simultaneously

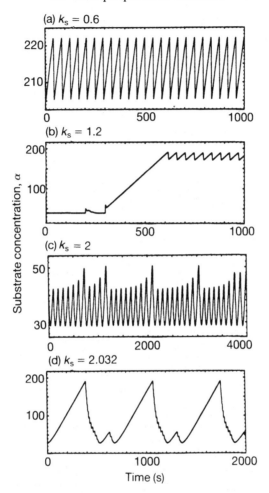

Fig. 4.3. Different modes of dynamic behaviour observed in the biochemical model with multiple regulation, for increasing values of parameter $k_s$ (in s$^{-1}$): (a) 0.6, simple periodic oscillations; (b) 1.2, hard excitation; (c) 2, chaos; (d) 2.032, complex periodic oscillations (bursting). Only the substrate concentration is represented as a function of time. The curves are obtained by numerical integration of eqns (4.1) for the parameter values of fig. 4.2 (Decroly & Goldbeter, 1982).

stable periodic regimes was obtained fortuitously (Decroly & Goldbeter, 1982). Sometimes the numerical integration of eqns (4.1) for different initial conditions leads to either one of two stable limit cycles. The boundary between the two cycles is extremely sharp: in fig. 4.4, where the system evolves towards either of the limit cycles, initial

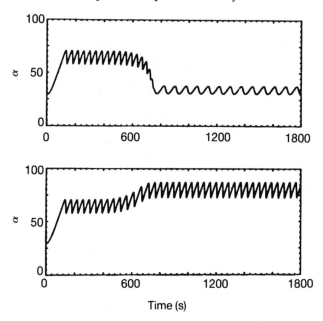

Fig. 4.4. Birhythmicity: two stable limit cycles coexist for the same parameter values. The curves are obtained by numerical integration of eqns (4.1) for the parameter values of fig. 4.2, with $k_s = 1.8 \text{ s}^{-1}$. The evolution towards either one of the two stable limit cycles depends on initial conditions; these differ in the fifth decimal place for $\alpha$: 32.02223 for the upper curve, and 32.02222 for the bottom one, while $\beta = 250$ and $\gamma = 0.25$ for the two curves. These initial conditions are close to an unstable periodic trajectory that the system transiently follows before settling on one or the other stable cycle (Decroly & Goldbeter, 1982).

conditions are identical for variables $\beta$ and $\gamma$, and differ only in the fifth decimal place for variable $\alpha$. Owing to the choice and proximity of the initial conditions, the system begins its evolution in the immediate vicinity of an unstable periodic trajectory along which it makes a number of cycles before escaping to either of the two stable limit cycles; the initial conditions thus determine the attraction basin into which the system eventually falls.

This phenomenon of **birhythmicity** is observed in two domains of values of parameter $k_s$, at a fixed value of parameter $v$, as indicated by the bifurcation diagram of fig. 4.2 and by table 4.1. The coexistence occurs for the stable cycles LC1 and LC2, or for LC2 and LC3. For larger values of constant $k_s$, the limit cycle LC2 can also coexist with solutions of period $2^n$ ($n = 2, 4, 8, 16, \ldots$) issued from LC3, or with chaos, to which this sequence of period-doubling leads. An example of such a coexis-

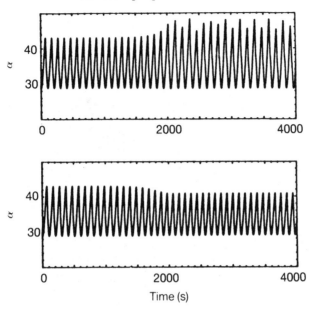

Fig. 4.5. Coexistence between a stable limit cycle and a regime of stable chaos. Parameter values are those of fig. 4.2, with $k_s = 1.99$ s$^{-1}$. Initial conditions are $\beta = 188.8$ and $\gamma = 0.3367$, with $\alpha = 29.19988$ for the upper curve, and $\alpha = 29.19989$ for the lower one. As in fig. 4.4, the system evolves along an unstable limit cycle before stabilizing in either of the two stable oscillatory regimes (Decroly & Goldbeter, 1982).

tence between stable periodic oscillations and chaos is presented in fig. 4.5. Here again, the boundary between the attraction basins of these two asymptotically stable solutions is formed by an unstable limit cycle over which the system briefly evolves before making its choice, when the initial conditions are close to the separatrix. The latter defines a sharp boundary between the attraction basins: in fig. 4.5, the initial values of $\beta$ and $\gamma$ are identical, while the initial substrate concentration varies in the fifth decimal place.

The closeness of the two distinct regions of birhythmicity that involve the coexistence between the limit cycle LC2 with either LC1 or LC3 suggests the possibility that the two domains may overlap for some values of the other parameters. This conjecture is indeed verified: by varying the rate of substrate injection, the bifurcation diagram of fig. 4.2 can transform into that represented schematically in fig. 4.6. The region of coexistence between LC2 and LC3 is here displaced towards lower values of parameter $k_s$ for which the limit cycle LC1 subsists. Thus, in a

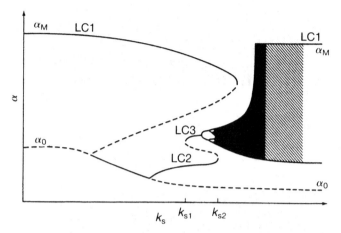

Fig. 4.6. Schematic bifurcation diagram showing the origin of trirhythmicity. For certain values of $k_s$, the three stable limit cycles LC1, LC2 and LC3 coexist; they are separated by unstable limit cycles. The black and the dashed domains correspond to chaos and bursting, respectively (Decroly & Goldbeter, 1985).

narrow range of $k_s$ values, the limit cycles LC1, LC2 and LC3 coexist; two unstable limit cycles separate these three stable, periodic trajectories.

**Trirhythmicity** is illustrated in fig. 4.7 where the three stable limit cycles are reached from distinct initial conditions, for the same set of parameter values (Decroly & Goldbeter, 1984a). Initial conditions in parts a, b and c only differ, in the first decimal place, by variable $\alpha$. Each of the stable cycles is attained after a brief excursion along an unstable limit cycle.

The transition between the three stable periodic regimes can also occur, as in the case of birhythmicity, in response to a perturbation such as the addition of an appropriate quantity of substrate, provided that the perturbation occurs at the adequate phase of each oscillation (fig. 4.8).

## 4.4 Chaos

The appearance of aperiodic oscillations beyond a point of accumulation of a cascade of period-doubling bifurcations is one of the best-known scenarios for the emergence of chaos (Feigenbaum, 1978; Bergé *et al.*, 1984). It is also along this way that chaos arrives in the multiply regulated enzyme model. Another example of this type of irregular

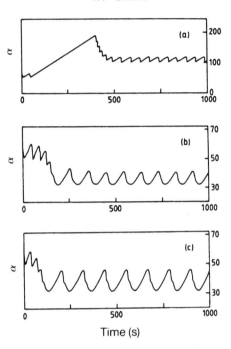

Fig. 4.7. Trirhythmicity: coexistence of three simultaneously stable limit cycles. The three situations differ in the initial value of the substrate, $\alpha$: (a) 61.5; (b) 61; (c) 60.5. In each case, the initial concentrations of the other variables are $\beta = 131$ and $\gamma = 0.25$. The curves are obtained by integration of eqns (4.1) for the parameter values of fig. 4.2, with $v = 0.4$ s$^{-1}$ and $k_s = 1.632$ s$^{-1}$ (Decroly & Goldbeter, 1984a).

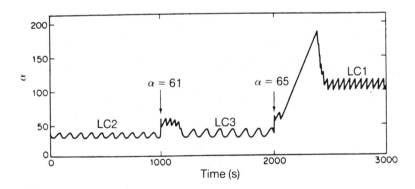

Fig. 4.8. Transitions between the three stable limit cycles of fig. 4.7, induced by successive additions of substrate (Decroly & Goldbeter, 1985).

oscillation, already illustrated in fig. 4.3c, is shown in fig. 4.9. This chaotic behaviour is deterministic (Lorenz, 1963) in that it is generated by the differential equations (4.1) in the absence of noise.

In the phase space, the trajectory followed by the system never passes again through the same point, but remains confined to a finite portion of this space (fig. 4.10): the system evolves towards a **strange attractor** (Ruelle, 1989). The unpredictability of the time evolution in the chaotic regime is associated with the sensitivity to initial conditions: two points, initially close to each other on the strange attractor, will diverge exponentially in the course of time.

Chaos can further be characterized by resorting to Poincaré sections. By determining, for example, the value $\alpha_n$ of the substrate corresponding to the $n$th peak, $\beta_n$, of product $P_1$ in the course of aperiodic oscillations, we may construct the one-dimensional return map giving $\alpha_{n+1}$ as a function of $\alpha_n$ (Decroly, 1987a; Decroly & Goldbeter, 1987). The continuous character of the curve thus obtained (fig. 4.11) denotes the deterministic nature of the chaotic behaviour.

Moreover, the values of parameter $k_s$ for which period-doubling bifurcations occur before the onset of aperiodic oscillations define a sequence (Decroly, 1987a,b) characterized by a value close to that obtained by Feigenbaum (1978) for one of the universal constants associated with the cascade of period-doubling bifurcations leading to chaos.

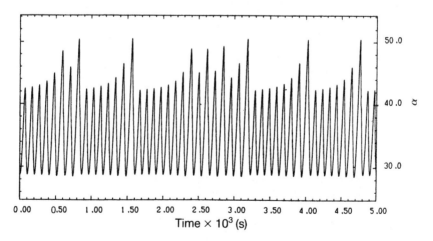

Fig. 4.9. Chaotic evolution of the biochemical model with multiple regulation. The curve is obtained by numerical integration of eqns (4.1) in the domain of chaos shown in fig. 4.2 (Goldbeter & Decroly, 1983).

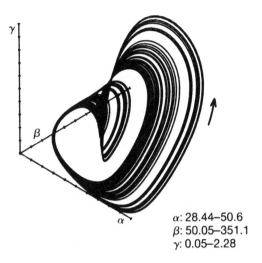

α: 28.44–50.6
β: 50.05–351.1
γ: 0.05–2.28

Fig. 4.10. Strange attractor corresponding to the chaotic dynamics of fig. 4.2c. The domain of variation of the three variables is indicated, as well as the direction of evolution on the attractor (Decroly & Goldbeter, 1982).

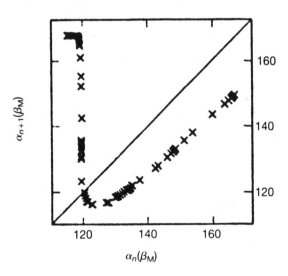

Fig. 4.11. Poincaré section for chaos. The construction of the return map $\alpha_{n+1}(\beta_M) = f\,[\alpha_n(\beta_M)]$ is described in the text (see also fig. 4.24). Parameter values are those of fig. 4.2, with $v = 0.25\,\text{s}^{-1}$ and $k_s = 1.537\,\text{s}^{-1}$ (Decroly & Goldbeter, 1987).

The cascade of period-doubling bifurcations is not restricted to the limit cycle LC2. Thus, for other values of the substrate injection rate, the bifurcation diagram as a function of $k_s$ takes the form indicated in fig. 4.12b (for the sake of comparison, fig. 4.2 is redrawn schematically in part a). In the lower diagram, the period-doubling cascade leading to chaos is observed not only for limit cycle LC2 but also, at larger values of $k_s$, for limit cycle LC1.

The domain in which chaos occurs in parameter space remains nevertheless reduced. To be convinced of this, we may determine the amplitude of variation of each parameter that is required for moving out of the domain of chaos when all other parameters retain the values giving rise to the aperiodic oscillations of fig. 4.9. The results are indicated in table 4.2. Although the domain of chaos is far from being negligible,

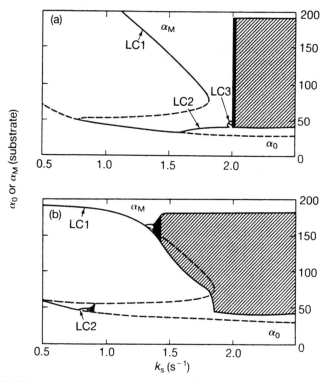

Fig. 4.12. Bifurcation diagram as a function of $k_s$ for two values of parameter $v$. The diagram shows, in a schematic manner, how chaos (black domain) can appear on the periodic branch LC3 ((a), same situation as in fig. 4.2), or from the limit cycles LC1 and LC2 (b). Diagram (b) is obtained for the parameter values of fig. 4.2, with $v = 0.25$ s$^{-1}$ (Decroly & Goldbeter, 1985).

Table 4.2. *Domain of parameter values for which eqns (4.1) produce chaotic behaviour, in the conditions of fig. 4.9*

| Parameter | Domain of chaos |
|---|---|
| $v$ (s$^{-1}$) | 0.446–0.451 |
| $k_s$ (s$^{-1}$) | 1.99–2.036 |
| $L_1$ | $4.96 \times 10^8 - 5.4 \times 10^8$ |
| $L_2$ | 99–100.3 |
| $q_1$ | 49.6–50.2 |
| $q_2$ | 0.0197–0.0201 |
| $\sigma_1$ (s$^{-1}$) | 9.51–10.10 |
| $\sigma_2$ (s$^{-1}$) | 9.97–10.20 |

Goldbeter & Decroly, 1983.

variation by a few % or less in the value of one or the other parameters suffices to abolish aperiodic behaviour.

## 4.5 Multiple oscillations, fractal boundaries and final state sensitivity

When the system admits several, simultaneously stable periodic solutions, the boundary that separates their attraction basins is not always as sharp as in the case of figs. 4.4 and 4.7. There, an abrupt threshold characterizes the evolution towards either one of the stable limit cycles. In certain cases, the structure of the attraction basins is more complex.

In the case of trirhythmicity, for example, when the initial values of $\beta$ and $\gamma$ are fixed and only the initial concentration of the substrate, $\alpha$, is changed, we observe (table 4.3) that the system evolves, alternatively, towards the cycle LC1 or LC2. The intervals of $\alpha$ values corresponding to these successive choices diminish exponentially as $\alpha$ increases, until an accumulation point is reached, beyond which the system evolves towards the third cycle, LC3. The origin of such behaviour, also observed in a different context by Takesue & Kaneko (1984), can be elucidated by means of one-dimensional maps (Decroly & Goldbeter, 1985; Decroly, 1987a). A prediction of this analysis, verified by numerical simulations, is that from certain initial conditions the system can evolve to one of the stable cycles after passing, successively, in a transient manner, by each of the two unstable cycles.

Another complex structure of the attraction basins is observed in certain cases of birhythmicity. Thus, the system can evolve towards either one of limit cycles LC1 and LC2, starting from proximate initial conditions, after oscillating in a chaotic manner (fig. 4.13). For these parameter values, the two stable limit cycles are separated by a regime of unstable chaos (Grebogi *et al.*, 1983b). Consequently, the attraction basins of limit cycles LC1 and LC2 possess a fractal boundary. This conclusion is illustrated in fig. 4.14. The initial values of $\beta$ and $\gamma$ being fixed, a variation of the initial value of $\alpha$ leads either to limit cycle LC2 (in which case a vertical line is drawn for that value of $\alpha$), or to the cycle LC1 (the space is then left blank). The evolution towards either one of the two cycles as a function of $\alpha$ is irregular as well as unpredictable. Moreover, the phenomenon possesses the property of self-similarity: any enlargement of a segment of the $\alpha$ axis reveals a structure of alternation between LC1 and LC2 similar to that of the initial segment (Decroly & Goldbeter, 1984b). The boundary between the two stable limit cycles has therefore a fractal nature.

The behaviour illustrated by figs. 4.13 and 4.14 provides an example of **final state sensitivity** (Grebogi *et al.*, 1983a). The evolution towards one or other final state is unpredictable when the unstable trajectory that defines the boundary of their attraction basins is a strange attractor rather than a simple limit cycle.

Another consequence of the fractal nature of the basins' boundaries

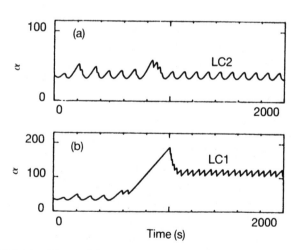

Fig. 4.13. Birhythmicity and final state sensitivity. The biochemical system with multiple regulation evolves to either one of two stable limit cycles after crossing a regime of unstable chaos. The curves are obtained by numerical integration of eqns (4.1) for parameter values of fig. 4.2, with $v = 0.39 \text{ s}^{-1}$ and $k_s = 1.6 \text{ s}^{-1}$. Initial conditions are $\beta = 128$, $\gamma = 0.2439$ with $\alpha = 35.6$ for (a) and $\alpha = 35.7$ for (b) (Decroly & Goldbeter, 1984b).

Table 4.3. *Complex structure of the attraction basins of limit cycles LC1 and LC2 in the case of trirhythmicity*

| Initial value of $\alpha$ | Attracting limit cycle |
| --- | --- |
| 58.10429 | LC2 |
| 58.95438 | LC3 |
| 60.28994 | LC2 |
| 60.55276 | LC3 |
| 60.91664 | LC2 |
| 60.98123 | LC3 |
| 61.06684 | LC2 |
| 61.08157 | LC3 |
| 61.10089 | LC2 |
| 61.10418 | LC3 |
| 61.10850 | LC2 |
| 61.10923 | LC3 |
| 61.11020 | LC2 |
| 61.11036 | LC3 |
| 61.11057 | LC2 |
| 61.110615 | LC3 |
| 61.110633 | LC2 |
| 61.110671 | LC3 |
| 61.1106822 | LC2 |
| 61.1106840 | LC3 |
| 61.11068637 | LC2 |
| 61.11068678 | LC3 |
| 61.110687309 | LC2 |
| 61.110687399 | LC3 |
| 61.110687516 | LC2 |
| 61.110687536 | LC3 |
| ⋮ | ⋮ |
| 61.110687575 | LC1 |

The initial concentrations of $\beta$ and $\gamma$ being fixed, the initial value of $\alpha$ is augmented continuously and the limit cycle reached by system (4.1) is indicated.
Decroly & Goldbeter, 1985.

is that the effect of a perturbation of each of the limit cycles becomes unpredictable. The addition of increasing amounts of $\alpha$ at a given phase of limit cycle LC2 leads first to cycle LC1 then to LC2, in an apparently random manner (fig. 4.15). This sensitivity of the final state nevertheless occurs in the model only in precise conditions that remain relatively rare in comparison to the case where a sharp boundary separates the attraction basins of multiple periodic solutions.

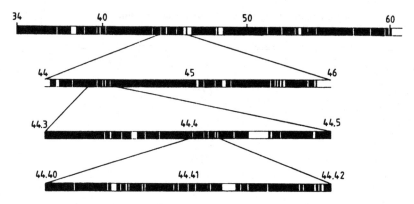

Fig. 4.14. Self-similar structure of the attraction basins of the two limit cycles in the case of final state sensitivity described in fig. 4.13. The initial values of $\beta$ and $\gamma$ being fixed, the initial value of $\alpha$ is varied in a continuous manner. A vertical line is traced when the system evolves towards limit cycle LC2. The white zones correspond to values for which the system evolves towards limit cycle LC1. Successive enlargements of the domains of variation of $\alpha$ illustrate the self-similarity of the random alternation between the two limit cycles, as a function of the initial substrate concentration (Decroly & Goldbeter, 1984b).

## 4.6 Complex periodic oscillations: bursting

Periodic behaviour of the 'bursting' type corresponds to flares of high-frequency oscillations regularly spaced from each other by phases of quiescence. Such oscillations are among the most common observed at the cellular level in electrically excitable systems. Two experimental examples of bursting are shown in fig. 4.16. Part a illustrates the autonomous electrical activity of the neuron R15 in *Aplysia*; this neuron, which is probably the most investigated nerve cell after the giant squid axon (Adams & Benson, 1985), represents the experimental prototype of this type of oscillation. Part b shows the bursting behaviour observed for the membrane potential of pancreatic β-cells (Perez-Armendariz *et al.*, 1985).

Numerous experimental (Alving, 1968; Junge & Stephens, 1973; Gola, 1974; Chaplain, 1976; Meech, 1979; Gorman & Hermann, 1982; Johnston & Brown, 1984, Adams & Benson, 1985; Alonso, Faure & Beaudet, 1994) and theoretical studies (Both, Finger & Chaplain, 1976; Plant & Kim, 1976; Connor *et al.*, 1977; Plant, 1978; Carpenter, 1979; Chay & Rinzel, 1985; Rinzel & Lee, 1986; Ermentrout & Kopell, 1986; Kopell & Ermentrout, 1986; Rinzel, 1987; Destexhe *et al.*, 1993; Bertram *et al.*, 1995; Destexhe & Sejnowski, 1995) have been devoted to

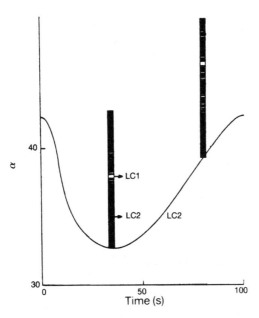

Fig. 4.15. Influence of the fractal structure of the attraction basins on the sensitivity to perturbations in the course of oscillations. The system is perturbed at two different phases of limit cycle LC2 by an instantaneous increase in $\alpha$. A short horizontal line is traced at that value of $\alpha$ when the system returns to limit cycle LC2; an empty space corresponds to a transition to limit cycle LC1 (Decroly & Goldbeter, 1984b).

the mechanism of neuronal oscillations of the bursting type. Bursting oscillations in the membrane potential of $\beta$-cells have also been studied in a theoretical manner (Chay & Keizer, 1983; Chay, 1993). In view of its importance and widespread occurrence, it is interesting to determine the origin of bursting behaviour and, in particular, the mechanism underlying the transition from simple to complex periodic oscillations. The three-variable, multiply regulated enzyme system provides a useful prototype for studying this transition. Even if most cases of bursting are observed experimentally in neurons whose oscillatory properties are related to the electrical excitability of the nerve membrane, we can hope to identify, in the biochemical model, mechanisms that will be of sufficiently broad significance to throw light, in a more general manner, on the transition from simple to complex oscillatory behaviour in other biological systems.

In the phase space, the trajectory associated with bursting oscillations

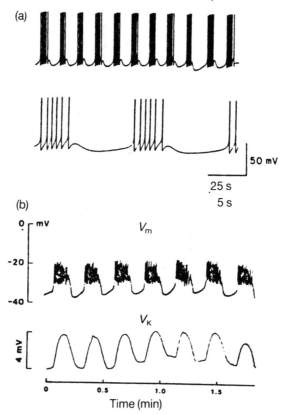

Fig. 4.16. Examples of bursting oscillations in neurophysiology. (a) Variation of the membrane potential in the neuron R15 from *Aplysia*, at two different time scales (Adams & Benson, 1985). (b) Complex periodic variation of the membrane potential ($V_m$) of pancreatic β-cells (upper curve), associated with a simple periodic variation of the extracellular potassium concentration ($V_K$) (Lebrun & Atwater, 1985).

takes the form of a limit cycle with several loops (fig. 4.17) corresponding to as many successive peaks of products $P_1$ (β) and $P_2$ (γ); movement along these loops takes place in the direction of a decrease in α, the substrate being consumed during synthesis of $P_1$. Moreover, the synthesis of a peak in γ is accompanied, for a similar reason, by a decrease in β. In contrast to the strange attractor of fig. 4.10, the curve in fig. 4.17 always passes again through a given point of the trajectory: the fact that the curve is closed reflects the periodic nature of these complex oscillations.

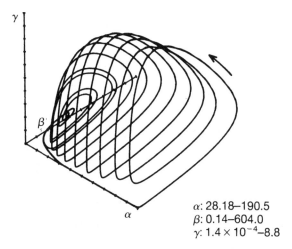

$\alpha$: 28.18–190.5
$\beta$: 0.14–604.0
$\gamma$: $1.4 \times 10^{-4}$–8.8

Fig. 4.17. Phase space trajectory associated with bursting in the biochemical model with multiple regulation. The curve corresponds to the complex periodic oscillations of fig. 4.3d (Decroly & Goldbeter, 1982).

### *Different modes of bursting*

That complex periodic behaviour in the model is equivalent to bursting of the type shown in fig. 4.16 becomes particularly clear when we consider the variation of product $P_1$ (or $P_2$) in the course of time, rather than the variation of the substrate. The pattern of bursting changes, however, as a function of the parameter values of the system.

Shown in fig. 4.18 are various types of periodic oscillation, simple or complex, obtained for different values of the kinetic constant $k_s$. In part a, the system oscillates in a simple periodic manner. In part b, rapid, minute oscillations are observed on top of the large-amplitude peaks in $\beta$; such a behaviour represents a first pattern of bursting. In part c, four peaks of large amplitude in $\beta$ are repeated in a periodic manner; this pattern of bursting will be represented by the notation $\pi(4)$. In part d, so as to better show the succession of four peaks followed by another group of four peaks in $\beta$ – all the peaks having a different amplitude – the value of $\alpha$ is represented instead of $\beta$; this pattern of bursting is of the type $\pi(4, 4)$. Finally, part e represents periodic oscillations of an even more complex nature, observed in the model for another value of parameter $v$; this pattern of bursting, of the type $\pi(11, 2, 3, 2)$, is constituted by the periodic repetition of a set of four groups containing, successively, 11, 2, 3 and 2 peaks in $\beta$ of distinct amplitude. The latter

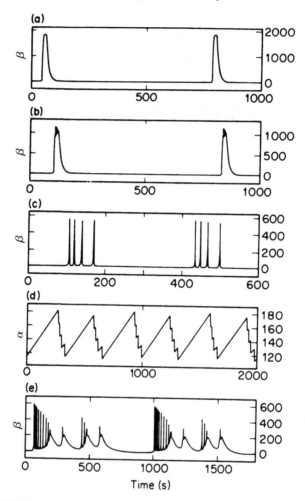

Fig. 4.18. Different types of simple or complex periodic behaviour observed in the model for the multiply regulated biochemical system, as a function of parameter $k_s$: (a) 15 s$^{-1}$, simple periodic oscillations; (b) 8 s$^{-1}$, bursting with a series of small, rapid oscillations on the top of slower oscillations of larger amplitude; (c) 1.53 s$^{-1}$, bursting oscillations with 4 peaks per period, of the type $\pi(4)$; (d) 1.534 s$^{-1}$, bursting pattern of the type $\pi(4, 4)$; (e) 2 s$^{-1}$, bursting oscillations of the type $\pi(11, 2, 3, 2)$. The curves are obtained by integration of eqns (4.1) for parameter values of fig. 4.2, with $v = 0.25$ s$^{-1}$ in (a)–(d) and $v = 0.445$ s$^{-1}$ in (e). Only the value of $\beta$ is shown, except in (d) where it is replaced by the value of $\alpha$ (Decroly & Goldbeter, 1987).

Table 4.4. *Different modes of bursting observed by numerical integration of eqns (4.1) as a function of parameter $k_s$*

| $k_s$ (s$^{-1}$) | Behavioural mode | $k_s$ (s$^{-1}$) | Behavioural mode |
|---|---|---|---|
| 1.3 | $\pi(1)$ | 1.87 | $\pi(11)$ |
| $1.34 \leq k_s < 1.4$ | Period doubling sequence | 1.88 | $\pi(10)$ |
| 1.4 | Chaos | 1.95 | $\pi(10)$ |
| 1.4473 | $\pi(4)$ | 1.96 | $\pi(9)$ |
| 1.532 | $\pi(4)$ | 2.23 | $\pi(9)$ |
| 1.533 | $\pi(4, 4)$ | 2.24 | $\pi(8)$ |
| 1.534 | $\pi(4, 4)$ | 2.91 | $\pi(8)$ |
| 1.535 | Chaos | 2.92 | $\pi(7)$ |
| 1.5355 | $\pi(4, 3)$ | 4.16 | $\pi(7)$ |
| 1.536 | Chaos | 4.18 | $\pi(6)$ |
| 1.537 | Chaos | 5.9 | $\pi(6)$ |
| 1.5378 | $\pi(4, 2)$ | 6.0 | $\pi(5)$ |
| 1.544 | $\pi(4, 2)$ | 8.5 | $\pi(5)$ |
| 1.545 | Chaos | 8.6 | $\pi(4)$ |
| 1.55 | $\pi(4, 1) = \pi(5)$ | 9.6 | $\pi(4)$ |
| 1.65 | $\pi(5)$ | 9.7 | $\pi(3)$ |
| 1.7 | $\pi(6)$ | 10.8 | $\pi(3)$ |
| 1.8 | $\pi(6, 2)$ | 10.9 | $\pi(2)$ |
| 1.81 | $\pi(7)$ | 12.5 | $\pi(2)$ |
| 1.85 | $\pi(7)$ | 12.6 | $\pi(1)$ |
| 1.86 | $\pi(11)$ | $\infty$ | $\pi(1)$ |

Decroly & Goldbeter, 1987.

pattern of complex bursting – rather rare in the model and even more in the experiments – occurs only in an extremely reduced domain of parameter space. Experimentally, the most commonly observed types of bursting are those of parts b and c.

Table 4.4 shows how the main patterns of bursting occur in the model when parameter $k_s$ moves across the domain of complex periodic oscillations. Bursting occurs after the system has passed a domain of chaotic behaviour, which is itself reached beyond a cascade of period-doubling bifurcations issued from a simple periodic solution. A few narrow windows of chaos separate the first patterns of bursting observed.

### Analysis of bursting by reduction to a two-variable system

The numerical study of eqns (4.1) shows that the behaviour of the biochemical system in the course of bursting can be decomposed into two phases: in the first, $\beta$ and $\gamma$ remain close to their steady-state values while variable $\alpha$ slowly accumulates. In the second phase, $\beta$ and $\gamma$ undergo rapid oscillations while $\alpha$ decreases in a relatively slow manner. The slower variation of $\alpha$ suggests that we can rewrite system (4.1) as:

$$\frac{d\alpha}{dt} = \epsilon \, (v' - \sigma'_1 \, \phi) \tag{4.4a}$$

$$\frac{d\beta}{dt} = q_1 \, \sigma_1 \, \phi - \sigma_2 \, \eta \tag{4.4b}$$

$$\frac{d\gamma}{dt} = q_2 \, \sigma_2 \, \eta - k_s \, \gamma \tag{4.4c}$$

where $\epsilon \, v' = v$ and $\epsilon \, \sigma'_1 = \sigma_1$.

In the limit $\epsilon \to 0$, the substrate concentration can be treated as a parameter governing the evolution of the rapid system (4.4b,c). Although this approximation is rather crude, it allows us to reach a better understanding of the transition from simple to complex periodic behaviour (Decroly & Goldbeter, 1987). This method, proposed by Rinzel (1987), has been successfully used in the theoretical analysis of bursting in certain neurons and in the $\beta$-cells of the pancreas (Rinzel & Lee, 1986). The following discussion of bursting in the multiply regulated enzyme system is based on the study of Decroly & Goldbeter (1987). An asymptotic analysis of bursting in that system has recently been proposed by Holden & Erneux (1993) within the limit of large allosteric constants.

The analysis of the subsystem (4.4b,c) permits us to establish the curve yielding the steady state of variable $\beta$ as a function of parameter $\alpha$ within the rapid system $(\beta, \gamma)$. This curve has the form of an S-shaped sigmoid, as indicated in fig. 4.19a. This result can be explained by the analysis developed in chapter 2 for the mono-enzyme model proposed for glycolytic oscillations. In that model, three steady-state values of product $\gamma$ are obtained for a given value of the substrate concentration $\alpha$ when the latter is held constant in an appropriate range (see also fig. 3.3, p. 95). Here, for certain values of parameter $\alpha$, we observe a similar phenomenon of bistability for $\beta$, the substrate of enzyme $E_2$ in the system $(\beta, \gamma)$. The median branch of the hysteresis loop yielding $\beta_0$ as a function of $\alpha$ is unstable; it almost coincides, in its lower part, with the bottom branch of the sigmoid curve.

The different diagrams giving the steady state of $\beta$ as a function of parameter $\alpha$ in fig. 4.19 are established for decreasing values of the kinetic constant $k_s$. In parts a, b, c and e, the trajectory followed by the complete three-variable system governed by eqns (4.1) is represented schematically. In order not to obscure the figure by superposing two curves, the schematic trajectory passes beyond the lower branch of the steady-state curve, which gives the false impression that the concentration $\beta$ may become negative!

At higher values of $k_s$ (fig. 4.19a), the upper and lower branches of the steady-state curve of $\beta$ are stable in the range of $\alpha$ values considered. As indicated above, $\alpha$ behaves in fact as a slow variable of the

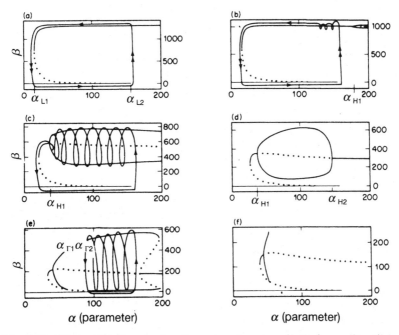

Fig. 4.19. Bifurcation diagram for the reduced system (4.4b,c) as a function of parameter $\alpha$. On the ordinate, the steady-state concentration of $\beta$ in that system is shown, as well as the maximum value reached by $\beta$ in the course of oscillations. The diagrams are obtained numerically by means of the program AUTO (Doedel, 1981), for decreasing values of parameter $k_s$ (in s$^{-1}$): (a) 10; (b) 7.78; (c) 4.5; (d) 2.7; (e) 1.7; (f) 1.2. Solid or dashed lines denote, respectively, stable and unstable (steady or periodic) regimes. The arrowed trajectories represent, schematically, the dynamic behaviour of the full, three-variable system (4.1a–c). The particular values of $\alpha$ relate to the limit points of the hysteresis curve ($\alpha_{L1}$, $\alpha_{L2}$), the Hopf bifurcation points ($\alpha_{H1}$, $\alpha_{H2}$), and the points corresponding to the appearance of homoclinic orbits ($\alpha_{\Gamma1}$, $\alpha_{\Gamma2}$) (Decroly & Goldbeter, 1987).

complete system (4.1). The dynamic behaviour of the latter system can be explained by fig. 4.19a as follows: starting from a low value of $\alpha$, the value of $\beta$ is low and enzyme $E_1$ is therefore predominantly in its inactive state. The net variation of $\alpha$ is then positive since the rate of substrate injection exceeds the rate of its consumption in the enzyme reaction. Consequently, 'parameter' $\alpha$ in fig. 4.19a slowly increases while variable $\beta$ follows the curve defined by the steady state of the rapid subsystem $(\beta, \gamma)$. When $\alpha$ reaches the limit point $\alpha_{L2}$ where the lower and median branches of the steady-state sigmoid merge, the rapid subsystem no longer admits any steady state on the lower branch; the system governed by eqns (4.1) then evolves towards the upper branch of the steady-state curve of $\beta$ (trajectory marked by arrowheads in fig. 4.19a).

On the upper branch, the value of $\beta$ is sufficiently large for enzyme $E_1$ to be fully activated by its product. The rate of the enzyme reaction then exceeds the rate of substrate injection, and the value of $\alpha$ begins to decrease in the course of time. The complete system (4.1) therefore progresses along the upper branch of the sigmoid in the direction opposite to that in which it moved along the lower branch. When $\alpha$ reaches the limit point $\alpha_{L1}$, where the upper and median branches of the sigmoid merge, the system falls back on to the lower branch where resides the only steady state now accessible to the rapid subsystem $(\beta, \gamma)$. The simple periodic behaviour thus described accounts for the oscillations shown in fig. 4.18a.

For a slightly lower value of $k_s$ (fig. 4.19b), the stability analysis of the rapid subsystem indicates that the upper branch of the curve yielding the steady state of $\beta$ as a function of parameter $\alpha$ becomes unstable for a value of $\alpha$ just above the limit point $\alpha_{L2}$. At this instability point $\alpha_{H1}$, a Hopf bifurcation occurs, beyond which a stable limit cycle arises; the envelope of the corresponding oscillations is represented in fig. 4.19b. Just below $\alpha_{H1}$, the steady state of the subsystem $(\beta, \gamma)$ is a stable focus around which damped oscillations occur. The behaviour of the full system (4.1) can be discussed as for part a. The only difference is that now, when the system jumps towards the upper branch of the sigmoid as soon as $\alpha$ exceeds the value $\alpha_{L1}$, damped oscillations of decreasing amplitude occur until the decrease in $\alpha$ is such that the steady state of the rapid subsystem becomes a stable node. The oscillations have thus a form similar to the pattern of bursting represented in fig. 4.18b: a short train of small, decreasing oscillations ride the crest of the large-amplitude peak in $\beta$.

In fig. 4.19c, a lower value of $k_s$ brings about a shift of the Hopf bifurcation point to the left, so that $\alpha_{H1}$ is now located between $\alpha_{L1}$ and $\alpha_{L2}$. Consequently, the steady state of the rapid subsystem is an unstable focus, surrounded by a limit cycle, between $\alpha_{L2}$ and $\alpha_{H1}$. When the complete system jumps towards the upper branch in $\alpha_{L2}$, large-amplitude oscillations occur; the latter continue until $\alpha$ has decreased below $\alpha_{H1}$ where they become damped, before the system returns to the bottom branch in $\alpha_{L1}$. Such a fully developed bursting is of the type represented in fig. 4.18c.

For an even lower value of $k_s$ (fig. 4.19d), a second Hopf bifurcation point appears just to the left of $\alpha_{L2}$. The dynamic behaviour of the system is similar to the one described above. When the value of $k_s$ decreases further, the envelope of the limit cycle in the rapid subsystem approaches and eventually reaches the median, unstable branch of the steady-state sigmoid curve. As indicated in fig. 4.18e, a **homoclinic orbit** thereby arises at both points $\alpha_{\Gamma 1}$ and $\alpha_{\Gamma 2}$. When the full system attains the upper branch in $\alpha_{L2}$, high-frequency oscillations proceed until $\alpha$ reaches the value $\alpha_{\Gamma 2}$: the system then falls back abruptly on to the bottom branch, and a new cycle of bursting resumes with the increase in the substrate level.

The creation of a homoclinic orbit thus gives rise to a sharp decline in the number of peaks within a burst. The abrupt transition from the bursting pattern with 11 peaks per period, $\pi(11)$, to the pattern with 7 peaks, $\pi(7)$, in table 4.4 originates from the appearance of such a homoclinic orbit. It should be noted that the situation depicted in fig. 4.19e creates conditions suitable for the coexistence between bursting oscillations and a limit cycle that would be stabilized around a value of $\alpha$ between $\alpha_{L1}$ and $\alpha_{\Gamma 1}$. Such a situation is indeed close to that described further in chapter 6 for the origin of birhythmicity of a similar kind in a model for the intercellular communication system of *Dictyostelium* amoebae.

Figure 4.19f shows the behaviour of the system for an even lower value of $k_s$. Here, the point $\alpha_{\Gamma 2}$ coincides with the limit point $\alpha_{L2}$, so that the only state accessible to the system between $\alpha_{\Gamma 1}$ and $\alpha_{L2}$ is the bottom branch of the steady state curve of the rapid subsystem. As soon as $\alpha$ exceeds $\alpha_{L2}$, the system evolves towards a large-amplitude limit cycle, not shown on the figure. Bursting disappears in such conditions.

Indicated in fig. 4.20 as a function of the rate constant $k_s$ and of parameter $\alpha$ in the rapid subsystem (4.4b,c) is the position of the limit points $\alpha_{L1}$ and $\alpha_{L2}$, as well as that of the Hopf bifurcation points $\alpha_{H1}$, $\alpha_{H2}$ and of the homoclinic orbit $\alpha_\Gamma$ (which can admit the two values $\alpha_{\Gamma1}$ and $\alpha_{\Gamma2}$ for a given value of $k_s$). This diagram was obtained numerically (Decroly & Goldbeter, 1987) by means of the program AUTO developed by Doedel (1981) for the automatic determination of steady and periodic solutions in systems of ordinary differential equations. At point D, the limit point $\alpha_{L1}$ coincides with the Hopf bifurcation point $\alpha_{H1}$ as well as with the point of emergence of a homoclinic orbit, $\alpha_{\Gamma1}$. The horizontal dashed lines refer to the situations a–f considered in fig. 4.19.

The above analysis allows us to explain a particular mode of bursting that differs from those more commonly described. The interval separating two successive peaks within a volley generally passes through a minimum in the course of bursting, so that the curve yielding the interval as a function of time possesses a parabolic shape. This type of 'parabolic' bursting (Ermentrout & Koppel, 1986; Rinzel, 1987) is that usually observed in the experiments and in the models. In the present system, we can observe another mode of bursting that is rather of an 'inverse parabolic' type. The time interval between two successive peaks within a volley indeed goes through a maximum (fig. 4.21). This phenomenon occurs in the vicinity of a homoclinic orbit: the system's evolution slows

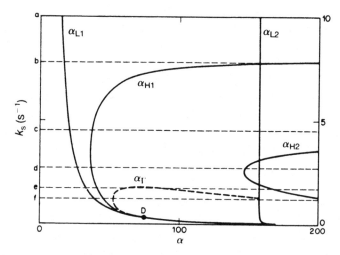

Fig. 4.20. Behaviour of the two-dimensional system (4.4b,c) in the parameter space $\alpha$–$k_s$. The values of $\alpha$ corresponding to the limit points of the hysteresis curve ($\alpha_{L1}$, $\alpha_{L2}$), to the Hopf bifurcation points ($\alpha_{H1}$, $\alpha_{H2}$) and to the points corresponding to the appearance of homoclinic orbits ($\alpha_{\Gamma1}$, $\alpha_{\Gamma2}$) have been obtained numerically by means of the program AUTO (Doedel, 1981); D is a triple point. Dashed horizontal lines refer to the values of $k_s$ considered in parts (a)–(f) of fig. 4.19 (Decroly & Goldbeter, 1987).

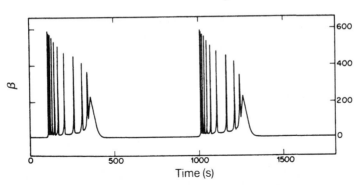

Fig. 4.21. 'Inverse parabolic' bursting obtained in system (4.1a–c) for $v = 0.25\ \mathrm{s}^{-1}$ and $k_s = 1.86\ \mathrm{s}^{-1}$, near the point of homoclinic tangency in the rapid, two-variable subsystem $(\beta, \gamma)$. Other parameter values are those of fig. 4.3 (Decroly & Goldbeter, 1987).

down in the middle of the bursting phase as the amplitude of the limit cycle in the rapid subsystem approaches the saddle branch of the hysteresis curve yielding the steady state as a function of parameter $\alpha$.

The preceding discussion suggests that the rate of substrate evolution governs the number of peaks in a burst. A faster progression of the system on the upper branch of the hysteresis curve in the situations described by fig. 4.19c–e should indeed result in reducing both the number of peaks and the period of oscillations. The numerical integration of eqns (4.4) for increasing values of parameter $\epsilon$ confirms this conjecture. As indicated by fig. 4.22, the number of loops of the limit cycle decreases as $\epsilon$ rises.

If it accounts for the transition from simple periodic behaviour to the patterns of bursting represented in fig. 4.18b and c, the analysis of the two-dimensional system does not provide an explanation for the occurrence of more complex patterns of bursting such as those of parts d and e. The study of a one-dimensional return map allows us to comprehend the origin of such complex modes of periodic behaviour in the multiply regulated enzyme model. Such maps represent a powerful tool for studying the dynamics of nonlinear systems (Collet & Eckmann, 1980; Gumowski & Mira, 1980) and have proved useful in the theoretical investigation of bursting or chaos in chemical systems (Rinzel & Troy, 1983; Bergé *et al.*, 1984) as well as in the β-cells of the pancreas (Chay & Rinzel, 1985).

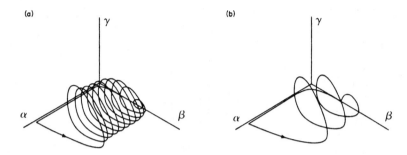

Fig. 4.22. Behaviour of system (4.4) in the phase space, for two values of parameter $\epsilon$: (a) 0.5 and (b) 2. The curves are obtained by integration of eqns (4.4) for the parameter values of fig. (4.12b) with $k_s = 5$ s$^{-1}$, $v' = 0.25$ s$^{-1}$, $\sigma_1' = 10$ s$^{-1}$. These values correspond to $v = 0.125$ s$^{-1}$, $\sigma_1 = 5$ s$^{-1}$ and $q_1 = 100$ in (a), and to $v = 0.5$ s$^{-1}$, $\sigma_1 = 20$ s$^{-1}$ and $q_1 = 25$ in (b). The value of $q_1$ is adjusted so that the product $q_1\sigma_1$ remains constant (Decroly & Goldbeter, 1987).

### Analysis of a one-dimensional return map

To characterize the different behavioural modes of the model, it is useful to simplify its dynamics by further reducing the dimension of the system. This can be done by constructing a one-dimensional return map. A Poincaré section previously described (see fig. 4.11) can be utilized to this end. When plotting the concentration $\alpha_{n+1}$ corresponding to the $(n + 1)$th peak of $\beta$ as a function of $\alpha_n$, we obtain, for the asymptotic regime attained by numerical integration of eqns (4.1), a certain number of points, finite or infinite, depending on the more or less complex nature of the oscillatory behaviour.

Thus, fig. 4.23a represents the result obtained when parameter values give rise to deterministic chaos. The continuous aspect of the curve reflects the chaotic nature of the evolution: an infinity of $\alpha$ values can correspond to the successive maxima in $\beta$, but these values of the substrate concentration are not disseminated at random, which fact translates into the continuous nature of the Poincaré section.

In this representation, simple periodic behaviour corresponds to a fixed point located on the bisectrix, given that such a behaviour is characterized by the equality $\alpha_{n+1} = \alpha_n$ for all values of $n$. A complex periodic solution will correspond to a finite number of points, none of which will be located on the bisectrix. Thus, the pattern of bursting $\pi(4)$, which contains four successive peaks of $\beta$, corresponds to four distinct points in the one-dimensional equivalent (fig. 4.23b). The construction of the Poincaré section for the similar situation of a pattern $\pi(3)$ is

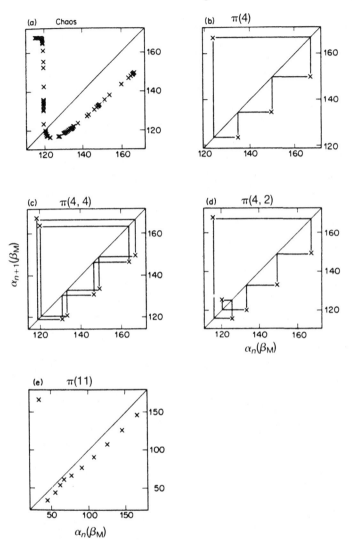

Fig. 4.23. Poincaré sections obtained in system (4.1) for different values of parameter $k_s$ (in s$^{-1}$): (a) 1.537; (b) 1.5; (c) 1.534; (d) 1.539; (e) 1.86. Situations (c)
and (e) correspond, respectively, to figs. 4.18d and 4.21. The continuous curve
in (a) corresponds to chaos; the simple or complex pattern of bursting obtained
in the other cases is indicated. The results are obtained by integration of eqns
(4.1). The construction of the return map $\alpha_{n+1} = f[(\alpha_n)]$ is explained in fig. 4.24
and in the text (Decroly & Goldbeter, 1987).

detailed schematically in fig. 4.24. More complex modes of bursting are reflected by a larger number of points in the corresponding section, as indicated by parts c, d and e obtained, respectively, for the bursting patterns $\pi(4, 4)$, $\pi(4, 2)$ and $\pi(11)$.

Although the results of fig. 4.23 are obtained for different values of parameter $k_s$, it becomes apparent that the points are distributed along an underlying curve whose shape is similar each time (see also fig. 4.24). As indicated by the continuous curve obtained for chaos (fig. 4.23a), the common dynamics that gives rise to these different results produces a return map $\alpha_{n+1} = f(\alpha_n)$ formed by three main parts: an upper, quasi-horizontal part, a part (on the right) parallel to the first bisectrix, and a third, central part linking the other two together. Such a map, obtained by numerical integration of eqns (4.1), is implicitly present in the other diagrams of fig. 4.23.

### A piecewise linear map for bursting

On the basis of these numerical results, an approximation of the unidimensional map by a piecewise linear function $x_{n+1} = f(x_n)$, can be constructed (Decroly, 1987a; Decroly & Goldbeter, 1987). This function, defined by eqns (4.5), is formed by three linear segments. The first, obeying eqn (4.5a), relates to the horizontal part of the map, where the value of $x$ (i.e. $\alpha$) reaches its maximum $M$; this particular value corre-

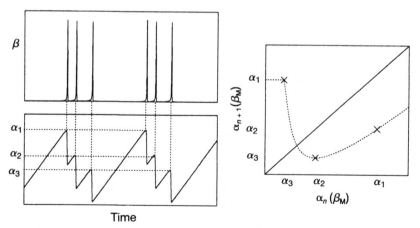

Fig. 4.24. Schematic construction of the return map $\alpha_{n+1} = f[\alpha_n(\beta_M)]$ used in fig. 4.23, from the oscillations generated by the model with multiple regulation (Decroly, 1987a).

sponds to the fact that the substrate accumulates up to the same, maximum level before the volley of oscillations begins. The successive peaks in the course of bursting produce a decrease of the substrate until the active phase of bursting comes to an end; the substrate, which has decreased down to a minimum value $m$, can then re-accumulate, and a new active phase of bursting begins when $\alpha$ reaches its maximum value. The two remaining segments, obeying eqns (4.5b,c), account for the other aspects of the oscillatory dynamics. In particular, the fact that the segment defined by relation (4.5c) is parallel to the first bisectrix and is located at a distance $a$ from it is equivalent to the observation that a quasi-constant quantity $a$ of substrate is consumed by the production of each of the successive peaks in $\beta$. An important simplification in this approximation is to avoid taking into account the curved nature of the numerically obtained map near the minimum of $\alpha_{n+1}$ (see fig. 4.23a). The consequences of this simplification are considered further below.

$$x_n \leq 1 : x_{n+1} = f_1(x_n) = M \tag{4.5a}$$

$$1 < x_n \leq m : x_{n+1} = f_2(x_n) = -b\,x_n + b + M \tag{4.5b}$$

$$m < x_n : x_{n+1} = f_3(x_n) = x_n - a \tag{4.5c}$$

The piecewise linear function defined by these three relations is represented in fig. 4.25. Three parameters control this map: the value $M$ corresponds to the maximum level of substrate reached at the beginning of the active, bursting phase, i.e. at the top of the first peak in $\beta$; parameter $a$ measures the quantity of substrate consumed by the synthesis of each peak of $\beta$; finally, the slope $b$ of the median portion of the linear map determines the value of the minimum $m$ that corresponds to the substrate level reached in the last peak of the volley, at the end of the active phase of bursting. Determining the intersection of the segments defined by eqns (4.5a,b) yields:

$$m = (a + b + M)/(b + 1) \tag{4.6}$$

The abscissa value of the intersection point of segments $f_1(x_n)$ and $f_2(x_n)$ is arbitrarily taken to be equal to unity. The fixed point $x^*$ where segment $f_2(x_n)$ intersects with the first bisectrix is unstable as long as the absolute value of the slope, $|-b|$, remains larger than unity.

Equations (4.5) give the value of $x_{n+1}$ as a function of $x_n$. Starting from

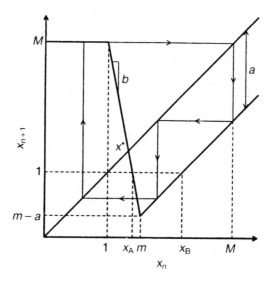

Fig. 4.25. Piecewise linear map constructed on the basis of the results of fig. 4.23 to account for complex periodic oscillations of the 'bursting' type. The one-dimensional return map $x_{n+1} = f(x_n)$ is defined by eqns (4.5) which contain the three parameters $M$, $a$ and $b$. The arrowed trajectory corresponds to the simple pattern of bursting with three peaks per period $\pi(3)$. Parameter values are $a = 6$, $b = 5$, $M = 11$ (Decroly & Goldbeter, 1987).

a given value, $x_1$, these equations thus permit us to generate the value of $x_2$; in the next iteration, the equations generate the value of $x_3$ corresponding to $x_2$, and so forth. Practically, we may start from the point of abscissa $x_1 = M$, located on the right branch of the map in fig. 4.25 (this point corresponds to the first peak of a bursting phase). The value of the next abscissa coordinate is obtained by determining the intersection $I_1$ of the first bisectrix with the horizontal passing through this first point: in this intersection, $x_2 = x_1$. The next point of the dynamics is therefore obtained by drawing the vertical through $I_1$ and by determining its intersection with one of the branches of the piecewise linear map.

The next abscissa value, $x_3$, is obtained similarly, by determining the intersection $I_2$ of the first bisectrix with the horizontal through the point of abscissa value $x_2$; the vertical through $I_2$ intersects with the map in the third point of the temporal sequence associated with the pattern of bursting. This process, thus repeated, generates a sequence of successive values of the abscissa coordinates $x_n$, and thereby provides a picture of the temporal dynamics of the system. A periodic orbit ends as soon as the value $x_i$ becomes equal to the starting value, $x_1$.

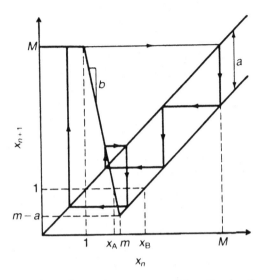

Fig. 4.26. Explanation of complex patterns of bursting by the piecewise linear map defined by eqns (4.5). The arrowed trajectory corresponds to the complex pattern of bursting $\pi(3, 2)$. Parameter values are $a = 4.3$, $b = 5$, $M = 11$ (Decroly & Goldbeter, 1987).

Two values of the abscissa coordinate $x_n$ play a particular role in the iterative dynamics of the system governed by eqns (4.5). These are the values denoted $x_A$ and $x_B$ in figs. 4.25 and 4.26. As soon as the value $x_n$ iterates to a value between $x_A$ and $x_B$, the next value of $x_n$ will be equal to the starting value, $M$. The fact that the first segment of the map is horizontal imposes the condition that sooner or later (as soon as $x_A \leq x_n \leq x_B$), the iteration will bring the system back to the initial value. As a consequence, by construction, the system admits **constrained periodic orbits** (CPOs). Such orbits would not necessarily exist if the branch $f_1(x_n)$ of the one-dimensional map were not strictly horizontal.

Depending on whether the representative point of the system iterates from the right branch $f_3(x_n)$ to the horizontal branch $f_1(x_n)$ immediately, or by making an excursion to the median branch $f_2(x_n)$, the constrained periodic orbit will be of simple or complex nature. A simple CPO corresponds to a pattern of bursting of the type $\pi(j)$, while a complex CPO will correspond to the bursting type $\pi(j, k, l \ldots)$ where the number of indices reflects the number of times the iteration will bring the system on to the median branch $f_2(x_n)$ before the constrained periodic orbit ends by iteration towards the horizontal branch $f_1(x_n)$. The number of different indices also corresponds to the number of distinct groups of spikes in the course of one period of bursting.

Two types of constrained periodic orbit, obtained for different values of parameter $a$, are represented in figs. 4.25 and 4.26. The situation in fig. 4.25 corresponds to a simple pattern of bursting, of the type $\pi(3)$. Starting from the value $x_1 = M$, the iteration of eqns (4.5) indeed twice lowers that value by the same amount $a$, before $x_3$ iterates again towards the starting value, so that $x_4 = x_1 = M$. Thus, to a number $N$ of peaks in $\beta$ within a burst corresponds, in the iterative dynamics, a number $(N-1)$ of reductions of the value of $x_n$ by the quantity $a$ (see also fig. 4.24). Figure 4.26 illustrates a constrained periodic orbit of the complex type, corresponding to the pattern of bursting $\pi(3, 2)$. Indeed, starting from the value $M$, the abscissa value diminishes twice by the amount $a$, before increasing again as a result of the excursion to the median branch of the map; then it diminishes again, once, by quantity $a$ corresponding to the synthesis of a new peak of $\beta$, before the next iteration returns the starting value, $M$.

The equations of the piecewise linear map allow us to determine the domains of existence of simple, constrained periodic orbits with $p$ peaks, $\pi(p)$, as a function of parameter $a$. The mode $\pi(p)$ is obtained as soon as condition (4.7) is satisfied:

$$x_A \le x_{p-1} \le x_B \tag{4.7}$$

where the abscissa values of the limit points $x_A$ and $x_B$ are given by eqns (4.8), while the value of $x_{p-1}$ is determined according to eqn (4.9):

$$x_A = (M + b - 1)/b, \; x_B = a + 1 \tag{4.8}$$

$$x_{p-1} = M - (p - 2) a \tag{4.9}$$

Note that, by construction, constrained periodic orbits can exist only if the ordinate $(m - a)$ of the minimum of the map is less than unity. Taking into account relation (4.6), this constraint on the parameters of the map is expressed by the inequality:

$$a > (M - 1)/b \tag{4.10}$$

Figure 4.27 shows how the variation of parameter $a$ influences the number of peaks of constrained periodic orbits of the simple type $\pi(p)$,

Fig. 4.27. Domains of values of *a* for which constrained periodic orbits of the simple type $\pi(p)$ exist in the piecewise linear map defined by eqns (4.5). Parameter values for the map are those of fig. 4.26 (Decroly & Goldbeter, 1987).

for the fixed values of the other parameters of the map, $b = 5$, $M = 11$. When *a* is sufficiently large ($a > 10$), the number of peaks of CPOs is always equal to 2: a single decrease by quantity *a* suffices then for condition (4.7) to be fulfilled, starting from *M*. A decrease in *a* causes an increase in the number of peaks in the bursting phase. For the limit value $a = 2$, condition (4.7) is satisfied for $p = 6$. Lower values of *a* do not lead to any CPO, since they violate condition (4.10). Figure 4.27 also indicates that the domain of values of *a* producing a simple pattern of bursting $\pi(p)$ diminishes as the number of peaks in the pattern increases.

What happens in the intervals of *a* values that do not produce any constrained periodic orbit of the simple type? The numerical simulation of eqns (4.5) shows that constrained periodic orbits of the complex type are then observed. Typical results of such simulations are represented in fig. 4.28, for $b = 7$, $M = 11$. The upper panel shows the successive values of *x*, obtained according to eqns (4.5), as a function of parameter *a* varying from 1.76 to 1.82. The number of values of *x* for a given value of *a* is equal to the total number of peaks in the pattern of bursting. Values marked *a–g* correspond, respectively, to the patterns $\pi(6, 2)$, $\pi(6, 2, 6)$, $\pi(6, 2, 5)$, $\pi(6, 2, 4)$, $\pi(6, 2, 3)$, $\pi(6, 2, 2)$ and $\pi(6, 3)$. These complex patterns of bursting nevertheless remain relatively simple when compared to those that occur in the separating intervals; the existence of the latter, extremely complex patterns is reflected in the huge number of values of *x* obtained for certain values of *a*.

When enlarging the scale of resolution, we obtain more and more complex patterns of bursting, but the structure as a function of parameter *a* remains identical, with an alternation between patterns of bursting that are more or less complex with respect to each other. Thus, the lower panel of fig. 4.28 presents results from the iteration of eqns (4.5) in the interval of *a* values between 1.778 and 1.785. This interval corresponds to the transition between the patterns $\pi(6, 2, 5)$ and $\pi(6, 2, 4)$ – i.e. between points *c* and *d* in the upper panel. Here, points *c* and *d*

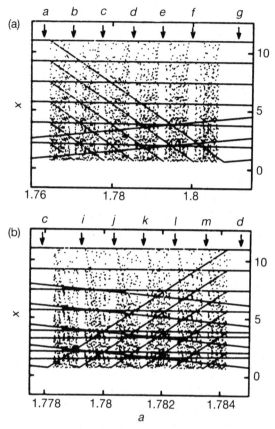

Fig. 4.28. Self-similarity of sequences of bursting patterns as a function of parameter *a* in the piecewise linear map defined by eqns (4.5). For a given value of *a*, the number of values of *x* equals the number of distinct peaks in the pattern of bursting. The variation of parameter *a* from 1.76 to 1.82 in (a) illustrates the passage from the pattern $\pi(6, 2)$ to the pattern $\pi(6, 3)$. In (b), the variation of *a* over the narrower range between 1.778 and 1.785 shows the transition between the more complex patterns $\pi(6, 2, 5)$ and $\pi(6, 2, 4)$. Each time, more complex patterns alternate with relatively simpler patterns of bursting. The results are obtained by iteration of eqns (4.5) for $b = 7$ and $M = 11$ (Decroly & Goldbeter, 1987).

correspond to the same patterns of bursting as previously, while points *i–m* correspond to the more complex patterns $\pi(6, 2, 4, 2)$, $\pi(6, 2, 4, 3)$, $\pi(6, 2, 4, 4)$, $\pi(6, 2, 4, 5)$ and $\pi(6, 2, 4, 6)$. The structure of the patterns of bursting as a function of parameter *a* in the piecewise linear map therefore possesses the property of self-similarity. The precise sequences of complex patterns can be generated in an analytical manner (Decroly, 1987a; Decroly & Goldbeter, 1987).

More and more complex patterns of bursting, similar to that shown in fig. 4.18e, can be obtained by iteration of eqns (4.5). As indicated by the results shown in figs. 4.27 and 4.28, however, the intervals of values of $a$ for which these extremely complex modes of bursting appear become, necessarily, smaller and smaller as the complexity of the oscillations increases. This could explain why such highly complex oscillations are very rare experimentally, and why they can be observed in the differential eqns (4.1) only in a minute domain of parameter space.

The piecewise linear map defined by eqns (4.5) thus allows us to explain the transitions between different simple or complex patterns of bursting as a function of the variation of parameters whose significance can be related to the properties of the biochemical model from which the map originates. Parameter $a$, for example, is linked to the quantity of substrate consumed by the production of a peak of product $P_1$. An increase in the maximum rate $\sigma_1$ of the reaction catalysed by enzyme $E_1$ should therefore correspond to an increase in parameter $a$ of the piecewise linear map. Likewise, a rise in the rate constant $k_s$ results in a decrease in the amount of product $P_2$ within the system; enzyme $E_2$, activated by $P_2$, should therefore become less active as $k_s$ increases. The amount of $P_1$, the substrate for enzyme $E_2$, will then tend to increase, owing to its diminished consumption in the second enzyme reaction. As enzyme $E_1$ is activated by $P_1$, the increased level of this product raises the rate of enzyme $E_1$, which results in an increased amount of substrate consumed during synthesis of a peak of $P_1$. Thus we can see how an increase in the rate constant $k_s$ in the enzyme model can also be associated with a larger value of parameter $a$ in the one-dimensional map studied for bursting.

The comparison of the results of table 4.4 (for $k_s > 1.85$) with those of fig. 4.27 confirms that the increase in these two parameters results, in the differential system as in the piecewise linear map, in the transition from the pattern of bursting $\pi(p)$ to the pattern $\pi(p + 1)$.

The piecewise linear map also allows us to comprehend how the variation of another parameter, such as $b$, can elicit the transition from complex periodic oscillations to simple periodic behaviour. For such a transition to occur, it suffices that the segment $f_2(x_n)$ acquire a less negative slope, so that the fixed point $x^*$ of the return map becomes stable.

### A nonlinear map for bursting and chaos

There is, however, one type of complex oscillatory behaviour that the piecewise linear map defined by eqns (4.5) cannot produce: chaos. For the latter phenomenon to occur, the one-dimensional map must be modified so that it admits a curved junction between the segments $f_2(x_n)$ and $f_3(x_n)$, as in the map derived numerically from the equations of the model (see fig. 4.23a).

The return map defined by eqns (4.11), and represented in the inset to fig. 4.29, takes this characteristic into account. This nonlinear map allows us to unify the various modes of simple or complex oscillatory behaviour observed in the differential system (4.1), including chaos (Decroly, 1987a; Decroly & Goldbeter, 1987):

$$x_n \le x_j : x_{n+1} = g_1(x_n) = M$$

$$x_n > x_j : x_{n+1} = g_2(x_n) = [x_n^2 - (A+1) x_n + B]/(x_n - 1) \qquad (4.11)$$

where $x_j$ is the value of $x$ such that $g_2(x_j) = M$. The three parameters of the map are now $A$, $B$ and $M$.

As shown in fig. 4.29, the iteration of this nonlinear map for different values of parameter $A$ yields a rather faithful image of the behaviour of the differential system (4.1). As $A$ increases, a simple periodic solution

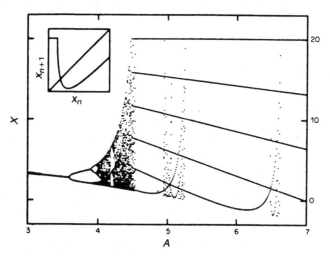

Fig. 4.29. The nonlinear return map represented in the inset and defined by eqns (4.11) accounts for simple and complex patterns of bursting, as well as for chaos, as a function of parameter $A$. Iterations of the return map are performed for $M = 20$ and $B = 7$ (Decroly & Goldbeter, 1987).

undergoes a cascade of period-doubling bifurcations leading to chaos; the latter is followed by an alternation of simple or complex periodic orbits. This is precisely what is observed (see table 4.4) upon increasing parameter $k_s$ in the biochemical model.

## 4.7 Domains of occurrence of the various modes of dynamic behaviour

In view of the rich repertoire of oscillatory behaviours characterizing the multiply regulated enzyme model, the question arises as to what is the relative importance of these behavioural modes in parameter space. To obtain insights into this, the study of the dynamics of system (4.1), carried out mainly, so far, with respect to parameter $k_s$, has been extended to the second important control parameter, the substrate injection rate, $v$.

The different domains of dynamic behaviour in the parameter space $v$–$k_s$ are represented in fig. 4.30. The diagram spans five orders of magnitude of constant $k_s$, and three orders of magnitude of parameter $v$. The domains marked 'stable' are those in which the single steady state admitted by the system is stable; outside these domains, this state is unstable. When instability of the steady state prevails, one or the other mode of temporal organization detailed above is observed.

The largest domain in this parameter space is clearly that of simple periodic oscillations. Second in importance is the domain of complex periodic oscillations. Then comes the domain of coexistence between a stable limit cycle and a stable steady state (dotted zone), which situation is associated with the phenomenon of hard excitation. Just below the latter domain are two regions of birhythmicity corresponding to the coexistence of limit cycles LC1 and LC2 on the one hand and LC2 and LC3 on the other. These two domains of birhythmicity partly overlap; their intersection defines the domain of trirhythmicity, where the three stable limit cycles LC1, LC2 and LC3 coexist. Near the domains of birhythmicity are three distinct regions of chaos (dark zones), whose size is relatively reduced.

The comparison of the different behavioural domains in parameter space shows that simple periodic oscillations remain, by far, the most common type of dynamic behaviour. Complex periodic oscillations of the 'bursting' type are also rather frequent, but much less than simple oscillations. The coexistence between a steady state and a limit cycle comes third by virtue of the importance of the domain in which such behaviour occurs in the $v$–$k_s$ plane. Birhythmicity and chaos come next

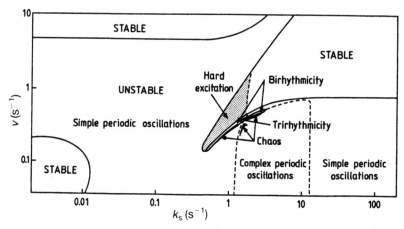

Fig. 4.30. Diagram showing the various modes of dynamic behaviour of the multiply regulated biochemical system in the parameter space $v$–$k_s$. The indications 'stable' and 'unstable' relate to the stability properties of the unique steady state admitted by eqns (4.1). The domain of coexistence of such a stable state with a stable limit cycle is represented by the dotted area. Two domains of birhythmicity are observed; their overlap gives rise to trirhythmicity. Three regions of chaos are represented in black, while the domains of simple or complex periodic oscillations occupy the rest of the space. The diagram is established for the parameter values of fig. 4.2 (Decroly & Goldbeter, 1982; Goldbeter, Decroly & Martiel, 1984).

as they occur much less frequently than simple or complex periodic oscillations and hard excitation (how other parameters of the model affect the extent of the chaotic domain has been addressed in section 4.4; see table 4.2 above). Trirhythmicity is the least common mode of dynamic behaviour. Most complex oscillatory phenomena occur in a range of $k_s$ values between 1 and 10 $s^{-1}$ (fig. 4.30), in which simple periodic oscillations are practically absent.

The appearance of complex modes of oscillatory behaviour such as chaos or bursting is due to the presence, within the same system, of two instability-generating mechanisms. It is when each of these mechanisms acquires a comparable importance that their interaction gives rise to complex dynamic phenomena. The fact that simple periodic behaviour nevertheless remains the most common suggests that for a large choice of parameter values, the uncoupling of the two mechanisms allows one of them to be active while the other remains 'silent'.

Despite the fact that they have been established for a specific model based on enzyme kinetics, these results are of more general significance in that they show how the simultaneous presence of two instability-generating mechanisms within the same chemical or biological system can lead from simple periodic behaviour to complex oscillations such as bursting and chaos. This conclusion is corroborated by the similarity of the one-dimensional map obtained here for bursting and chaos (figs. 4.25 and 4.26) with that obtained by Chay & Rinzel (1985) in a model for similar oscillatory phenomena in the β-cells of the pancreas.

The complex oscillations predicted by the model can be related to those sometimes observed, at low values of the substrate injection rate, in glycolysing yeast extracts. Such complex glycolytic oscillations (fig. 4.31) could represent chaos resulting from the interaction between oscillating phosphofructokinase and a second instability-generating reaction, catalysed by another glycolytic enzyme, in a small range of values of the substrate injection rate.

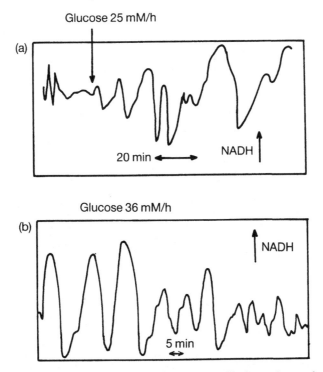

Fig. 4.31. Two examples of complex glycolytic oscillations observed at low values of the substrate injection rate in yeast extracts. The curves are due to (a) Pye (1969) and (b) Hess & Boiteux (1973).

As indicated by the analysis of the multiply regulated biochemical system, birhythmicity occurs in a domain of $v$ values close to those producing chaos. Of interest, in this respect, is the preliminary evidence obtained by Markus & Hess (1990) for birhythmicity in glycolysing yeast extracts. These authors showed the existence of a hysteresis loop between two types of oscillations upon increasing and then decreasing the value of the substrate injection rate. A direct test of the transition between two different rhythms at the same value of the substrate input remains to be performed.

As shown in the following chapters, the study of simple and complex oscillations in the signalling system used by *Dictyostelium discoideum* amoebae for intercellular communication confirms the general significance of the results established for the multiply regulated biochemical system. This model thus represents a three-variable prototype for analysing the transition from simple to complex oscillations in biochemical and cellular systems.

# Part III

## Oscillations of cAMP in *Dictyostelium* cells

# Models for the periodic synthesis and relay of cAMP signals in *Dictyostelium discoideum* amoebae

The amoeba *Dictyostelium discoideum* (Raper, 1935) is one of the most studied organisms in developmental biology (Bonner, 1967; Loomis, 1975, 1982). The main reason underlying the attractiveness of these amoebae is twofold. On one hand, they can switch from a unicellular to a multicellular stage during their life cycle (Raper, 1940a,b) by resorting to a chemical mechanism of intercellular communication (Bonner, 1947; Shaffer, 1956, 1962; Devreotes, 1982; Gerisch, 1982, 1987), which presents some similarities with hormonal communications in higher organisms (Newell, 1977). On the other hand, once the multicellular stage is reached, the amoebae differentiate into at least two distinct cell types, thus providing a simple model for the study of pattern formation (Nanjundiah & Saran, 1992) and cell differentiation in eukaryotes (Takeuchi, Noce & Tasaka, 1986; Gross, 1994). Recent advances in the manipulation of the *Dictyostelium* genome reinforce the importance of this organism in the study of the molecular bases of development.

It is for yet another reason that *Dictyostelium* amoebae draw our attention here. The mechanism of intercellular communication that governs the transition between the isolated and collective phases of their life cycle possesses a periodic nature (Gerisch, 1968, 1971; Shaffer, 1962). After reviewing the experimental facts that make this phenomenon a prototype for biochemical oscillations in cell biology, theoretical models are developed in this chapter to account for this biochemical rhythm. A particularly interesting aspect of the *Dictyostelium* example, which will be highlighted here, is that it illustrates well the endless feedback between modelling and experimental studies. As the latter progress, sometimes stimulated or guided by the motivation of testing theoretical predictions, alterations to the initial model become necessary. The new version of the model based on new information gathered from the experiments aims to account for hitherto unexplained observations, and leads to new theoretical predictions. The latter will face the test of further investigations and, from this continuing interaction between the two complementary approaches, a better understanding of the dynamic phenomenon will be gained in an asymptotic manner.

The possible occurrence in the *Dictyostelium* system of complex oscillatory phenomena such as birhythmicity, bursting and chaos is discussed in chapter 6. The appearance of periodic behaviour in the course of development provides a model for the ontogenesis of biological rhythms. This aspect is treated in chapter 7. Finally, an additional interest of the intercellular communication system of *Dictyostelium* amoebae is that it allows us to address the question of the physiological function of the periodic phenomenon. This question is dealt with in chapter 8, where the discussion is extended to the role of pulsatile hormone secretion in higher organisms.

### 5.1 Oscillations of cAMP and the life cycle of *Dictyostelium* amoebae

The life cycle of each amoeba (see Bonner, 1967; Loomis, 1975, 1982) begins with the germination of a spore (fig. 5.1). The cell that emerges from it grows and divides as long as it finds in the surrounding medium bacteria on which it can feed. This phase of growth stops when food becomes scarce. Then begins a strikingly different phase of development: following starvation, the amoebae collect to form an aggregate that may contain up to 100 000 cells; hence the name **social amoebae** given to this species of cellular slime mould. The resulting aggregate takes the form of a slug, and finally differentiates, after a phase of migration, into a fruiting body (fig. 5.2). This structure consists of a mass of spores surmounting a stalk that can measure up to a few millimetres. Spore dispersal leads back to the initial phase of the life cycle (Raper, 1940a,b; Konijn & Raper, 1961; Bonner, 1967; Shaffer, 1962; Darmon & Brachet, 1978; Devreotes, 1982, 1989; Gerisch, 1968, 1982, 1987; Klein *et al.*, 1988; Gross, 1994).

### *Waves of cAMP and the periodic aggregation of* D. discoideum

In most amoebae of the *Dictyostelium* family of slime moulds, and in particular in the most studied species, *D. discoideum* (Raper, 1935), aggregation proceeds in a periodic manner. This phenomenon is particularly apparent when we observe the aggregation of the amoebae on a

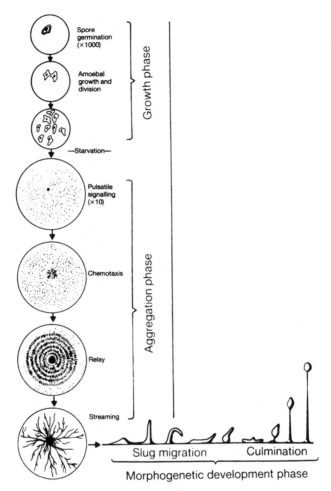

Fig. 5.1. Life cycle of *D. discoideum* amoebae showing the successive phases of growth, aggregation, and fruiting body formation (Newell, 1977).

solid support such as agar (Shaffer, 1956, 1962; Gerisch, 1971). Time lapse film of aggregation (Gerisch, Kuczka & Heunert, 1963) demonstrates the occurrence of waves of amoebae moving with a periodicity of the order of 5 to 10 min (Durston, 1974a).

Everything happens as if the amoebae were collecting around cells behaving as aggregation centres that periodically emit a chemical signal to attract the cells present in their environment (Shaffer, 1962). The signal was given the generic name of **acrasin** (Shaffer, 1956), and the name

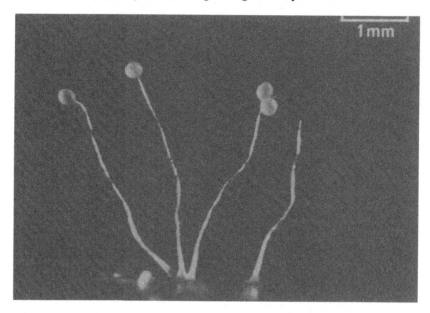

Fig. 5.2. Fruiting bodies formed by *D. discoideum* amoebae after the phases of aggregation and migration that follow starvation (Loomis, 1975).

**acrasiae** was given to the amoebae that utilize this mode of chemical communication to pass from the unicellular to the multicellular stage.

To explain the fact that a given centre controls the aggregation of a large number of cells on territories whose dimensions can reach up to 1 cm, Shaffer (1962) further suggested, on the basis of microscopic observations, that amoebae are capable of relaying the signals emitted by a centre located in their vicinity. Given that the amoebae respond chemotactically to the signals emitted by the centres (Bonner, 1947), such a relay, combined with the periodic release of the chemical signal, would explain the wavelike nature of aggregation, over territories whose dimensions are macroscopic compared with cellular size.

Typical aggregation territories are shown in fig. 5.3. The structure in concentric or spiral waves is clearly visible. As illustrated by fig. 5.4, clear bands correspond to chemotactically responding amoebae moving towards aggregation centres; dark bands correspond to immobile amoebae in a refractory phase.

The chemical analysis of the chemotactic signal utilized by *D. discoideum* amoebae in the course of aggregation showed (Konijn *et al.*, 1967) it to be cyclic AMP (cAMP). This molecule (fig. 5.5) is known to play an important role in higher organisms, where stimulation by

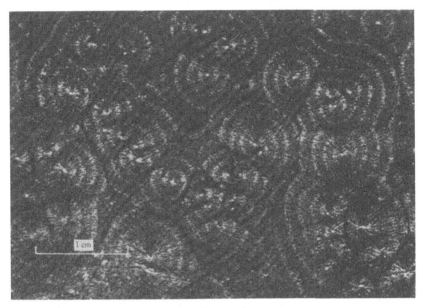

Fig. 5.3. Waves observed during aggregation of *D. discoideum* amoebae on agar. Light bands represent amoebae moving chemotactically in response to signals emitted by aggregation centres. Dark bands correspond to amoebae that are immobile due either to refractoriness or to the absence of chemotactic signal (Alcantara & Monk, 1974).

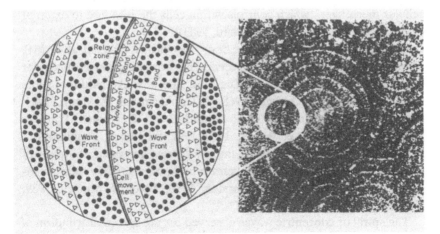

Fig. 5.4. Scheme showing the behaviour of *D. discoideum* amoebae within light and dark bands observed in pictures of aggregation on agar. The wavefront of the relayed signal and its movement are indicated, as well as cells moving towards the centre and those that are temporarily at rest. The scheme on the left represents an enlargement of part of one of the aggregation trerritories shown in fig. 5.3, where the spatial scale is indicated (Newell, 1977).

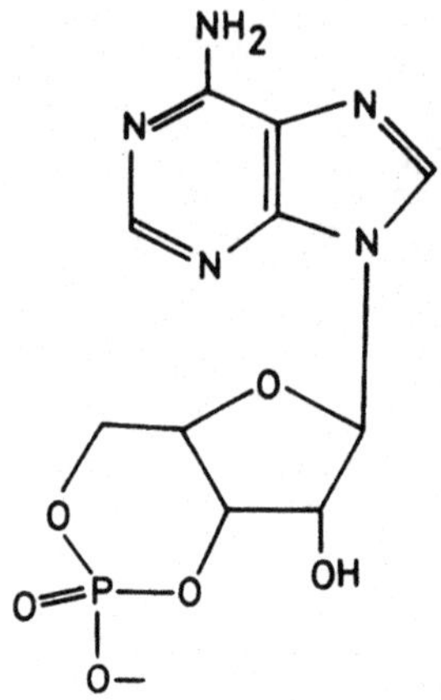

Fig. 5.5. Cyclic AMP, the molecule used as intercellular communication factor by *D. discoideum* amoebae.

hormones or neurotransmitters often elicits the synthesis of this 'intra-cellular messenger', which controls within cells the response to external stimulation (Greengard, 1978; Cohen, 1983).

By means of a fluorescent antibody specifically directed at cAMP, Tomchik & Devreotes (1981) succeeded in correlating waves of cellular movement with waves of cAMP (fig. 5.6). The wavelike nature of slime mould aggregation (Gerisch, 1968, 1971; Alcantara & Monk, 1974; Gross, Peacey & Trevan, 1976; Newell, 1977; Devreotes, Potel & Mackay, 1983; Siegert & Weijer, 1989, 1991; Steinbock, Hashimoto & Müller, 1991) therefore reflects the underlying presence of waves of cAMP produced by aggregation centres and propagated by cells consti-tuting an excitable medium (Durston, 1973; Tyson & Murray, 1989; Tyson *et al.*, 1989; see section 5.9).

The spiral or concentric waves observed for the spatial distribution of cAMP (fig. 5.6) present a striking analogy with similar wavelike phe-nomena found in oscillatory chemical systems, of which the Belousov–Zhabotinsky reaction (fig. 5.7) provides the best-known example (Winfree, 1972a).

The transition from the isolated state to the collective phase has

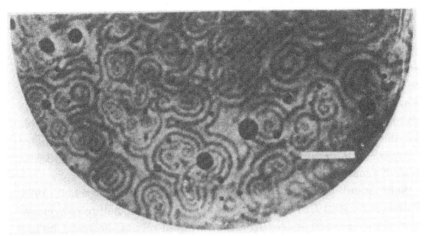

Fig. 5.6. Waves of cAMP observed by immunofluorography in the course of *D. discoideum* aggregation. These waves superimpose on those of chemotactically moving cells. Calibration bar, 1 cm (Tomchik & Devreotes, 1981).

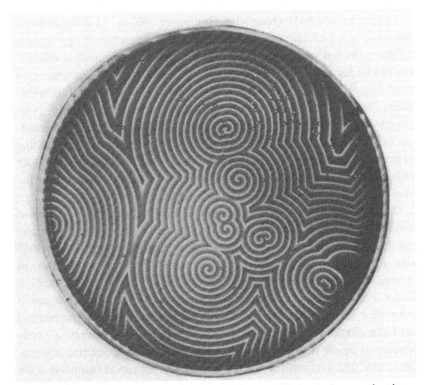

Fig. 5.7. Chemical waves observed in the Belousov–Zhabotinsky reaction in an unstirred medium (Winfree, 1972a).

become a classical problem in theoretical biology. In the first analysis devoted to this question, Keller & Segel (1970) thus proposed that aggregation represents a pattern of spatial organization resulting from an instability of the homogeneous spatial distribution of amoebae, beyond a critical value of acrasin secretion. This analysis applies particularly well to species that aggregate in a nonperiodic manner: the absence of periodic secretion and relay of the signal is then responsible for the relative smallness of aggregation territories, hence the name *Dictyostelium minutum* given to the best-known species that aggregates in such a manner (Gerisch, 1968).

Subsequent analyses (Cohen & Robertson, 1971; Mackay, 1978; Parnas & Segel, 1978) took into account the periodic nature of aggregation in species such as *D. discoideum*, without relying, however, on any precise mechanism explaining the origin of the periodic phenomenon. More recent theoretical studies (Tyson, 1989; Tyson & Murray, 1989; Tyson *et al.*, 1989; Höfer *et al.*, 1995; see also section 5.9) obviate this shortcoming by incorporating the model discussed below for the oscillatory synthesis of cAMP (Martiel & Goldbeter, 1987a). Another theoretical analysis of cAMP wave propagation in the course of aggregation has been carried out by Monk & Othmer (1990), who used an alternative model for the periodic generation of cAMP signals. The azimuthal instability responsible for the formation of streams at the end of aggregation has also been investigated theoretically (Nanjundiah, 1973; Levine & Reynolds, 1991). Most computer simulations treat the amoebae as immobile, as a first approximation, and focus on the propagating waves of cAMP; some theoretical studies, however, aim at incoporating both diffusion of extracellular cAMP and chemotactic movement of the cells (Mackay, 1978; Levine, 1994; Höfer *et al.*, 1995).

The periodic nature of aggregation is observed in other species of slime moulds whose acrasin can again be cAMP, or some other molecule (Schaap & Wang, 1984). In some species, periodic signals of cAMP play a role at a later stage of development, in the coordination of chemotactic movements during fruiting body formation (Schaap & Wang, 1984). As most experimental studies pertain to *D. discoideum*, it is for that species that the mechanism of intercellular communication has been elucidated in greatest detail. The aim of the theoretical study developed below is to propose models, based on experimental observations, that will account for the oscillatory phenomenon responsible for the periodic nature of aggregation in *D. discoideum*.

### *Oscillations of cAMP*

An important step in the the experimental characterization of the periodic phenomenon was made when Gerisch & Hess (1974) turned to studying the behaviour of *D. discoideum* amoebae in continuously stirred cell suspensions. By recording light scattering within such suspensions, these authors showed that spontaneous oscillations occur several hours after the beginning of starvation. These oscillations in light scattering have a period of the order of 5 to 10 min (fig. 5.8). The periodicity of the behaviour in suspensions is thus identical with that of the waves observed during aggregation on a solid support. The shape of the oscillations in suspensions changes in the course of time (Gerisch & Hess, 1974; Wurster, 1988); after an initial series of spikes, the oscillations become more sinusoidal (fig. 5.8). The periodic variation in light scattering reflects changes in cellular morphology that are brought about by the response of the cells to the chemotactic signal secreted within the stirred suspension.

The initial study of Gerisch & Hess (1974) did not propose any molecular basis for the oscillations in light scattering in suspensions of *D. discoideum* amoebae. However, in that same study, the authors showed that extracellular cAMP is capable of inducing an immediate phase shift of the oscillations (fig. 5.9). This observation suggests an important role for the chemotactic signal in the control of the oscillatory phenomenon.

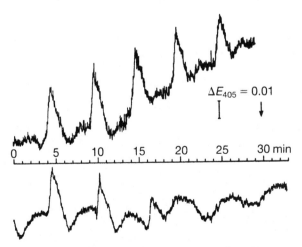

$\Delta E_{405} = 0.01$

Fig. 5.8. Spontaneous oscillations of light scattering in a suspension of *D. discoideum* amoebae (Gerisch *et al.*, 1979).

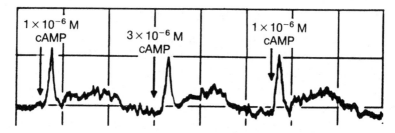

Fig. 5.9. Phase shift of light scattering oscillations by cAMP pulses in suspensions of *D. discoideum* (Gerisch & Hess, 1974).

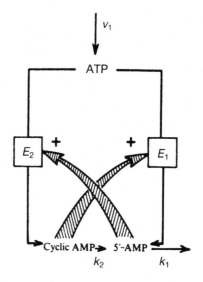

Fig. 5.10. Model for the synthesis of cAMP in *D. discoideum* (Goldbeter (1975), based on the experimental observations of Rossomando & Sussman (1973). This regulatory scheme is probably incorrect, given that the initial observations apparently suffered from an experimental artefact in the determination of cAMP levels.

Building on these observations, Goldbeter (1975) proposed a model (fig. 5.10) based on the regulatory properties of the system synthesizing cAMP in *D. discoideum*, as characterized by Rossomando & Sussman (1973). This model predicts that the synthesis of cAMP possesses an oscillatory character (fig. 5.11) resulting from the instability of the steady state brought about by the cross-catalytic regulation observed in the experiments.

The existence of cAMP oscillations concomitant with the periodic variation in light scattering was confirmed shortly thereafter by the

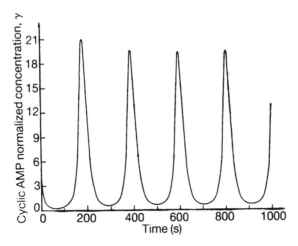

Fig. 5.11. Oscillations of cAMP predicted by the three-variable model schematized in fig. 5.10 (Goldbeter, 1975).

experiments of Gerisch & Wick (1975). As shown in fig. 5.12, intracellular cAMP oscillates in phase with light scattering, with a period close to 9 min; its peak precedes by some 10 s that of extracellular cAMP. Just as with the oscillations in light scattering, the oscillations in intra- and extracellular cAMP can be phase shifted by a pulse of extracellular cAMP (Gerisch *et al.*, 1979). As indicated by fig. 5.13, however, the effect of the perturbation depends on the phase at which the addition of cAMP occurs. A delay of the next peak is obtained when the stimulation takes place just after a peak, when cAMP is still close to maximum. Thereafter the stimulation gives rise to the synthesis of a precocious peak of cAMP. Combining these results on phase shift as a function of phase of perturbation yields a **phase response curve** (fig. 5.13, upper part) characteristic of a number of biological oscillators (see also the theoretical phase response curves obtained for glycolytic and calcium oscillations in figs. 2.19 (p. 60) and 9.16 (p. 376), respectively).

### Relay of cAMP signals

The advantage of studies in cell suspensions is that they allow for the synchronization of the amoebae through the cAMP secreted into the extracellular medium. Since it is easy to record continuously light scattering within such suspensions, this method is commonly used for the study of the oscillations, rather than the necessarily discontinuous,

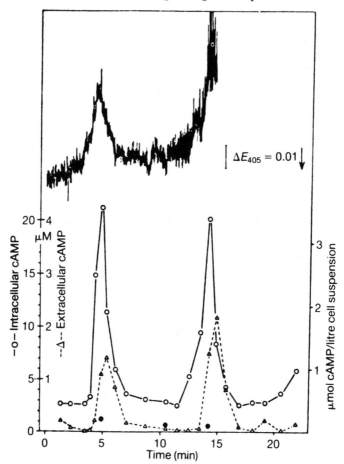

Fig. 5.12. Oscillations of intracellular and extracellular cAMP in suspensions of *D. discoideum* amoebae. These oscillations are in phase with those of light scattering (Gerisch & Wick, 1975).

chemical analysis of the cAMP content of the cells and of the external medium. The latter method was nevertheless used successfully by Roos *et al.* (1975) and by Shaffer (1975) to reveal a second type of dynamic behaviour, namely the relay of cAMP signals whose amplitude exceeds a threshold. This behaviour is illustrated in fig. 5.14; in the absence of any oscillation, the addition of a sufficient amount of extracellular cAMP (e.g. $5 \times 10^{-8}$ M) elicits the synthesis of a peak of intracellular cAMP that rises from 1 to 20 mM, before returning to its stable, steady level. The cAMP signal also elicits the synthesis of a pulse of cyclic GMP (Janssens & Van Haastert, 1987); the latter response, which is not

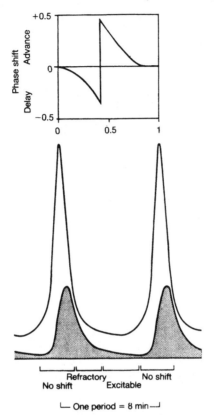

Fig. 5.13. Phase response curve showing the phase shift induced by pulses of cAMP applied in the course of oscillations. One period of the oscillations divides into a phase of weak or no response, a refractory phase corresponding to a delay, and an 'excitable' phase corresponding to a phase advance of the oscillations. The bottom part of the figure shows a schematic representation of the time course of intracellular and extracellular (shaded area) cAMP (Gerisch *et al.*, 1979).

considered below, is involved in chemotaxis through a mechanism involving myosin regulation (Liu & Newell, 1994).

The above observations account for the periodic secretion and relay of cAMP signals in the course of aggregation of amoebae on agar. The question which now arises is how to determine the molecular mechanism responsible for these two modes of dynamic behaviour. The theoretical study of experimentally based models clarifies the origin of relay and oscillations and permits us to explain why certain cells behave as centres while other cells act just as relays in the course of aggregation.

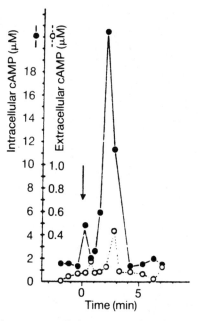

Fig. 5.14. Relay of an extracellular cAMP signal in a suspension of *D. dis-coideum* cells. In response to a $5 \times 10^{-8}$ M cAMP signal, cells amplify this per-turbation by synthesizing a peak of intracellular cAMP, followed by its transport into the extracellular medium (Roos *et al.*, 1975).

## 5.2 Allosteric model for cAMP oscillations

The first model proposed for the mechanism of intercellular communi-cation in *D. discoideum* (fig. 5.10) successfully predicted the existence of sustained oscillations of cAMP (Goldbeter, 1975). The experimental data on which that model was based, however, turned out to be partly inexact due to an experimental artefact. Although the theoretical pre-diction of cAMP oscillations was soon corroborated by the observations of Gerisch & Wick (1975), the source of the oscillations lay in another regulatory mechanism that remained to be determined.

An important feature of the signalling system, suggested by the observations of Roos *et al.* (1975) and Shaffer (1975), is that the stimu-lation by extracellular cAMP enhances the synthesis of intracellular cAMP. This activation of adenylate cyclase, the enzyme catalysing the production of cAMP from ATP, follows the binding of extracellular cAMP to a specific receptor located on the outer face of the cell mem-brane (Roos & Gerisch, 1976). Synthesis of cAMP in *D. discoideum* therefore is a self-amplifying process (fig. 5.15): the more cAMP accu-

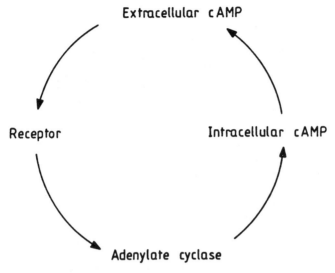

**Extracellular cAMP**

**Receptor**

**Intracellular cAMP**

**Adenylate cyclase**

Fig. 5.15. Self-amplification creates a positive feedback loop in the synthesis of cAMP in *D. discoideum*.

mulates in the external medium, the more it binds to the receptor and activates the intracellular production of cAMP; transport of the latter into the extracellular medium results in a positive feedback loop that renders the whole process autocatalytic.

The analysis of the model for glycolytic oscillations based on the regulatory properties of phosphofructokinase (chapter 2) highlighted the importance of autocatalysis as a source of instability leading to periodic behaviour. The synthesis of cAMP in *D. discoideum* provides us with a second example of autocatalytic regulation in biochemistry. The hypothesis on which the models proposed below are based is that the autocatalytic regulation of adenylate cyclase plays a primary role in the origin of cAMP oscillations.

The model proposed by Goldbeter & Segel (1977) for the dynamic behaviour of the cAMP signalling system of *D. discoideum* is based on such a positive feedback. This model (fig. 5.16) takes into account the binding of extracellular cAMP to a membrane receptor (Henderson, 1975; Gerisch & Malchow, 1976) and the subsequent activation of adenylate cyclase (Roos & Gerisch, 1976), which is located on the inner face of the cell membrane (Farnham, 1975); moreover, cAMP is transported into the extracellular medium, where it is hydrolysed by the membrane-bound and extracellular forms of phosphodiesterase. The

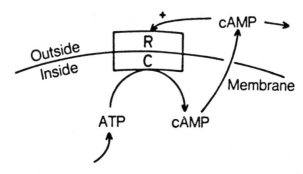

Fig. 5.16. Allosteric model for cAMP synthesis in *D. discoideum* amoebae. Extracellular cAMP binds to the receptor (R) and activates adenylate cyclase (C), which transforms substrate ATP into intracellular cAMP. Arrows denote the synthesis of ATP, the transport of cAMP across the membrane into the extracellular medium, and cAMP hydrolysis by phosphodiesterase (Goldbeter & Segel, 1977).

role of this enzyme secreted by the amoebae is to erase any trace of the chemotactic signal so as to prevent flooding of the cells and to permit them to respond maximally to the next signal emitted by the aggregation centre (Darmon, Barra & Brachet, 1978; Faure *et al.*, 1989).

At the time this model was proposed, no precise information was available on the mechanism by which the cAMP–receptor complex activates adenylate cyclase. Therefore a simple hypothesis was retained, namely that the receptor and adenylate cyclase form the regulatory (R) and catalytic (C) subunits of an allosteric complex, which bind extracellular cAMP and intracellular ATP, respectively; moreover, it is assumed that this complex possesses a tetrameric structure of the type $R_2C_2$. The input of the substrate ATP – but not the level of this metabolite – was considered to be constant in the course of time.

The three variables in that model are intracellular ATP ($\alpha$), intracellular cAMP ($\beta$), and extracellular cAMP ($\gamma$). We assume that the complex obeys the allosteric model of Monod *et al.* (1965), i.e. each regulatory or catalytic subunit can exist in two states, one of which has a larger affinity towards its ligand or, in the case of the catalytic subunit, is more active than the other; the transition between these two states is concerted. When we assume, moreover, that the system remains spatially homogeneous (which corresponds to the conditions in continuously stirred cell suspensions), the time evolution of the system is governed by the three ordinary differential equations:

$$\frac{d\alpha}{dt} = v - \sigma \phi$$

$$\frac{d\beta}{dt} = q\sigma \phi - k_t \beta$$

$$\frac{d\gamma}{dt} = (k_t \beta/h) - k \gamma \tag{5.1}$$

where the rate function $\phi$ for adenylate cyclase is given by the relation:

$$\phi = \frac{\alpha (1 + \alpha) (1 + \gamma)^2}{L + (1 + \alpha)^2 (1 + \gamma)^2} \tag{5.2}$$

In these equations, $v$ and $\sigma$ denote, respectively, the constant rate of synthesis of ATP and the maximum velocity of adenylate cyclase, both divided by the Michaelis constant $K_m$ of the substrate for the active form of the enzyme; $\alpha$ represents the concentration of ATP divided by $K_m$; $\beta$ and $\gamma$ are the concentrations of intracellular and extracellular cAMP, divided by the dissociation constant $K_P$ of extracellular cAMP for the receptor ($\alpha$, $\beta$ and $\gamma$ are thus dimensionless); $q = K_m/K_P$; $L$ is the allosteric constant of the receptor–enzyme complex; $k_t$ and $k$ denote, respectively, the apparent first-order kinetic constants for the transport of cAMP into the extracellular medium (Dinauer, MacKay &

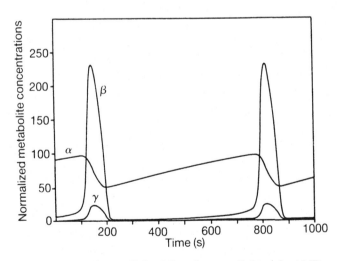

Fig. 5.17. Oscillations of intracellular ($\beta$) and extracellular ($\gamma$) cAMP, accompanied by a significant periodic variation of ATP ($\alpha$) in the allosteric model for the synthesis of cAMP in *D. discoideum*. The curves are obtained by numerical integration of eqns (5.1) for the following parameter values: $v = 0.1$ s$^{-1}$, $\sigma = 1.2$ s$^{-1}$, $k = k_t = 0.4$ s$^{-1}$, $L = 10^6$, $q = 100$, $h = 10$ (Goldbeter & Segel, 1977).

Devreotes, 1980a), and for the hydrolysis of extracellular cAMP by phosphodiesterase; finally, $h$ is a dilution factor equal to the ratio of extracellular to total intracellular volume.

The stability analysis of the unique steady state admitted by eqns (5.1) yields parameter values for which this state becomes unstable (see also Odell, 1980). The numerical integration of the kinetic equations then shows that the system acquires periodic behaviour in the course of time (fig. 5.17): large-amplitude oscillations are observed for intracellular cAMP ($\beta$); their spike shape is reminiscent of that of the experimentally observed oscillations (fig. 5.12). Oscillations of extracellular cAMP ($\gamma$), of reduced amplitude, closely follow the periodic evolution of variable $\beta$. Moreover, these oscillations are accompanied by a significant variation of the ATP concentration ($\alpha$).

## 5.3 Link between relay and oscillations of cAMP

### *Excitability: pulsatile amplification of suprathreshold cAMP signals*

The numerical simulations of eqns (5.1) also permit the demonstration of another type of dynamic behaviour, in a region of the parameter space where the steady state of the system is stable. In certain conditions close to those that give rise to sustained oscillations, the system in this type of stable steady state is indeed capable of amplifying in a pulsatory manner small perturbations corresponding to an instantaneous rise in extracellular cAMP. Such behaviour is illustrated in fig. 5.18: a slight increase in $\gamma$ above the steady-state level, corresponding to the addition of a $10^{-7}$ M pulse of cAMP, gives rise to the synthesis of a peak of intracellular cAMP; the latter rises from 1 to 20 mM before returning to its pre-stimulus level. As for the oscillations, the peak of intracellular cAMP is followed by a peak of extracellular cAMP and is accompanied by a sizeable variation in the ATP level.

Such pulsatory amplification of the extracellular cAMP pulse accounts for the phenomenon of relay of cAMP signals observed during aggregation of the amoebae on agar (Robertson & Drage, 1975) as well as in experiments in cell suspensions (Roos *et al.*, 1975; see fig. 5.14). Table 5.1 shows that the qualitative and quantitative predictions of the model regarding relay behaviour yield satisfactory agreement with experimental observations.

When the response of the system (fig. 5.18), measured by the ratio $\beta_M/\beta_0$ of the maximum intracellular cAMP level at the height of the

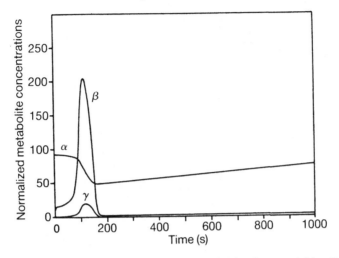

Fig. 5.18. Relay of a signal of extracellular cAMP in the three-variable allosteric model. The curves are obtained by integration of eqns (5.1) for the parameter values of fig. 5.17, with $v = 0.04$ s$^{-1}$. Initial conditions are $\alpha = 92.366$, $\beta = 10$, and $\gamma = 2$. These conditions correspond to taking the steady-state values for $\alpha$ and $\beta$, and to doubling the steady-state concentration of extracellular cAMP ($\gamma$) (Goldbeter & Segel, 1977).

pulse divided by the steady-state value of $\beta$, is plotted as a function of the magnitude of stimulation measured by the ratio $\gamma_i / \gamma_0$ of the initial value of extracellular cAMP divided by the value of $\gamma$ at steady state, we obtain the curve represented in fig. 5.19. The peculiarity of this dose–response curve is to possess an extremely abrupt threshold. In the case considered, the transition between a weak and a quasi-maximum response occurs when $\gamma_i$ exceeds a value between 1.846 and 1.848. The time taken to reach the peak of the response decreases as the amplitude of the stimulation augments. This behaviour reflects the **excitability** of the system synthesizing cAMP in *D. discoideum*. Excitability indeed consists, as in nerve cells (Fitzhugh, 1961), in the pulsatory amplification of a suprathreshold perturbation followed by the return of the system to the stable steady state that prevailed before the onset of stimulation.

Numerical simulations thus indicate that eqns (5.1) can produce either sustained oscillations or excitable behaviour, in slightly different conditions, i.e. for closely related values of the parameters. Phase plane analysis allows us to determine the conditions in which either phenomenon occurs; this analysis further shows that the two behavioural modes are necessarily linked (Goldbeter, Erneux & Segel, 1978). The same link between excitability and oscillations (see also Hahn *et al.*, 1974) is

Table 5.1. *Comparison of the predictions of the allosteric model with experimental observations on the relay of cAMP signals*

| Relay | Relative amplitude[a] | Time for maximum relay (s) | Half width (s) | Delay between extra- and intra-cellular cAMP (s) | $\text{Min[ATP]}/\text{max[cAMP]}_i$ | $\text{Max[cAMP]}_i/\text{max[cAMP]}_e$ | Steady-state ratio $\text{[ATP]}/\text{[cAMP]}_i$ |
|---|---|---|---|---|---|---|---|
| Experiment | 10–25 | 100–120 | 60 | 30–40 | 50 | 20–50 | $10^3$ |
| Model | 20 | 113 | 53 | 3 | 25 | 10 | $9.24 \times 10^2$ |

[a] Relative amplitude of relay is given as the maximum of the intracellular cAMP peak divided by the steady-state level of intracellular cAMP.

The comparison bears on (col. 1) the relative amplitude of the response, defined as the ratio of the maximum and steady-state values of intracellular cAMP, (2) the time taken to reach the maximum of the response, (3) the half-width of the cAMP peak (in s), (4) the delay between intra- and extracellular cAMP, and (5–7) the relative concentrations of ATP and intra- and extracellular cAMP. Goldbeter & Segel, 1977.

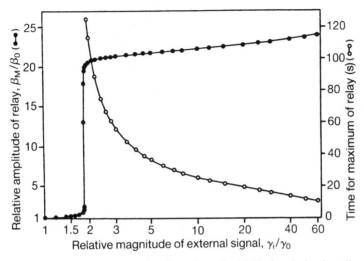

Fig. 5.19. Dose–response curve for the relay of cAMP signals in the allosteric model. The response is measured by the maximum $\beta_M$ of the peak of intracellular cAMP (fig. 5.18), divided by the steady-state value, while the stimulation is measured by the initial value of extracellular cAMP, $\gamma_i$, divided by the steady-state value, $\gamma_0$. Also indicated is the time required to reach the maximum of the response (Goldbeter & Segel, 1977).

observed for nerve cells (Fitzhugh, 1961) as well as for oscillatory chemical reactions (Winfree, 1972a; De Kepper, 1976).

### *Phase plane analysis of a two-variable approximation*

To apply phase plane analysis to the three-variable system (5.1), it is first desirable to seek a reduction of the number of variables down to two. System (5.1) can be written as:

$$\frac{d\alpha}{dt} = v - \sigma\,\phi$$

$$\left(\frac{1}{q}\right)\frac{d\beta}{dt} = \sigma\,\phi - (k_t\,\beta/q)$$

$$\frac{d\gamma}{dt} = (k_t\,\beta/h) - k\,\gamma \tag{5.3}$$

where the rate function $\phi$ remains defined by relation (5.2).

The value of parameter $q$ is of the order $10^3$, in view of the smallness of the dissociation constant of the cAMP receptor (approx. $10^{-7}$ M) with respect to the Michaelis constant of adenylate cyclase (approx. $10^{-4}$ M) (see table 5.2). In the limit $(1/q) \to 0$, assuming that the ratio $(k_t /q)$ remains finite, we can adopt a quasi-steady-state hypothesis for the 'rapid' (i.e. rapidly changing) variable $\beta$. A satisfactory approximation of the initial system (5.1) is then provided by the two-variable system (5.4) in which the value of $\beta$ is given by the algebraic relation (5.4a) while the constant $\lambda$ is defined as the ratio $(q\sigma/h)$:

$$\beta = q\sigma/k_t \, \phi \tag{5.4a}$$

$$\frac{d\alpha}{dt} = v - \sigma\phi \tag{5.4b}$$

$$\frac{d\gamma}{dt} = k \, (\lambda \, \phi - \gamma) \tag{5.4c}$$

This two-variable system (Goldbeter *et al.*, 1978) presents the additional advantage of being formally identical with the system of eqns (2.7) studied in chapter 2 for glycolytic oscillations. This similarity stems from the basic structure common to the two models: a substrate, injected at a constant rate, is transformed in a reaction catalysed by an allosteric enzyme activated by the reaction product. In the cAMP-synthesizing system in *D. discoideum*, activation is indirect as extracellular cAMP enhances the synthesis of intracellular cAMP, which is then transported into the extracellular medium. However, the hypothesis of a quasi-steady state for intracellular cAMP is tantamount to considering that the variation of $\beta$ is so fast that the enzyme is, *de facto*, activated directly by its apparent product, extracellular cAMP.

All results established in chapter 2 for the analysis of eqns (2.7) in the phase plane therefore also apply to the two-variable system (5.4b,c). In particular, the two nullclines of the system remain defined by eqns (2.21). When parameter $\lambda$ exceeds a critical value, the product nullcline $\gamma = \lambda \, \phi$ takes the form of a sigmoid possessing a region in which the slope $(d\alpha/d\gamma)$ is negative in the phase plane $(\alpha, \gamma)$. When the substrate nullcline $v = \sigma \, \phi$ intersects the product nullcline in that region, in such a manner that condition (2.26) is satisfied, the steady state, which lies at the intersection of the two nullclines, is unstable and the system evolves towards a limit cycle corresponding to sustained oscillations (fig. 2.13). Such behaviour translates here into oscillations of ATP and extracellular cAMP, to which are associated oscillations in intracellular cAMP because of relation (5.4a), which links $\beta$ to the 'slower' variables $\alpha$ and $\gamma$.

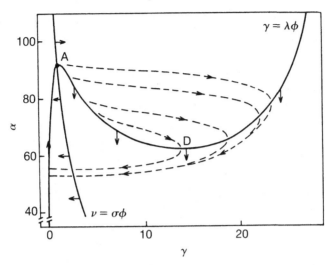

Fig. 5.20. Excitability in the phase space of the two-variable system (5.4b,c) describing cAMP synthesis in *D. discoideum*. The substrate and product null-clines are represented; their intersection A defines the steady state. The dashed curves represent the trajectories followed by the system, starting from different initial conditions, before returning to the stable steady state (Goldbeter *et al.*, 1978).

Different behaviour is observed when the substrate nullcline inter-sects the product nullcline just to the left of the domain of negative slope (fig. 5.20). The steady state of the system (denoted by A in fig. 5.20) is then stable. Let us see how the system now responds to the addition of a limited amount of extracellular cAMP. In the phase plane, such a stimulation corresponds to an instantaneous displacement of variable $\gamma$ towards the right, on the horizontal through the steady state. Integration of eqns (5.4) shows that the system reacts to the stimulation by undergoing a large excursion in the phase plane, in the course of which it reaches the right limb of the sigmoid nullcline before slowly returning to the stable steady state. This excursion corresponds to the synthesis of a peak in both intra- and extracellular cAMP.

If phase plane analysis thus accounts for 'relay' behaviour, can it also explain the existence of an abrupt threshold for the amplification of perturbations? When the steady state is located at A as in fig. 5.20, a tiny displacement of $\gamma$ towards the right will suffice to produce the excitable response. However, when the substrate nullcline is located more to the left (for example, for smaller values of parameter $v$), the steady state moves further away from the region of negative slope. A displacement of $\gamma$ towards the right will not produce any relay response as long as the initial condition resulting from the perturbation is located on the left of the median part of the curve $(d\gamma/dt) = 0$. This curve, indeed, separates the region where the derivative of $\gamma$ is positive (region above the nullcline) from the region where this derivative is negative (below the nullcline). The point where the horizontal through the steady state intersects with the median branch of the product null-cline defines the threshold that must be exceeded for the perturbation to be amplified by the excitable system.

This analysis predicts that the excitability threshold augments when the steady state moves further away from the region of negative slope. This prediction is verified by numerical simulations of the model. How the position of the steady state and the excitability threshold vary when the substrate injection rate $v$ decreases in the full, three-variable system (5.1) is indicated in fig. 5.21. The dashed curve represents a typical tra-jectory for relay, while the inset shows the time evolution of intracellu-lar cAMP during such a response. A comparison with fig. 5.20 indicates that the reduction to two variables yields a satisfactory picture of the dynamics of the three-variable system; moreover, this reduction allows us to explain the existence of the excitability threshold and its increase when the variation of a control parameter displaces the system away from the oscillatory domain.

The analysis of eqns (5.4b,c) in the phase plane leads to the impor-tant conclusion that oscillatory behaviour and excitability are necessari-ly linked in the cAMP signalling system of *D. discoideum*. Both modes of dynamic behaviour result from the existence of a sigmoid nullcline possessing a region of negative slope associated with instability of the steady state. Depending on whether the steady state lies in the domain of instability on the nullcline or just to its left, the system will behave as a spontaneous oscillator or will be capable only of amplifying suprathreshold perturbations. Therefore, there is no need to invoke the existence of two distinct biochemical mechanisms, one for oscillations and the other for relay. Similar consequences of the form of nullclines in the phase plane were also stressed by Cohen (1977) in his qualitative discussion of cAMP signalling in *Dictyostelium*.

An alternative view has been proposed (Geller & Brenner, 1978) on the basis of experimental observations. Thus, the addition of dinitro-phenol to *D. discoideum* amoebae oscillating in suspensions suppresses the oscillations but preserves the capability of cells to amplify extracel-lular cAMP signals (Geller & Brenner, 1978). Far from indicating nec-

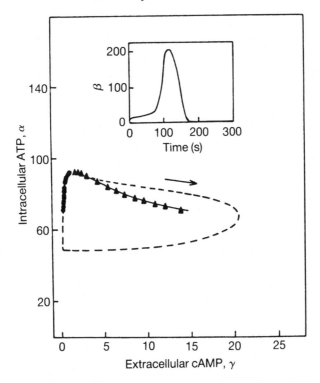

Fig. 5.21. Excitability in the three-variable system (5.1). The trajectory followed by this system in the conditions of fig. 5.18 is projected (dashed line) onto the plane ($\alpha$, $\gamma$); the corresponding evolution of $\beta$ is represented in the inset. Shown as a function of the position of the steady state (filled circles) is the initial value of $\gamma$ (filled triangles) that must be exceeded to obtain the pulsatory amplification of the perturbation. The locus of these threshold values represents the 'threshold separatrix' (Fitzhugh, 1961) whose position is close to the median branch of the product nullcline in the two-variable system (fig. 5.20) (Goldbeter *et al.*, 1978).

essarily the silencing of an oscillatory mechanism that would permit the expression of a second mechanism responsible for relay, these observations can be interpreted, in terms of the present analysis, as indicating the passage from one type of dynamic behaviour (oscillations) to another (relay), following the variation of one or more control parameters brought about by dinitrophenol. The same biochemical mechanism can thus manifest itself by one or other mode of dynamic behaviour, for closely related values of the parameters.

The phase plane analysis developed here for the cAMP-synthesizing system of *D. discoideum* is analogous to that presented by Fitzhugh

(1961) in his classic study of the excitable and oscillatory properties of the nerve membrane. In that study, Fitzhugh demonstrated the link between these two properties and brought to light, by means of phase plane analysis based on the sigmoid shape of one of the two nullclines, the existence of an abrupt threshold for excitability; his analysis further explained the origin of the relative and absolute refractory periods that characterize excitable behaviour. From this viewpoint, beyond the differences in underlying mechanisms, *D. discoideum* amoebae can be regarded as being analogous to neurons: the two types of cell indeed possess similar dynamic properties, namely the capability of oscillating in an autonomous manner or of amplifying suprathreshold perturbations as a result of their excitability.

### 5.4 Necessity of amending the allosteric model for cAMP signalling

To understand the reasons why it is necessary to modify the hypotheses on which the allosteric model for cAMP signalling is based, it is useful to define more precisely the mechanism that underlies oscillations in this model.

Any rhythm can be comprehended as a sequence of two phases that merge and follow each other in an autonomous manner. The first phase, of **amplification**, corresponds here to the onset of a peak in cAMP synthesis. The second phase, of **limitation**, corresponds to the decrease in cAMP synthesis, once the maximum is passed. Autonomous oscillations occur because of the regenerative process that leads to the renewal of the self-amplification phase after cAMP has reached its minimum level. To understand the mechanism of oscillations is tantamount to identifying this regenerative process as well as those responsible for self-amplification and for its limitation.

#### *Substrate availability as a factor limiting self-amplification*

In the allosteric model of fig. 5.16, the amplification of cAMP synthesis originates from the positive feedback loop (fig. 5.15) resulting from the activation of adenylate cyclase, following binding of extracellular cAMP to the receptor. This self-amplification would lead to a 'runaway' phenomenon and to a biochemical 'explosion' were it not for the existence of processes limiting such an autocatalysis. The limiting effect results here from the drop in substrate level during synthesis of a peak of cAMP. As indicated by fig. 5.17, the ATP level ($\alpha$) decreases significantly when $\beta$ increases; as the substrate is supplied at a finite, constant rate, and because of its enhanced consumption, the level of substrate cannot indefinitely support elevated synthesis of cAMP.

The discussion of the two-variable system (5.4b,c) helps to clarify the sequence of events. When the term $k\lambda\phi$, which denotes the rate of cAMP synthesis in eqn (5.4c), is exceeded by the term $k\gamma$, which denotes cAMP hydrolysis by phosphodiesterase, the level of extracellular cAMP begins to decrease. There results a further decrease in the rate of synthesis, because of the diminished binding of cAMP to the receptor and of the subsequent reduction in adenylate cyclase activation; this further accelerates the drop in cAMP. When cAMP reaches its minimum, adenylate cyclase functions at a basal rate (in the absence of activation by the receptor free from cAMP); the substrate can re-accumulate, since its rate of input then exceeds its rate of consumption in the reaction catalysed by adenylate cyclase. When extracellular cAMP has itself accumulated so as to reach a level where its binding to the receptor begins to significantly activate adenylate cyclase, self-amplification resumes and a new cycle of oscillations begins.

The effect limiting self-amplification in this model is therefore linked to the sizeable decrease in ATP that accompanies the synthesis of cAMP. The model predicts that the level of ATP must oscillate with a significant amplitude in the course of cAMP oscillations, even though the ratio between the maximum and minimum values of $\alpha$ remains smaller than 2, i.e. well below the corresponding ratio predicted for intracellular cAMP, $\beta$ (see fig. 5.17). Experimental observations nevertheless indicate (Roos, Scheidegger & Gerisch, 1977) that the level of ATP remains practically unchanged in the course of cAMP oscillations, around the value of 1.2 mM, even if a slight decrease is perceptible during the synthesis of a cAMP peak (fig. 5.22).

This experimental observation does not exclude, however, the possibility of a limiting role for ATP if the latter were compartmentalized in at least two pools, one of which, located near the membrane, would serve as source of substrate for adenylate cyclase. Experimentally, any significant variation of ATP in such a pool would be masked by the constancy of a larger compartment that would not be involved in the synthesis of cAMP.

### Role of receptor desensitization in the decrease in cAMP synthesis

Results of another kind (Theibert & Devreotes, 1983) are proof that the limitation of self-amplification does not result from the consumption of

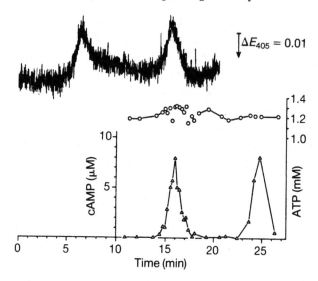

Fig. 5.22. Variation of ATP in the course of cAMP oscillations. These experimental observations in suspensions of *D. discoideum* indicate that ATP (circles) varies only slightly in the course of large-amplitude oscillations of cAMP (triangles), and remains buffered in the domain 1.15–1.25 mM (Roos *et al.*, 1977).

ATP in the course of cAMP synthesis. An important observation in this respect (Brenner & Thoms, 1984) is that caffeine inhibits cAMP oscillations in *D. discoideum* (fig. 5.23). Theibert & Devreotes (1983) took advantage of this observation in a series of experiments conducted under conditions where the cAMP level is maintained constant in the extracellular medium. This type of experiment permits the quantitative determination of the various aspects of the response of the cAMP signalling system to a variation of extracellular cAMP of known profile and magnitude (Devreotes & Steck, 1979; Dinauer *et al.*, 1980a; Dinauer, Steck & Devreotes, 1980b,c). As indicated further below, such experimental conditions are artificial in that they suppress the self-amplification that is one of the most conspicuous properties of the cAMP signalling system. The study of the system in these controlled conditions nevertheless yields extremely useful insights into the signalling mechanism.

The most salient results of the experiments of Theibert & Devreotes (1983) are summarized in fig. 5.24. The dashed curve denotes the imposed variation in extracellular cAMP, while the solid curves relate to the response of the system, in the form of cAMP synthesis. When the

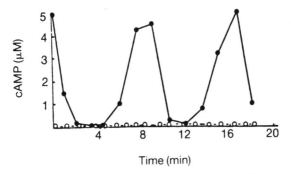

Time (min)

Fig. 5.23. Inhibition of cAMP oscillations by caffeine. The curves show the evolution of the extracellular cAMP concentration in the absence (filled circles) and presence (empty circles) of 2 mM caffeine in a *D. discoideum* suspension (Kimmel, 1987).

extracellular cAMP concentration undergoes a step increase such that the cAMP level instantaneously reaches a higher, constant value, cells respond to such a stimulation by synthesizing cAMP; however, although the stimulus is maintained at a higher level, the response declines and after a few minutes cAMP synthesis returns to the low level it had prior to stimulation (fig. 5.24a and b). This phenomenon illustrates the property of **adaptation** to constant stimuli. Most sensory systems possess such a property, the molecular basis of which is discussed in more detail in section 5.7 as well as in chapter 8.

If caffeine is added 8 min before the onset of stimulation and is removed when stimulation by cAMP begins (fig. 5.24c and d), the response is practically normal. If, on the contrary, pretreatment by caffeine is performed in the presence of cAMP (parts e and f), subsequent stimulation by cAMP fails to elicit any intracellular synthesis of cAMP. The latter observation indicates that adaptation to the external stimulation is not brought about by a decrease in the level of ATP, since caffeine has the effect of preventing the activation of adenylate cyclase. From these results Theibert & Devreotes (1983) concluded that the phenomenon of adaptation originates from a process located on the way from the receptor to the cyclase. A plausible hypothesis, corroborated by the studies of Klein (1979) on the possible modification of the cAMP receptor, is that adaptation results from the phenomenon of **desensitization** of the receptor in the presence of its specific ligand, cAMP.

Receptor desensitization consists in a decrease of the physiological

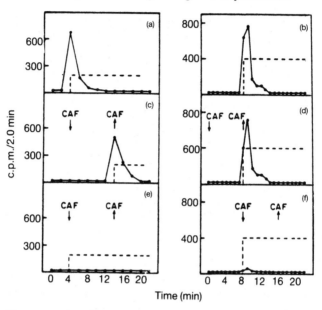

Fig. 5.24. Demonstration, by means of caffeine (CAF), that adaptation to the cAMP stimulus takes place in the absence of ATP consumption (see text), which suggests that desensitization occurs at the receptor level or on the path leading from the receptor to adenylate cyclase (Theibert & Devreotes, 1983).

effect elicited by the binding of the specific ligand, following the continuous contact of the latter with the receptor. This phenomenon is widely known for hormones as well as for neurotransmitters. An example is provided by desensitization of the acetylcholine receptor, which results from the transition of the allosteric receptor to a conformational state possessing a high affinity for the ligand but a reduced capacity for inducing the cellular response (Changeux, 1981). Such a process was at the basis of the model proposed by Katz & Thesleff (1957) to account for the decrease in the effect of acetylcholine in the course of time, following preincubation of the receptor with its ligand.

Besides purely conformational changes, desensitization often results from a covalent modification of the receptor in the presence of its ligand. Thus, adaptation of the chemotactic response of bacteria is due to the reversible methylation of membrane receptors, following the binding of specific ligands (Koshland, 1979; Springer, Goy & Adler, 1979).

In *D. discoideum*, it appears, likewise, that the desensitization responsible for adaptation to constant cAMP stimuli results from covalent modification. Such a modification takes here the form of a

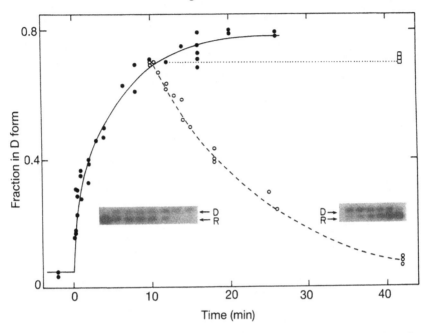

Fig. 5.25. Reversible modification of the cAMP receptor induced by an extracellular stimulus of cAMP in *D. discoideum* (filled circles). The D state is a phosphorylated form of the cAMP receptor. When the stimulation ceases, the receptor returns to its basal state R through dephosphorylation (open circles) (Devreotes & Sherring, 1985).

reversible phosphorylation (Lubs-Haukeness & Klein, 1982; Devreotes & Sherring, 1985; Meier & Klein, 1988; Vaughan & Devreotes, 1988), as in the case of the β-adrenergic receptor (Lefkowitz & Caron, 1986). Experiments indicate that, besides enhancing the synthesis of intracellular cAMP, stimulation by extracellular cAMP induces the phosphorylation of the receptor (fig. 5.25). The latter is dephosphorylated in a few minutes as soon as stimulation ceases. The fraction of receptor phosphorylated increases with the magnitude of the extracellular signal (fig. 5.26). It appears that the phosphorylated form of the cAMP receptor is less apt at inducing the synthesis of intracellular cAMP, hence the association of receptor modification with adaptation to constant stimuli. In the course of cAMP oscillations, the receptor varies, with the same periodicity, between the phosphorylated and dephosphorylated states (fig. 5.27); the maximum of cAMP oscillations slightly precedes the maximum in the nonphosphorylated receptor state (Klein, P. *et al.*, 1985; Gundersen *et al.*, 1989).

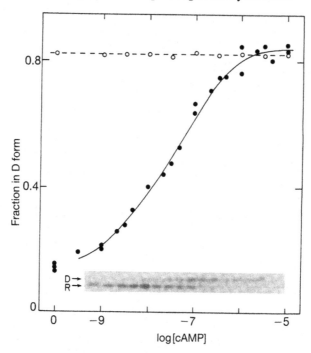

Fig. 5.26. Steady-state level of receptor phosphorylation in the presence of continuous stimulation by cAMP. The phosphorylation level increases with the amplitude of the stimulus from $10^{-9}$ M on, until it reaches a maximum for a cAMP concentration close to $10^{-5}$ M (Devreotes & Sherring, 1985).

In view of these observations, it is clear that the allosteric model of Goldbeter & Segel (1977) must be amended so as to replace the primary role of the substrate in limiting the autocatalysis by a corresponding role for receptor desensitization. Theoretical studies in this direction (Goldbeter & Martiel, 1980, 1983; Martiel & Goldbeter, 1984) preceded the experimental demonstration that the phenomenon of adaptation is due to phosphorylation of the receptor (Devreotes & Sherring, 1985; Klein, C. *et al.*, 1985). The acquisition of precise experimental data allowed the construction of a detailed model for the synthesis of cAMP in *D. discoideum*, based on desensitization of the receptor through reversible phosphorylation (Martiel & Goldbeter, 1987a).

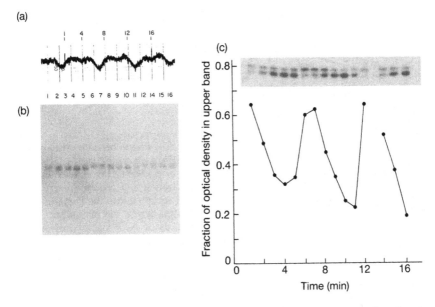

Fig. 5.27. Alternation between the phosphorylated and nonphosphorylated forms of the receptor ((b),(c)) in the course of cAMP oscillations in *D. discoideum*. The two forms of the cAMP receptor correspond to distinct bands observed on protein gels after purification ((b), and upper part of (c)). cAMP oscillates in phase with light scattering within the cell suspension (a). The fraction of receptor in the upper band (corresponding to the desensitized form) reaches its maximum close to 65% just after the phase of active synthesis of cAMP, which suggests a causal relationship between the inactivation of adenylate cyclase and receptor modification (Klein, P. *et al.*, 1985).

## 5.5 Model based on desensitization of the cAMP receptor

### *Description of the model*

The model based on desensitization of the cAMP receptor (Martiel & Goldbeter, 1984, 1987a) is represented schematically in fig. 5.28. This model retains a certain number of elements of the preceding model (fig. 5.16). Thus, extracellular cAMP binds to the receptor, but the latter now exists in two states, one of which is active (R) and the other desensitized (D); only the complex formed by the receptor in the R state with cAMP is capable of activating adenylate cyclase (C). In contrast with the preceding model, the latter enzyme is not linked here to the recep-

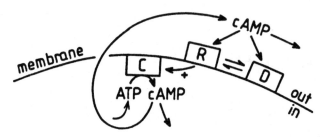

Fig. 5.28. Model for the synthesis of cAMP in *D. discoideum* based on desensitization of the cAMP receptor. Extracellular cAMP binds to the active (R) and desensitized (D) forms of the receptor. Binding of cAMP to the R state elicits the activation of adenylate cyclase (C), via the action of a G-protein (not represented), as well as the reversible transition of the receptor into the D state. The intracellular cAMP thus synthesized is transported across the membrane into the extracellular medium. Arrows denote synthesis of ATP and hydrolysis of cAMP by the intra- and extracellular forms of phosphodiesterase (Martiel & Goldbeter, 1984, 1987a).

tor in the same allosteric complex; moreover, the kinetics of the enzyme is considered to be Michaelian. As in the model outlined in fig. 5.16, intracellular cAMP is transported into the extracellular medium where it is hydrolysed by phosphodiesterase. Also taken into account here is the activity of an intracellular form of the enzyme. In terms of covalent modification of the receptor (Devreotes & Sherring, 1985), the D state represents the form desensitized through phosphorylation, while the R state is the active, dephosphorylated form.

The model corresponds to the detailed sequence (5.5) formed by the following reaction steps *a–j*:

*Step a:* Transitions between the active and desensitized states of the free receptor. The kinetic constants $k_1$ and $k_{-1}$ relate to the reactions catalysed, respectively, by the kinase and phosphatase responsible for the phosphorylation and dephosphorylation of the receptor.

*Steps b and c:* Binding of extracellular cAMP (P) to the two states of the receptor, with affinities that may or may not differ, depending on the relative values of the dissociation constants $K_R = d_1/a_1$ and $K_D = d_2/a_2$.

*Step d:* Transitions between the R and D states of the receptor bound to cAMP. The kinetic constants $k_2$ and $k_{-2}$ relate, respectively, to the kinetic activity of the kinase and phosphatase.

*Step e:* Transformation of free adenylate cyclase (C) into its active form (E) through the binding of two molecules of the complex formed by the receptor in the R state with cAMP. This step, represented here in a global manner, probably involves a GTP-binding protein (often referred to as a G-protein), as observed in the activation of adenylate cyclase in higher organisms (Janssens & Van Haastert, 1987). The effect of incorporating the action of G-proteins is discussed at the end of this chapter.

*Steps f and g:* Binding of the substrate ATP (S) to the forms E and C of adenylate cyclase and transformation into intracellular cAMP ($P_i$) in a Michaelian process. The form E possesses a larger affinity for ATP and/or a higher catalytic activity than the free form C of adenylate cyclase.

*Step h:* Hydrolysis of cAMP by the intracellular form of phospho-diesterase.

*Steps i:* Transport of cAMP into the extracellular medium, followed by its hydrolysis by the extracellular and membrane-bound forms of phosphodiesterase. In agreement with the experiments (Dinauer *et al.*, 1980a), these two steps, as well as step *h*, are supposed to obey apparent first-order kinetics.

*Steps j:* Production of ATP at a constant rate, and utilization of ATP by the cell in reactions unrelated to cAMP synthesis.

$$(a) \quad R \underset{k_{-1}}{\overset{k_1}{\rightleftharpoons}} D$$

$$(b) \quad R + P \underset{d_1}{\overset{a_1}{\rightleftharpoons}} RP$$

$$(c) \quad D + P \underset{d_2}{\overset{a_2}{\rightleftharpoons}} DP$$

$$(d) \quad RP \underset{k_{-2}}{\overset{k_2}{\rightleftharpoons}} DP$$

$$(e) \quad 2RP + C \underset{d_3}{\overset{a_3}{\rightleftharpoons}} E$$

$$(f) \quad E + S \underset{d_4}{\overset{a_4}{\rightleftharpoons}} ES \overset{k_4}{\longrightarrow} E + P_i$$

$$(g) \quad C + S \underset{d_5}{\overset{a_5}{\rightleftharpoons}} CS \overset{k_5}{\longrightarrow} C + P_i$$

$$(h) \quad P_i \overset{k_i}{\longrightarrow}$$

$$(i) \quad P_i \overset{k_t}{\longrightarrow} P \overset{k_e}{\longrightarrow}$$

$$(j) \quad \overset{v_i}{\longrightarrow} S \overset{k'}{\longrightarrow} \tag{5.5}$$

The nonlinearity of the activation step $e$ of adenylate cyclase by two molecules of the RP complex turns out to be necessary to account for excitable and oscillatory behaviour. This hypothesis, not yet corroborated by experimental observations in *D. discoideum*, is based on the fact that the activation of certain cellular responses, such as that triggered by the binding of gonadotropin-releasing hormone (GnRH) to pituitary cells (Conn, Rogers & McNeil, 1982; Blum & Conn, 1982; Blum, 1985), requires similar coupling with two molecules of the receptor–ligand complex.

In another version of the model (Martiel & Goldbeter, 1984), the nonlinearity is displaced from step $e$ to steps $b$ and $c$ of cAMP binding to the two states of the receptor, which is then supposed to be dimeric. Some studies support the existence of such a type of positive cooperativity in *D. discoideum* (Coukell, 1981); however, other results point rather to the occurrence of negative cooperativity, although they might also be interpreted in terms of multiple forms of the receptor (Mullens & Newell, 1978). The two sources of nonlinearity yield similar results regarding the dynamic behaviour of the model. We shall focus here on the effect of a nonlinear activation of adenylate cyclase, which might also reflect, in an implicit manner, an amplification due to the interaction of the enzyme with the G-protein, which itself becomes active through collision with the complex RP. This question is re-examined at the end of this chapter where the role of G-proteins is discussed in more detail.

### Kinetic equations and reduction to a four-variable system

No fewer than 11 variables appear in the reaction scheme (5.5). These are the concentrations of the metabolites and of the free and complexed forms of the receptor and adenylate cyclase. Because of the existence of conservation relations for the total amount of adenylate cyclase and of receptor, the number of independent variables can be reduced to nine. In the conditions of spatial homogeneity corresponding to the experiments in well-stirred cellular suspensions, the time evolution of the system is governed by the system of nine ordinary differential equations (Martiel & Goldbeter, 1987a; see Appendix, p. 234):

$$\frac{d\rho}{dt} = k_1(-\rho + L_1\delta) + d_1(-\rho\gamma + x)$$

$$\frac{d\delta}{dt} = k_1(\rho - L_1\delta) + d_2(-\delta c\gamma + y)$$

$$\frac{dx}{dt} = k_2(-x + L_2y) + d_1(\rho\gamma - x) + (2\mu/h)d_3(-\epsilon x^2\,\bar{c} + \bar{e})$$

$$\frac{d\bar{c}}{dt} = d_3(-\epsilon x^2\,\bar{c} + \bar{e}) + (d_5 + k_5)(-\bar{c}\alpha\theta + \overline{cs})$$

$$\frac{d\bar{e}}{dt} = d_3(\epsilon x^2\,\bar{c} - \bar{e}) + (d_4 + k_4)(-\bar{e}\alpha + \overline{es})$$

$$\frac{d\overline{es}}{dt} = (d_4 + k_4)(\bar{e}\alpha - \overline{es})$$

$$\frac{d\alpha}{dt} = v - k'\alpha - \sigma(\overline{es} + \lambda\overline{cs}) + \theta_E[(d_4 + k_4)$$
$$(-\bar{e}\alpha + \overline{es}) + (d_5 + k_5)(-\bar{c}\alpha\theta + \overline{cs})]$$

$$\frac{d\beta}{dt} = q\sigma(\overline{es} + \lambda\overline{cs}) - (k_i + k_t)\beta$$

$$\frac{d\gamma}{dt} = (k_t\beta/h) - k_e\gamma + \eta[d_1(-\rho\gamma + x)$$
$$+ d_2(-\delta c\gamma + y)] \tag{5.6}$$

In these equations, the fractions of enzyme or of receptor are defined as follows (see Appendix for further details):

$$\rho = R/R_T, \ \delta = D/R_T, \ x = RP/R_T, \ c = C/R_T, \ \bar{cs} = CS/R_T,$$
$$\bar{e} = E/E_T, \ \bar{es} = ES/E_T \tag{5.7}$$

where $R$, $D$, $C$ and $E$ denote the concentrations of receptor in R or D state, activated receptor–enzyme complex and free enzyme, respectively, while $CS$ and $ES$ denote the corresponding enzyme–substrate complexes; $R_T$ and $E_T$ denote the total quantities of receptor and of adenylate cyclase. The normalized metabolic variables are defined by:

$$\alpha = S/K_m, \ \beta = P_i/K_R, \ \gamma = P/K_R \tag{5.8}$$

where $S$, $P_i$ and $P$ denote the concentration of substrate, intracellular cAMP and extracellular cAMP, respectively. The different parameters appearing in eqns (5.6)–(5.8) are defined in table 5.3, as a function of the kinetic constants of the reaction scheme (5.5).

The study of the nine-variable system (5.6) is of course rather cumbersome. It would be useful to simplify it by reducing the number of variables. Such a reduction is often possible because certain variables evolve more rapidly than others. A quasi-steady-state hypothesis adopted for the fast variables then allows us to transform the corresponding kinetic equations into algebraic relations. Such an approach was followed in chapter 2 for the reduction of an 11-variable system, obtained in the case of a dimeric enzyme model for glycolytic oscillations, to the form of system (2.7), which contains only two variables.

The reduction of system (5.6) turns out to be much more delicate because the variables cannot be readily separated into 'slow' and 'fast'. Each of the kinetic equations (5.6) indeed contains both slow and fast terms. Thus, in the equation for the fraction of free, active receptor $\rho$, the first term $k_1(-\rho + L_1 \delta)$, which relates to the reversible modification of the receptor, is relatively slow compared to the processes of cAMP binding to the receptor, expressed by the term $d_1(-\rho\gamma + x)$. To reduce the number of variables, it is first necessary to define new variables by linear combination of some of the original ones. After rearranging and separating slow and fast variables, this procedure leads to system (5.9), which contains four variables, including ATP (see the Appendix to this chapter, and Martiel & Goldbeter (1987a) for further details on this reduction):

$$\frac{d\rho_T}{dt} = -f_1(\gamma)\,\rho_T + f_2(\gamma)\,(1 - \rho_T) \tag{5.9a}$$

$$\frac{d\alpha}{dt} = v - k'\,\alpha - \sigma\,\phi\,(\rho_T, \gamma, \alpha) \tag{5.9b}$$

$$\frac{d\beta}{dt} = q\sigma \, \phi \, (\rho_T, \gamma, \alpha) - (k_i + k_t) \, \beta \qquad (5.9c)$$

$$\frac{d\gamma}{dt} = (k_t\beta/h) - k_e \, \gamma \qquad (5.9d)$$

where:

$$f_1(\gamma) = \frac{k_1 + k_2 \, \gamma}{1 + \gamma}, \; f_2(\gamma) = \frac{k_1 L_1 + k_2 L_2 c\gamma}{1 + c\gamma},$$

$$\phi \, (\rho_T, \gamma, \alpha) = \frac{\alpha \, (\lambda\theta + \epsilon Y^2)}{1 + \alpha\theta + \epsilon Y^2 \, (1 + \alpha)}, \; Y = \frac{\rho_T \, \gamma}{1 + \gamma} \qquad (5.10)$$

In these equations, as before, $\alpha$, $\beta$ and $\gamma$ denote the normalized concentrations of ATP, intracellular cAMP and extracellular cAMP, whereas $\rho_T$ represents the total fraction of receptor in the active state (i.e. the sum of the concentration of the free form R and those of the complexes RP, E, and ES, divided by $R_T$).

### Reduction to three variables

The numerical study of the four-variable system (5.9a–d) reveals that it is capable of sustained oscillatory behaviour. These results, developed in further detail in the following section, also indicate that ATP remains practically constant in the course of cAMP oscillations (fig. 5.30, below). Thus, in contrast with the allosteric model considered above, the model based on receptor desensitization can account for the experimental observation (fig. 5.22) on the limited variation of ATP in the course of cAMP oscillations. Once we have established that the model predicts this characteristic feature of the experimental system, we may consider, as a first approximation, that ATP remains constant in time at the value given by eqn (5.11), in view of the relative smallness of the maximum rate of adenylate cyclase compared to the accumulative rate of ATP utilization in other pathways:

$$\alpha = v/k' \qquad (5.11)$$

The cAMP-synthesizing system in *D. discoideum* is thus finally governed by the system of three differential equations:

$$\frac{d\rho_T}{dt} = -f_1(\gamma) \, \rho_T + f_2(\gamma) \, (1 - \rho_T)$$

$$\frac{d\beta}{dt} = q\sigma \, \phi \, (\rho_T, \, \gamma, \, \alpha) - (k_i + k_t) \, \beta$$

$$\frac{d\gamma}{dt} = (k_t\beta/h) - k_e\gamma \tag{5.12}$$

In certain conditions discussed below, additional hypotheses permit a further reduction to only two variables; the advantage of such a reduction is to facilitate the analysis of relay and oscillations in the phase plane. However, to be valid, these hypotheses require, for certain parameters, values that differ significantly from those that are experimentally available. Therefore the study of oscillations and relay will be carried out mainly in the three-variable system (5.12), or in the four-variable system (5.9a–d) from which it is derived.

## 5.6  Role of receptor desensitization in oscillations and relay of cAMP signals

The synthesis of cAMP in *D. discoideum* possesses an autocatalytic nature. Only the limitation by receptor desensitization prevents this synthesis from becoming 'explosive'. It is the combination of these antagonistic effects that is responsible for the two most conspicuous dynamic properties of these amoebae, namely oscillations and relay of cAMP signals.

### *Autonomous oscillations of cAMP*

The systems of eqns (5.9a–d) and (5.12) have been subjected to linear stability analysis, in order to determine the conditions in which the regulation of the cAMP-synthesizing system gives rise to an instability of the steady state followed by sustained oscillations (Martiel & Goldbeter, 1987a). Two examples of a typical stability diagram established as a function of parameters $L_1$ and $L_2$ are shown in fig. 5.29. These parameters denote here the ratio between the kinetic constants of dephosphorylation and phosphorylation of the receptor, free or complexed with the ligand, respectively (see table 5.3 for further details on the definition of the parameters).

The two parts of fig. 5.29 are obtained as a function of $L_1$ and $L_2$, for two distinct values of the coefficient $c$, which measures the differential affinity of cAMP for the R and D states of the receptor. In region I, sustained oscillations occur around an unstable steady state. In region II,

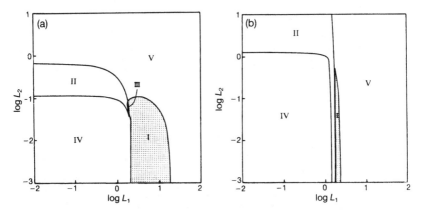

Fig. 5.29. Stability diagrams showing the various modes of dynamic behaviour of the model for cAMP synthesis based on receptor desensitization. The diagrams, obtained by linear stability analysis of the steady states of the three-variable system (5.12), are established as a function of parameters $L_1$ and $L_2$ for two values of coefficient $c$ characterizing the nonexclusive binding of cAMP to the receptor: (a) 10, and (b) 0.1. Domain I denotes instability of a unique steady state and corresponds to sustained oscillations of the limit cycle type. Two stable steady states coexist in II; they correspond to a weak and an elevated level of extracellular cAMP. In domain III, three steady states are obtained, only one of which is stable; a stable limit cycle can surround an unstable steady state corresponding to a high level of cAMP. Outside these regions, system (5.12) admits a unique, stable steady state corresponding to a low (IV) or elevated (V) level of extracellular cAMP. Parameter values are given in table 5.3 (Martiel & Goldbeter, 1987a).

two stable steady states coexist; they correspond to a high or a low level of cAMP, and are separated by an unstable steady state. Three distinct steady states are obtained in region III, only one of which is stable; a coexistence between this state and a stable limit cycle surrounding one of the unstable states can then be observed. Outside these domains, the system admits a single steady state corresponding to a low (domain IV) or high (domain V) level of cAMP.

Although oscillations remain possible for values of coefficient $c$ smaller than 1, the comparison of the two diagrams shows that the oscillatory domain widens when the value of $c$ increases above unity; cAMP then has a larger affinity for the desensitized state of the receptor. Notwithstanding the absence of oscillations in that system, such a situation is observed for the acetylcholine receptor, whose desensitized state possesses a higher affinity for its ligand (Changeux, 1981).

The numerical integration of the kinetic equations for parameter values corresponding to a point located in domain I confirms that periodic behaviour occurs when the unique steady state admitted by the system is unstable. Typical oscillations obtained in the four-variable system (5.9a–d) are shown in fig. 5.30. In part a, the variation in ATP ($\alpha$) is indicated, together with that in intracellular ($\beta$) and extracellular ($\gamma$) cAMP.

As mentioned above and demonstrated in fig. 5.30, ATP varies only slightly in the course of large-amplitude cAMP oscillations. What is then the factor responsible for the decrease in cAMP once the latter has reached its maximum? The evolution of the total fraction of receptor in the active state ($\rho_T$) shows that this fraction begins to decrease from the onset of the acceleration of cAMP synthesis; this decrease results in a drop in adenylate cyclase activation and, subsequently, in the level of cAMP. Another factor that contributes to this drop in cAMP is the action of the intracellular and extracellular forms of phosphodiesterase.

Figure 5.30b shows again, on a larger scale, the variation of extracellular cAMP. Also represented in this figure are the variation of the total fraction of receptor in the desensitized state ($\delta_T$), and the receptor saturation function ($\bar{Y}$). This function, defined by relation (5.13), takes into account the binding of cAMP to the two forms of the receptor:

$$\bar{Y} = \rho_T \frac{\gamma}{1+\gamma} + (1-\rho_T) \frac{c\gamma}{1+c\gamma} \tag{5.13}$$

The phase difference between the saturation function of the receptor and extracellular cAMP depends on the value of parameter $c$, which measures the relative affinity of the ligand for the two receptor states. A negligible phase shift should be observed as long as the R state possesses a higher affinity for cAMP ($c < 1$); in the opposite case, the saturation function should reach its maximum after the peak in cAMP. A detailed experimental study, not yet available, would permit us to obtain information on that point and could indirectly yield indications as to the value of coefficient $c$. The curves in fig. 5.30a and b also show that for the parameter values chosen (and, in particular, for $c = 10$), the maximum of the fraction of active, dephosphorylated receptor shortly precedes the maximum in cAMP, while the increase in the fraction of desensitized receptor follows the peak in cAMP.

The fact that the periodic variation in cAMP is accompanied by an alternation between the phosphorylated and nonphosphorylated forms of the receptor accounts well for the observations of Klein, P. *et al.* (1985) represented in fig. 5.27. In both the experimental and theoretical curves, the peak in modified receptor follows shortly after the peaks in

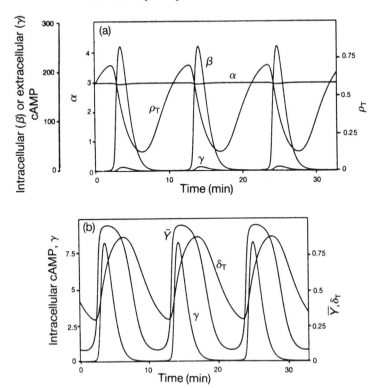

Fig. 5.30. Sustained oscillations of cAMP in the model based on receptor desensitization. (a) The evolution of intracellular cAMP ($\beta$), ATP ($\alpha$), the total fraction of receptor in active state ($\rho_T$), and extracellular cAMP ($\gamma$). The latter is represented, on an enlarged scale, in (b) together with the total fraction of receptor in the desensitized state ($\delta_T$) and the saturation function ($\bar{Y}$) measuring binding of cAMP to the two receptor states. The curves are obtained by numerical integration of the four-variable system (5.9) for the parameter values indicated in table 5.3; most of these values are those determined experimentally and available in the literature (see table 5.2). Similar curves are obtained by integration of the three-variable system (5.12) when ATP is maintained constant at the value $\alpha = 3$ (Martiel & Goldbeter, 1987a).

intra- and extracellular cAMP. A similar result is obtained by integration of eqns (5.12), which govern the three-variable system, when ATP is maintained constant at the value prescribed by relation (5.11).

The agreement of the theoretical predictions with the experimental results is not only qualitative but also, in a large measure, quantitative. Thus, the oscillations obtained (fig. 5.30) match in period and in amplitude those observed in *D. discoideum* suspensions, when parameters are given the values determined in these experiments. Table 5.2

Table 5.2. *Experimental values of parameters in the model for cAMP synthesis in* D. discoideum *based on receptor desensitization*

| Parameter | Definition | Experimental range | Reference |
|---|---|---|---|
| $K_R$ | Dissociation constant of cAMP–receptor complex in R state[a] | $10^{-7}$ M | Klein (1969); Mullens & Newell (1978) |
| | | $3.6 \times 10^{-8}$ M | Henderson (1975) |
| | | $10^{-8}$ M to $1.4 \times 10^{-7}$ M | Coukell (1981) |
| | | $1.5 \times 10^{-9}$ M | Devreotes & Sherring (1985) |
| $K_D$ | Dissociation constant of cAMP– receptor complex in D state[a] | $9 \times 10^{-9}$ M | Mullens & Newell (1978) |
| | | $3 \times 10^{-9}$ M to $9 \times 10^{-9}$ M | Klein (1979) |
| | | $3 \times 10^{-9}$ M | Devreotes & Sherring (1985) |
| $k_1$ | Rate constant for modification step $R \to D$ | $0.012$ min$^{-1}$ | Devreotes & Sherring (1985) |
| $k_{-1}$ | Rate constant for demodification step $D \to R$ | $0.104$ min$^{-1}$ | Devreotes & Sherring (1985) |
| $k_2$ | Rate constant for modification step $RP \to DP$ | $0.222$ min$^{-1}$ | Devreotes & Sherring (1985) |
| $k_{-2}$ | Rate constant for demodification step $DP \to RP$ | $0.055$ min$^{-1}$ | Devreotes & Sherring (1985) |
| cAMP receptor | | $10^5$–$10^6$ molecules/cell | Gerisch & Malchow (1976) |
| Adenylate cyclase | | $2 \times 10^4$ molecules/cell | Gerisch & Malchow (1976) |
| Cell density | | $10^7$ to $2 \times 10^8$ cells/ml | Gerisch & Malchow (1976) |
| Intracellular volume | | $10^{-12}$ litre | Gerisch & Malchow (1976) |
| | | $7.5 \times 10^{-13}$ litre | Europe-Finner & Newell (1984) |
| Activity of adenylate cyclase | Basal rate | $5.7 \times 10^{-7}$ M/min | Loomis (1979) |
| | At maximum activation | $2.3 \times 10^{-5}$ M/min | Gerisch & Malchow (1976) |
| $K_{m,\,cyclase}$ | Michaelis constant of adenylate cyclase | $0.4$ mM | Gerisch & Malchow (1976) |
| | | $0.2$–$0.5$ mM | Klein (1976) |
| | | $17$ μM and $0.4$ mM | de Gunzburg et al. (1980) |

| Parameter | Definition | Experimental range | Reference |
|---|---|---|---|
| $k_i$ | Apparent first-order rate constant for intracellular phosphodiesterase | $1.73 \text{ min}^{-1}$ | Dinauer *et al.* (1980a) |
| $V_{max,PDE}$ | Maximum activity for extracellular phosphodiesterase | $10^{-5} \text{ M/min}$ $5 \times 10^{-5} \text{ M/min}$ | Klein & Darmon (1976) Yeh *et al.* (1978) |
| $K_{m,PDE}$ | Michaelis constant of extracellular phosphodiesterase for cAMP | $4 \times 10^{-6} \text{ M}$ | Gerisch & Malchow (1976) |
| $k_t$ | Apparent first-order rate constant for cAMP transport into extracellular medium | $0.34\text{–}0.94 \text{ min}^{-1}$ | Dinauer *et al.* (1980a) |
| ATP | | $1\text{–}1.5 \text{ mM}$ | Gerisch & Malchow (1976) Roos *et al.* (1977) |

[a] Some $K_R$ or $K_D$ values might pertain to cAMP receptors not related to adenylate cyclase. Values measured by Van Haastert (1984) in kinetic binding experiments on a time scale of seconds probably pertain to the formation of distinct receptor complexes during the phase of activation of adenylate cyclase, rather than to the later phase of adaptation mediated by receptor modification. Martiel & Goldbeter, 1987a.

presents the values available for the various parameters, while table 5.3 contains the values retained for numerical simulations of the model, as well as a definition of the parameters in terms of the kinetic constants related to the reaction steps (5.5).

The agreement between theory and experiment extends to the amplitude of the variations in fraction $\rho_T$. In fig. 5.30, the fraction of desensitized receptor varies periodically from 30% up to 85%, while the corresponding fraction of phosphorylated receptor in the experiments (Klein, P. *et al.*, 1985) oscillates between 25% and 65% (fig. 5.27).

As indicated in table 5.3, only the values of the phosphorylation and dephosphorylation constants $k_1$, $k_{-1}$, $k_2$, $k_{-2}$ chosen for simulations differ from the values determined experimentally by Devreotes & Sherring (1985). To obtain a period of the order of 10 min for the oscillations of cAMP, which is that observed in the experiments (fig. 5.12), it is

Table 5.3. *Parameter values considered in numerical simulations of oscillations and relay of cAMP in the model based on receptor desensitization*

| Parameter | Definition or expression in terms of the parameters of scheme (5.5) | Experimental range (see table 5.2) | Numerical value considered |
|---|---|---|---|
| $K_R$ | $d_1/a_1$ | $1.5 \times 10^{-9}$ to $1.4 \times 10^{-7}$ M | $10^{-7}$ M |
| $K_D$ | $d_2/a_2$ | $3 \times 10^{-9}$ to $9 \times 10^{-9}$ M | $10^{-8}$ M[a] |
| $c$ | $K_R/K_D$ | 0.15–50 | 10 |
| $k_1$ | | 0.012 min$^{-1}$ | 0.036 min$^{-1}$ |
| $k_{-1}$ | | 0.104 min$^{-1}$ | 0.36 min$^{-1}$ |
| $k_2$ | | 0.222 min$^{-1}$ | 0.666 min$^{-1}$ |
| $k_{-2}$ | | 0.055 min$^{-1}$ | 0.00333 min$^{-1}$ |
| $L_1$ | $k_{-1}/k_1$ | 8.67 | 10 |
| $L_2$ | $k_{-2}/k_2$ | 0.25 | 0.005 |
| $K_m$ | $K_{m,cyclase} = (d_4 + k_4)/a_4$ | 17 μM to 0.5 mM | 0.4 mM |
| $q$ | $K_m/K_R$ | 120–3.3 × 10$^5$ | 4 × 10$^3$ |
| $\sigma$ | $V_{max,cyclase}/K_m = k_4 E_T/K_m$ | 0.05–1.35 min$^{-1}$ | 0.6 min$^{-1}$ |
| $k_i$ | | 1.73 min$^{-1}$ | 1.7 min$^{-1}$ |
| $k_e$ | $V_{max,PDE}/K_{m,PDE}$[b] | 2.5–12.5 min$^{-1}$ | 5.4 min$^{-1}$ |
| $k_t$ | | 0.34–0.94 min$^{-1}$ | 0.9 min$^{-1}$ |
| $\alpha$ | ATP/$K_m$ | 2.4–70 | 3 |
| $h$ | Ratio of extracellular to intracellular volumes | 5–100[c] | 5 |
| $E_T$ | Total intracellular concentration of adenylate cyclase | $3 \times 10^{-8}$ M[d] | $3 \times 10^{-8}$ M |
| $R_T$ | Total extracellular concentration of cAMP receptor | $1.5 \times 10^{-9}$ to $3 \times 10^{-7}$ M[c] | $3 \times 10^{-8}$ M |
| $\mu$ | $E_T/R_T$ | 0.1–10 | 1 |
| $\eta$ | $R_T/K_R$ | 0.02–200 | 0.3 |
| $\theta_E$ | $E_T/K_m$ | $2 \times 10^{-3}$ to $6 \times 10^{-5}$ | Neglected |
| $\theta$ | $K_m/K'_m$ (ratio of Michaelis constants for E and C forms of adenylate cyclase) | Not available | 0.01 |
| $\lambda$ | $k_5/k_4$ (ratio of catalytic constants of forms C and E of adenylate cyclase) | Not available | 0.01 |
| $\epsilon$ | $R_T^2 (a_3/d_3)$; coupling constant for activation of C by 2RP (step $e$) | Not available | 1 |

[a] For $c \neq 10$, $K_D = 10^{-7}$ M/$c$.
[b] This parameter includes the effect of the membrane-bound and extracellular forms of phosphodiesterase, which both act on extracellular cAMP.
[c] For a cell density ranging from $2 \times 10^8$ to $10^7$ cells/ml.
[d] Established for an intracellular volume of $10^{-12}$ litre.
Martiel & Goldbeter, 1987a.

necessary to multiply these values by a factor of 3. Without this manipulation, the period of the oscillations is of the order of 30 min, i.e. three times too large. While this departure from published experimental data can seem to present a defect of the model that should be cured by some modification of the hypotheses, another view, just as legitimate, is to consider the predictive value of such a departure. It now appears that the second interpretation, in terms of the predictive power of the model, is more valid. Additional experiments have indeed allowed Vaughan & Devreotes (1988) to refine the values of the receptor phosphorylation and dephosphorylation rate constants. In agreement with the predictions of the model, the new values of these parameters are multiplied by a factor ranging from 2 to 5 with respect to the values mentioned in tables 5.2 and 5.3.

### *Mechanism of oscillations*

The results shown in fig. 5.30 and the determination of the fluxes between the various states of the receptor in the course of time (Martiel & Goldbeter, 1987a) allow us to reconstruct the sequence of events whose spontaneous blending gives rise to periodic behaviour. That sequence of interacting processes constitutes the molecular mechanism of sustained oscillations. When starting from a low level of extracellular cAMP, the receptor is predominantly in the R state, given the high value of $L_1$ equal to the ratio $R/D$ in the absence of ligand (see table 5.3). Adenylate cyclase operates at its basal rate, and the level of extracellular cAMP slowly rises. Binding of cAMP to the receptor in the R state elicits activation of the enzyme. As soon as the level of extracellular cAMP reaches the level where autocatalysis becomes effective ($\gamma \approx 1$), binding of the ligand to the receptor results in a marked acceleration of cAMP synthesis: this corresponds to the abrupt, rising phase of the cAMP peak.

As the concentration of the ligand increases, the receptor passes from the active into the desensitized state, because of the low value of constant $L_2$, which measures the ratio $RP/DP$ at saturation by the ligand. The receptor is now predominantly in its phosphorylated state, and its ability to activate adenylate cyclase reaches a minimum; a decrease in the synthesis of cAMP ensues. Extracellular cAMP thus decreases as a result of this diminished synthesis but also because of its hydrolysis by phosphodiesterase. Owing to the drop in cAMP, the receptor progressively returns towards its active state, by dephosphorylation catalysed by the receptor phosphatase. A cycle of the oscillations is completed and the next one begins when the initial value of the ratio $R/D$ is restored.

This mechanism allows us to understand why the kinetic constants of the kinase and of the phosphatase implicated in the covalent modification of the receptor control the period of the oscillations. The major part of a period is occupied by the interval between two peaks of cAMP (fig. 5.30). The process that predominates during that phase is the return, through dephosphorylation, of the D to the R form of the receptor; the velocity of this step is governed by the rate constant $k_{-1}$. For a given value of the ratios $L_1 = k_{-1}/k_1$ and $L_2 = k_{-2}/k_2$, multiplication of the kinetic constants of the kinase and phosphatase by a factor of 3, for example, will have the approximate effect of dividing, by the same factor, the period of the oscillations.

### *Relay of cAMP signals: excitability*

Like the allosteric model analysed in section 5.2 above, the model based on receptor desensitization is capable of describing the amplification of cAMP pulses by the signalling system when the latter is initially in a stable steady state, provided that the amplitude of the stimuli exceeds a threshold.

This excitable behaviour is illustrated in fig. 5.31. At time zero, the value of $\gamma$ is increased by 0.3 unit, which corresponds to the instantaneous addition of a $3 \times 10^{-8}$ M pulse of extracellular cAMP. The system responds to this perturbation by producing a peak of intracellular cAMP before moving towards the steady state. If a second stimulus of the same magnitude is applied 10 min after the first signal, a second response is triggered, but its amplitude is weaker than that of the preceding response. A third, identical stimulus, applied 7.5 min later, results in an even more reduced response. The curve showing the evolution of $\rho_T$ as a function of time indicates that the decrease in the response to successive signals is due to the fact that the transition of the receptor into the active state is still incomplete when the second and third stimuli occur.

When the amplitude of the response, measured by the ratio $\beta_M/\beta_0$ of the maximum of the peak in intracellular cAMP divided by the steady state level of $\beta$, is plotted as a function of the stimulus, measured by the ratio $\gamma_i/\gamma_0$ of the initial value of extracellular cAMP divided by the steady state level of $\gamma$, we obtain the dose–response curve for relay rep-

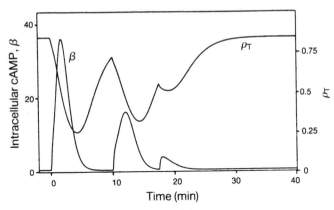

Fig. 5.31. Relay of cAMP signals. Starting from a stable steady state, the system is perturbed by an increase in $\gamma$ by 0.3 unit (which corresponds to the addition of a $3 \times 10^{-8}$ M pulse of extracellular cAMP for a dissociation constant $K_R = 10^{-7}$ M; see tables 5.2 and 5.3). The curve for $\beta$ shows the amplification of this perturbation, in the form of synthesis of a peak of intracellular cAMP. A second stimulus of similar magnitude applied in $t = 10$ min produces a response of smaller amplitude, while the response to a third, identical stimulus in $t = 17.5$ min is even weaker. The curve showing the corresponding evolution of the fraction of active receptor is also given. The curves are obtained by integration of eqns (5.12) for $\sigma = 0.57$ min$^{-1}$, $k_e = 3.58$ min$^{-1}$, $k_i = 0.958$ min$^{-1}$, $\epsilon = 0.108$; the values of the other parameters are as in table 5.3 (Martiel & Goldbeter, 1987a).

resented in fig. 5.32. This curve demonstrates the existence of an abrupt threshold below which cAMP signals fail to induce any cellular response; beyond the threshold, the amplitude of the relay response approaches a plateau. Given that for the parameter values considered $\gamma_0 = 0.0284$, and that the dissociation constant $K_R$ is equal to $10^{-7}$ M (see tables 5.2 and 5.3), the value of the threshold predicted by the model, close to $\gamma = 3\gamma_0$, corresponds to an extracellular cAMP concentration of the order of $9 \times 10^{-9}$ M. This value compares with the experimental value, which falls between $2 \times 10^{-9}$ and $10^{-8}$ M, which was determined for relay during aggregation on agar (Robertson & Drage, 1975; Grutsch & Robertson, 1978). No dose–response curve for relay is available for experiments in cell suspensions.

The existence of the abrupt threshold for the relay response is linked to the excitable properties of the cAMP-synthesizing system. Like the oscillations, excitability is due mainly to the self-amplifying nature of cAMP production in the slime mould. As indicated by the curve yielding the saturation function of the receptor as a function of the cAMP signal in fig. 5.32, no threshold is detectable for binding of the ligand to

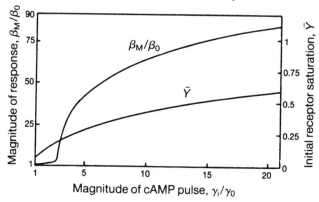

Fig. 5.32. Existence of a threshold for the relay of cAMP signals. The amplitude of the response, measured by the ratio of the maximum of the peak of intracellular cAMP divided by the steady-state level of $\beta$, is represented as a function of the initial value of extracellular cAMP divided by the steady-state level of $\gamma$, under the conditions of fig. 5.31. The dose–response curve for the synthesis of cAMP possesses an abrupt threshold. In contrast, no threshold is apparent in the saturation function of the cAMP receptor, $\bar{Y}$ (Martiel & Goldbeter, 1987a).

the receptor. Nor does the threshold for relay originate from the non-linear coupling between the receptor and adenylate cyclase, as demonstrated by the results obtained in the absence of self-amplification, which are detailed below (fig. 5.39). The occurrence of a sharp threshold is thus due solely to a nonlinear dynamic property of the signalling system, namely excitable behaviour.

Besides the existence of a threshold for stimulation, excitable systems are characterized by the existence of a refractory period during which the response to further stimulation is reduced or even totally absent (Fitzhugh, 1961). As shown in fig. 5.33, the cAMP signalling system of *D. discoideum* shares this property. Represented in this figure is the ratio $\beta_{M2}/\beta_{M1}$ of the maxima of two successive peaks of intracellular cAMP, as a function of the time interval separating the second stimulus from the maximum of the first response, which defines time zero. For 4 min, no cAMP synthesis can be elicited by the second stimulus. This phase defines an **absolute refractory period**. Thereafter, the response to the second stimulus increases gradually, until the second maximum reaches the value of the first, after some 15 min. This second phase defines a **relative refractory period**. The two types of refractory period are known in other excitable systems, for example in nerve cells (Fitzhugh, 1961).

The curve yielding the time evolution of the fraction $\rho_T$ after the first

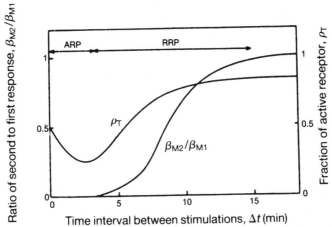

Fig. 5.33. Absolute and relative refractory periods for relay of cAMP signals. The ratio $\beta_{M2}/\beta_{M1}$ of the maximum of the second peak of intracellular cAMP divided by the maximum of the first response in fig. 5.31 is plotted as a function of the time separating the second stimulus from the peak of the first response. For 4 min, no second response can be measured, which indicates the existence of an absolute refractory period (ARP). Then, the amplitude of the second response progressively increases, until it reaches the amplitude of the initial response. This second phase represents a relative refractory period (RRP). The corresponding variation of the fraction of active receptor in response to the first stimulation is also shown (Martiel & Goldbeter, 1987a).

stimulation (fig. 5.33) helps to clarify the molecular basis of the refractory period for relay. The absolute refractory period corresponds to the phase of decrease of the fraction of active receptor after the first cAMP pulse. The slow rise of this fraction up to its value prior to the first stimulation corresponds to the relative refractory period during which a second stimulus, identical with the preceding one, elicits a response of reduced amplitude.

### The two-variable core of the cAMP signalling mechanism

Reducing the dynamics of a complex system to that of a two-variable system is the goal pursued in most studies devoted to periodic or excitable behaviour, in chemistry as in biology. The main impetus for such a reduction is that it allows us to study these phenomena by means of phase plane analysis. The latter clarifies the origin of the two modes of dynamic behaviour, and highlights basic features common to different systems. This approach was followed in the study of excitability and oscillations in the Belousov–Zhabotinsky reaction (Tyson, 1977), and in a number of models based on the equations of Hodgkin & Huxley (1952) for neuronal dynamics. Several two-variable reductions of the latter, four-variable system have been analysed (Kokoz & Krinskii, 1973; Krinskii & Kokoz, 1973; Rinzel, 1985), and other two-variable versions, not directly related to the Hodgkin–Huxley equations, have been proposed as simple models for the generation of the action potential (Fitzhugh, 1961; Hindmarsh & Rose, 1982).

Thus the question arises as to whether a further reduction of system (5.12) is possible, which would allow the description of the dynamics of cAMP signalling in terms of two variables only. For sufficiently large values of parameters $q$, $k_i$ and $k_t$, the variation of $\beta$ in the course of time is much faster than that of $\rho_T$ and $\gamma$ in the three-variable system (5.12). In such conditions, a quasi-steady-state hypothesis could be justified for $\beta$, whose kinetic equation would then reduce to the algebraic relation:

$$\beta = (q\sigma\,\phi)/(k_i + k_t) \tag{5.14}$$

In such conditions, the dynamics of the cAMP signalling system is governed by the two differential equations:

$$\frac{d\rho_T}{dt} = -f_1(\gamma)\,\rho_T + f_2(\gamma)\,(1 - \rho_T)$$

$$\frac{d\gamma}{dt} = q'\sigma\,\phi\,(\rho_T,\,\gamma) - k_e\,\gamma \tag{5.15}$$

with:

$$q' = \frac{qk_t}{h\,(k_i + k_t\,)}$$

As indicated in table 5.2, if the value of parameter $q$ is sufficiently high to justify the hypothesis of a quasi-steady state for variable $\beta$, the situation is somewhat different for the experimental values of $k_i$ and $k_t$. The latter are too low for the hypothesis to hold rigorously. It is for this reason that the preceding results on oscillations and relay were obtained by means of the three-variable system (5.12). It is nevertheless interesting to study the behaviour of the two-variable system (5.15), in order to provide an estimate for the error made upon reducing the number of variables to only two. This error could well turn out to be minute, and thus acceptable, in comparison with the advantage provided by the possibility of resorting to the phase plane analysis of a two-variable system.

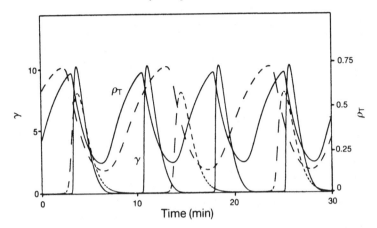

Fig. 5.34. Comparison of cAMP oscillations in the two-variable system (5.15) (solid line) and in the three-variable system (5.12) (dashed line). The curves, obtained for the parameter values of table 5.3, are drawn so that the first peak of cAMP coincides in the two systems (Martiel & Goldbeter, 1987a).

Compared in fig. 5.34 are the results obtained for the oscillations in $\gamma$ and $\rho_T$ by integrating the three-variable system (5.12) (dashed line) and the two-variable system (5.15) (solid line). The curves are drawn so that the first maximum in $\gamma$ occurs at the same time in both systems. Although, mathematically, the reduction to two variables is not rigorously justified for the parameter values considered, it nevertheless produces results that remain relatively satisfactory, from both a qualitative and a quantitative point of view. Whereas in the case considered the period is of 10.7 min in the three-variable system, it drops down to 7.5 min in the reduced system, while the amplitude of $\gamma$ passes from 7.8 to 10.3. In fig. 5.34, the error on the period and amplitude resulting from the reduction to two variables is thus of the order of 25%, which remains acceptable for a number of applications.

One of these applications consists in introducing system (5.15) as the mechanism generating cAMP signals in a model that takes into account the diffusion of this substance in the unstirred reaction medium, so as to simulate the behaviour of the amoebae in the course of aggregation (see section 5.9). Such a study (Tyson, 1989; Tyson & Murray, 1989; Tyson *et al.*, 1989; Höfer *et al.*, 1995) predicts the occurrence of concentric or spiral waves of cAMP, which are in good agreement with those observed (Tomchik & Devreotes, 1981) during aggregation on agar.

In addition to its greater simplicity, the two-variable system allows us to demonstrate the close link that connects excitable and oscillatory

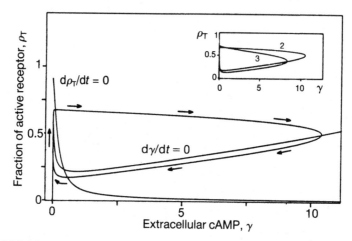

Fig. 5.35. Phase portrait for cAMP oscillations in the two-variable system (5.15). The two nullclines in the ($\rho_T$, $\gamma$) phase plane are indicated, as well as the limit cycle surrounding the unstable steady state. The inset shows the comparison of this cycle with that obtained in the three-variable system (5.12). Parameter values are those of fig. 5.34 (Martiel & Goldbeter, 1987a).

behaviour in the phase plane. A phase portrait corresponding to the occurrence of sustained oscillations is shown in fig. 5.35. Here, as in the case of the allosteric model, the nullcline of product $\gamma$ has a sigmoidal shape and possesses a region of negative slope. The steady state is unstable when the nullcline of variable $\rho_T$ intersects with the sigmoid nullcline in that region, provided the slope ($d\rho_T/d\gamma$) is sufficiently negative. The system then evolves towards a stable limit cycle surrounding the steady state. In the inset of fig. 5.35 are shown, for comparative purpose, the limit cycles obtained in the ($\rho_T$, $\gamma$) plane for the two- and three-variable systems. In both cases, the general shape of the limit cycle is similar, although the amplitude of extracellular cAMP is slightly larger in the reduced system.

For closely related values of the parameters, the nullcline of $\rho_T$ cuts the nullcline of $\gamma$ to the left of the instability domain (fig. 5.36). The stable steady state is then excitable: a horizontal displacement towards the right, corresponding to the addition of a sufficient amount of extracellular cAMP, gives rise to the amplification of this perturbation and to the subsequent production of a peak of extracellular cAMP. Of course, this response is also accompanied by the synthesis of a peak of intracellular cAMP, as stipulated by relation (5.14).

The enlargement in the inset of fig. 5.36 shows the existence of an abrupt threshold for the excitable response. The threshold is dictated by

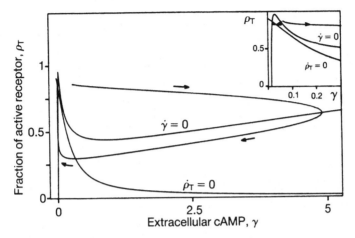

Fig. 5.36. Phase portrait for relay of suprathreshold cAMP signals in the two-variable system (5.15). The two nullclines in the phase plane ($\rho_T$, $\gamma$) are indicated, as well as the trajectory corresponding to the amplification of a perturbation away from the stable steady state in the form of an instantaneous increase in extracellular cAMP. The inset shows the response to two perturbations below and above the threshold, respectively. The curves are obtained for $\sigma = 0.57$ min$^{-1}$, $k_e = 4$ min$^{-1}$, $k_i = 1$ min$^{-1}$, $\epsilon = 0.15$; other parameter values are those of fig. 5.34 (Martiel & Goldbeter, 1987a). Dot notation has been used for clarity, e.g. $\dot{\gamma} = d\gamma/dt$.

the position of the median branch of the sigmoid nullcline. Any perturbation corresponding to an initial condition located to the left of this 'threshold separatrix' (Fitzhugh, 1961) will be damped, while perturbations of larger magnitude, corresponding to an initial value of $\gamma$ located to the right of the separatrix, will be amplified owing to the excitable properties of the cAMP signalling system.

Periodic and excitable behaviour thus both originate from the nullcline structure of the two-variable system in the phase plane. The two types of dynamic behaviour are necessarily linked, and arise from a common biochemical mechanism. The nature of the observed behaviour will depend on the values of the parameters. This conclusion corroborates that obtained in the allosteric model analysed above. Here, the fraction of active receptor replaces the substrate ATP as second variable, besides extracellular cAMP.

The two-variable system (5.15) represents the true core of the mechanism of cAMP relay and oscillations in *D. discoideum*. This reduced form, indeed, accounts for the essential properties of the cAMP signalling system, namely the capability to oscillate in an autonomous manner and to relay suprathreshold pulses of cAMP.

The qualitative and, in a a large measure, quantitative agreement of this model with experimental observations should not hide the fact that it is based on a certain number of simplifying assumptions or as yet unverified conjectures as to the precise mechanism of some steps of reaction scheme (5.5). Thus, one or more G-proteins play a role in the activation or inhibition of adenylate cyclase after binding of extracellular cAMP to the active and desensitized receptor states (Van Haastert, 1984; Janssens & Van Haastert, 1987; Snaar-Jagalska & Van Haastert, 1990); this issue is discussed further below in section 5.9. $Ca^{2+}$ could play a role in the control of adenylate cyclase and could contribute to the termination of a cAMP pulse by inhibiting the enzyme; this as yet unverified conjecture is at the basis of an alternative model proposed for cAMP relay and oscillations in *D. discoideum* (Othmer, Monk & Rapp, 1985; Rapp, Monk & Othmer, 1985; Monk & Othmer, 1989). The latter model, however, does not take into account the phenomenon of desensitization of the cAMP receptor, which plays an essential role here.

A certain number of experimental observations suggest that besides cAMP oscillations, other periodic phenomena could be at work during aggregation. Thus, oscillations of light scattering in *D. discoideum* suspensions have at first a spike shape before becoming more sinusoidal (Gerisch & Hess, 1974; Wurster, 1988). A mutant has even been isolated that seems to present spike-shaped oscillations in the absence of any detectable variation in cAMP (Wurster & Mohn, 1987). Furthermore, V. Nanjundiah (personal communication) has shown that the repeated dilution of amoebae in suspensions does not suppress oscillatory behaviour and influences only slightly the period of the phenomenon, even if cell density is reduced by three orders of magnitude (we cannot exclude, however, the presence of small aggregates in the suspension, within which cell density would be higher).

These observations suggest the possible existence of a second oscillator that would be intracellular and would not depend on activation by external cAMP while retaining the capability of being phase shifted by it (Nanjundiah & Wurster, 1989). Models based on such an activation cease to oscillate when the cell density is reduced by more than two orders of magnitude. The possibility of a second intracellular oscillator therefore exists; the latter could be based on the direct activation of adenylate cyclase by intracellular cAMP (Goldbeter & Segel, 1977). An attractive possibility is that the second oscillator involves the autonomous generation of spikes of intracellular $Ca^{2+}$ (Dupont &

Goldbeter, 1989). $Ca^{2+}$ oscillations, which often have a period of the order of minutes, have been observed in a number of different cell types; this oscillatory phenomenon is discussed in detail in chapter 10. A direct link between $Ca^{2+}$ oscillations and cAMP would exist because inositol 1,4,5-trisphosphate ($IP_3$), whose synthesis is elicited by the cAMP stimulus (Europe-Finner & Newell, 1987), controls the mechanism of $Ca^{2+}$ spiking (see chapter 10). Oscillations of extracellular $Ca^{2+}$ have been recorded in *D. discoideum* suspensions (Bumann, Malchow & Wurster, 1986; Wurster, 1988), but periodic variations of intracellular $Ca^{2+}$ have not yet been observed.

That the activation of adenylate cyclase by extracellular cAMP plays an essential role in the mechanism of cAMP oscillations is nevertheless demonstrated by the inhibition of periodic behaviour by adenosine, which interferes with cAMP binding to the receptor (Newell, 1982; Newell & Ross, 1982), and by caffeine (Theibert & Devreotes, 1983; Brenner & Thoms, 1984; Kimmel, 1987), which uncouples the receptor from adenylate cyclase.

Despite its necessarily incomplete nature, it appears that the model analysed above for the synthesis of cAMP in *D. discoideum* amoebae is based on the most conspicuous properties of the signalling system, namely self-amplification due to the activation of adenylate cyclase that follows the binding of cAMP to the receptor, and desensitization of this receptor through cAMP-induced phosphorylation.

## 5.7 Adaptation to constant stimuli

As the model based on desensitization of the cAMP receptor accounts well for excitable and oscillatory behaviour in the synthesis of cAMP in *D. discoideum*, it is necessary to test its predictive power for the types of response observed in other experimental conditions. In particular, it remains to be shown whether the model can account for the response of cells subjected to constant stimulation by cAMP.

The instructive series of experiments carried out by Devreotes and coworkers (Devreotes & Steck, 1979; Dinauer *et al.*, 1980a,b,c; Theibert & Devreotes, 1983) has permitted the determination of the time evolution of cAMP synthesis in the absence of the positive feedback loop that is normally exerted by extracellular cAMP (compare fig. 5.37b and c). Although such unphysiological conditions do not occur during the life cycle of *D. discoideum*, they nevertheless yield highly useful information on the various factors that control the synthesis of cAMP in the

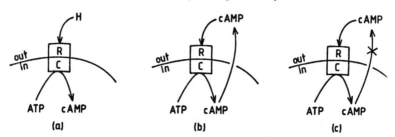

Fig. 5.37. Different types of activation of adenylate cyclase (C) elicited by the binding of an extracellular signal to the receptor (R). (a) Stimulation by a hormone H (a G-protein, not represented, participates in the activation of the enzyme); (b) autocatalytic control by cAMP in *D. discoideum*; (c) the positive feedback present in (b) is suppressed when the level of extracellular cAMP is held constant, so that the situation becomes equivalent to that considered in (a) (Goldbeter *et al.*, 1984).

course of the relay response. As already mentioned, this type of experiment has demonstrated, for example, the primary role played by receptor desensitization in the decrease of cAMP synthesis upon prolonged stimulation by the ligand.

Another interesting aspect of maintaining the level of extracellular cAMP constant (fig. 5.37c) is that the cAMP signalling system of *D. discoideum* then becomes analogous to traditional hormonal systems. In the latter (fig. 5.37a), indeed, a hormone (H) binds to a membrane receptor and, through the intermediary action of a G-protein (not represented in the scheme of the figure), activates adenylate cyclase. The cAMP thus synthesized is referred to as a **second messenger**, while the extracellular hormone represents the **first messenger**. The remarkable property of the slime mould *D. discoideum* is that in this system *cAMP plays the double role of first as well as second messenger* (Konijn, 1972). From this coincidence arises the positive feedback responsible for excitability and oscillations. Suppressing this autocatalytic regulation by holding extracellular cAMP at a constant level (fig. 5.37c) is tantamount to transforming the situation represented in fig. 5.37b into that schematized in part a. The results obtained for the cAMP-synthesizing system of *D. discoideum* in these conditions therefore provide insights for the study of proper hormonal systems, as discussed in chapter 8.

To determine the behaviour of the model in conditions where extracellular cAMP is maintained at a fixed value, it is necessary to treat variable $\gamma$ in system (5.12) as a parameter. This system then reduces to another two-variable system:

$$\frac{d\rho_T}{dt} = -f_1(\gamma)\,\rho_T + f_2(\gamma)\,(1 - \rho_T)$$

$$\frac{d\beta}{dt} = q\sigma\,\phi\,(\rho_T, \gamma, \alpha) - (k_i + k_t)\,\beta \qquad (5.16)$$

Figure 5.24a and b (see above) illustrate the main result of the experimental response to a step increase in extracellular cAMP: although the level of extracellular cAMP remains constant at a high value, the synthesis of intracellular cAMP is biphasic, as it decreases to its prestimulus level after passing through a maximum. This phenomenon of **adaptation** results from the phosphorylation of the receptor, accompanied by desensitization. The latter begins as soon as stimulation by cAMP is initiated (fig. 5.25), and the level of receptor phosphorylation established at steady state depends on the magnitude of the signal (fig. 5.26).

The simulations of system (5.16) are in agreement with these observations. In fig. 5.38, the response of the system to three distinct step

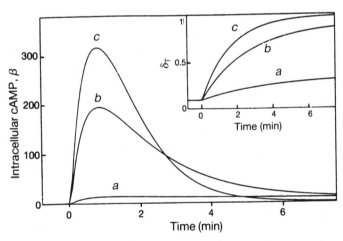

Fig. 5.38. Adaptation to constant stimuli. At time zero, the system is subjected to an increase in extracellular cAMP, $\gamma$, whose level rises from zero to 0.1 (curve a), 1 (curve b) and 10 (curve c). For $K_R = 10^{-7}$ M, these stimuli correspond to the passage from the cAMP level from zero to $10^{-8}$ M, $10^{-7}$ M, and $10^{-6}$ M, respectively. The synthesis of intracellular cAMP resulting from these stimulations is indicated; the responses are characterized by the phenomenon of adaptation. In inset, the evolution of fraction $\delta_T$ indicates that adaptation correlates with receptor modification. The magnitude of receptor desensitization increases with the intensity of stimulation. The curves are obtained by numerical integration of the two-variable system (5.16) for the parameter values given in table 5.3 (Martiel & Goldbeter, 1987a).

increases in γ, corresponding to three constant stimuli of increasing magnitude, is indicated, while the inset shows the accompanying variation in the fraction of phosphorylated receptor. The model accounts for the property of adaptation to constant stimuli and shows that the decrease in cAMP synthesis is brought about by an increase in the level of phosphorylated, desensitized receptor. The synthesis of cAMP returns to a minimum level when phosphorylation reaches a steady state. Furthermore, in agreement with the experiments, the level of receptor phosphorylated at the steady state progressively rises with the amplitude of the external signal.

As in the case of bacterial chemotaxis, where receptor methylation acts as a counterbalance to changed external conditions and thus allows the cells to adapt to constant stimulation (Koshland, 1980), phosphorylation of the cAMP receptor provides the counterbalance thanks to which *D. discoideum* amoebae adapt to constant cAMP stimuli.

Represented in fig. 5.39 is the dose–response curve giving the maximum of the peak of cAMP synthesis, $\beta_M$, as a function of the amplitude of stimulation. In contrast to the corresponding curve of fig. 5.32 obtained for system (5.12), this curve is not characterized by any threshold. The same is true for the curve yielding receptor saturation by cAMP. This observation is in agreement with the experimental results (Devreotes & Steck, 1979), which indicate the absence of threshold for

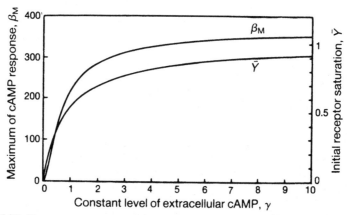

Fig. 5.39. Dose–response curve for constant cAMP stimuli. The maximum of the peak in intracellular cAMP obtained in fig. 5.38 is plotted as a function of the final, constant level of extracellular cAMP. The abrupt threshold that characterized the relay of pulsatile signals (fig. 5.32) has disappeared. The saturation function of the receptor, $\bar{Y}$, is also represented (Martiel & Goldbeter, 1987a).

relay of cAMP signals when the level of extracellular cAMP is maintained constant.

Self-amplification, suppressed in the latter conditions, is thus the process that is responsible for the existence of an abrupt threshold when the level of extracellular cAMP is allowed to vary freely. The fact that the response in fig. 5.39 does not display any threshold confirms that the latter owes its existence, in fig. 5.32, to the dynamic property of excitability linked to self-amplification, and not to the nonlinear coupling between the receptor and the cyclase, since the latter process is also at work in system (5.16).

The model also allows us to reproduce (fig. 5.40) the results of the experiments with caffeine (fig. 5.24). When system (5.16) is submitted to an elevation of the level of extracellular cAMP at time zero in the presence of an agent such as caffeine, which uncouples the receptor from the cyclase – this effect is obtained in the simulations by transiently setting the coupling constant $\epsilon$ equal to zero – no significant response

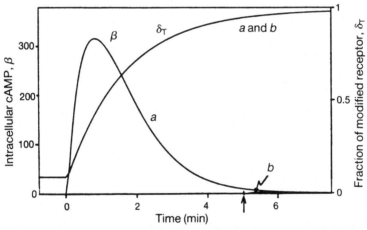

Fig. 5.40. Adaptation in the absence of activation of adenylate cyclase. In response to the stimulus $\gamma = 0 \rightarrow 10$, the system synthesizes a peak of intracellular cAMP (curve $a$); this curve is identical with curve $c$ in fig. 5.38. In curve $b$, the system is stimulated in an identical manner from the beginning, but the coupling between the receptor and adenylate cyclase has been suppressed for 5 min, by setting the parameter $\epsilon$ equal to zero. No response is observed when the coupling is re-established in the presence of stimulation. The curve for the fraction of desensitized receptor shows that adaptation to the stimulus has occurred despite uncoupling of the enzyme from the receptor. These results account for the experiments of fig. 5.24. The curves are obtained by integration of eqns (5.16) for the parameter values of fig. 5.38; fraction $\delta_T$ is obtained by taking the relation $\delta_T = 1 - \rho_T$ into account (Martiel & Goldbeter, 1987a).

is obtained when the coupling is restored in the presence of an identical stimulation (fig. 5.40, curve b). As indicated in fig. 5.40, the evolution of the total fraction of desensitized receptor, $\delta_T$, is the same in the absence (curve b) as in the presence (curve a) of the coupling between the receptor and the enzyme. Adaptation depends only on the degree of phosphorylation of the receptor; preincubation of the latter with a certain amount of extracellular cAMP necessarily leads to adaptation to that signal and, hence, to a lack of responsiveness when an identical signal is applied once the coupling to the enzyme is restored.

Yet another type of experiment allows the comparison of the theoretical predictions of the model with the experiments. Devreotes & Steck (1979) showed that adaptation to constant stimuli can be avoided when subjecting the amoebae to a sequence of successive step increases in stimulus such that the level of extracellular cAMP is doubled every 90 s, starting from an initial value of $10^{-12}$ M up to a final value of $10^{-5}$ M. The cumulated level of cAMP synthesized by the cells then rises until it reaches a plateau (fig. 5.41b). Simulations of eqns (5.16) in similar conditions furnish comparable results (fig. 5.41a) when taking for the parameters the values yielding good agreement with the experiments on cAMP oscillations. In the theoretical curve, the cumulated concentration of cAMP reaches half of its final value when the level of extracellular cAMP is close to $8 \times 10^{-8}$ M, 27 min after the beginning of stimulation, starting from a cAMP concentration of $10^{-12}$ M. The corresponding values in the experimental curve are, respectively, $10^{-7}$ M and 25 min. A significant difference is, however, that the theoretical curve is sharper than the experimental one, which extends over a wider range of stimuli. This indicates that the model is not as sensitive as it should be to stimuli of low magnitude, below $10^{-9}$ M (see also section 5.8).

### 5.8  Further extension of the model: incorporation of G-proteins into the mechanism of signal transduction

While the simulations of figs. 5.40 and 5.41 show reasonable agreement with experimental observations, there is one set of experiments for which the model fails to yield satisfactory results. These experiments, carried out by Devreotes & Steck (1979), pertain to the effect of increasing the level of extracellular cAMP from zero up to $10^{-6}$ M in four successive phases lasting 225 s each. In response to such sequential increments, the cells synthesize and excrete cAMP in four well-separated peaks of varying magnitude (fig. 5.42a). In these conditions, numeri-

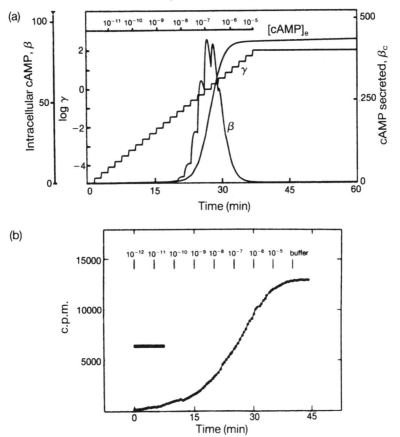

Fig. 5.41. Continuous activation of adenylate cyclase by successive doublings of the constant cAMP stimuli. (a) The level of extracellular cAMP is increased from $1.193 \times 10^{-12}$ M up to $10^{-5}$ M by 25 successive doublings; each level of stimulation is maintained for 90 s, as in the experiments of Devreotes & Steck (1979). The time evolution of intracellular cAMP ($\beta$) is represented, as well as that of the cumulated quantity of cAMP synthesized, $\beta_c$, and the level of stimulation ($\gamma$). The corresponding scale for extracellular cAMP is indicated, for the value $K_R = 10^{-7}$ M. The curves are obtained by integration of eqns (5.16) for the parameter values of fig. 5.38 (Martiel & Goldbeter, 1987a). (b) Experimental results of Devreotes & Steck (1979) to which the simulations relate. The curve represents the accumulated amount of cAMP synthesized by the cells; vertical bars indicate the level of extracellular cAMP. The heavy bar indicates the total duration (~ 8 min) of the response to a single, instantaneous $0–10^{-5}$ M step.

cal simulations of the model based on receptor desensitization fail to yield well-separated peaks: as previously noted by Monk & Othmer (1990), the synthesis of cAMP begins only at the second increment from $10^{-9}$ M to $10^{-8}$ M, and the three subsequent responses remain glued to

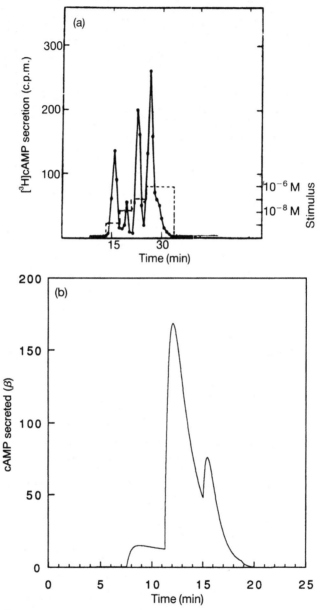

Fig. 5.42. Response to four successive increments in extracellular cAMP. The level of extracellular cAMP is brought successively from zero to $10^{-9}$ M, $10^{-8}$ M, $10^{-7}$ M, and $10^{-6}$ M in four steps lasting 225 s each. (a) Experimental results (first part of Fig. 8 of Devreotes & Steck, 1979) (b) Theoretical predictions from the model based on receptor desensitization; the curve is obtained by integration of eqns (5.16) for the parameter values of fig. 5.38.

each other (fig. 5.42b), in contrast to what is observed in the experiments.

Thus, the model fails on two accounts with regard to this particular experiment, which, even if it corresponds to nonphysiological conditions, is nevertheless of great interest as it throws light on the signalling mechanism and sets constraints on any theoretical models. First, the sensitivity at low levels of extracellular cAMP is not sufficient, as the system fails to respond to the first increase from zero to $10^{-9}$ M cAMP. Second, the decrease in cAMP synthesis after an increment in stimulation is not sufficiently rapid; hence, the lack of clear separation between the successive responses, in contrast to what is seen in the experiments.

To account for the observations of Devreotes & Steck, it is therefore necessary to amend the model both with respect to enhanced sensitivity

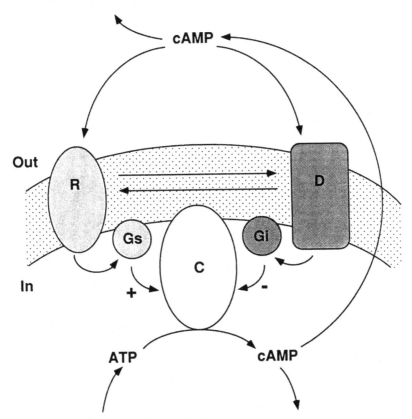

Fig. 5.43. Extended model for cAMP signalling in *D. discoideum*, incorporating receptor desensitization as well as G-proteins (see text) (Halloy, 1995; Halloy & Goldbeter, 1995).

and to clear-cut separation between successive responses. To that end, the model based on receptor desensitization has been extended to take into account explicitly the role of G-proteins in signal transduction. The model schematized in fig. 5.43 is based on the experimental results of Snaar-Jagalska & Van Haastert (1990); these results indicate that, as in other organisms (Gilman, 1984), cAMP binding to the active state of the receptor triggers the activation of adenylate cyclase via an activating protein, $G_s$, while cAMP binding to the desensitized state of the receptor contributes to the inactivation of adenylate cyclase via an inhibitory protein, $G_i$. Both $G_s$ and $G_i$ would be members of the family of GTP-binding proteins, also called G-proteins. These proteins play a ubiquitous role in signal transduction (Gilman, 1987; Birnbaumer, Abramowitz & Brown, 1990); they become activated through the binding of GTP following stimulation by a ligand–receptor complex, and thereafter inactivate, following hydrolysis of GTP into GDP by the

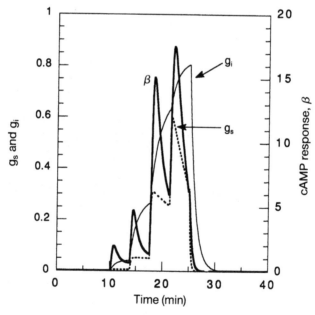

Fig. 5.44. Typical response generated by the model incorporating G-proteins and receptor desensitization, when the cAMP signalling system is subjected to four successive increments in extracellular cAMP from $10^{-9}$ M to $10^{-6}$ M; these conditions match the experimental conditions of fig. 5.42a. Shown is the time course of intracellular cAMP ($\beta$), the fraction of active $G_s$ protein (dashed line), and the fraction of active $G_i$ protein (thin solid line) (Halloy, 1995; Halloy & Goldbeter, 1995).

GTPase activity carried by G-proteins (Bourne, Sanders & McCormick, 1990).

When incorporating the role of $G_s$ and $G_i$ into the model based on receptor desensitization, the two variables of system (5.16) are supplemented by two additional variables, namely, the fraction of active protein $G_s$ and the fraction of active $G_i$, denoted by $g_s$ and $g_i$. The dynamics of the signalling system is then governed by a set of four nonlinear differential equations (Halloy, 1995; Halloy & Goldbeter, 1995) whose response to the experimental protocol of fig. 5.42a is shown in fig. 5.44. In agreement with experimental observations, the system already responds to the first increase in extracellular cAMP, from zero up to $10^{-9}$ M; moreover, the subsequent responses are well separated. As shown in fig. 5.44, the latter property is clearly due to the increase in the level of the inhibitory protein $G_i$, which accompanies increased stimulation. The analysis of the model indicates that of crucial importance for the right time course in cAMP synthesis are the relative rates of inactivation of $G_s$ and $G_i$ through GTP hydrolysis. A related model for adaptation to constant stimuli, incorporating G-proteins, has also been studied by Tang & Othmer (1994a).

When the positive feedback loop is restored, i.e. when extracellular cAMP becomes a variable – as in physiological conditions – the signalling system is governed by a set of five differential equations. Shown in fig. 5.45 is a typical example of sustained oscillations generated by such an extended model. In addition to extracellular cAMP, the figure shows the periodic variation of the fractions of active $G_s$ and $G_i$ proteins in the course of time. Here again oscillations occur only when some step in the pathway for activation of adenylate cyclase by extracellular cAMP possesses a nonlinear nature. Such nonlinearity could occur at the level of the receptor, e.g. in the form of microaggregation – as observed, for example, in the case of the GnRH receptor (Conn *et al.*, 1982) – that would lead here to the activation of G-proteins, or at the level of the coupling between G-proteins and adenylate cyclase.

Incorporation of the role of G-proteins in transduction of the cAMP signal therefore allows us to account both for the results of experiments on oscillations and on adaptation to constant stimuli. With the exception mentioned above (see fig. 5.42), most experimental results can, however, be accounted for in a satisfactory manner when the role of G-proteins is not explicitly incorporated into the model for cAMP signalling. The model based on receptor desensitization thus permits us to unify the different types of dynamic behaviour observed under various

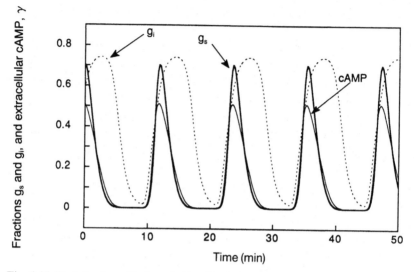

Fig. 5.45. Sustained oscillations in the five-variable model for cAMP signalling in *D. discoideum*, incorporating receptor desensitization as well as the two types of G-protein (Halloy, 1995; Halloy & Goldbeter, 1995).

experimental conditions. When the external signal is maintained constant, the system possesses the capability of adapting to such stimuli. When the self-amplification loop prevails, as in the natural conditions encountered during the life cycle, in the course of the aggregation that follows starvation, the cAMP signalling system acquires the capability of oscillating in a spontaneous manner or of relaying the cAMP stimuli whose amplitude exceeds a threshold.

## 5.9 Link between oscillations and waves of cAMP

Closely related to cAMP oscillations and relay is the generation of cAMP waves in the course of *D. discoideum* aggregation on agar. The theoretical study of how spatial patterns arise in aggregating *Dictyostelium* cells dates back to the pioneering work of Keller & Segel (1970) who showed that the formation of distinct aggregation territories in homogeneous populations of amoebae can be seen as the transition to a spatial structure of the Turing type, as a result of instability with respect to diffusion. What was lacking from this description, however, was the spatiotemporal nature of aggregation in *D. discoideum*. The analysis nevertheless applies well to the aggregation of other slime mould species such as *Dictyostelium minutum*, which lacks the oscillatory and relay mechanism at the time of aggregation (Gerisch, 1968).

Subsequent studies aimed at incorporating the spatiotemporal aspects of *D. discoideum* aggregation. Such studies relied on one- (Parnas & Segel, 1978) and two-dimensional (Mackay, 1978) computer simulations in which an *ad hoc* mechanism was used to generate the periodic cAMP signal, which was then relayed by chemotactically responding cells. The two-dimensional simulations reproduced many features of wavelike aggregation including concentric (target) and spiral patterns. The transition from target patterns to streams of amoebae observed at later stages of aggregation was also accounted for theoretically in terms of an azimuthal instability (Nanjundiah, 1973); a similar study was performed more recently (Levine & Reynolds, 1991) using the model based on receptor desensitization discussed earlier in this chapter. Other recent studies were devoted to patterns observed at still later stages of differentiation. Thus, Siegert & Weijer (1992) have modelled the wavelike chemotactic movement of cells within the slug; one result of that study is the characterization of a transition from a scroll wave in the tip to a planar wave in the adjacent, anterior region of the slug. A recent extension of that study, based on the receptor desensitization model discussed here, shows that cAMP waves could direct cell movement and gene expression in various portions of the slug (Bretschneider, Siegert & Weijer, 1995). Present modelling efforts, also based on the oscillatory dynamics of the receptor desensitization model, aim at incorporating cell movement during the aggregation phase, in addition to the oscillatory cAMP dynamics (Höfer *et al.*, 1995; S. Panfilov, personal communication; see also Levine, 1994).

With respect to wavelike aggregation, previous two-dimensional studies have incorporated a detailed mechanism for the periodic generation and relay of cAMP signals, but only cAMP diffusion was taken into account while the slower chemotactic movement of cells was disregarded as a first approximation. Thus, incorporation of the desensitization model discussed above into a grid representing such two-dimensional fields of amoebae resulted in concentric and spiral waves (Tyson, 1989; Tyson & Murray, 1989; Tyson *et al.*, 1989), while similar results were obtained using an alternative signalling mechanism based on the putative inhibition of adenylate cyclase by $Ca^{2+}$ (Monk & Othmer, 1990).

Typical spatial patterns obtained by means of the model based on receptor desensitization are shown in fig. 5.46. In isotropic conditions, a

(a)

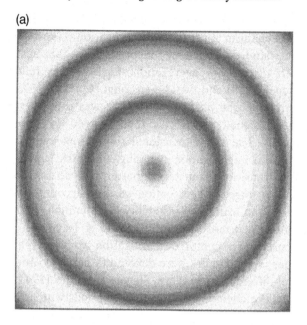

(b)

Fig. 5.46. Spatial patterns obtained in the three-variable model for cAMP signalling governed by eqns (5.12) supplemented with a term for the diffusion of
extracellular cAMP. (a) Concentric (target) patterns. (b) Spiral patterns. The
square area of 1 cm$^2$ is represented by a mesh of 160 × 160 points in (a) and
100 × 100 points in (b). In (a) a pacemaker, represented by four points at the

pacemaker cell placed at the centre of a field of excitable cells readily generates a concentric wave of cAMP (fig. 5.46a). Spiral waves can also be obtained (fig. 5.46b), but special, somewhat artificial initial conditions are needed (Tyson *et al.*, 1989), as is the case for most studies of spirals in excitable media (Winfree, 1991b; Mikhailov, Davydov & Zykov, 1994). Thus, a typical initial condition leading to spiral formation consists in cutting a concentric pattern in two halves and retaining only one of these; the remaining part readily generates spirals at its ends, as shown in fig. 5.46b. A similar procedure was previously used by others (Tyson & Murray, 1989; Tyson *et al.*, 1989; Monk & Othmer, 1990) to obtain spiral waves in the cAMP signalling system.

The question arises, however, as to the possibility of obtaining the spontaneous formation of spiral waves starting from more natural initial conditions; spiral waves are known, indeed, to occur in fields of aggregating amoebae without any artificial perturbation (see fig. 5.6). Some heterogeneity in initial conditions is of course needed to generate spiral waves in an excitable medium. Several types of initial heterogeneity have been tested, such as random distribution of cAMP concentration in the excitable medium, or random local changes in some parameter values such as adenylate cyclase or phosphodiesterase activity. Generally, however, target patterns always develop in these conditions instead of spiral patterns. The search for natural initial conditions leading to spiral waves of cAMP indicates (J. Halloy, J. Pontès & A. Goldbeter, unpublished data) that orthogonal gradients of appropriate, reduced magnitude in the fraction of active receptor ($\rho_T$) and extracellular cAMP ($\gamma$) can give rise to the formation of a cAMP spiral. Another, more natural initial condition consists in initiating a planar wave from one extremity of the two-dimensional system and then

centre of the mesh, emits autonomous oscillations of cAMP as in fig. 5.30. Cells in other points of the mesh are excitable as they have a slightly different value of parameter $\sigma$, which is equal to 0.36 min$^{-1}$. Other parameter values are as in fig. 5.30 with $k_1 = 0.09$, $k_{-1} = 0.9$, $k_2 = 1.665$, $k_{-2} = 0.008325$ (in min$^{-1}$). The system is taken initially at the stable steady state for excitable cells, and on a point of the limit cycle for the central pacemaker cells. In (b), all cells are excitable, with $\sigma = 0.34$ min$^{-1}$; other parameter values are as in (a). To obtain the spiral wave, a planar wave is first produced by stimulating the system along one side. The resulting wave is then cut into two pieces of which only half is retained; initial (steady-state) conditions are then set in the remaining part of the system, and the spiral wave is seen to develop from the remaining portion of the initially planar wave (J. Halloy & A. Goldbeter, unpublished results). Results similar to those shown in (b) have been obtained by Tyson & Murray (1989) and, with another model for cAMP signalling, by Monk & Othmer (1990).

applying a pulse of cAMP that initiates a wave in the orthogonal direction; this procedure is based on that used (Davidenko *et al.*, 1990) to trigger spiral waves of electrical activity in cardiac tissue.

Spiral waves and streaming patterns have recently been obtained in a computer simulation study which incorporates both the receptor desensitization model for cAMP signalling and cell chemotactic movements (Höfer *et al.*, 1995). The results of that study indicate that the chemotactic response of the amoebae favours the spontaneous occurrence of spiral patterns in the course of *D. discoideum* aggregation.

### Appendix: Model for oscillations and relay of cAMP signals in *Dictyostelium* cells: reduction to a three-variable system†

The time evolution of the concentration of the various species appearing in the kinetic scheme (5.5) is governed by the differential equations (5.6):

$$\frac{d\rho}{dt} = k_1(-\rho + L_1\delta) + d_1(-\rho\gamma + x)$$

$$\frac{d\delta}{dt} = k_1(\rho - L_1\delta) + d_2(-\delta c\gamma + y)$$

$$\frac{dx}{dt} = k_2(-x + L_2 y) + d_1(\rho\gamma - x) + (2\mu/h)d_3(-\epsilon x^2\,\bar{c} + \bar{e})$$

$$\frac{d\bar{c}}{dt} = d_3(-\epsilon x^2\bar{c} + \bar{e}) + (d_5 + k_5)(-\bar{c}\alpha\theta + \bar{cs})$$

$$\frac{d\bar{e}}{dt} = d_3(\epsilon x^2\bar{c} - \bar{e}) + (d_4 + k_4)(-\bar{e}\alpha + \bar{es})$$

$$\frac{d\bar{es}}{dt} = (d_4 + k_4)(\bar{e}\alpha - \bar{es})$$

$$\frac{d\alpha}{dt} = v - k'\alpha - \sigma(\bar{es} + \lambda\bar{cs}) + \theta_E[(d_4 + k_4)$$

$$\times (-\bar{e}\alpha + \bar{es}) + (d_5 + k_5)(-\bar{c}\alpha\theta + \bar{cs})]$$

$$\frac{d\beta}{dt} = q\sigma(\bar{es} - \lambda\bar{cs}) - (k_i + k_t)\beta$$

$$\frac{d\gamma}{dt} = (k_t\beta/h) - k_e\gamma + \eta[d_1(-\rho\gamma + x)$$

$$+ d_2(-\delta c\gamma + y)] \tag{A1}$$

† This appendix is from Martiel & Goldbeter (1987a).

In these equations $\rho = R/R_T$, $\delta = D/R_T$, $x = RP/R_T$, $\bar{e} = E/E_T$, $\bar{es} = ES/E_T$, $c = C/R_T$, and $\bar{cs} = CS/R_T$, where $R_T$ and $E_T$ represent the total amount of receptor and of adenylate cyclase; $\beta$ and $\gamma$ denote respectively the concentrations of intracellular and extracellular cAMP divided by the dissociation constant $K_R = d_1/a_1$; $\alpha$ is the intracellular level of ATP normalized by the Michaelis constant $K_m = (d_4 + k_4)/a_4$. Moreover, $L_1 = k_{-1}/k_1$ is the equilibrium ratio of the states R and D in the absence of ligand, whereas $L_2 = k_{-2}/k_2$; $c = K_R/K_D$ where $K_D = d_2/a_2$ is the nonexclusive binding coefficient of extracellular cAMP for the two receptor states; $v = v_i/K_m$; $h$ is the dilution factor (see Table 5.3 for definition of other parameters).

In deriving eqns (A1), we made use of the two conservation relations for the receptor and for adenylate cyclase:

$$R_T = R + D + RP + DP + (2/h)(E + ES)$$

$$E_T = E + ES + C + CS \tag{A2}$$

The assumption that $R_T$ and $E_T$ remain constant holds in first approximation, given that the time scale for the variation of these parameters is much longer than the time scale for relay and oscillations. The conservation relations (A2) take into account the fact that the receptor and adenylate cyclase concentrations are defined with respect to the extracellular and intracellular volumes, respectively. Equations (A2) yield the following expressions for $y$ and $\bar{cs}$, which supplement eqns (A1):

$$y = 1 - \rho - \delta - x - (2\mu/h)(\bar{e} + \bar{es})$$

$$\bar{cs} = 1 - \bar{c} - \bar{e} - es \tag{A3}$$

In the limit of fast binding of extracellular cAMP to both forms of the receptor and fast association between RP and C, C and S, E and S, the following inequality on the rate constants for the reaction steps 1 holds:

$$(k_1, k_{-1}, k_2, k_{-2}, k_i, k_t, k_e, \sigma, k')$$

$$<<(a_1, d_1, a_2, d_2, a_3, d_3, a_4, d_4, a_5, d_5) \tag{A4}$$

As eqns (A1) contain both fast binding and slow modification terms, we introduce new variables to separate the nine differential eqns (A1) into two sets, one associated with the slower time scale governing the interconversion of the receptor forms, and another associated with the faster time scale governing the binding reactions.

Let us define $\rho_T$ and $\delta_T$ as the total fractions of the receptor in the active and inactive (modified) states, $\bar{Y}$ as the total fraction of the receptor forms bound to cAMP, $A$ as the total concentration of intracellular ATP (free plus bound to the two forms of adenylate cyclase), and $\Gamma$ as the normalized total concentration of extracellular cAMP (free plus bound to the two receptor states). These new variables are expressed as a function of the old ones by eqns (A5):

$$\rho_T = 1 - \delta_T = \rho + x + (2\mu/h)(\bar{e} + \bar{es})$$

$$\bar{Y} = x + y + (2\mu/h)(\bar{e} + \bar{es})$$

$$A = \alpha + \theta_E(\bar{cs} + \bar{es})$$

$$\Gamma = \gamma + \eta(1 - \rho - \delta) = \gamma + \eta\bar{Y} \tag{A5}$$

The evolution equations for these new variables can now be obtained by taking their time derivative from eqns (A5) and inserting into the resulting relations the relevant equations from eqns (A1). This procedure yields the new set of eqns (A6) in which the first four equations govern the evolution of slower variables, while the remaining equations relate to fast variables (system (A6) can be complemented by one of the three eqations for $\rho$, $\delta$, or $x$ in (A1) but these equations contain both fast and slow terms and will therefore not be used in the subsequent reduction):

$$\frac{d\rho_T}{dt} = k_1(-\rho + L_1\delta) + k_2(-x + L_2y)$$

$$\frac{dA}{dt} = v - k'\alpha - \sigma(\bar{es} + \lambda\bar{cs})$$

$$\frac{d\beta}{dt} = q\sigma(\bar{es} + \lambda\bar{cs}) - (k_i + k_t)\beta$$

$$\frac{d\Gamma}{dt} = (k_t/h)\beta - k_e\gamma$$

$$\frac{d\bar{Y}}{dt} = d_1(\rho\gamma - x) + d_2(\delta c\gamma - y)$$

$$\frac{d\bar{c}}{dt} = d_3(-\epsilon x^2 \bar{c} + \bar{e}) + (d_5 + k_5)(-\bar{c}\alpha\theta + \bar{cs})$$

$$\frac{d\bar{e}}{dt} = d_3(\epsilon x^2 \bar{c} - \bar{e}) + (d_4 + k_4)(-\bar{e}\alpha + \bar{es})$$

$$\frac{d\bar{es}}{dt} = (d_4 + k_4)(\bar{e}\alpha - \bar{es}) \tag{A6}$$

where $y$ and $\bar{c}$ are still given by eqns (A3).

We now require that, after an initial transient phase, the differential equations for the fastest variables $\bar{Y}$, $\bar{c}$, $\bar{e}$, and $\bar{es}$ reduce to algebraic equations corresponding to the quasi-steady-state hypothesis for these receptor and enzymic forms. This condition leads to:

$$d_1(\rho\gamma - x) + d_2(\delta c\gamma - y) = 0$$

$$\bar{e}\alpha - \bar{es} = 0$$

$$\bar{c}\alpha\theta - \bar{cs} = 0$$

$$\epsilon x^2 \bar{c} - \bar{e} = 0 \tag{A7}$$

In the first of these equations each of the two terms in parentheses vanishes since, when the quasi-steady-state regime holds, the set relations (A7) must remain independent of the actual values of $d_1$ and $d_2$. The five algebraic relations obtained from (A7), plus the four kinetic equations for the slow variables in (A6), correspond to the nine degrees of freedom of the initial system (A1).

Taking into account the conservation relations (A3) we obtain eqn (A8) for $\bar{c}$ as a function of $\alpha$ and $x$; similar relations are obtained for $\bar{e}$, $\bar{es}$, and $\bar{cs}$:

$$\bar{c} = [1 + \alpha\theta + \epsilon x^2(1 + \alpha)]^{-1} \tag{A8}$$

The evolution equations for the remaining slow variables can now be transformed according to relations (A7) and (A8), yielding the four-variable differential system:

$$\frac{d\rho_\Gamma}{dt} = k_1(-\rho + L_1\delta) + k_2(-\rho\gamma + L_2\delta c\gamma)$$

$$\frac{dA}{dt} = v - k'\alpha - \sigma\alpha(\lambda\theta + \epsilon\rho^2\gamma^2)/$$

$$[1 + \alpha\theta + \epsilon\rho^2\gamma^2(1 + \alpha)]$$

$$\frac{d\beta}{dt} = q\sigma\alpha(\lambda\theta + \epsilon\rho^2\gamma^2)/$$

$$[1 + \alpha\theta + \epsilon\rho^2\gamma^2(1 + \alpha)] - (k_i + k_t)\beta$$

$$\frac{d\Gamma}{dt} = (k_t/h)\beta - k_e\gamma \tag{A9}$$

To express in these equations the old variables in terms of the new ones, we use the definitions (A5) which take the form:

$$\rho_T = \rho(1 + \gamma) + (2\mu/h)[\epsilon\rho^2\gamma^2(1 + \alpha)]/$$

$$[1 + \alpha\theta + \epsilon\rho^2\gamma^2(1 + \alpha)]$$

$$A = \alpha + \theta_E[(\alpha\theta + \epsilon\rho^2\gamma^2)]/[1 + \alpha\theta + \epsilon\rho^2\gamma^2(1 + \alpha)]$$

$$\Gamma = \gamma + \eta(\rho\gamma + \delta c\gamma) + (2\mu/h)[\epsilon\rho^2\gamma^2(1 + \alpha)]/$$

$$[1 + \alpha\theta + \epsilon\rho^2\gamma^2(1 + \alpha)] \qquad (A10)$$

As experimentally observed (see Tables 5.2 and 5.3), the parameters $(\mu/h)$, $\theta_E$, and $\eta$ are much smaller than unity. Neglecting the terms multiplied by these factors in eqns (A10), we obtain the simpler expressions for the new variables $\rho_T$, $A$, and $\Gamma$ as a function of the original ones $\rho$, $\alpha$, and $\gamma$:

$$\rho_T = \rho(1 + \gamma); \; A = \alpha; \; \Gamma = \gamma \qquad (A11)$$

Inserting the expressions (A7) into (A11) and taking into account the conservation relations (A3), we get:

$$\rho = \rho_T/(1 + \gamma); \; \delta = (1 - \rho_T)/(1 + c\gamma) \qquad (A12)$$

The four-variable system (A9) takes the final form (A13):

$$\frac{d\rho_T}{dt} = -\rho_T[(k_1 + k_2\gamma)/(1 + \gamma)]$$

$$+ (1 - \rho_T)[(k_1L_1 + k_2L_2c\gamma)/(1 + c\gamma)]$$

$$\frac{d\alpha}{dt} = v - k'\alpha - \sigma\phi(\rho_T, \gamma, \alpha)$$

$$\frac{d\beta}{dt} = q\sigma\phi(\rho_T, \gamma, \alpha) - (k_i + k_t)\beta$$

$$\frac{d\gamma}{dt} = (k_t/h)\beta - k_e\gamma \qquad (A13)$$

with $\phi(\rho_T, \gamma, \alpha) = \alpha(\lambda\theta + \epsilon Y^2)/[1 + \alpha\theta + \epsilon Y^2(1 + \alpha)]$; $\quad Y = \rho_T\gamma/(1 + \gamma)$. These equations are identical with the four-variable system (5.9) analysed in the text. System (5.9a–d) further reduces to the three-variable system (5.12) when we consider that the ATP level ($\alpha$) does not vary in time.

Equations (A13) are valid in the limit of negligible ($\mu/h$), $\theta_E$, and $\eta$ (see eqns (A11)). The values of the two first parameters are generally smaller than that of the last, which may reach 200 (see table 5.3). It may therefore be of interest to determine the behaviour of the system when the cAMP bound to the receptor is taken into account (this amounts to keeping the term $\eta(\rho\gamma + \delta c\gamma)$ in eqns (A10)). In the limit ($\mu/h$) $\to 0$, $\theta_E \to 0$ and finite $\eta$ we obtain, instead of eqns (A11):

$$\rho_T = \rho(1 + \gamma); \quad A = \alpha; \quad \Gamma = \gamma + \eta(\rho\gamma + \delta c\gamma) \tag{A14}$$

The differential system (A13) takes the form

$$\frac{d\rho_T}{dt} = -\rho_T f_1(\gamma) + (1 - \rho_T)f_2(\gamma)$$

$$\frac{d\alpha}{dt} = v - k'\alpha - \sigma\phi(\rho_T, \gamma, \alpha)$$

$$\frac{d\beta}{dt} = q\sigma\phi(\rho_T, \gamma, \alpha) - (k_i + k_t)\beta$$

$$\frac{d\Gamma}{dt} = (k_t/h)\beta - k_e\gamma \tag{A15}$$

with $\phi(\rho_T, \gamma, \alpha) = \alpha(\lambda\theta + \epsilon Y^2)/[1 + \alpha\theta + \epsilon Y^2(1 + \alpha)]$; $Y = \rho_T\gamma/(1 + \gamma)$; $f_1(\gamma) = (k_1 + k_2\gamma)/(1 + \gamma)$; $f_2(\gamma) = (k_1 L_1 + k_2 L_2 c\gamma)/(1 + c\gamma)$; $\Gamma = \gamma + \eta[\rho_T\gamma/(1 + \gamma) + (1 - \rho_T)c\gamma/(1 + c\gamma)]$. $\gamma$ is given by the unique positive root of the third-degree polynomial:

$$-c\gamma^3 + (c\Gamma - 1 - c - \eta c)\gamma^2$$

$$+ [(1 + c)\Gamma - 1 - \eta c - \eta\rho_T(1 - c)]\gamma + \Gamma = 0 \tag{A16}$$

which must be solved in the course of the numerical integration of eqns (A15). When $K_R = 10^{-7}$ M, the value of $\eta$ is 0.3 (see table 5.3) and the approximation (A13) can be made without resorting to the more complicated expressions (A14) and (A16), as indicated by comparative integration of eqns (A13) and (A15).

# 6

# Complex oscillations and chaos in the cAMP signalling system of *Dictyostelium*

In a somewhat surprising manner, numerical simulations of the model for cAMP signalling in *D. discoideum* revealed the possibility of complex modes of oscillatory behaviour similar to those studied in chapter 4 in the model for a multiply regulated enzyme system with two autocatalytic reactions coupled in series.

While the presence of two instability-generating mechanisms gives rise, in the latter model, to two endogenous oscillatory processes whose interaction results in complex oscillatory phenomena, the fact that such phenomena also occur in the model for cAMP signalling is, at first, rather surprising. This model, indeed, contains but a single instability-generating mechanism based on the positive feedback exerted by extracellular cAMP on its own synthesis, through the binding of cAMP to the receptor and the ensuing activation of adenylate cyclase. Before addressing the origin of complex oscillatory behaviour, we shall first examine the different types of complex phenomena observed in the model for cAMP synthesis in *D. discoideum*.

## 6.1 Complex oscillations in a seven-variable model for cAMP signalling

The first indication of complex oscillations was that of the complex periodic behaviour of the 'bursting' type shown in fig. 6.1 (Martiel & Goldbeter, 1985). These oscillations resemble those obtained in the model for the multiply regulated enzyme system analysed in chapter 4.

These complex periodic oscillations were observed in the model for cAMP signalling in *D. discoideum* by numerical integration of the differential equations, before reduction of the number of variables. As indicated in chapter 5, excitable and oscillatory behaviour occur in this model as a result of self-amplification in cAMP synthesis. The nonlin-

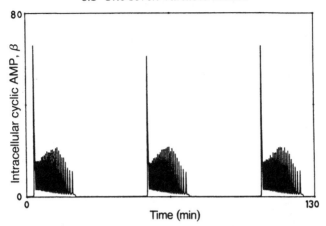

Fig. 6.1. Complex periodic oscillations (bursting) obtained in the model for cAMP synthesis in *D. discoideum* based on receptor desensitization. Only the evolution of intracellular cAMP ($\beta$) is represented. The curve is obtained by numerical integration of the seven-variable system (6.2) for the following parameter values: $\theta = \lambda = 0.01$; $\epsilon = 0.2$; $\eta = \mu = 0.1$; $h = 5$; $L_1 = 300$; $L_2 = L_1/c^2$ with $c = 100$; $k_1 = 0.15 \text{ s}^{-1}$, $k_2 = 0.06 \text{ s}^{-1}$; $\sigma = 0.1 \text{ s}^{-1}$; $k_e = 0.8 \text{ s}^{-1}$; $k_i = 0.6 \text{ s}^{-1}$; $k_t = 0.4 \text{ s}^{-1}$; $d_1 = d_2 = d_3 = 10 \text{ s}^{-1}$, $v = 7.5 \times 10^{-3} \text{ s}^{-1}$, $q = 4000$ and $K_R^{1/2} = 10^{-7} \text{ M}$ (Martiel & Goldbeter, 1985a).

earity of self-amplification can take different forms, depending on whether the step of activation of adenylate cyclase by the cAMP–receptor complex is bi- or trimolecular. Introduction of the explicit coupling of the receptor to G-proteins and of the latter to adenylate cyclase will further modify the form of the nonlinearity, as discussed at the end of chapter 5. Whereas simple periodic oscillations have also been obtained with these equations (see fig. 5.45), the study of complex oscillatory behaviour is currently being carried out in this extended model. The results reported below pertain to the model analysed in greatest detail in chapter 5, prior to the introduction of G-proteins.

In the model analysed in section 5.5, positive cooperativity originated from the activation of adenylate cyclase by two molecules of the cAMP–receptor complex. Similar results are obtained when this co-operativity is replaced by that associated with the binding of two molecules of cAMP to an allosteric receptor, while the activation of adenylate cyclase occurs through the coupling of the enzyme with a single molecule of the cAMP–receptor complex. The reaction scheme (5.5) is then replaced by the sequence of reaction steps:

$$R \underset{k_{-1}}{\overset{k_1}{\rightleftharpoons}} D,$$

$$R + 2P \underset{d_1}{\overset{a_1}{\rightleftharpoons}} RP_2; \quad D + 2P \underset{d_2}{\overset{a_2}{\rightleftharpoons}} DP_2,$$

$$RP_2 \underset{k_{-2}}{\overset{k_2}{\rightleftharpoons}} DP_2; \quad RP_2 + C \underset{d_3}{\overset{a_3}{\rightleftharpoons}} E,$$

$$E + S \underset{d_4}{\overset{a_4}{\rightleftharpoons}} ES \overset{k_4}{\longrightarrow} E + P_i,$$

$$C + S \underset{d_5}{\overset{a_5}{\rightleftharpoons}} CS \overset{k_5}{\longrightarrow} C + P_i,$$

$$P_i \overset{k_i}{\longrightarrow}; \quad P_i \overset{k_t}{\longrightarrow} P; \quad P \overset{k_e}{\longrightarrow} \tag{6.1}$$

The time evolution of the concentrations of the substrate ATP ($\alpha$), of intracellular ($\beta$) and extracellular ($\gamma$) cAMP, and of the different complexes formed by adenylate cyclase and by the cAMP receptor is then governed by a system of nine differential equations, as in the slightly different model studied in chapter 5. When a quasi-steady-state hypothesis is adopted for the enzyme–substrate complexes formed by adenylate cyclase in its free (C) and activated (E) states, the dynamics is described by the system of seven differential equations (6.2). In these equations, variables and parameters are defined as in eqns (5.6) (see table 5.3), but for dimensional reasons, $\beta$ and $\gamma$ represent the concentrations of intracellular and extracellular cAMP divided by $K_R^{1/2}$; moreover, $c = (K_R/K_D)^{1/2}$ and $\varphi = (1 + \alpha)^{-1}$ (Martiel & Goldbeter, 1984; Goldbeter, Decroly & Martiel, 1984).

$$\frac{d\rho}{dt} = k_1(-\rho + L_1\delta) + d_1(-\rho\gamma^2 + x)$$

$$\frac{d\delta}{dt} = k_1(\rho - L_1\delta) + d_2[1 - \rho - \delta(1 + c^2\gamma^2) - x - \mu(1 - \bar{c}(1 + \alpha\theta))]$$

$$\frac{dx}{dt} = k_2\{L_2[1 - \rho - \delta - \mu(1 - \bar{c}(1 + \alpha\theta))] - (1 + L_2)x\}$$
$$\quad + d_1(\rho\gamma^2 - x) + d_3\mu[\varphi - \bar{c}(\epsilon x + \varphi(1 + \alpha\theta))]$$

$$\frac{d\bar{c}}{dt} = d_3[\varphi - \bar{c}(\epsilon x + \varphi(1 + \alpha\theta))],$$

$$\frac{d\alpha}{dt} = v - \sigma\alpha\varphi[1 - \bar{c}(1 + \alpha\theta(1 - \lambda) - \lambda\theta)]$$

$$\frac{d\beta}{dt} = q\sigma\alpha\varphi[1 - \bar{c}(1 + \alpha\theta(1 - \lambda) - \lambda\theta)] - (k_i + k_t)\beta$$

$$\frac{d\gamma}{dt} = (k_t\beta/h) - k_e\gamma + 2\eta\{d_1(-\rho\gamma^2 + x)$$

$$+ d_2[1 - \rho - \delta(1 + c^2\gamma^2) - x - \mu(1 - \bar{c}(1 + \alpha\theta))]\} \qquad (6.2)$$

The complex periodic oscillations of fig. 6.1 were obtained by integration of eqns (6.2). The number of peaks of cAMP in the active phase of bursting depends on parameter values in the model. Thus, a variation in parameter $L_1$, which measures the relative rates of dephosphorylation and phosphorylation of the receptor in the free state (see chapter 5), controls the number of cAMP spikes over a period. This number can decrease progressively until a single peak subsists; the system then recovers its simple periodic behaviour.

## 6.2 Complex oscillatory phenomena in a three-variable model for cAMP signalling

The reduction of system (6.2) to only four variables can be performed as in the slightly different model analysed in the preceding chapter (details of this reduction are given in the Appendix to chapter 5; see also Martiel & Goldbeter, 1987a). As previously, the four variables retained in the reduced system are the total fraction of receptor in active state ($\rho_T$), the concentration of the substrate ATP ($\alpha$), as well as the concentrations of intracellular ($\beta$) and extracellular ($\gamma$) cAMP.

As in the case of the four-variable system (5.9) obtained when the source of cooperativity lies in the activation of adenylate cyclase rather than in the binding of cAMP to the receptor, the level of ATP varies only slightly in the course of oscillations. This observation allows us to reduce the number of variables by considering that the level of ATP remains fixed in the course of time. For system (5.9), this simplification leads to the three-variable system (5.12). Such an assumption is not retained here: indeed, as soon as the ATP concentration is held constant, all manifestations of complex oscillatory behaviour disappear in the model. The reasons for such a phenomenon are elucidated below.

In order to reduce the number of variables down to three, it is thus desirable to eliminate some variable other than ATP, if we wish to retain the possibility of complex oscillations. As indicated in chapter 5, a quasi-steady-state hypothesis for variable $\beta$, justified by the large value of parameter $q$, allows the transformation of the kinetic equation for $\beta$ into an algebraic relation. It is precisely such a reduction that led to the transformation of system (5.9) into the two-variable system (5.15), when the ATP level is maintained constant in the course of time. When the latter constraint is not imposed, system (6.2) reduces to the three-variable system (Goldbeter & Martiel, 1985; Martiel & Goldbeter, 1987b):

$$\frac{d\alpha}{dt} = v - \sigma \, \phi \, (\alpha, \rho_T, \gamma)$$

$$\frac{d\rho_T}{dt} = f_2(\gamma) - \rho_T \, [f_1(\gamma) + f_2(\gamma)]$$

$$\frac{d\gamma}{dt} = q' \, \sigma \, \phi \, (\alpha, \rho_T, \gamma) - k_e \, \gamma \qquad (6.3)$$

where:

$$q' = \frac{q k_t}{h \, (k_i + k_t)},$$

$$f_1 \, (\gamma) = \frac{k_1 + k_2 \, \gamma^2}{1 + \gamma^2}, \quad f_2 \, (\gamma) = \frac{k_1 \, L_1 + k_2 \, L_2 \, c^2 \, \gamma^2}{1 + c^2 \, \gamma^2},$$

and:

$$\phi \, (\gamma) = \frac{\alpha \left( \lambda \, \theta + \dfrac{\epsilon \, \rho_T \, \gamma^2}{1 + \gamma^2} \right)}{(1 + \alpha \, \theta) + \left( \dfrac{\epsilon \, \rho_T \, \gamma^2}{1 + \gamma^2} \right) (1 + \alpha)} \qquad (6.4)$$

### Bifurcation diagrams as a function of $v$ and $k_e$

The richness of behavioural modes in the model is best illustrated by means of bifurcation diagrams in parameter space (Li *et al.*, 1992a). The diagrams following are established for parameter values close to those available from the experimental literature (these values are listed in table 5.2). Shown in fig. 6.2 is such a diagram as a function of two key parameters, $v$ and $k_e$ (parameter $v$ is scaled as $v^* = v/\mu$, where $\mu$ is a small, fixed quantity). The main domains, separated by the Hopf bifurcation boundary (solid lines), are those where the system evolves toward a stable steady state (dotted regions), simple periodic behaviour (domain 1), or complex periodic oscillations of the bursting type (domain 2). The upper dotted line separating simple from complex

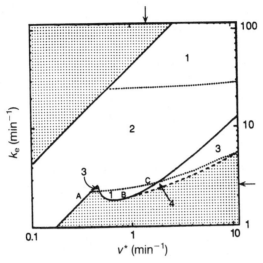

Fig. 6.2. Bifurcation diagram for the three-variable system (6.3) governing cAMP synthesis in a homogeneous cell population. The diagram is established as a function of extracellular phosphodiesterase activity ($k_e$) and of the scaled input rate of ATP, $v^* = v/\mu$. In the dotted domain the system admits a single stable steady state. The unique stable steady state coexists with simple periodic oscillations in domain 4 and with complex oscillations (periodic or chaotic) in domain 3. The dotted line between domains 1 and 2 indicates the instability of the simple periodic oscillations (see text and fig. 6.3); complex oscillations occur in domain 2. Point C ($k_e = 2.625$ min$^{-1}$, $v^* = 1.407$ min$^{-1}$) denotes the chaotic behaviour considered in the simulations of figs. 6.21 and 6.22. Points A and B correspond to numerical experiments on the mixing of two oscillatory populations, described in section 6.6. The upper vertical arrow refers to the value of $v^*$ ($= 1.4073825$ min$^{-1}$) chosen for the bifurcation diagram established as a function of $k_e$ in fig. 6.3. The horizontal arrow at the right-hand side represents the fixed value of $k_e$ ($= 2.625$ min$^{-1}$) used in obtaining the bifurcation diagram as a function of $v^*$ in fig. 6.5. The diagram has been established by means of linear stability analysis and numerical simulations; the lower dotted and dashed lines were obtained by means of AUTO (Doedel, 1981). Parameter values are $k_1 = 1.125$ min$^{-1}$, $k_2 = 0.45$ min$^{-1}$, $k_i = 4.5$ min$^{-1}$, $k_t = 3$ min$^{-1}$, $\sigma = 0.75$ min$^{-1}$, $L_1 = 316.2277$, $L_2 = 10^{-4} L_1$, $\theta = \lambda = 0.01$, $\epsilon = 0.2$, $c = 100$, $q = 4000$, $h = 5$, while the small parameter $\mu$ is fixed at the value $3.125 \times 10^{-3}$ (Li *et al.*, 1992a).

oscillations denotes the first period-adding bifurcation in which two peaks of cAMP ($\gamma$) per period are observed; further decrease in $k_e$ will lead to a progressive increase in the number of cAMP peaks during each period of bursting (see also fig. 6.3a). The lower dotted line in fig. 6.2 denotes the first period-doubling bifurcation, which usually precedes the evolution to chaos in this model. The dashed line at the bottom is the locus of limit points defining the boundary of the domain of hard excitation where a stable steady state coexists with oscillations that are either complex (domain 3) or simply periodic (domain 4). In the region of the kink on the lower Hopf bifurcation line, complex phenomena including birhythmicity occur (see legend to fig. 6.2). The occurrence of chaos has not been determined in full detail but should in principle appear in a region located above but very close to the lower dotted line, e.g. at point C, where chaos appears after a sequence of period-doubling bifurcations.

To refine the bifurcation structure of the model, it is useful to make a vertical section through the diagram of fig. 6.2 in the $v–k_e$ parameter space by fixing the substrate input rate $v$ at a particular value (vertical arrow at the top of fig. 6.2).

The 'behaviour spectrum' of a homogeneous population as a function of parameter $k_e$ (fig. 6.3) reveals a rich variety of dynamic behavioural modes of the cAMP signalling system. Starting from a low initial value of $k_e$ (fig. 6.3a), the system evolves toward a stable steady state, represented by the value of the extracellular cAMP concentration, $\gamma_0$. Around $k_e = 2.4$ min$^{-1}$ (fig. 6.3b), a subcritical Hopf bifurcation occurs beyond which the steady state becomes unstable (dashed line): in a range roughly extending from $k_e = 2.2$ to $2.4$ min$^{-1}$ in the conditions of fig. 6.3, the system thus admits a coexistence between a stable steady state and a stable limit cycle represented by the upper solid line showing the maximum cAMP level in the course of oscillations, $\gamma_M$; these two stable solutions are separated by an unstable limit cycle (dashed line).

Figure 6.3a indicates that the stable limit cycle undergoes a series of complex changes that are clarified in the successive enlargements in parts b and c. Thus, fig. 6.3c shows that the stable periodic solution transforms into chaos (black area) through a series of period-doubling bifurcations as $k_e$ increases. At larger values of $k_e$, the oscillations abruptly take the form of complex patterns of bursting (dotted domains in fig. 6.3b and c) in which phases of quiescence separate active phases characterized by a first, large amplitude peak followed by a series of smaller peaks (see fig. 6.3a). How these complex oscillations transform into chaos upon a decrease in $k_e$ has not been investigated in further detail.

As indicated in fig. 6.3a, further increase in $k_e$ beyond a value of between 4 and 5 min$^{-1}$ leads to a progressive decrease in the number of small peaks that follow the large-amplitude spike at the beginning of the bursting phase; moreover, the amplitude of successive peaks in that

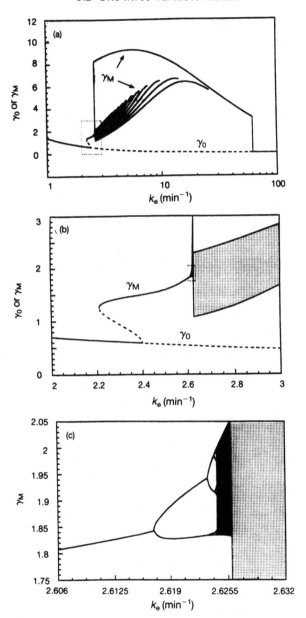

Fig. 6.3. Bifurcation diagram for eqns (6.3) showing the steady state ($\gamma_0$) or maximum value of the concentration of extracellular cAMP ($\gamma_M$) as a function of parameter $k_e$. (b) and (c) Successive enlargements of the area within the dotted box in (a). The diagram represents a vertical section through the $k_e$–$v^*$ bifurcation diagram of fig. 6.2, at the value of $v/\mu = 1.4073825$ min⁻¹ marked by the vertical arrow in that figure (Li *et al.*, 1992a).

phase progressively increases. The first periodic, bursting pattern recognizable in fig. 6.3a comprises 10 small spikes of increasing magnitude following a large peak. Bursting oscillations comprising more spikes were encountered in our numerical simulations; these oscillations are represented by the area that looks black as a result of the crowding of too many lines. Such a bursting pattern can be denoted by $\pi(1, n)$ where $n$ represents the number of small spikes. Therefore, as $k_e$ increases, $n$ progressively diminishes in a process where one spike is lost at a time. Clearly visible in the bifurcation diagram of fig. 6.3a are the transitions from $\pi(1, 7)$ to $\pi(1, 1)$. Eventually, limit cycle oscillations of the relaxation type, with a single peak and an extremely long period, are obtained. Beyond a second Hopf bifurcation point of the supercritical type, the steady state recovers its stability.

If we consider the sequence of bifurcations in the opposite direction, i.e. starting from a large initial value of $k_e$ close to 100 min$^{-1}$, fig. 6.3a indicates that a single limit cycle transforms into bursting and leads to chaos by a period-adding route. Oscillations with complex bursting patterns then merge with the aperiodic oscillations located to the right of the period-doubling cascade shown in fig. 6.3c.

Shown in fig. 6.4 are the different types of bursting obtained in the three-variable model. Besides the type already discussed with regard to fig. 6.3, and exemplified by the pattern $\pi(1, 8)$ in fig. 6.4c, we can also observe qualitatively different patterns of bursting such as those shown in fig. 6.4a and b. The latter two types of complex periodic oscillation occur in the vicinity of point B in fig. 6.2. The origin of such bursting patterns can be comprehended by resorting to a discussion of the dynamics of the fast subsystem $\rho_T$–$\gamma$ in which variable $\alpha$ is treated as a slowly varying parameter (see below).

With regard to the physiological situation, the domain of interest is that of chaos and of simple periodic oscillations located to the left of the region of chaos in fig. 6.3. Indeed, only there is the period of the order of minutes, as observed in the experiments, whereas the period increases up to hours at larger values of $k_e$ (see fig. 6.4). However, the domain of bursting is also considered in the following, for the sake of completeness, even if it probably represents unphysiological behaviour because of the very long periods with which it is associated. It should be noted, however, that doublet or triplet waves are sometimes observed in the course of aggregation on agar (Gottmann & Weijer, 1986); such waves could be related to the bursting properties of aggregation centres which

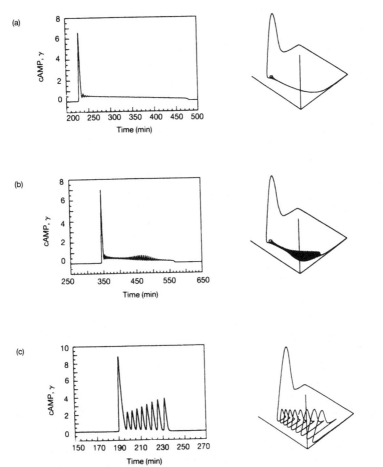

Fig. 6.4. Different patterns of bursting obtained in the model for cAMP signalling. The time course and corresponding phase space trajectories are obtained by numerical integration of eqns (6.3) by means of Gear's method (NAG subroutine D02EBF) for the parameter values of fig. 6.2; from top to bottom, the values of $k_e$ and $v^* = v/\mu$ are (in min$^{-1}$) (1.8, 2, 4) and (0.18, 0.2, 0.4) (Li *et al.*, 1992b).

would emit two or three pulses of cAMP followed by a quiescent phase over a period.

With respect to parameter $v$, a somewhat similar bifurcation diagram is obtained when fixing $k_e$. For the value $k_e = 2.625$ min$^{-1}$ in fig. 6.2, upon decreasing $v$ from a value corresponding to a stable steady state we successively observe the coexistence of such a state with a stable limit cycle

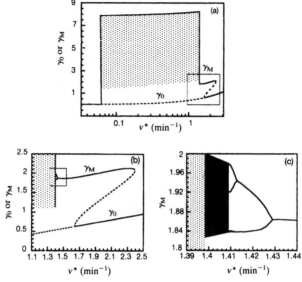

Fig. 6.5. Bifurcation diagram showing the steady state ($\gamma_0$) and maximum concentration of extracellular cAMP as a function of parameter value $v^* = v/\mu$. As in fig. 6.3, (b) and (c) represent successive enlargements of the boxed areas in (a) and (b), respectively. The black domain in (c) represents chaos. The diagram represents a horizontal section through the diagram of fig. 6.2, with $k_e = 2.625 \text{ min}^{-1}$ (Li *et al.*, 1992a).

and a sequence of period-doubling bifurcations leading to chaos (fig. 6.5a–c). At smaller values of $v$, complex patterns of bursting are observed. These patterns of bursting disappear abruptly at still lower values of $v^*$; the sharpness of this transition remains to be clarified.

### Birhythmicity

As indicated by the above bifurcation diagrams, the three-variable system (6.3) is capable of displaying different modes of simple or complex oscillatory behaviour. One additional mode is that of birhythmicity: for certain values of the parameters, eqns (6.3) admit a coexistence between two simultaneously stable periodic regimes. In the phase plane ($\rho_T$, $\alpha$, $\gamma$), these two regimes correspond to two limit cycles, one of which possesses a smaller amplitude and the second the folded appearance characteristic of bursting (fig. 6.6).

The passage from one type of oscillation to the other can be achieved by chemical perturbation. Thus, the addition of a sufficient amount of

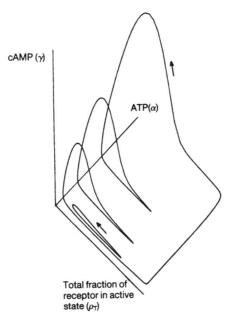

cAMP (γ)

ATP(α)

Total fraction of
receptor in active
state (ρ_T)

Fig. 6.6. Birhythmicity in the phase space of the model for cAMP signalling. The curves were obtained by integration of the three-variable system (6.3) for the following parameter values: $L_1 = 85.507$; $L_2 = 8.551 \times 10^{-3}$; $k_e = 0.35 \text{ s}^{-1}$; $v = 1.4125 \times 10^{-4} \text{ s}^{-1}$; other parameter values are as in fig. 6.1 (Goldbeter & Martiel, 1985).

extracellular cAMP at the appropriate phase induces the transition from the small to the large limit cycle, while a similar perturbation of the latter cycle allows the return to the small-amplitude periodic regime (fig. 6.7).

The comparative study of the sensitivity of the two oscillatory regimes towards perturbations indicates that the passage from the small to the large cycle is much easier than the reverse transition (Goldbeter & Martiel, 1985). Indeed, as for birhythmicity in the two-variable auto-catalytic model with product recycling into substrate (chapter 3), the passage from the small to the large limit cycle occurs as soon as the addition of product exceeds a threshold. In contrast, to return to the small-amplitude limit cycle, it is necessary that the quantity of product added lies between two critical values, given that the addition of too large an amount of product would bring the system across the attraction basin of the small cycle into the other side of the initial basin.

In fig. 6.7, the threshold corresponding to the first transition (first

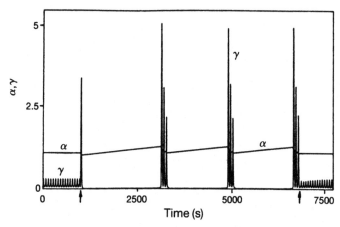

Fig. 6.7. Reversible transitions between two stable limit cycles in the model for cAMP signalling. Small-amplitude oscillations (fig. 6.6) are perturbed in $t = 1000$ s (when $\rho_T = 0.167$ and $\alpha = 1.113$) by an instantaneous increase from $\gamma = 0.318$ to 0.418. The system then switches to the large-amplitude limit cycle. The latter is perturbed in $t = 6800$ s (when $\rho_T = 0.774$ and $\alpha = 1.111$) by an increase in $\gamma$ from 0.048 to 0.058; the system then returns to the limit cycle of reduced amplitude. The curves are obtained for the situation of fig. 6.6, in the three-variable system (6.3) (Goldbeter & Martiel, 1985).

arrow) amounts to a 30% increase in cAMP. The passage from the large to the small limit cycle (second arrow) occurs only when the instantaneous concentration of cAMP is increased by 15–30% (Goldbeter & Martiel, 1985).

### 6.3 Analysis of bursting and birhythmicity in a two-variable system

The phenomena of bursting and birhythmicity can be analysed in system (6.3) by means of a reduction to two variables, based on the observation that the substrate ATP varies much more slowly than the other two variables (Martiel & Goldbeter, 1987b). This approach, analogous to that developed in chapter 4 for the analysis of bursting in the multiply regulated enzyme system, is inspired by a method proposed by Rinzel for the theoretical investigation of bursting in nerve or pancreatic β-cells (Rinzel & Lee, 1986; Rinzel, 1987).

The dynamics of the fast subsystem ($\rho_T$, $\gamma$), in which $\alpha$ is now a parameter, is governed by:

$$\frac{d\rho_T}{dt} = f_2(\gamma) - \rho_T \left[ f_1(\gamma) + f_2(\gamma) \right]$$

$$\frac{d\gamma}{dt} = \sigma^* \, \phi \, (\rho_T, \alpha, \gamma) - k_e \, \gamma \tag{6.5}$$

where functions $f_1(\gamma)$, $f_2(\gamma)$, and $\phi \, (\rho_T, \alpha, \gamma)$ remain defined by relations (6.4), while the normalized maximum rate of adenylate cyclase is given by:

$$\sigma^* = \frac{\sigma \, k_t \, q}{h \, (k_i + k_t)} \tag{6.6}$$

The situation to be analysed by means of this reduction differs slightly from that of fig. 6.6. While the small limit cycle was contained within the large cycle there, it is located outside the latter cycle in the case considered below. The two oscillatory regimes that coexist in the phase plane in fig. 6.8 are represented as a function of time in fig. 6.9. Here again the large cycle is of the 'bursting' type, with only two peaks per period.

The curves of figs. 6.8 and 6.9 are obtained by numerical integration of the three-variable system (6.3) for the same set of parameter values. The origin of this birhythmicity and of bursting is clarified when we determine the bifurcation diagram of the fast, two-variable system (6.5) as a function of the substrate concentration (Martiel & Goldbeter, 1987b). This bifurcation diagram is schematized in fig. 6.10. On the ordinate are shown the steady-state concentration of extracellular cAMP, $\gamma_0$, as well as its mean value $<\gamma>$ over a period of the oscillations. The abscissa gives the substrate concentration, $\alpha$, considered as control parameter.

The curve yielding $\gamma_0$ as a function of $\alpha$ shows the existence of a phenomenon of bistability in the reduced system when the substrate concentration is held constant. Three distinct values of $\gamma_0$ are obtained in fig. 6.10 in the range $\alpha''_1 < \alpha < \alpha''_2$. The linear stability analysis of eqns (6.5) reveals that the steady state on the lower branch of the hysteresis curve is always stable, while it is unstable on the median branch, between points $S_1$ and $S_2$. On the upper branch, the steady state is unstable in the domain $\alpha_1 < \alpha < \alpha_2$. Two families of periodic solutions, denoted $\Gamma_1$ and $\Gamma_2$, appear through a Hopf bifurcation at the points $H_1$ and $H_2$; they disappear at the points $H'_1$ and $H'_2$, of abscissae $\alpha'_1$, $\alpha'_2$, when the amplitude of the limit cycle is such that the latter reaches the

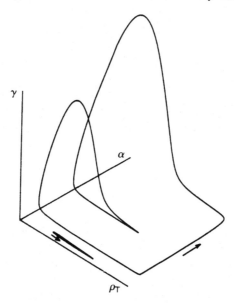

Fig. 6.8. Another case of coexistence between two simultaneously stable limit cycles in the three-variable system (6.3). Here, in contrast to the situation of fig. 6.6, the small-amplitude cycle is outside the large-amplitude one. Parameter values are those of fig. 6.6, except $L_2 = 8.5507 \times 10^{-3}$, $k_1 = 0.141$ s$^{-1}$, $k_2 = 0.0564$ s$^{-1}$ and $v = 3.15 \times 10^{-2}$ s$^{-1}$. Appropriate initial conditions for the small and large limit cycles are, respectively: $\alpha = 1.24$; $\rho_T = 0.951$; $\gamma = 1.84 \times 10^{-2}$; and $\alpha = 1.095$; $\rho_T = 0.308$; $\gamma = 0.118$ (Martiel & Goldbeter, 1987b).

unstable, median branch of the hysteresis curve, thus giving rise to two homoclinic orbits.

The dynamics of the full, three-variable system (6.3) can be comprehended in terms of the bifurcation diagram obtained for the reduced system, as soon as the substrate is considered as a slow variable rather than as a parameter whose value would remain fixed in the course of time. When the unique steady state admitted by the equations is unstable, the complete system moves along the lower branch of the hysteresis curve in fig. 6.10, starting from a low level of extracellular cAMP. Since adenylate cyclase then operates at a reduced activity, the net rate of substrate input (equal, here, to the difference between the rate of synthesis of ATP and the rate of its utilization in reactions other than that catalysed by adenylate cyclase) exceeds the rate of substrate consumption in the enzyme reaction; thus $v > \sigma \phi$ in the evolution equation of $\alpha$ in system (6.3). The substrate therefore slowly accumulates, and the sys-

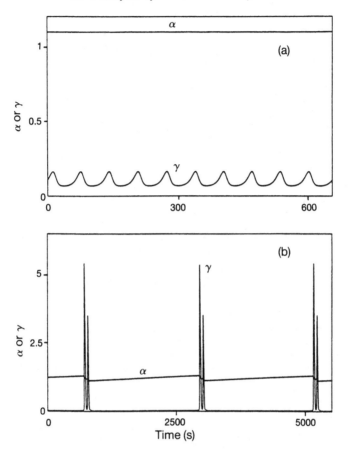

Fig. 6.9. Sustained oscillations corresponding to the two simultaneously stable limit cycles of fig. 6.8 (Martiel & Goldbeter, 1987b).

tem proceeds from the origin towards the point $S_2$, along the stable branch of the steady-state curve of the reduced system.

When the limit point $S_2$ is reached, an increase in $\alpha$ elicits the abrupt transition towards the upper branch of the hysteresis curve, given that the lower branch has now vanished. But the upper branch of the steady-state curve of the $(\rho_T, \gamma)$ system is unstable, and the analysis of the reduced system predicts that oscillations belonging to the branch of periodic solution $\Gamma_2$ should occur when $\alpha$ is close to the value $\alpha''_2$. These oscillations correspond to the active phase of bursting represented in fig. 6.9b.

Since the concentration of $\gamma$ on the upper branch of the steady-state

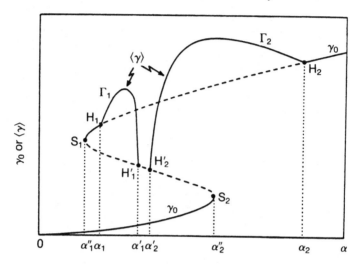

Fig. 6.10. Origin of bursting and birhythmicity. The bifurcation diagram for the two-variable system (6.5), established as a function of $\alpha$ considered as a parameter, shows the steady state $\gamma_0$ and the mean value $\langle\gamma\rangle$ of cAMP in the course of one period of the oscillations, on the two branches of periodic solutions $\Gamma_1$, $\Gamma_2$. The particular values of $\alpha$ indicated on the abscissa are defined in the text. For the sake of clarity, the diagram, obtained by means of the program AUTO (Doedel, 1981), is presented in a schematic manner: the precise values of the critical points are: $\alpha''_1 = 1.074$; $\alpha_1 = 1.084$; $\alpha'_1 = 1.101$; $\alpha'_2 = 1.126$; $\alpha''_2 = 1.287$; $\alpha_2 = 3.934$ (Martiel & Goldbeter, 1987b).

curve is sufficiently large to fully activate adenylate cyclase, each spike of cAMP consumes a certain quantity $\Delta\alpha$ of substrate; the movement of the system along the periodic branch $\Gamma_2$ therefore proceeds towards the left. This decrease in $\alpha$, accompanied by spikes in $\gamma$, continues until the system reaches a substrate concentration such that $\alpha \le \alpha'_2$: in that point, the stable periodic solution $\Gamma_2$ has disappeared, because of the existence of the homoclinic orbit, and the system abruptly returns to the only stable state accessible, on the lower branch of the steady-state hysteresis curve. There, the cycle of substrate accumulation from $\alpha'_2$ to $\alpha''_2$ resumes.

The preceding discussion accounts for the mechanism of large-amplitude oscillations of the 'bursting' type in figs. 6.7 and 6.9b. What then about the origin of the birhythmicity demonstrated in fig. 6.8? As in fig. 6.7, the passage to the small limit cycle occurs when an adequate amount of cAMP is added at the appropriate phase of the large-amplitude oscillations, namely just after the last peak of the active phase of bursting. The bifurcation diagram of the reduced system allows us to understand this result obtained from numerical simulations.

When the system is in the vicinity of point $H'_2$ in fig. 6.10, i.e. at the end of the active phase of bursting on the large-amplitude limit cycle, the addition of a quantity $\Delta \gamma$ of cAMP elicits the transient activation of adenylate cyclase and, subsequently, the consumption of a quantity $\Delta \alpha$ of substrate. If this quantity is larger than $(\alpha'_2 - \alpha'_1)$, the system will be captured by the periodic solution $\Gamma_1$ which exists in the interval $\alpha_1 < \alpha < \alpha'_1$.

In certain conditions, a corresponding periodic solution can appear in the complete system (6.3), for which $\alpha$ varies only weakly in that interval. This second periodic solution corresponds to the small limit cycle of fig. 6.7 and to the oscillations of fig. 6.9a. The reduced amplitude of oscillations in $\gamma$ is such that the small amount of substrate consumed by each spike of product is compensated for by the constant input of substrate. Thus, the level of $\alpha$ in the course of these oscillations does not fall below $\alpha_1$; in the opposite case, the small cycle would disappear and the oscillations on the large cycle would resume.

The bifurcation diagram of the reduced system also explains the necessity of finely tuning the perturbation required for switching from the large to the small limit cycle. Indeed, in $H'_2$, too small an addition of $\gamma$ will produce an insufficient decrease in substrate; the resulting value of $\alpha$ will remain larger than $\alpha'_1$, and bursting will resume. Similarly, too large an addition of $\gamma$ will result in such a high consumption of substrate that $\alpha$ will drop below the value $\alpha_1$, and the system, passing from $H_1$ to $S_1$, will fall back on to the stable branch of the hysteresis curve (as long as the three-variable system does not admit a stable steady state between $H_1$ and $S_1$), before moving again towards the limit point $S_2$, thus initiating a new cycle of large-amplitude oscillations.

The homoclinic orbits disappear in the reduced system $(\rho_T, \gamma)$ as soon as $H'_1$ coincides with $H'_2$ and the branches $\Gamma_1$ and $\Gamma_2$ merge. The analysis predicts that the number of spikes in the bursting regime should then abruptly rise, given that the range of variation of $\alpha$, suddenly enlarged, now extends from $\alpha''_1$ to $\alpha''_2$. At the same time, birhythmicity should disappear, or the pattern of birhythmicity could change from that shown in fig. 6.8 to one of the type represented in fig. 6.6.

The good agreement between the predictions based on the bifurcation diagram of the two-variable system and the dynamics of the three-variable system originates from the fact that the homoclinic orbits responsible here for birhythmicity occur for closely related parameter values in the two versions of the model. This is not always the case. Thus, in the situation of fig. 6.6, for example, the homoclinic orbits are present in the three-variable system (6.3) whereas they have disappeared in the reduced system (6.5) (Martiel & Goldbeter, 1987b). Nevertheless, the analysis of bursting and birhythmicity in terms of a fast, two-variable subsystem governed by the dynamics of the third variable treated as a slowly changing parameter allows us to throw light, in this model as in that analysed in chapter 4, on the origin of complex dynamics in the full, three-variable system.

### 6.4 Aperiodic oscillations of cAMP: chaos

As indicated by the bifurcation diagrams established as a function of parameters $v$ $(= \mu v^*)$ and $k_e$, chaos can occur in the three-variable model for cAMP signalling. The phenomenon is also seen in the seven-variable version of the model: for parameter values close to those producing bursting in fig. 6.1, the numerical integration of eqns (6.2) indeed shows the existence of aperiodic oscillations (fig. 6.11). The irregularity of these sustained oscillations shows up in the varying amplitude of the cAMP peaks, but mostly in the intervals between successive peaks. As

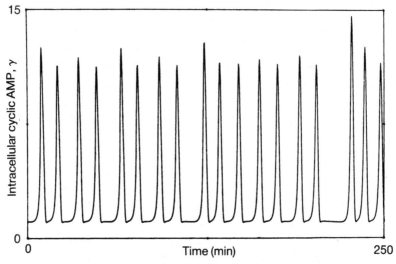

Fig. 6.11. Aperiodic oscillations (chaos) in cAMP synthesis predicted by the model based on receptor desensitization. The chaotic behaviour is obtained by numerical integration of the seven-variable system (6.2), for $v = 7.545 \times 10^{-4} \, s^{-1}$; other parameter values are those of fig. 6.1, divided by 10 for constants expressed in $s^{-1}$ (Martiel & Goldbeter, 1985a).

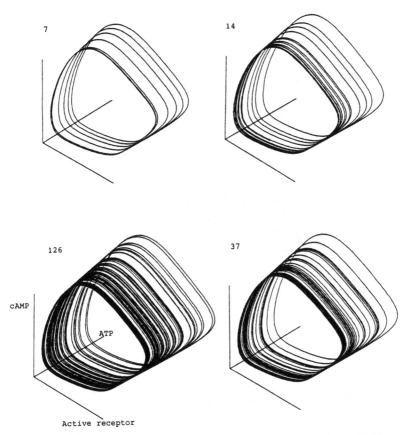

Fig. 6.12. Strange attractor corresponding to the aperiodic oscillations of fig. 6.11. The curve is obtained by projection of the trajectory of the seven-variable system (6.2) onto the space ($\rho$, $\alpha$, $\gamma$). The four figures show the successive states of the attractor after 7, 14, 37 and 126 cycles (J.L. Martiel & A. Goldbeter, unpublished results).

indicated below, several indices suggest that this aperiodic behaviour represents deterministic chaos (Martiel & Goldbeter, 1985; Goldbeter & Martiel, 1987).

In the phase plane ($\rho$, $\alpha$, $\beta$), the projection of the trajectory followed by the seven-variable system (6.2) takes the form of a strange attractor of which four successive states are shown in fig. 6.12. The system remains confined within a portion of the phase space, but the curve it follows in the course of oscillations never passes twice through any given point. This is one way by which this behaviour differs from complex periodic oscillations. Some cycles on the strange attractor are

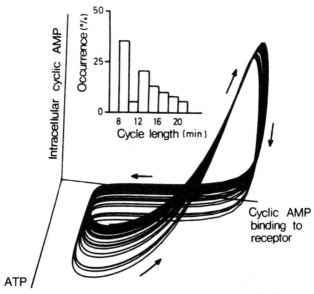

Fig. 6.13. Strange attractor corresponding to the chaotic behaviour of fig. 6.11. The curve is a projection in the space $(\alpha, \beta, Y)$ where $Y$ represents the total fraction of receptor bound to extracellular cAMP. The range of variation of ATP, intracellular cAMP and fraction $Y$ extends from 0 to 15, 1.545 to 1.585, and 0.15 to 0.9, respectively. The window shows a histogram of cycle lengths for 500 successive cycles on the strange attractor. The time interval between two successive peaks of cAMP is measured, and the percentage of cycle lengths within 2 min intervals is plotted (Martiel & Goldbeter, 1985a).

longer than others; the longer cycles correspond to a larger decrease in the level of substrate, and to the synthesis of a cAMP spike of greater magnitude, as is also confirmed by examination of fig. 6.11.

Shown in fig. 6.13 is another projection of the strange attractor in the $(\alpha, \beta, Y)$ space, where $Y$ represents the function of receptor saturation by cAMP. The inset of the figure portrays a histogram of cycle lengths for some 500 successive cycles on the attractor. While nearly 35% of the cycles have a length of between 8 and 10 min, fewer than 10% have a length of between 10 and 12 min; the rest of cycle lengths range from 12 to 22 min. This rather broad histogram differs from that obtained for periodic behaviour, which, by definition, is centred on a unique value of the interval between two cAMP spikes.

When the maximum of a peak of intracellular cAMP in fig. 6.11 is plotted as a function of the maximum of the preceding peak, we obtain a return map yielding $\beta_{n+1}$ as a function of $\beta_n$. This one-dimensional

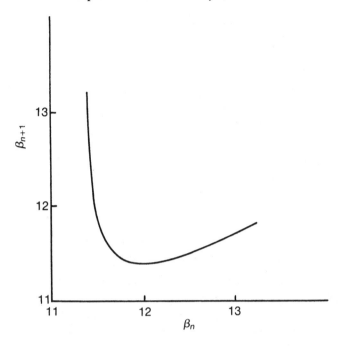

Fig. 6.14. Return map for the aperiodic behaviour of fig. 6.11. The maximum of the peak in intracellular cAMP is plotted as a function of the maximum of the preceding peak. The continuous nature of the curve is an indication of the chaotic nature of the system's evolution (Goldbeter & Martiel, 1987).

map, shown in fig. 6.14, is continuous, which property confirms that these oscillations represent deterministic chaos.

Another indication of the occurrence of chaos is given by the route leading to this irregular oscillatory behaviour in parameter space. Aperiodic oscillations indeed arise in the model after a cascade of period-doubling bifurcations as a function of parameter $v$, which measures the net rate of ATP supply to the adenylate cyclase reaction site. The successive values of parameter $v$ corresponding to these bifurcations obey the universal scheme described by Feigenbaum (1978) for the onset of chaos. Thus, the values of $v$ (in $s^{-1}$) associated with the first three period doublings, from period 1 to period 8, are as follows:

$$v_1 = 7.804 \times 10^{-4}$$

$$v_2 = 7.630 \times 10^{-4}$$

$$v_3 = 7.594 \times 10^{-4} \tag{6.7}$$

These values yield the fraction:

$$\frac{v_1 - v_2}{v_2 - v_3} = 4.83 \qquad (6.8)$$

which is close to the value of 4.669... found by Feigenbaum (1978) for the universal constant characterizing the cascade of period-doubling bifurcations leading to chaos.

Whereas the above data were obtained in the seven-variable version of the model, chaos has also been found in the reduced version (6.3) containing only three variables, namely $\rho_T$, $\alpha$ and $\gamma$. More details about these results are given at the end of this chapter.

### 6.5 The aperiodic aggregation of the *Dictyostelium* mutant *Fr 17*: an example of autonomous chaos at the cellular level?

Experimentally, the mechanism of intercellular communication by cAMP pulses in the course of *D. discoideum* aggregation is characterized by its periodicity. The latter is reflected by the wavelike movement of amoebae towards the aggregation centres, as a result of the periodic pulses of cAMP that the latter emit at regular intervals (Durston, 1974a). The periodic behaviour of the model based on receptor desensitization accounts for the periodic secretion of cAMP by aggregation centres, whereas excitable behaviour accounts for the relay of cAMP pulses by cells that amplify the suprathreshold signals emitted by the centres.

While relay and periodic oscillations are experimental properties accounted for by the model, the theoretical prediction of complex oscillatory behaviour raises the question of the occurrence of such phenomena in *Dictyostelium* amoebae. There is as yet no experimental evidence for birhythmicity; this phenomenon in fact remains to be established in biology, although recent results of Markus & Hess (1990) suggest its possible occurrence in glycolysis (see section 2.10). Birhythmicity has been reported to occur in chemical systems (Alamgir & Epstein, 1983; Lamba & Hudson, 1985; Citri & Epstein, 1988) following its prediction in the multiply regulated biochemical model analysed in chapter 4 (Decroly & Goldbeter, 1982). As regards bursting, a mutant studied by Gottmann & Weijer (1986) could provide an example of such complex periodic oscillations. This mutant indeed aggregates around centres by forming doublet or triplet waves, i.e. periodically spaced groups of two or three bands of chemotactically moving amoebae. This phenomenon could be interpreted in terms of bursting: centres would emit two or three spikes of cAMP separated from another group of spikes by a silent phase, as in the case illustrated in figs. 6.7 or 6.9b.

The behaviour of another mutant studied by Durston (1974b) could well be interpreted in terms of chaos (Martiel & Goldbeter, 1985; Goldbeter & Martiel, 1987). In his experiments on the wavelike aggregation of amoebae on agar, Durston measured the time intervals separating the successive waves of cells moving towards the aggregation centres, in the wild type *NC-4* as well as in two mutants. The histogram of intervals is relatively narrow in the wild type, where it is centred around a value close to 7 min, despite the fact that the data were collected from a large number of distinct experiments and concerned some 300 waves (first column in fig. 6.15). In the mutant *Fr17*, in contrast, the histogram of intervals is much broader and extends from 4 to 28 min (second column, fig. 6.15). In a given aggregation territory, the waves follow each other in *Fr17* at irregular intervals; for this reason Durston (1974b) suggested that this mutant be called **aperiodic**.

Further study of cell suspensions of the thermosensitive mutant *HH201* derived from *Fr17* revealed (Coukell & Chan, 1980) the existence of 'erratic' oscillations of intracellular cAMP (fig. 6.16a). This preliminary result (cAMP was assayed every minute, during four successive cycles only) and that obtained by Durston under other experimental conditions suggest that the irregular oscillations observed in these mutants might represent chaotic behaviour, of the type predicted by the model for cAMP signalling (fig. 6.11). While the synthesis of cAMP in the wild type would proceed in a periodic manner, cAMP synthesis in the mutants *Fr17* and *HH201* would possess a chaotic nature as a result of the variation in a control parameter consecutive to some mutation.

The mutations of *Fr17* and *HH201* affect the metabolism of cAMP (Kessin, 1977), and give rise to the precocious appearance of adenylate cyclase (fig. 6.16b) in the course of development after starvation (Coukell & Chan, 1980). In the model, accordingly, chaotic behaviour occurs for relatively large values of parameter $\sigma$, which measures the maximum activity of adenylate cyclase (Martiel & Goldbeter, 1985).

If the behaviour of the mutants *Fr17* and *HH201* did represent aperiodic oscillations, it would provide the first example of autonomous chaos at the cellular level, as well as an example of **dynamic disease** (Mackey & Glass, 1977) in a unicellular organism. While the transition to chaos in *Dictyostelium* would result from some genetic mutation, the

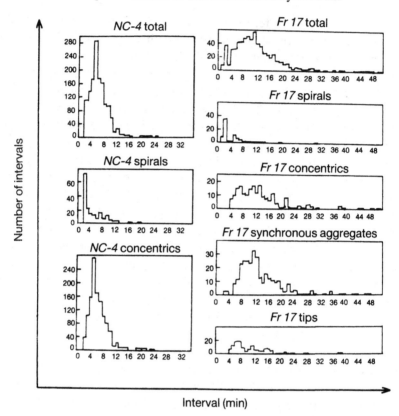

Fig. 6.15. Histogram of time intervals between successive waves in the course of aggregation on agar for the wild type *NC-4* and the aperiodic mutant *Fr17* of *D. discoideum*. The histograms are relatively narrow in the wild type, whose aggregation is periodic (the period of the waves, like that of cAMP oscillations, nevertheless changes in the course of development). In contrast, the histogram for *Fr17* is much wider, because of the 'aperiodic' aggregation of this mutant (Durston, 1974b).

addition of drugs has been reported to elicit the transition from periodic to chaotic oscillations in a molluscan neuron (Holden, Winlow & Haydon, 1982).

The model for cAMP signalling suggests that the addition of an appropriate amount of exogenous phosphodiesterase should transform chaos into periodic behaviour. This prediction could be tested during aggregation of the mutants *Fr17* or *HH201* on agar, in order to determine whether the broad histogram of intervals between successive waves becomes narrower in the presence of the enzyme, so as to

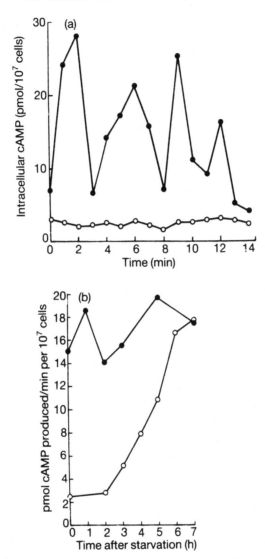

Fig. 6.16. (a) Irregular oscillations of intracellular cAMP in suspensions of the mutant *HH201* of *D. discoideum*, derived from the mutant *Fr17*. (b) Evolution of adenylate cyclase during the hours that follow starvation. Open and filled circles refer to the wild type and to the *HH201* mutant, respectively (Coukell & Chan, 1980).

become analogous to the histogram that characterizes the regular oscillatory behaviour of the wild type. Such experiments are, however, arduous; the study of the mutants in cell suspensions could provide a simpler approach that might help in tackling the issue.

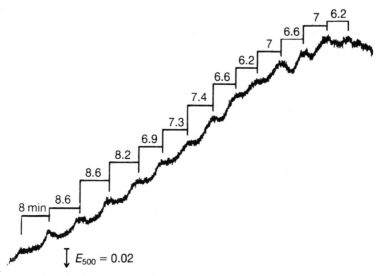

Fig. 6.17. Relatively regular oscillations of light scattering in suspensions of the mutant *HH201* of *D. discoideum*, derived from the putatively chaotic mutant *Fr17* (Goldbeter & Wurster, 1989).

In order to test in cell suspensions the nature of oscillations in the putatively chaotic mutants, Goldbeter & Wurster (1989) recorded light scattering in suspensions of the mutant *HH201* a few hours after starvation. Somewhat surprisingly, the oscillations observed in a series of experiments (fig. 6.17) do not differ significantly from those observed in the wild type. The interval between two successive peaks varies in a relatively narrow range as compared to the much broader spread observed during aggregation of the mutant *Fr17* (fig. 6.15). The drift of the interval that decreases from approximately 8 to 6 min along the 13 successive cycles is also observed in the wild type (Wurster, 1988). The behaviour in fig. 6.17 could thus be interpreted in terms of a periodic behaviour subjected to a slow variation in one or more control parameters that would affect the period of the oscillations. As may be seen in chapter 7, the fact that the biochemical parameters of the signalling system do evolve in the course of time supports this hypothesis.

The time series of fig. 6.17 is too short to allow for an unambiguous characterization of the oscillations in terms of chaos. Nevertheless, the variation of the interval between successive peaks remains reduced; this result differs from those pertaining to the 'aperiodic' (Durston, 1974b) or 'erratic' (Coukell & Chan, 1980) behaviour of the mutants *Fr17* and

HH201. Such a conclusion does not exclude the possibility of chaotic behaviour in these mutants, for at least two reasons. First, chaos generally occurs in a narrow range of parameter values, as indicated by the analysis of both the present model and that considered in chapter 4. Depending on growth conditions, the mutant could therefore evolve, after starvation, to a domain of periodic rather than chaotic behaviour, given the smallness of the domain of chaos in parameter space as compared with the much wider domain of periodic oscillations. This hypothesis is corroborated by the observation (Durston, 1974b) that aperiodic oscillations can become spontaneously periodic in the course of aggregation of the mutant *Fr17*. This phenomenon could be due to the exit from the chaotic domain and the concomitant entry into a domain of periodic behaviour, following variation of a control parameter such as the activity of adenylate cyclase or phosphodiesterase.

Another explanation of the rather regular nature of oscillations in fig. 6.17 relates to the behaviour of cells in continuously stirred suspensions. Observations of the wild type (Gerisch & Hess, 1974; see figs. 5.8 and 5.12) suggest that the amoebae, in these conditions, are tightly coupled through extracellular cAMP. Within the suspension, the amoebae oscillate synchronously because the faster entrain other cells that are either excitable or oscillate with a frequency close to that of the 'pacemakers'. If the cells oscillate in a chaotic manner, the question arises as to what will be the behaviour of the amoebae in the suspension. Chaos is indeed characterized by sensitivity to initial conditions, and two cycles on the same strange attractor within two distinct cells could have widely different durations, as shown by the histogram in fig. 6.13. Either the coupling is so strong that cells will display synchronous, chaotic oscillations, or different cells will independently release the cAMP signal in a chaotic manner; since other cells in the suspension could be excitable or capable of being phase shifted, the response of amoebae in these conditions could acquire a regularity that would be dictated by the value of the refractory period for relay.

Yet another possibility is that part of the cells within a continuously stirred suspension would oscillate in a chaotic manner if left on their own, while the remaining cells would oscillate periodically in the absence of the chaotic population. The coupling of the two populations within the same suspension could well suppress any manifestation of chaos by conferring upon the coupled system a global, regular behaviour.

### 6.6 Coupling chaotic and periodic behaviour in cell suspensions

The model developed for the cAMP signalling system can be used to test some of the hypotheses presented above for explaining the observation of regular oscillations instead of chaos in suspensions of the putatively chaotic mutant *HH201*. The basic question is to determine what happens upon coupling two populations of cells, one of which oscillates periodically and the other chaotically, or both oscillating chaotically on the same strange attractor but with different initial conditions. If chaos were to be destroyed by such a coupling, experiments in suspensions would lead only to regular oscillations and would therefore be less conclusive regarding the existence of the phenomenon than observations of aperiodic behaviour during slime mould aggregation on agar.

To examine the coupling of periodic cells with chaotic amoebae in a mixed suspension, let us consider the simplest case of mixing two populations, one periodic and the other chaotic (the results generalize to the mixing of a larger number of distinct populations). The two 'pure' suspensions are both described by the same group of three ordinary differential equations (6.3) but differ, in the simplest case, by the value of a single parameter. As a result, the two populations behave differently in the course of time. We shall first focus on the case where one population is chaotic and the other periodic. The mixing experiment is schematized in fig. 6.18, where cells from the two populations are represented in black and white, respectively.

The parameter that is chosen to differ for the two 'pure' populations is either the net rate of ATP supply within the cells ($v$), or the total activity of phosphodiesterase ($k_e$). The values of the two parameters for populations 1 and 2 will be denoted by $v_1$, $k_{e1}$ and $v_2$, $k_{e2}$, respectively.

The mixed suspension has the same volume and total number of cells as the starting suspensions containing the two homogeneous populations (see fig. 6.18 for a schematic representation of the mixing experiment). The fraction of cells from the purely chaotic and periodic populations in the final suspension are denoted by $F_1$ and $F_2$, respectively ($F_1 + F_2 = 1$). The dynamics of the mixed suspension is described by five kinetic equations given by eqns (6.9); these govern the evolution within each type of cell of the two intracellular variables, namely ATP, which serves as the substrate for cAMP synthesis, and the fraction of active cAMP receptor. The fifth equation governs the evolution of extracellular cAMP through which coupling occurs between the two

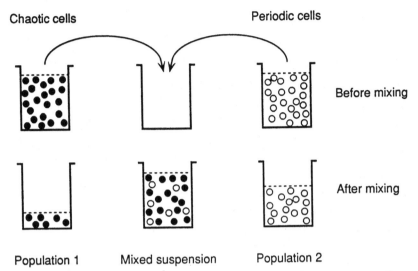

Population 1     Mixed suspension     Population 2

Fig. 6.18. Schematic representation of the mixing of two homogeneous cell suspensions exhibiting chaotic and periodic oscillations (Li *et al.*, 1992b).

populations present within the mixture; the production and degradation of extracellular cAMP depends on the fractions $F_1$ and $F_2$, which reflect the proportions of the two cell populations in the mixture:

$$\frac{d\alpha_j}{dt} = v_j - \sigma_j\, \phi_j\, (\alpha_j, \rho_{Tj}, \gamma)$$

$$\frac{d\rho_{Tj}}{dt} = f_{2j}(\gamma) - \rho_{Tj}\, [f_{1j}(\gamma) + f_{2j}(\gamma)] \qquad (j = 1, 2)$$

$$\frac{d\gamma}{dt} = F_1\, (q'_1\sigma_1\, \phi_1 - k_{e1}\, \gamma) + F_2\, (q'_2\sigma_2\, \phi_2 - k_{e2}\, \gamma) \qquad (6.9)$$

where parameter $q'$ and functions $f_1(\gamma)$, $f_2(\gamma)$ and $\phi(\gamma)$ remain defined by eqns (6.4) (Halloy *et al.*, 1990).

The global coupling in this study differs from the diffusive coupling that is generally adopted in studies of coupled oscillators (Landahl & Licko, 1973; Alamgir & Epstein, 1984; Bar-Eli, 1984; Aronson, Doedel & Othmer, 1987; Boukalouch *et al.*, 1987; Lengyel & Epstein, 1991). Here, the coupling results from the fact that cells secrete cAMP into the extracellular medium; this extracellular signal is shared by all cells in the suspension. Feedback of extracellular cAMP on the intracellular dynamics of each cell occurs through binding of cAMP to the cell surface receptor.

### *Period-doubling route to chaos as a function of the relative proportions of chaotic and periodic cells*

For simplicity, let us consider first that the two populations differ only by parameter $v_i$ which measures the net, constant input of ATP into the adenylate cyclase reaction site within the cells. To determine the effect of coupling chaotic with periodic behaviour, we consider the situation where population 1 has a value of $v_1$ corresponding to chaotic behaviour while the second population has a value of $v_2$ corresponding to periodic oscillations. The question that arises is how the dynamics of the mixed suspension varies as a function of the relative proportions $F_1$ and $F_2 = 1 - F_1$ of the two populations.

The behaviour in phase space at the extreme values $F_1 = 0$ ($F_2 = 1$) and $F_1 = 1$ ($F_2 = 0$) is shown in figs. 6.19a and d, respectively. As indicated above, parameter values have been chosen so that a homogeneous population 2 undergoes periodic behaviour of the limit cycle type, while a homogeneous population 1 oscillates in an aperiodic manner and evolves towards a strange attractor. When $F_1$ is progressively increased from zero to unity, the mixed system's behaviour switches from periodic to chaotic through a sequence of period-doubling bifurcations. This sequence, observed up to period 8, is shown in fig. 6.20 where the domain of chaos is schematized by the black region. The phase space trajectories corresponding to period-2 and period-4 oscillations are represented in fig. 6.19b and c, respectively.

Thus, when the proportion of chaotic cells in the mixed suspension is progressively increased from 0 to 100%, the dynamics changes from periodic to chaotic through a sequence of period-doubling bifurcations. What is of interest is that such a well-known route to chaos is encountered here as a function of the relative proportions of periodic and chaotic cells upon mixing such populations in a common suspension.

### *Suppression of chaos by periodic oscillations*

An intriguing property illustrated by the bifurcation diagram of fig. 6.20 is that the behaviour of the mixed suspension is strongly tilted towards periodic behaviour. To examine this phenomenon in more detail, it is useful to focus on the simple case where the two populations differ only

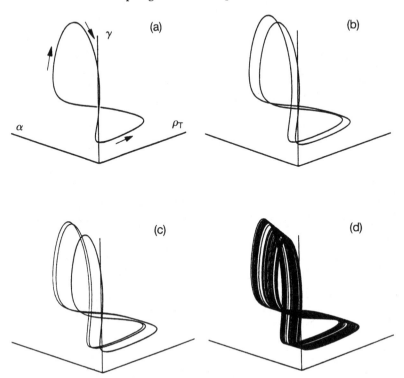

Fig. 6.19. Evolution in the phase space at different values of the relative proportions of (initially) chaotic and periodic cell populations in a mixed suspension containing various amounts of the two types of cells. (a) Oscillations of the limit cycle type obtained for $v_2 = 4.5 \times 10^{-3}$ min$^{-1}$ when the suspension contains only cells of periodic population 2 ($F_1 = 0$, $F_2 = 1$); arrows show the direction of movement and the trajectory has been broken to indicate the part that comes behind (the portion of the curve in front corresponds to a decrease in all three variables after a peak in cAMP). (b) Period-2 oscillations obtained upon adding to periodic population 2 cells from the chaotic population 1, for which $v_1 = 4.396875 \times 10^{-3}$ min$^{-1}$; the value of the fraction of the (initially), chaotic population is $F_1 = 0.5$. (c) Period-4 oscillations obtained when $F_1$ is increased up to 0.86; notice that two of the loops of the trajectory over a period are very close to each other, which is also apparent in the bifurcation diagram of fig. 6.20. (d) Chaotic behaviour corresponding to a strange attractor when the suspension contains only cells of population 1 ($F_1 = 1$). The curves are obtained by numerical integration of eqns (6.9) for the above-indicated values of $v_1$ and $v_2$; other parameter values, which hold for the two populations, are as in fig. 6.2 . Variables $\rho_T$ and $\alpha$ relate to population 2 in (a)–(c), and to the homogeneous population 1 in (d); variable $\gamma$ is shared by the two populations. Ranges of variation for $\rho_T$, $\alpha$ and $\gamma$ are 0–1, 0.65–0.68 and 0–2.2, respectively. Initial conditions were $\alpha = 0.6729$ and $\rho_T = 0.2446$ for both populations, while $\gamma = 1.7033$. The curves were obtained after a transient of 500–1000 min. The period of the oscillations shown in (a)–(c) is of the order of 8–10 min; thus for $F_1 = 0.3$ and $F_2 = 0.7$ the period is equal to 8.7 min (Halloy *et al.* 1990).

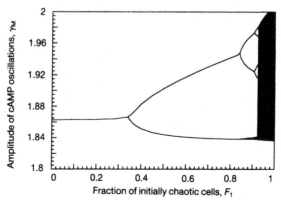

Fig. 6.20. Sequence of period-doubling bifurcations leading to chaos, obtained upon adding to a suspension of periodic cells progressively increasing amounts of cells behaving chaotically. The bifurcation diagram indicates the dynamic behaviour of the suspension after mixing the two populations. Only the maximum (or maxima) of $\gamma$ in the course of oscillations is plotted as a function of $F_1$, which denotes the fraction of cells within the mixed suspension that initially oscillate in a chaotic manner. The diagram is established as indicated in fig. 6.19; parameter values for the periodic and chaotic populations are those of fig. 6.19a and d, respectively (Halloy *et al.* 1990).

by parameter $k_e$ ($k_{e1} \neq k_{e2}$). Then it can be shown (Li *et al.*, 1992a) that the two populations will eventually synchronize in time; i.e. for sufficiently long times, system (6.9) will evolve to a time-dependent state in which $\alpha_1$ and $\rho_{T1}$ will approach $\alpha_2$ and $\rho_{T2}$, respectively.

Thus, when the two populations differ only in the rate of cAMP hydrolysis by extracellular phosphodiesterase ($k_e$) while the values of all the other parameters (including $v$) are identical for the two populations, the dynamics of the mixed suspension, governed by the five independent differential equations (6.9), asymptotically reduces to that of a 'pure' suspension, governed by three independent differential equations, with an effective value of $k_e$ denoted by $k_{eff}$:

$$k_{eff} = F_1 k_{e1} + F_2 k_{e2} \qquad (6.10)$$

In these conditions, the asymptotic behaviour of the mixed suspension can thus be predicted solely on the basis of the relative proportions $F_1$ and $F_2$, and of the 'behaviour spectrum' of a homogeneous suspension, such as that illustrated by the bifurcation diagram of fig. 6.3. The mixing will result simply in a shift in the dynamics of each of the two cell populations on to a common type of intermediate behaviour, since $k_{e1} < k_{eff} < k_{e2}$, as can be easily seen in eqn (6.10). From a mathematical point of view, the mixing maps the dynamics of the two homogeneous cell populations into the dynamics of another homogeneous cell population. This map is parameterized by the fraction $F_1$ (or $F_2 = 1 - F_1$). In the simplest case discussed above, the dynamics of the mixed population can be obtained in an exactly predictable way provided that the bifurcation diagram is known.

The effect of mixing a chaotic with a periodic population is illustrated in figs. 6.21 and 6.22, in a case where the two populations differ only by parameter $k_e$ (Li *et al.*, 1992a,b). The chaotic behaviour of population 1 and the periodic behaviour of population 2 prior to mixing are repre-

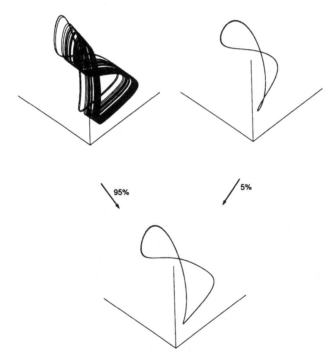

Fig. 6.21. Mixing 5% periodic cells with 95% chaotic cells results in periodic cAMP oscillations in the final suspension. Shown are the phase space trajectories for the chaotic population 1 (upper left), the homogeneous, periodic population 2 (upper right), and the final mixed suspension, which oscillates on a limit cycle (lower part of figure) differing from that of population 2. The curves are obtained by integrations of eqns (6.3) for the pure population 1 or 2, and of eqns (6.9) for the mixed suspension. For the chaotic population 1, the parameter values correspond to point C in fig. 6.2 ($k_{e1}$ = 2.625 min$^{-1}$), while for the periodic population 2, $k_{e2}$ = 2.4257 min$^{-1}$; in both cases, $v^* = v/\mu = 1.407$ min$^{-1}$ (Li *et al.*, 1992a).

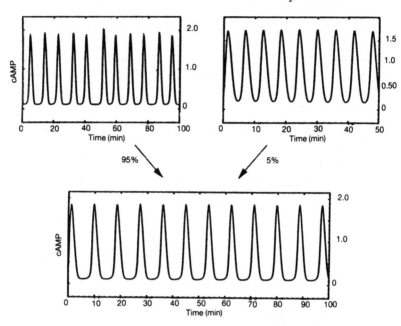

Fig. 6.22. Mixing 5% periodic cells (upper right) with 95% chaotic cells (upper left) results in periodic cAMP oscillations in the final suspension (bottom). The curves show the temporal evolution of extracellular cAMP ($\gamma$) corresponding to the three parts of fig. 6.21 (Li *et al.*, 1992b).

sented in the upper parts of figs. 6.21 and 6.22 in the phase space and as a function of time, respectively. As can be expected from eqn (6.10), the dynamic behaviour of the mixed suspension depends markedly on the relative proportions of cells from the two populations. Most striking is the result that a tiny proportion of cells from population 2 can impose its periodic properties on a suspension containing a large majority of cells from chaotic population 1.

In the specific case considered in figs. 6.21 and 6.22, the mixing of 5% of periodic cells with 95% of chaotic cells results in periodic oscillations of cAMP (lower part of figures); the strange attractor associated with chaos thus transforms into a limit cycle. The periodic oscillations in the mixed suspension correspond to the behaviour predicted by the bifurcation diagram of fig. 6.3 for the effective value of parameter $k_e$ given by eqn (6.10). Qualitatively similar results are also obtained by numerical simulations when the two populations differ only by their intracellular supply of ATP, namely, $v_1 \neq v_2$.

The question arises as to what is the value of the minimum fraction of

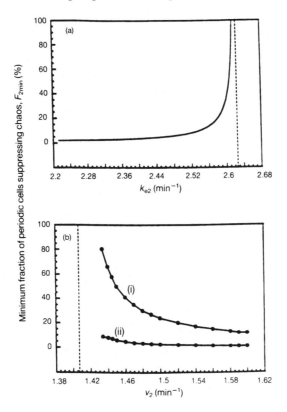

Fig. 6.23. Minimum fraction of periodic cells suppressing chaos. (a) The value of the minimum fraction of cells from periodic population 2 in the final suspension, $F_{2min}$, capable of transforming chaos of population 1 into simple periodic oscillations is plotted as a function of parameter $k_{e2}$; the chaotic behaviour of population 1 is that shown in fig. 6.21 (upper panel, left). The curve was generated according to eqn (6.11) with $k_{e1}^{ch} = 2.6257$ min$^{-1}$ and $k_{e2}^{pd} = 2.6167$ min$^{-1}$. (b) Plotted as a function of $v_2$ is the value of the minimum fraction of periodic cells from population 2 needed to transform the chaotic behaviour of population 1 into (i) simple periodic oscillations, and (ii) oscillations of period 8, with eight peaks of cAMP per period. The results are obtained numerically by integration of eqns (6.9) (dots) as described in the legend to fig. 6.21. The values $k_{e1} = 2.625$ min$^{-1}$ and $v_1/\mu = 1.4073825$ min$^{-1}$ are taken for the chaotic population; other parameter values are as in fig. 6.2. The vertical, dashed line indicates the value of $k_{e1}$ or $v_1$ giving rise to chaos in population 1 (Li *et al.*, 1992a).

cells from the periodic population capable of suppressing chaos in the mixed suspension. This minimum fraction, $F_{2min}$, depends on the parameters of the model. In the two panels of fig. 6.23 parameters $k_{e2}$ or $v_2$ are varied over the range yielding simple periodic oscillations in population 2. The data in fig. 6.23 show that in both situations, when population 2 is sufficiently far away from the domain of aperiodic oscillations, a tiny fraction of periodic cells, sometimes as small as a few per cent, suffices to suppress chaos in the mixed suspension.

As shown in fig. 6.23a, $F_{2min}$ decreases from a value above 90% to less than 5% as the values of parameters $k_{e2}$ and $v_2$ of the periodic population move further away from the values $k_{e1}$ and $v_1$ which give rise to chaos in population 1 (the latter values correspond to the vertical dashed lines in fig. 6.23).

Unlike the effect of $v_2$ which was established solely by numerical simulations, the effect of $k_{e2}$ could be predicted on the basis of the 'behaviour spectrum' in fig. 6.3. When the two populations differ by the amount of phosphodiesterase acting on extracellular cAMP ($k_e$), mixing will eventually lead to their full synchronization, and the mixed suspension will behave as a homogeneous population for which the effective value of $k_e$ will be equal to that given by eqn (6.10) where $F_1$ and $F_2$ denote the fractions of populations 1 and 2 in the final suspension; this argument readily generalizes to the case of mixing $n$ populations. The curve in fig. 6.23b yielding the minimum fraction $F_{2min}$ of cells needed to transform chaos into simple periodic oscillations was thus obtained from eqn (6.10) by means of the relation:

$$F_{2min} = \frac{k_{e1}^{ch} - k_{e2}^{pd}}{k_{e1}^{ch} - k_{e2}} \tag{6.11}$$

where $k_{e1}^{ch}$ and $k_{e2}^{pd}$ refer, respectively, to the value of $k_e$ producing chaos in population 1 and to the largest limiting value of this parameter yielding oscillations of period 1 in population 2 (see fig. 6.3c). Simple periodic oscillations, i.e. oscillations of period 1 with a single peak per period, occur over a finite range of $k_{e2}$ when $k_{e2} < k_{e2}^{pd}$; increasing $k_{e2}$ beyond the latter value leads to chaos through a sequence of period-doubling bifurcations. As shown by the comparison of the two curves established as a function of $v_2$ (lower part of fig. 6.23), the value of $F_{2min}$ needed to transform chaos into complex periodic behaviour, e.g. oscillations of period 8 with eight peaks per period, can be much smaller than that required for transforming chaos into oscillations of period 1.

The above discussion focused on the suppression of chaos by periodic oscillations. As shown by the bifurcation diagram in fig. 6.3, other dynamic transitions may be brought about by the coupling of two populations endowed with distinct dynamic properties; the global behaviour of the mixed suspension will, as in the cases discussed above, depend on the new, effective value of $k_e$ according to eqn (6.10). Thus the mixing of two oscillating populations or of two populations initially in two different steady states can produce simple periodic oscillations, bursting or even chaos, for suitable proportions $F_1$ and $F_2$ (Li *et al.*, 1992a).

## 6.7 Suppression of chaos by the periodic forcing of a strange attractor

Another way of seeing the suppressive effect of the periodic population on chaos is to view the role of the small fraction of periodic cells as equivalent to that of a small-amplitude periodic forcing of chaos (Li *et al.*, 1992a,b). To test this hypothesis, the system of eqns (6.3) governing cAMP oscillations in a homogeneous population of *Dictyostelium* cells is considered under conditions where this three-variable system produces chaos (fig. 6.24). A sinusoidal term representing the forcing of the chaotic population by a small-amplitude periodic input of cAMP is then added to the kinetic equation for extracellular cAMP, so that the dynamics of the cAMP signalling system is governed by the nonautonomous system (6.3) in which the last equation is replaced by:

$$\frac{d\gamma}{dt} = q' \, \sigma \, \phi \, (\alpha, \rho_\text{T}, \gamma) - k_e \, \gamma + \frac{A}{2} \left[ 1 + \sin \left( \frac{2\pi t}{T} \right) \right] \qquad (6.12)$$

where $A$ and $T$ are, respectively, the amplitude and the period of the forcing signal, while the various functions remain defined by eqns (6.4). The numerical integration of these equations shows that a periodic forcing of minute amplitude suffices to transform the strange attractor of fig. 6.22 into a limit cycle (fig. 6.24). In the case considered, the amplitude of the sinusoidal input is of the order of 2% of the maximum cAMP concentration in the course of aperiodic oscillations, while the period of the forcing input is close to that of the limit cycle considered in the upper right panel in fig. 6.22.

## 6.8 Origin of complex oscillations in the model for cAMP signalling in *Dictyostelium*

What is the origin of complex oscillatory phenomena in the cAMP signalling system of *D. discoideum* amoebae? There exists but a single self-amplification process in this system, whereas it is the interplay between two instability-generating mechanisms, each based on a

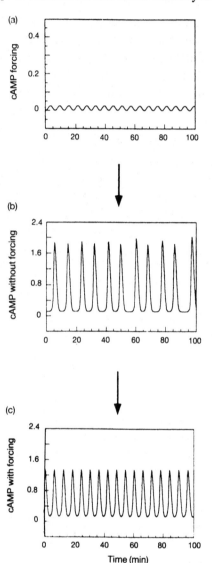

Fig. 6.24. Suppression of chaos by a small-amplitude, periodic input of cAMP. The chaotic oscillations of cAMP (b) are the same as those considered in fig. 6.21 (top, left part). The system is subjected to a sinusoidal input of cAMP (a), as described by eqn (6.3d). Such forcing of the strange attractor leads to periodic oscillations of cAMP (c); the latter are obtained by numerical integration of the first two equations of system (6.3) and eqn (6.12) for $A = 0.025$ and $T = 6$ min (Li *et al.*, 1992b).

self-amplification feedback, which leads to bursting and chaos in the multiply regulated enzyme system studied in chapter 4.

Although the model for cAMP signalling based on receptor desensitization contains but a single feedback loop, in fact it hides two distinct mechanisms, each of which can, on its own, produce sustained oscillations. These two mechanisms, coupled in parallel, share the same process of self-amplification, namely the activation of adenylate cyclase that follows binding of cAMP to the receptor, but the two mechanisms differ by the process limiting this autocatalysis (fig. 6.25). In the first oscillatory mechanism, the limitation arises from the passage of the active receptor into the desensitized state at high concentrations of extracellular cAMP. In the second mechanism, it is the limitation by

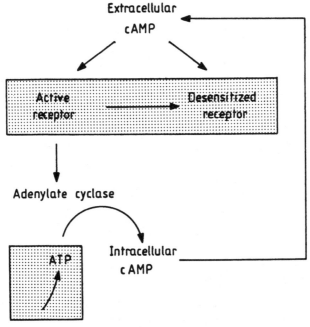

Fig. 6.25. Origin of complex oscillations in the cAMP signalling system of *D. discoideum*. Complex behaviour (birhythmicity, bursting and chaos) originates from the interaction of two endogenous oscillatory mechanisms that are coupled in parallel. The two mechanisms share the same feedback loop of self-amplification by extracellular cAMP, via the binding of the latter to the receptor, and differ by the process responsible for limiting autocatalysis (dotted area): the first limiting process is based on receptor desensitization, and the second on substrate availability at the adenylate cyclase reaction site (Goldbeter & Martiel, 1987).

substrate availability that curtails the self-amplification in cAMP synthesis. The two limiting processes are represented by the dotted boxes in fig. 6.25.

Complex oscillations such as bursting or chaos arise when these two effects acquire comparable importance, so that an interplay occurs between the two, simultaneously active, oscillatory mechanisms. In contrast, when parameter values are such that one of the mechanisms is active while the other remains 'silent', oscillations have a simple periodic character.

In line with this explanation, complex periodic oscillations, birhythmicity and chaos disappear in the model when the concentration of the substrate is held constant in the course of time. The system then admits a unique oscillatory mechanism based on the coupling between self-amplification in cAMP synthesis and its sole limitation by receptor desensitization.

To address the origin of regular oscillatory behaviour of the putatively chaotic mutants *Fr17* and *HH201* in cell suspensions, the model for the cAMP signalling system was used to investigate the dynamic phenomena brought about by the coupling of two populations of *Dictyostelium* cells in which periodic or chaotic oscillations of cAMP occur. Rather than being coupled through transport processes between two distinct reactors undergoing oscillations (Landahl & Licko, 1973; Bar Eli, 1984; Aronson *et al.*, 1987; Boukalouch *et al.*, 1987; Crowley & Epstein, 1989), the two populations of oscillating cells are coupled within the same stirred suspension through the sharing of a common biochemical signal molecule released by all cells present in the medium. Upon binding to a cell surface receptor, this extracellular signal of cAMP in turn influences the intracellular dynamics of cells from the two populations. Such a strong chemical coupling presents some similarities with the experimental system in which two oscillatory chemical reactions are coupled electrically (Crowley & Field, 1986) or through the sharing of a common intermediate (Alamgir & Epstein, 1984), and with the case of nerve or pancreatic cells synchronized by ionic coupling, e.g. by extracellular $K^+$ (Rinzel, Sherman & Stokes, 1992; Sherman & Rinzel, 1992; Stokes & Rinzel, 1993).

When the two cell populations differ only in the amount of extracellular phosphodiesterase, measured by parameter $k_e$, the mixed suspension will eventually synchronize and adopt the behaviour of a homogeneous population of cells characterized by an effective value of $k_e$, which is a linear combination of the values of this parameter for each population, weighted by the fraction of each cell type in the final suspension. The knowledge of the detailed bifurcation diagram as a function of $k_e$ in the three-variable system describing the behaviour of a homogeneous population then suffices to predict the dynamic behaviour of the mixed suspension, regardless of the relative proportions of cells from the two populations. Similar results are obtained when the two populations differ by other parameters such as $v$, which measures the net input of ATP within cells. Furthermore, the results can be generalized to the case where the mixed suspension contains cells from an arbitrary number of cell populations (in the limit, all cells might be different).

Of all modes of dynamic behaviour examined here, only simple periodic oscillations of cAMP have been clearly established in the experiments (Gerisch & Wick, 1975). While the aperiodic signalling properties observed for the mutant *Fr17* (Durston, 1974a) might represent chaos, bursting might account for the occasional occurrence of doublet or triplet waves in the course of aggregation (Gottmann & Weijer, 1986). Long-term bursting of the kind shown in figs. 6.1 and 6.4 remains a theoretical prediction that could be checked in cell suspensions if the activity of extracellular phosphodiesterase could be set at the appropriate (nonphysiological) level.

Besides the demonstration of chaos in the model for cAMP signalling, the main topic of this chapter concerns the coupling between chaos and periodic oscillations. When a periodic cell population and a chaotic population are coupled through extracellular cAMP in a mixed suspension, a tiny proportion of periodic cells (population 2) sometimes suffices to suppress the chaotic behaviour of the majority of cells (population 1) present in the medium. The suppression of chaos was characterized by obtaining the minimum fraction $F_{2min}$ of periodic cells transforming the chaotic behaviour of population 1 into simple or complex periodic oscillations. The value of $F_{2min}$ was determined as a function of the parameters in two conditions, namely when the two cell populations differ by $v$ or by $k_e$. In each case, it appears that chaos in the mixed suspension can be suppressed by the presence of only a small percentage of periodic cells when the relevant parameter of the periodic population is far away from the value producing chaos in the other population.

The reason why chaos is easily suppressed by periodic oscillations in this model is attributable to the relative smallness of the domain of chaos in parameter space as compared to the domain of periodic behaviour. When the two populations differ by parameter $k_e$, their mixing results in establishing an effective, new value of $k_e$ equal to $F_1 k_{e1} + F_2 k_{e2}$. Given that a minute change in $k_{e1}$ suffices to move out of the region of chaos, it is intuitively clear why a tiny proportion of periodic cells (corresponding to a value of $F_2$ of only a few per cent) will transform aperiodic into periodic oscillations. A similar line of reasoning holds for the similar results obtained numerically when the coupled populations differ by parameter $v$, although for this parameter we lack an analytical expression in the form of a linear combination of $v_1$ and $v_2$, as found when the two populations differ by $k_e$.

The question arises as to the generality of the results obtained for the suppression of chaos by periodic oscillations. The relative ease in suppressing chaos in this model results directly from the smallness of the domain of chaos in parameter space. If the domain of autonomous chaos were larger than the domain of periodic behaviour, an inverse conclusion should hold, namely a small proportion of chaotic cells could well suffice to transform periodic into aperiodic oscillations. In all models of biochemical significance investigated so far, particularly in this book (see also Glass & Malta, 1990), the domain of autonomous chaos was significantly more reduced than that of periodic oscillations. This conclusion might be specific to these models and/or to the range of parameters investigated so far.

In models studied in other fields, such as the Lorenz model, chaos occupies a large domain in parameter space. In agreement with a more general study of the interaction between chaotic oscillations (Pikovsky, 1984), diffusive coupling of strange attractors in the Lorenz model (Lorenz, 1991) and in a modified version of the Sel'kov model (Badola, Kumar & Kulkarni, 1991) was shown to produce chaos. Because the mixing of two population differing by $k_e$ always results in a shift to an intermediate value of this parameter, the mixing of two chaotic populations in the present model will produce chaos unless the effective value of $k_e$ falls within a periodic window in the chaotic domain. Due to the relative smallness of these windows, however, the emergence of periodic behaviour from the coupling of two strange attractors should remain a rare event. In a physical context, a case where the coupling between chaotic attractors produces periodic oscillations has recently been described (Rul'kov *et al.*, 1992) for two coupled electronic circuits.

With respect to the possible occurrence of chaos in *Dictyostelium* cells, the present results provide one plausible explanation for the occurrence of rather regular oscillations in suspensions of the putatively chaotic mutant *HH201* (Goldbeter & Wurster, 1989) (see section 6.5 for a discussion of an alternative explanation). Strong coupling of amoebae in such suspensions could suppress chaos if a small proportion of cells were to behave in a periodic manner, even if the majority of cells on their own behaved chaotically. Such a suppression of chaos in mixed suspensions would not occur on agar, where any chaotic centre would be free to generate cAMP signals in an aperiodic manner over its aggregation territory. This would explain the apparently paradoxical observation (Durston, 1974a) of aperiodic signalling during aggregation of the mutant *Fr17* on agar.

Besides oscillations of cAMP in *Dictyostelium*, the physiological significance of the present results bears on the robustness of regular biological rhythms, since they show that chaos can be easily suppressed by coupling with periodic oscillations. Such results could be applicable to electrically excitable cells, e.g. in cardiac or neural tissues. Of interest in this regard is the finding (Destexhe & Babloyantz, 1991) that the coupling of 2% of cells from a neuronal network to the periodic input from the thalamus suffices to enhance coherent behaviour in the network, which otherwise displays spatiotemporal chaos.

The present results provide another way of 'controlling' chaos (Ott *et al.*, 1990; Petrov *et al.*, 1993; Shinbrot *et al.*, 1993). Chaos can be viewed as containing an infinity of unstable periodic orbits. Besides physical systems, the control of chaos by stabilization of one such periodic orbit through finely tuned perturbation in a parameter has been achieved in the Belousov–Zhabotinsky reaction (Peng, Petrov & Showalter, 1991; Petrov *et al.*, 1993) and for ouabain-induced aperiodic behaviour in an isolated piece of rabbit ventricular tissue (Garfinkel *et al.*, 1992); a similar claim was recently made for the control of chaos in the brain (Schiff *et al.*, 1994). Here the stabilization of the periodic orbit is achieved by the coupling of chaos with periodic behaviour. Similar to the elimination of chaos by application of weak periodic forcing in theoretical studies of the periodically driven pendulum (Braiman & Goldhirsch, 1991) and of the Lorenz model (Guidi, Halloy & Goldbeter, 1995), the suppression of chaos by a tiny proportion of periodic *Dictyostelium* cells turns out to be equivalent to the suppressive effect of a low-amplitude, sinusoidal forcing of a strange attractor.

# 7

# The onset of cAMP oscillations in *Dictyostelium* as a model for the ontogenesis of biological rhythms

Biological rhythms occur only under precise conditions, and variations in a control parameter can bring about their disappearance. In a symmetrical manner, the variation of such a parameter can lead to the appearance of a rhythm in the course of development. There is no example as yet where the molecular basis of the ontogenesis of a biological rhythm is known in detail. The rhythm of intercellular communication in the slime mould *Dictyostelium discoideum* provides us with a prototype for the study of this question.

## 7.1 Appearance of excitability and oscillations in the course of development of *Dictyostelium* amoebae

Studies of aggregation of *D. discoideum* amoebae on agar have shown that cells are capable of relaying an artificial cAMP signal applied iontophoretically by means of a pipette, before being able to produce sustained oscillations in an autonomous manner (Robertson & Drage, 1975; Gingle & Robertson, 1976). These spontaneous oscillations of cAMP occur only a few hours after the beginning of starvation.

Studies of *D. discoideum* cells in suspensions confirm this observation (Gerisch *et al.*, 1979). At the beginning of starvation, the amoebae are unable to amplify cAMP signals. Some 2 h later, the amoebae amplify cAMP pulses whose amplitude exceeds a threshold. Four hours after the beginning of starvation, spontaneous oscillations arise and are maintained for 3 h before giving way to excitable behaviour (fig. 7.1). During the hours preceding starvation, the amoebae thus undergo successive transitions that modify the dynamic properties of the signalling system that controls intercellular communication. The sequence of developmental transitions leads from the absence of excitability to relay, and from relay to autonomous oscillations of cAMP.

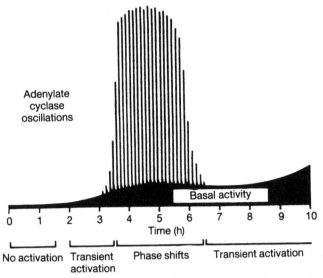

Fig. 7.1. Evolution of the system of intercellular communication by cAMP signals in *D. discoideum* in the course of development. Time is measured in hours after the beginning of starvation. The black zone at the bottom of the figure shows the evolution of the basal activity of adenylate cyclase. Until 2 h after the beginning of starvation, the enzyme fails to be activated by cAMP signals; the transient activation, corresponding to the phenomenon of relay, is observed thereafter, followed by the onset of autonomous oscillations of cAMP. The latter spontaneously appear about 4 h after the beginning of starvation and continue for about 3 h, before the system recovers its excitable behaviour (Gerisch *et al.*,1979).

Identifying the molecular basis of these transitions would permit us to comprehend the ontogenesis of the intercellular communication rhythm in *Dictyostelium*. At the same time, the clarification of this sequence would provide an explanation for the differences between centres and relay cells in the course of aggregation. The theoretical models studied above for cAMP synthesis in *D. discoideum* allow us to obtain elements of response to these questions, in terms of changes in dynamic behaviour due to the passage through a bifurcation point in the course of development (Goldbeter & Segel, 1980; Goldbeter, 1981; Martiel & Goldbeter, 1988).

As explained at the end of this chapter, although specifically formulated for the case of *Dictyostelium* cells, the ideas presented below generalize to other biological systems in which dynamic properties such as excitable or oscillatory behaviour arise in the course of development.

## 7.2 Developmental path in parameter space: a molecular basis for the ontogenesis of cAMP oscillations

For a first theoretical approach of the transitions between relay and oscillations, it is useful to return to the allosteric model proposed for the mechanism of cAMP synthesis in *D. discoideum* (Goldbeter & Segel, 1980) (see section 5.2 and fig. 5.16 for a scheme of that model). Two key parameters in any model for cAMP synthesis in *D. discoideum* are the activity of adenylate cyclase, which catalyses the production of cAMP from ATP, and the activity of phosphodiesterase, which hydrolyses the signal in the extracellular medium. In the allosteric model governed by eqns (5.1), parameters $\sigma$ and $k$ measure, respectively, the maximum rate of the cyclase and of the phosphodiesterase.

The study of the stability properties of the unique steady state admitted by eqns (5.1) permits us to establish the stability diagram of fig. 7.2 in which the dashed area C denotes the domain of instability of the steady state, where sustained oscillations of the limit cycle type occur. In domain B, the steady state is stable but excitable, as the system amplifies, in a pulsatory manner, a cAMP signal whose given amplitude exceeds a threshold. Everywhere else in the diagram the steady state is stable but nonexcitable.

To account for the transitions **no relay → relay → oscillations**, it is desirable that the system, in the $\sigma - k$ plane, follow the path marked by an arrow. This trajectory, starting in A from a nonexcitable state characterized by a low level of cAMP, would enter the excitable domain B, where relay would occur, before moving into domain C where cAMP oscillations would begin spontaneously. Finally, the passage to a domain located to the left of region C would bring an end to the oscillations by leading the system into a stable state characterized by an elevated level of cAMP.

Discussing the dynamics of the cAMP synthesizing system as a function of the stability diagram in the $\sigma - k$ parameter space is justified as long as the changes in the values of these parameters remain slow compared to the changes in the metabolic variables of the model. This approximation is reasonable, given that the period of oscillations is of the order of 5 to 10 min, while the variation of the enzyme activities, resulting from protein synthesis, extends over several hours. To each pair of values of parameters $\sigma$ and $k$ we may thus associate a certain mode of dynamic behaviour; the sequence of values taken by the two parameters in the course of development specifies a trajectory with which a particular sequence of behavioural modes is associated.

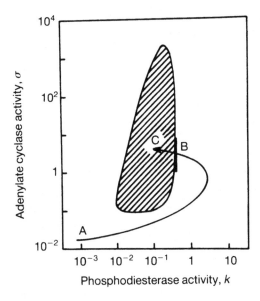

Fig. 7.2. Developmental path of the cAMP signalling system in the parameter space formed by adenylate cyclase and phosphodiesterase activity. The stability diagram is established by linear stability analysis of the steady state admitted by the three-variable system (5.1) governing the dynamics of the allosteric model for cAMP signalling in *D. discoideum* (see section 5.2). In domain C sustained oscillations occur around an unstable steady state. In domain B, the steady state is stable but excitable as it amplifies in a pulsatile manner a suprathreshold per-turbation of given amplitude. Outside these domains the steady state is stable and not excitable. The arrow crossing successively domains A, B and C repre-sents the developmental path that the system should follow in that parameter space to account for the observed sequence of developmental transitions no relay → relay → oscillations (Goldbeter, 1980).

The **developmental path** (Goldbeter & Segel, 1980) in the $\sigma - k$ para-meter space suggested above accounts for the transitions observed for the dynamics of the cAMP signalling system. In the space of the para-meters considered, this path corresponds to an initial increase in the activity of adenylate cyclase and phosphodiesterase, followed by a decrease in the activity of the latter enzyme. How does this prediction, based on the variation of the two enzyme activities after starvation, compare with experimental observations?

The beginning of starvation corresponds in *D. discoideum* to the initi-ation of a new developmental programme that leads to the formation of a multicellular aggregate and to its final transformation into a fruiting body. The initial phase of this developmental programme is characterized by an intense protein synthesis activity (Loomis, 1979). Cells must acquire all the biochemical equipment that they will need for aggregation. In particular, they must build up the mechanism of intercellular communication that will allow the centres to emit cAMP signals and the other cells to respond chemotactically to these stimuli. During the hours that follow starvation, accordingly, we observe a marked increase in the activity of adenylate cyclase (fig. 7.3a) and of the intra- and extracellular forms of phosphodiesterase (fig. 7.3b). In parallel, the level of cAMP binding to the cell increases (fig. 7.3c), as a result of enhanced receptor synthesis. Each of the elements of the machinery of intercellular communication thus takes its place after the beginning of starvation, and undergoes a sigmoidal increase in the course of time until its level reaches a maximum some 6 h later when the aggregation phase begins.

The initial increase predicted by the developmental path of fig. 7.2 corresponds well to this time evolution of adenylate cyclase and phosphodiesterase, but the agreement does not extend to the later decrease in phosphodiesterase activity, which is not corroborated by experimental observations (fig. 7.3b). A modification of eqns (5.1) allows us to amend this shortcoming of the model. When taking into account the consumption of ATP in reactions other than that catalysed by adenylate cyclase, an additional term $(-k'\alpha)$ enters the kinetic equation for ATP $(\alpha)$ in system (5.1). The stability diagram in the $\sigma - k$ parameter space remains practically unchanged for low values of constant $k'$. For non-negligible values of this parameter, the diagram is deformed, as indicated in fig. 7.4 (Goldbeter & Segel, 1980).

The hatched domain C represents, as previously, the domain of sustained, autonomous oscillations of cAMP around a unique, unstable steady state. In domain B, the unique steady state is stable and excitable; in this region of parameter space, the system amplifies a cAMP signal of given magnitude. In domain E, three steady states are obtained in the model, two of which – or only one – are stable. Everywhere else, the unique steady state is stable and nonexcitable, i.e. relay does not occur for the signal of given magnitude; in A and D, this state corresponds, respectively, to a low or high level of extracellular cAMP. In the diagram of fig. 7.4, as in that of fig. 7.2, the boundary of the domain of sustained oscillations C is sharp in that it represents the locus of bifurcation points beyond which a stable limit cycle forms around an unstable steady state. In contrast, the boundary of the relay domain B does not correspond to any bifurcation; its extension increases with the magnitude of the cAMP signal.

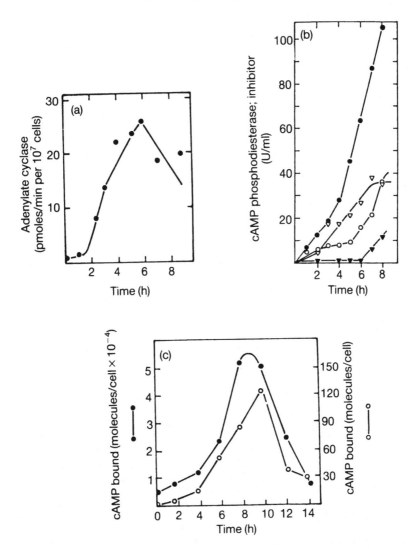

Fig. 7.3. Evolution of the biochemical parameters of the mechanism of intercellular communication during the hours that follow starvation in *D. discoideum*. The variation in the activity of (a) adenylate cyclase and (b) intra- and extracellular phosphodiesterase; and (c) the evolution of the quantity of cAMP receptor measured in experiments performed with two different levels of cAMP. In (b), circles and triangles refer to total phosphodiesterase and to a protein inhibitor of the enzyme, respectively; closed symbols relate to cells treated by pulses of cAMP resulting in $10^{-7}$ M cAMP every 5 min, whereas open symbols relate to untreated cells (data collected from various authors by Loomis, 1979).

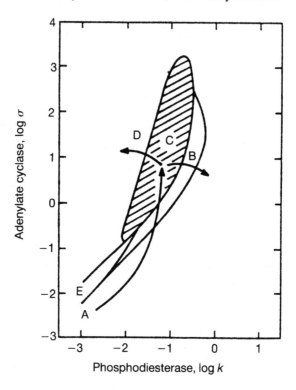

Fig. 7.4. Developmental path for the mechanism of intercellular communication by cAMP signals in *D. discoideum*, in agreement with the variations observed for the activity of adenylate cyclase and phosphodiesterase after starvation. The diagram is constructed as indicated in fig. 7.2, for system (5.1) to which the term $(-k'\alpha)$ has been added in the evolution equation for variable $\alpha$, to take into account the utilization of ATP to ends other than cAMP synthesis. In these conditions, the developmental path accounting for the sequential transitions of fig. 7.1 corresponds to the increase in the two enzyme activities that is observed in the hours that follow starvation (fig. 7.3). Domains A, B and C have the same meaning as in fig. 7.2; domain D corresponds to a stable steady state characterized by an elevated level of cAMP, while two stable steady states can coexist in E. The signal considered for amplification in B is $\gamma = 10$. Parameter values are $v = 0.04 \text{ s}^{-1}$, $k_t = 0.4 \text{ s}^{-1}$, $k' = 10^{-3} \text{ s}^{-1}$, $h = 10$, $L = 10^6$, $q = 100$ (Goldbeter & Segel, 1980).

With respect to fig. 7.2, fig. 7.4 presents the peculiarity of possessing a domain of bistability. Such a phenomenon has not yet been observed in *D. discoideum*, although certain experimental observations on the coexistence of a basal activity state with an activated state of adenylate cyclase (Juliani & Klein, 1978) could be interpreted in terms of a transition between two simultaneously stable steady states. Of more immediate relevance to the discussion of the transitions observed in the course of development is the displacement of the excitablity domain B, which, from being located to the right of the oscillatory domain C in fig. 7.2, now extends below that domain in fig. 7.4. Here, the developmental path accounting for the transitions no relay → relay → oscillations thus corresponds to a trajectory with which a continuous rise in adenylate cyclase and phosphodiesterase is associated. The late decrease in the latter enzyme, which was necessary to obtain the right behavioural sequence in fig. 7.2, is therefore not required in the modified diagram.

The continuous rise in the activity of the two enzymes catalysing, respectively, the production and destruction of cAMP – which rise is observed (Klein & Darmon, 1975; Klein, 1976) during the hours that follow starvation (fig. 7.3a and b) – thus allows us to explain the discontinuous transitions in the dynamic behaviour of the cAMP signalling system. The final transition that brings the system from the oscillatory regime into an excitable state could originate from a late variation in phosphodiesterase activity; the experimental curves (fig. 7.3b) indeed suggest the existence of a second peak of synthesis of this enzyme.

The predictions of the model could be tested by means of mutants. Thus, a mutant of *D. discoideum* deficient in phosphodiesterase should follow a quasi-vertical developmental path, starting from a low activity of adenylate cyclase. In fig. 7.4, such a trajectory would pass to the left of the oscillatory domain. The addition of an adequate amount of exogenous phosphodiesterase should then bring the system into the domain of spontaneous oscillations, without prior passage through the domain of excitability. Such a prediction is in agreement with the experimental observations of Darmon *et al.* (1978), who successfully induced periodic aggregation in a mutant lacking phosphodiesterase, after providing the mutant cells with the latter enzyme.

An early increase in adenylate cyclase should, however, bring the cells more quickly into the oscillatory domain. This is indeed observed in the mutant *Fr17* (discussed in chapter 6) for which the oscillations appear precociously (see fig. 6.16) as a result of the fact that adenylate cyclase accumulates more rapidly and at a higher level than in the wild type (Coukell & Chan, 1980). Finally, the mutant *91A* studied by Durston (1974b) is characterized by a rate of emergence of autonomous centres that is 20 times lower than in the wild type. This mutant seems blocked in the domain of relay behaviour. The model suggests that the amount of phosphodiesterase or of adenylate cyclase in the mutant *91A* could be too low to permit the cells to proceed along the developmental path up to the domain of oscillations; the amoebae would thus be blocked in the relay domain.

## 7.3  On the nature of centres and relays

During aggregation, what are the factors that induce a given cell to become a centre or to remain capable only of relaying the incoming signals of cAMP? Figure 7.4 provides an answer (Goldbeter & Segel, 1980).

An aggregation territory controlled by a centre can contain up to $10^5$ amoebae. At the beginning of starvation, these cells are not all perfectly synchronized; indeed they differ by their exact content of various enzymes, including adenylate cyclase and phosphodiesterase. Cells thus start on their developmental path, symbolized by the arrowed trajectory in fig. 7.4, from slightly different initial conditions in the space of parameters $\sigma$ and $k$. As a result of this heterogeneity, certain cells, more advanced than others, will be the first to penetrate into the domain of instability of the steady state. These amoebae will become the aggregation centres, as they become capable of spontaneously generating cAMP signals in a periodic manner.

At the moment when the first amoebae reach the oscillatory domain C, the other cells are less advanced on their developmental path. Those which are most behind are still in a nonexcitable state, while others have reached the domain B in which they acquire the property of relaying signals of sufficient magnitude. If we remove from an aggregation territory the cells that have become centres, we observe that other cells will replace them in that capacity; once these new centres are removed, others take up their function (Shaffer, 1962). This process can be repeated over several hours until no more cells can become centres.

These observations can be explained in terms of the evolution of the amoebae along the developmental path: if the first cells that entered the domain of autonomous oscillations are removed from the medium, other cells that follow them will eventually enter the domain and will thereby perform the function of aggregation centres. Removing these cells would permit the cells that have just reached the excitability domain to proceed in turn up to the domain of spontaneous oscillations. This process could be repeated until the variations of adenylate cyclase and phosphodiesterase would be such that the developmental path would lead the cells out of the domain of autonomous oscillations.

The suggestion that all cells have the capacity to become a centre is in agreement with experimental observations (Glazer & Newell, 1981). In fact, at the time of aggregation, a significant part of the cells could well behave as aggregation centres by synchronously emitting the periodic signals, while the other cells would act as relays. In cell suspensions, such a situation cannot be distinguished from that where a unique centre produces periodic signals amplified by the other, excitable amoebae.

For simplicity, the preceding discussion considered two control parameters only. If adenylate cyclase and phosphodiesterase are among the most important parameters of the mechanism of cAMP synthesis, nevertheless it remains clear that the developmental path meanders in an $n$-parameter space. Genetic studies (Coukell, 1975; Williams & Newell, 1976) have shown that nearly 50 genes are required for aggregation. Among these genes, a certain number are involved in the chemotactic response or in the establishment of cell contacts within the aggregate, while it is probable that the products of at least 10 genes play a role in the mechanism of periodic generation of the cAMP signals. The discussion in terms of two parameters can be generalized without conceptual difficulty to such a situation, and to any parameter space in which different types of dynamic behaviour can be demonstrated.

Thus, the successive transitions between a nonexcitable state, relay and cAMP oscillations also occur in the stability diagram of fig. 5.29 obtained for the model based on desensitization of the cAMP receptor. This diagram is established as a function of parameters $L_1$ and $L_2$ linked to the activities of the protein kinase and phosphatase that control the level of receptor phosphorylation. A variation in the kinase or phosphatase activity in the course of time could give rise to the observed transitions. In each parameter space, the path predicted theoretically to account for the behavioural transitions should be compared with the variations observed experimentally for these parameters, as done above for adenylate cyclase and phosphodiesterase.

The preceding discussion is based on stability (or behavioural) diagrams in parameter space, and incorporates the temporal variations of these parameters in an implicit manner only. We see below how the developmental path in a four-parameter space can lead to the dynamic transitions when we incorporate explicitly into the model the changes observed for these parameters.

## 7.4 Incorporating the variation of the parameters into the kinetic equations of the model

Figure 7.4 was obtained in a modified version of the allosteric model that takes into account the utilization of ATP other than in cAMP synthesis. Analogous diagrams are obtained in any model accounting for the phenomena of relay and oscillations. Still, we need to verify whether the predictions concerning the parameter changes required for explaining the sequential transitions are corroborated by experimental observations. The discussion of dynamic transitions in the allosteric model in terms of the stability diagram of fig. 7.4, moreover, possesses an implicit character. For these reasons, it would be useful to test the concept of a developmental path in the model based on receptor desensitization, and to determine the evolution of dynamic behaviour predicted by that model when we explicitly incorporate the temporal variation of the control parameters into the kinetic equations.

Given that in this model the receptor and adenylate cyclase are separate entities, it is possible to take into account the independent variation of these two parameters. Moreover, the model based on receptor desensitization takes explicitly into account the activity of the two forms of phosphodiesterase. For each of these four parameters, namely the activity of adenylate cyclase, of intra- and extracellular phosphodiesterase, and the quantity of receptor present at the surface of the membrane, the variation observed in the hours that follow starvation takes the form of a sigmoidal increase as a function of time (fig. 7.3a–c). The evolution of each of these parameters can thus be described, to a first approximation, by an equation of the logistic type (Goldbeter & Martiel, 1988; Martiel, 1988).

Under these conditions, the evolution of the mechanism of synthesis of cAMP signals is governed by the three-variable system (5.12), to which are added four differential equations for the variation of parameters $R_T$, $\sigma$, $k_e$, and $k_i$; these parameters will be taken as proportional to a variable $X$, whose time evolution is itself governed by the logistic equation:

$$\frac{\mathrm{d}X}{\mathrm{d}t} = kX(X_{max} - X) \tag{7.1}$$

Parameters $k$ and $X_{max}$ are chosen in such a manner that the evolution of $X$ and, consequently, that of the biochemical parameters, takes place over relatively long characteristic times, so that $X$ reaches its maximum value, $X_{max}$, after 6 h. Furthermore, the initial value of $X$ is taken as equal to unity, and the maximum is taken as equal to 50, so that $X$ will vary by a factor of 50 in the course of the 6 h interval considered; the variations observed for adenylate cyclase and phosphodiesterase in the hours following starvation are of this order (fig. 7.3a,b). To account for the fact that the amounts of receptor and of the different enzymes do not increase exactly at the same time, parameters $R_T$, $\sigma$, $k_e$ and $k_i$ are taken as proportional to $X (t - \tau)$ where $\tau$ represents a delay that may differ for each parameter. The fraction $R_T/(R_T)_f$, where $(R_T)_f$ represents the level of receptor reached after 6 h, is denoted by $f_R$. The dynamic behaviour of the signalling system after starvation is thus governed by the following evolution equations:

$$\frac{d\rho_T}{dt} = -f_1(\gamma)\rho_T + f_2(\gamma)(f_R - \rho_T)$$

$$\frac{d\beta}{dt} = q \, \sigma \, \phi \, (\rho_T, \gamma, \alpha, X) - (k_i + k_t) \, \beta$$

$$\frac{d\gamma}{dt} = (k_t \, \beta/h) - k_e \, \gamma$$

$$\frac{dX}{dt} = k \, X \, (X_{max} - X)$$

$$\sigma = 0.012 \, X(t)$$

$$f_R = \frac{X(t - \tau_1)}{X_{max}}$$

$$k_e = 0.108 \, X \, (t - \tau_2)$$

$$k_i = 0.034 \, X \, (t - \tau_2) \tag{7.2}$$

where:

$$f_1 (\gamma) = \frac{k_1 + k_2 \, \gamma}{1 + \gamma}, \, f_2 (\gamma) = \frac{k_1 L_1 + k_2 L_2 c \gamma}{1 + c \gamma},$$

$$\phi (\rho_T, \gamma, \alpha) = \frac{\alpha \, (\lambda\theta + \epsilon Y^2)}{1 + \alpha\theta + \epsilon Y^2 \, (1 + \alpha)}, \, Y = \frac{f_R \, \gamma}{1 + \gamma} \tag{7.3}$$

The sigmoidal evolution of the four control parameters governed by these equations is represented in fig. 7.5a. The response of the system to this slow variation in the parameters is shown in fig. 7.5b. The numerical integration of eqns (7.2) indicates that the level of extracellular cAMP

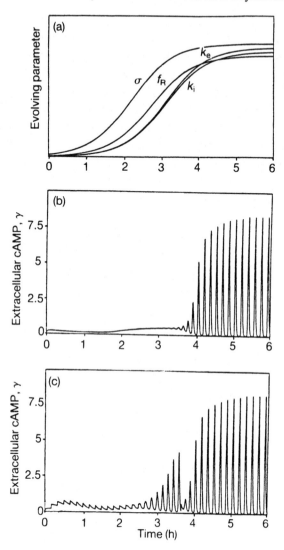

Fig. 7.5. Incorporating the variation of biochemical parameters into the description of the evolution of the mechanism of intercellular communication in *D. discoideum*. The sigmoidal increase observed during the 6 h that follow the beginning of starvation for adenylate cyclase ($\sigma$), the intracellular ($k_i$) and extracellular ($k_e$) forms of phosphodiesterase, and the quantity of cAMP receptor ($f_R$), is incorporated into the model based on receptor desensitization. The variation of these four parameters in the system now ruled by the enlarged set of equations (7.2), is represented in (a). The fraction $f_R$ denotes the receptor concentration divided by the level reached after 6 h. The response of the system to such a variation in the parameters is shown in (b): autonomous oscillations of cAMP occur after 4 h. (c) The response of the system to perturbations of extra-

progressively rises, in a monotonous manner, until autonomous oscillations set in some 4 h after the beginning of starvation. The rhythm thus arises, spontaneously, from the continuous variation of the control parameters.

Does the variation of the parameters also account for the sequential transitions no relay → relay → oscillations observed during this phase of development? To determine whether the system is capable of amplifying a cAMP signal, a pulse of extracellular cAMP of $3 \times 10^{-8}$ M is applied every 10 min, from time zero to the beginning of spontaneous oscillations, in the form of an instantaneous increase in variable $\gamma$ by 0.3 unit. Figure 7.5c shows that, at the beginning, the system fails to amplify the signal: the extracellular cAMP level indeed falls immediately after stimulation. Some 3 h after the onset of starvation, the system acquires the capability of amplifying the cAMP signals in a pulsatory manner. The system therefore becomes excitable before being able to generate periodic signals of cAMP.

The results shown in fig. 7.5 indicate that the incorporation of the continuous changes in four of the main control parameters allows us to account for the sequence of transitions in dynamic behaviour observed after starvation for the intercellular communication system of *D. discoideum* amoebae.

The results can be extended to take into account the variation of an even larger number of parameters; thus, it appears that a two-fold increase in the activity of the kinase and phosphatase involved in the reversible phosphorylation of the cAMP receptor would preserve the sequence of developmental transitions shown in fig. 7.5, while causing a progressive shortening of the period of oscillations. Such an acceleration of the cAMP rhythm is suggested by experiments in cell suspensions (see chapters 5 and 6, and Wurster, 1988).

It is interesting to compare the case of *D. discoideum* to that of the species *Dictyostelium minutum*, in which oscillations are also observed, but at a later stage of aggregation (Schaap, Konijn & Van Haastert, 1984). The changes in the activity of phosphodiesterase and in the cAMP receptor observed in this species after starvation are shown in fig. 7.6. The rise in the enzyme and in the receptor is much slower than

cellular cAMP applied from time zero until the beginning of spontaneous oscillations, every 10 min, in the form of an increase by 0.3 unit in $\gamma$; relay of these signals begins some 3 h after the beginning of starvation, just before cells become capable of autonomous oscillations (Goldbeter & Martiel, 1988).

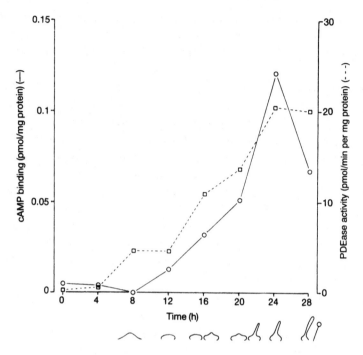

Fig. 7.6. Evolution of phosphodiesterase (PDEase) and of the cAMP receptor during the hours that follow starvation in *Dictyostelium minutum*. The diagram at the bottom shows the corresponding morphogenetic changes in this species of slime mould (Schaap *et al.*, 1984).

in *D. discoideum*, as indicated by comparison with fig. 7.3. In agreement with the preceding discussion, the oscillations in *D. minutum* occur only 12 to 16 h after the beginning of starvation (Schaap *et al.*, 1984).

These results are of general significance for the study of biological rhythms as they show how the continuous variation of certain control parameters can lead to the emergence of a rhythm in the course of development of an organism. Here, the level of certain proteins augments once the amoebae begin to synthesize the components of their intercellular communication system after starvation. As soon as the concentration of the cAMP receptor and the activity of enzymes such as adenylate cyclase and phosphodiesterase reach a critical value, oscillations appear spontaneously. Rinzel & Baer (1988) have shown, however, that a certain delay separates the time at which the parameters cross their bifurcation values and the moment at which oscillations begin; this delay increases with the time spent by the system in the stability region, before its entrance into the oscillatory domain.

Whereas in *Dictyostelium* the onset of oscillations would be due to the synthesis of the cAMP receptor and of a number of enzymes, other biological rhythms could appear in an analogous manner. Thus, neuronal rhythms could arise from the synthesis of certain ionic channels, once the conductances reach a critical value corresponding to a Hopf bifurcation (Holden & Yoda, 1981). A parameter controlling these oscillations could also be the activity of a protein kinase capable of modulating a conductance by altering the degree of phosphorylation of the ionic channel (Levitan & Benson, 1981; Levitan, 1985, 1994). Like the progressive accumulation of an ionic channel within the membrane, the continuous variation of the activity of such an enzyme could lead to the onset of periodic behaviour.

An analogous explanation at the molecular level can be proposed for the spontaneous emergence of the cardiac rhythm at a precise stage of embryonic development. This appearance of one of the most remarkable biological rhythms results from the fact that in the course of development cells in nodal tissues of the heart acquire the capability to generate periodically, in an autonomous manner, the electrical signal that initiates cardiac contraction (Fukii, Hirota & Kamino, 1981). As for the neurons, the variation of certain ionic conductances would endow these cells with the property of oscillating spontaneously, while other cardiac cells would only be excitable (DeHaan, 1980).

Finally, in oscillatory neural networks (Friesen & Stent, 1978; Rose & Benjamin, 1981; Getting, 1983, 1989; Selverston & Moulins, 1985; Friesen, 1989; Grillner *et al.*, 1989; Jacklet, 1989), periodic behaviour could result from the appearance of rhythmic properties in one or several cells of the network or, alternatively, from the establishment of inhibitory or activating intercellular connections (Kling & Szekely, 1968).

# Part IV

## From cAMP signalling in *Dictyostelium* to pulsatile hormone secretion

# 8

# Function of the rhythm of intercellular communication in *Dictyostelium*: link with pulsatile hormone secretion

Whereas the function of glycolytic oscillations, at least in yeast, remains puzzling, the same does not hold for the role of cAMP oscillations in *Dictyostelium* amoebae. After reviewing the different aspects of the function of these oscillations and the molecular bases for the efficiency of pulsatile signalling, a link is established here with the rhythmic, pulsatile secretion observed for an increasing number of hormones. Besides their circadian modulation (Van Cauter & Aschoff, 1989), many hormones are indeed secreted in the form of ultradian, high-frequency pulses (Crowley & Hofler, 1987; Negro-Vilar *et al.*, 1987). Particular attention is devoted below to the pulsatile release of gonadotropic hormones by the pituitary in response to pulses of the gonadotropin-releasing hormone (GnRH), emitted by the hypothalamus with a frequency close to one pulse per hour.

A model for the response of target cells to pulsatile stimulation in the presence of receptor desensitization allows us to investigate, in a general manner, the encoding of the pulsatile signal in terms of its frequency. This analysis, applied to the cAMP signalling system and to the pusatile release of GnRH, suggests the existence, in both cases, of a frequency capable of inducing an optimum response. The model shows how this optimum frequency is related to the kinetics of reversible desensitization in target cells. The validity of the results obtained in the general model is corroborated by the fact that similar results are recovered in the specific model for cAMP signalling based on receptor desensitization.

Beyond the case of GnRH, the conclusions on the existence of an optimum frequency of pulsatile signalling probably extend to other hormones, as is discussed for the cases of insulin and growth hormone. This chapter ends with a brief discussion of mechanisms underlying the generation of pulsatile signals in intercellular communication.

## 8.1 Function of cAMP oscillations in *Dictyostelium*

The primary function of the rhythm of intercellular communication in *D. discoideum* amoebae is to allow the centres to control the aggregation of a large number of cells. The existence of oscillations is indeed closely related to that of relay, i.e. excitability, as shown by the results presented in chapter 5. The fact that certain cells emit the chemotactic signal periodically permits, thanks to its relay by surrounding cells, rapid propagation of the chemical signal over distances of up to 1 cm; this scale is macroscopic with respect to cellular dimensions, which are of the order of 50 μm (Shaffer, 1962; Alcantara & Monk, 1974). This interpretation is corroborated by the existence of other species, such as *Dictyostelium minutum*, in which the mechanism of relay and oscillations is absent in the course of aggregation (Gerisch,1968); centres then emit the chemotactic factor in a monotonous manner. In the absence of relay, the information is propagated by simple diffusion, so that the centres control only small territories containing but a small number of amoebae; hence the name *D. minutum* given to that species of slime mould.

Besides the role of periodic signals in the control of the chemotactic response during aggregation, there is another, no less important, function. Experiments indicate (Darmon, Brachet & Pereira da Silva, 1975; Juliani & Klein, 1978) that certain mutants of *D. discoideum* submitted to periodic stimulation at the frequency of one cAMP pulse every 5 min, aggregate normally, whereas they fail to do so in the absence of stimulation or when they are subjected to a constant cAMP signal in the course of time.

Similar experiments carried out with the wild type (Gerisch *et al.*, 1975) showed that periodic signals of cAMP, in contrast to constant stimuli, accelerate cell differentiation by inducing the precocious synthesis of proteins involved in the aggregation process. Initial experiments demonstrated the induction of proteins necessary for the establishment of intercellular contacts; similar experiments (Klein & Darmon, 1977; Juliani & Klein, 1978; Chisholm, Hopkinson & Lodish, 1987; Kimmel, 1987; Mann & Firtel, 1987; Mann, Pinko & Firtel, 1988) established that the periodic signals of cAMP induce a differentiation programme in which other proteins such as adenylate cyclase, the cAMP receptor, and phosphodiesterase are synthesized. These developmental responses depend on the periodic nature of the stimulation and do not occur when the cAMP stimulus remains constant.

What is even more remarkable is that the effect of the periodic cAMP signals depends on their frequency. Thus, Wurster (1982) showed that the cAMP stimulation loses its efficacy when applied at the frequency of one pulse every 2 min; it then becomes as ineffectual as constant signals or pulsatile signals applied in a random manner (Nanjundiah, 1988).

## 8.2 Molecular bases of the efficiency of periodic signalling in *Dictyostelium*

The model based on desensitization of the cAMP receptor allows us to propose an explanation for these observations (Goldbeter, 1987a,b; 1988a; Martiel & Goldbeter, 1987a). Since the experiments of Wurster Wurster (1982) were conducted at an early stage of development, when cells are not yet excitable, we can determine the effect of periodic stimulation by extracellular cAMP under conditions where the latter metabolite remains controlled, as in the experiments of Devreotes & Steck (1979); the self-amplification by cAMP is then suppressed (see fig. 5.37c). The behaviour in the presence of self-amplification is not much different, but the existence of an absolute refractory phase when cells become excitable somewhat complicates the interpretation of the results (Martiel, 1988), although most conclusions remain valid under such conditions (Li & Goldbeter, 1990). The case where no self-amplification is present is of more general significance as it also applies to the patterns of pulsatile hormone secretion considered below.

When the external cAMP concentration is treated as a control parameter, the cAMP signalling system is governed by eqns (5.16) in which the two variables are intracellular cAMP ($\beta$) and the total fraction of receptor in the nondesensitized state ($\rho_T$). To determine the effect of periodic stimulation, the simplest is to consider a square-wave signal (fig. 8.1). During a lapse of time $\tau_1$, the level of extracellular cAMP is raised up to a constant value $\gamma_1$, before being decreased and maintained at the basal level $\gamma_0$ during a time interval $\tau_0$ (see also fig. 8.9 below). These two phases are repeated regularly so that the periodicity of the signal is equal to ($\tau_0 + \tau_1$). In the numerical simulations shown in fig. 8.1, the basal level of extracellular cAMP is nil whereas the maximum level in the course of stimulation is equal to $10^{-6}$ M.

Figure 8.1a shows the response of the signalling system to cAMP signals of 5 min duration, spaced at 5 min intervals. The response of the system to stimuli of similar amplitude and duration, when successive

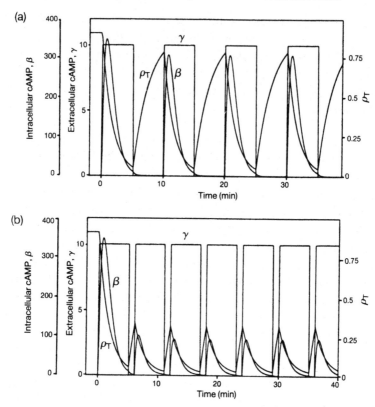

Fig. 8.1. Response to pulsatile stimulation by extracellular cAMP in the model for cAMP synthesis based on receptor desensitization. Equations (5.16) of the two-variable model are integrated in the case where periodic stimulation by extracellular cAMP ($\gamma$) takes the form of a square wave. In (a), the stimulus consists in raising $\gamma$ from 0 to 10 for 5 min at 5 min intervals. In (b), the same pulse is applied at 1 min intervals. In each case the variation of intracellular cAMP ($\beta$) is represented, as well as the variation of the total fraction of active receptor ($\rho_T$). Parameter values are those of fig. 5.38 (Martiel & Goldbeter, 1987a).

pulses are separated by 1 min only, is shown in fig. 8.1b. In each case the response of the system rapidly stabilizes in the course of time: indeed the synthesis of intracellular cAMP reaches a constant amplitude after a few stimuli. Moreover, this amplitude is less than that of the response to the first signal. When the interval between successive stimuli is 5 min, the amplitude of the response decreases only slightly in the course of stimulation. In contrast, when the interval reduces to 1 min, the amplitude of the response markedly decreases.

The above differences in dynamic behaviour can be explained by following the evolution of the fraction of active receptor, $\rho_T$. Figure 8.1a shows that the receptor begins to desensitize at the onset of each stimulation phase. The interval between two successive stimuli is, however, sufficiently large for the receptor to resensitize before the beginning of the next stimulation. Figure 8.1b shows that this is not the case when the interval is shorter: the receptor has not enough time to resensitize sufficiently between two stimuli; consequently, the amplitude of the response elicited by the cAMP signals can only be reduced.

These results suggest that the efficacy of the periodic signal is closely related to the phenomenon of receptor desensitization in the presence of prolonged stimuli. The model provides an explanation for the observation that constant cAMP signals have no effect on the differentiation of *D. discoideum* cells, whereas periodic signals applied every 5 min succeed in inducing the synthesis of proteins required for their development (Darmon *et al.*, 1975; Gerisch *et al.*, 1975). In the presence of constant stimulation, as shown by the experiments of Devreotes & Steck (1979) and by simulations of the model (Martiel & Goldbeter, 1984, 1987a), the receptor can only desensitize, thus leading to the reduction, or even disappearance, of cAMP synthesis after a few minutes.

The theoretical analysis further allows us to explain the specificity of the frequencies capable of inducing differentiation. As shown by Wurster (1982), cAMP signals spaced by 2 min fail to promote differentiation and are therefore as inefficient as constant stimuli. The results given in fig. 8.1 suggest that the cAMP receptor, which becomes phosphorylated in the course of stimulation, has not enough time, in 2 min, to be significantly dephosphorylated, so that the receptor remains primarily trapped in its desensitized state in the course of successive stimulations. There results a marked decrease in the synthesis of cAMP; the latter is probably too low to elicit the physiological effect induced by pulses applied at sufficiently long intervals.

The model allows the identification of the parameter that controls the relative efficiency of pulsatile stimuli of different periods. Indeed, numerical simulations indicate that the main process governing the response of the system to such stimuli is the rate of dephosphorylation, which determines the rapidity at which the receptor resensitizes between successive stimuli (this point is elaborated further in section 8.5). In *D. discoideum* amoebae, the rate of resensitization is thus governed by the activity of a phosphatase; the kinetic constant and the concentration of that enzyme are such that dephosphorylation takes place in a few minutes (fig. 5.25). This duration corresponds well to the intervals that permit a quasi-maximal response to periodic pulses of cAMP.

## 8.3 Link with pulsatile hormone secretion: the example of gonadotropic hormones

Intercellular communication by periodic, pulsatile signals is not observed only in *Dictyostelium* amoebae. In recent years, it has become clear that an analogous phenomenon occurs for a large number of hormones whose physiological effect is linked to their periodic secretion into the blood circulation (Crowley & Hofler, 1987; Wagner & Filicori, 1987; Leng, 1988). The mechanism of communication by periodic signals in the slime mould throws light on the function of pulsatile patterns of hormone secretion, and suggests that a major role of these rhythms is to provide an optimal mode of intercellular communication.

Pulsatile (or episodic) hormone secretion has been observed, among other examples, for insulin and glucagon (Goodner *et al.*, 1977; Lang *et al.*, 1979, 1982; Lefèbvre *et al.*, 1987), growth hormone (Tanenbaum & Martin, 1976; Borges *et al.*, 1984; Bassett & Gluckman, 1986), cortisol and adrenocorticotropic hormone (ACTH) (Van Cauter, 1987), as well as for corticotropin-releasing hormone (CRH), which liberates ACTH (Avgerinos *et al.*, 1986). In each of these cases, the hormone best exerts its effects on target cells when it reaches them in the form of ultradian pulses whose periodicity is close to that observed in physiological conditions.

The prototype for periodic hormone secretion remains, however, that of GnRH released by the hypothalamus at the frequency of one pulse per hour in humans and in the rhesus monkey (Dierschke *et al.*, 1970; Carmel, Araki & Ferin, 1976; Crowley *et al.*, 1985), and at a higher frequency in the rat (see Rasmussen, 1993). This hormone induces the release by the pituitary of the gonadotropic hormones LH (luteinizing hormone) and FSH (follicle-stimulating hormone) (Schally *et al.*, 1971), hence the name of LHRH (LH-releasing hormone) also given to GnRH (gonadotropin secretion by pituitary cells has recently been shown to involve GnRH-induced high-frequency $Ca^{2+}$ oscillations (Stojilkovic *et al.*, 1993; Tse *et al.*, 1993); see chapter 9). LH and FSH control gonadal development in men and women, as well as ovulation in the course of the menstrual cycle. A typical example of rhythmic secretion of GnRH accompanied by pulsatile release of LH and FSH is shown in fig. 8.2.

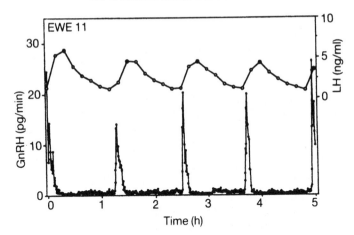

Fig. 8.2. Rhythmic, pulsatile secretion of GnRH inducing the pulsatile release of the gonadotropic hormones LH and FSH. The data, obtained in the ewe by Moenter *et al.* (1992), show a frequency close to one pulse per hour; a similar frequency is observed in rhesus monkey and humans.

Knobil and his coworkers (Belchetz *et al.*, 1978; Knobil, 1980; Wildt *et al.*, 1981; Pohl *et al.*, 1983) have performed a series of remarkable experiments in rhesus monkeys subjected to hypothalamic lesions. Because of these lesions, the monkeys were unable to generate autonomously the GnRH signal and, consequently, lacked the gonadotropic hormones since the latter were no longer secreted by the pituitary. When the GnRH hormone is injected into these monkeys in a constant manner, the level of LH and FSH in the blood remains low (fig. 8.3a). In contrast, when the hormone is injected at the physiological frequency of one 6 min pulse per hour, the normal level of gonadotropic hormones is progressively restored (fig. 8.3b). This transition is reversible (fig. 8.4): the passage from a pulsatile to a constant injection markedly reduces the level of LH and FSH, while the return to periodic injection permits the restoration of normal levels.

Even more remarkably, the level of gonadotropic hormones depends on the frequency of pituitary stimulation by periodic signals of GnRH (fig. 8.5). When the frequency of stimulation by GnRH rises up to two or three pulses per hour, the level of LH and FSH drops significantly. When, conversely, the frequency is lower than normal, the usual ratio between LH and FSH levels is strongly diminished. The latter observation can be explained by the difference in clearance rate of the two hormones within the circulation. Work on the effect of GnRH pulse

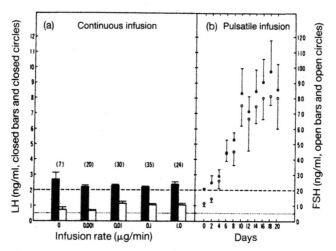

Fig. 8.3. Pituitary response to continuous or pulsatile stimulation by the hormone GnRH in the rhesus monkey. In (a), the continuous infusion of GnRH at four different rates fails to induce proper secretion of LH and of FSH. In contrast, pulsatile infusion of GnRH (b) in the form of 1 mg/min for 6 min every hour induces appropriate secretion of the gonadotropic hormones by the pituitary (Belchetz *et al.*, 1978).

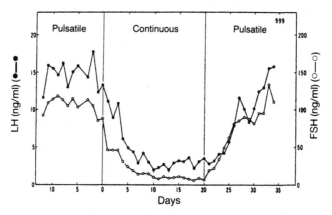

Fig. 8.4. Continuous injection (1 μg/min from day 0) of the hormone GnRH suppresses the secretion of the gonadotropic hormones LH and FSH by the pituitary in the rhesus monkey. Normal secretion of LH and FSH is progressively re-established when the GnRH decapeptide is again administered in a pulsatile fashion, as a 6 min pulse of 1 μg/min every hour. As in fig. 8.3, the rhesus monkey is unable to generate the GnRH signal autonomously, owing to a specific lesion of the hypothalamus (Belchetz *et al.*, 1978).

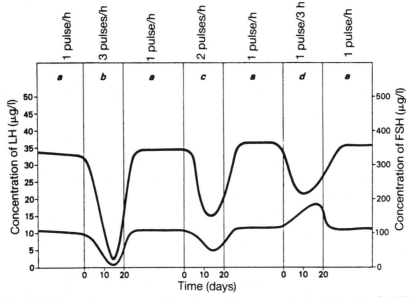

Fig. 8.5. Schematic summary of the effect of frequency of stimulation by GnRH on the secretion of the gonadotropic hormones LH (upper curve) and FSH (lower curve) in the rhesus monkey subjected to a hypothalamic lesion impairing its ability to generate GnRH (Thalabard *et al.*, 1984, based on data from Pohl *et al.*, 1983).

frequency on gonadotropin secretion has recently been extended to the level of gonadotropin mRNA production (Ishizaka *et al.*, 1992).

The studies of Knobil on the rhesus monkey demonstrate the importance of the frequency of stimulation for the physiological response. The hormonal control of the reproductive system relies on the frequency encoding of the GnRH signal: as emphasized by Knobil (1981), the temporal pattern of the hormonal signal becomes as important as, if not more important than, the concentration of the hormone in the blood; this view represents a truly innovative concept in endocrinology. The central nervous system in fact controls fertility (Karsch, 1987), as well as the onset of puberty (Wildt, Marshall & Knobil, 1980), by alterations in the frequency of GnRH secretion by the hypothalamus.

The clinical applications of these observations have led in recent years to remarkable developments, not only for hormones involved in the control of reproduction, but also for other hormones whose pulsatile secretion has been uncovered. A hormone whose normal pattern of pulsatile secretion is perturbed possesses increased efficiency when injected in a pulsatile rather than continuous manner; moreover the response of target cells improves when the stimulation frequency approaches the physiological value, as demonstrated for insulin and glucagon (Matthews *et al.*, 1983; Weigle, Koerker & Goodner, 1984; Komjati, Bratusch-Marrain & Waldhaüsl, 1986; Weigle & Goodner, 1986; Lefèbvre *et al.*, 1987; Paolisso *et al.*, 1989, 1991) and for growth hormone (Jansson *et al.*, 1982; Clark *et al.*, 1985; Hindmarsh *et al.*, 1990).

Several reproductive disorders involving impaired GnRH secretion are known in human physiology (Crowley *et al.*, 1985). Some of these disorders are now being treated by means of pulsatile GnRH injection in women (Santoro, Filicori & Crowley, 1986; Filicori, 1989) and men (Wagner, 1985; Wagner *et al.*, 1989). Therapeutic programmes based on the pulsatile injection of this hormone by means of a pump programmed at the appropriate frequency lead to the restoration of ovulation, thus allowing pregnancies to occur in previously infertile women (Leyendecker, Wildt & Hansmann, 1980; Reid, Leopold & Yen, 1981).

A close link exists between the efficiency of pulsatile patterns of hormone secretion and the efficacy of periodic signals of cAMP used by *Dictyostelium* amoebae during aggregation. The study of the latter system could shed light on the function of pulsatile hormone secretion, a phenomenon that is observed for an increasing number of hormones. In each case, the periodic secretion of the signal permits the repetitive occurrence of a quasi-maximal response from a target cell that retains the capacity of adapting to constant stimuli as a result of receptor desensitization (Goldbeter, 1987a,b, 1988a).

The results obtained by Wurster (1982) in *Dictyostelium* and by Knobil (1980) in the rhesus monkey show, nevertheless, that the frequency of stimulation is as important as the periodic nature of the signal in eliciting an adequate response. The results shown in fig. 8.1 obtained from the model for cAMP synthesis based on receptor desensitization in *D. discoideum* provide a theoretical basis for these observations. Confirming that the efficiency of the signal is a function of its frequency, these results suggest that the optimum frequency will be dictated by the relative values of the interval between successive stimuli and the time required for receptor resensitization.

To determine in more detail the link between the frequency of stimulation and the cellular response in the case of *Dictyostelium* as in that of hormonal signals, it is desirable to address in sufficiently general terms the effect of pulsatile stimulation on a receptor subjected to reversible desensitization. A simple, general model appropriate for such an analysis is presented below.

## 8.4 Response of a target cell to pulsatile stimulation: a general model based on receptor desensitization

The first model proposed for the phenomenon of adaptation to constant stimuli, due to Katz & Thesleff (1957), sought to explain the desensitization of the acetylcholine receptor following prolonged incubation of the receptor with its ligand. This model was based on the existence of two conformational states of the receptor, one of which is desensitized and possesses a reduced capacity to induce the cellular response. Experimental data confirm the existence of multiple conformational states of the acetylcholine receptor (Changeux, 1981), although these data reveal that the situation is more complex than that considered in the initial model.

A mathematical model for the adaptation of sensory systems to constant stimuli has been developed in a series of experimental and theoretical studies devoted to bacterial chemotaxis, a sensory response for which the phenomenon of adaptation is also observed (Macnab & Koshland, 1972; Koshland, 1977; Goldbeter & Koshland, 1982b). This model is based on the desensitization of the sensory receptor through conformational changes or some covalent modification. The latter mechanism underlies adaptation of the cellular response in bacteria and in *D. discoideum* amoebae, given that in these organisms the phenomenon is brought about, respectively, by methylation and phosphorylation of the receptor (Koshland, 1979; Springer *et al.*, 1979; Devreotes & Sherring, 1985; Klein, C. *et al.*, 1985; Meier & Klein, 1988; Vaughan & Devreotes, 1988).

Segel *et al.* (1985, 1986) have re-examined this model and extended it to account for the property of **exact adaptation** to constant stimuli. In contrast to **partial adaptation**, the former phenomenon, observed for bacterial chemotaxis and cAMP secretion induced by cAMP signals in *D. discoideum*, denotes the passage of the response through a maximum and its return to the same basal steady-state level regardless of the amplitude of the external stimulation. This model was applied to exact adaptation to constant stimuli in bacterial chemotaxis and in cAMP synthesis in the slime mould (Knox *et al.*, 1986). Li & Goldbeter (1989b) used the same model to determine its response to pulsatile stimulation as a function of the frequency and shape of the signal. The results of this study, presented below, apply both to the secretion of cAMP signals in *Dictyostelium* and to the pulsatile release of GnRH; in both cases the analysis allows us to clarify the molecular bases of the efficiency of the periodic signal.

### *A general receptor model based on reversible desensitization*

The general model considered for the response to external stimulation (fig. 8.6) is that of a receptor existing in two forms, active (R) and desensitized (D), which, upon binding the ligand (L), give rise to the complexes X ≡ RL and Y ≡ DL. The reversible transition between the active and desensitized states may correspond to a simple conformational change or to some covalent modification. The kinetic constants appearing in fig. 8.6 relate to the binding and dissociation of the ligand, for the two forms of the receptor, and to the reversible transitions between these forms. A similar model has been considered by Swillens & Dumont (1977) for the desensitization of hormone receptors; exact adaptation was, however, not examined in that study.

So as to retain a frame as general as possible for the analysis, we assume that the response of the system, elicited by binding of the ligand to the receptor, is proportional to a certain **activity** $A(t)$ whose value is a function of the relative populations of the four forms of the receptor, namely R, D, X and Y. More specifically, it is assumed that the activity is a linear combination of the concentrations of these four receptor

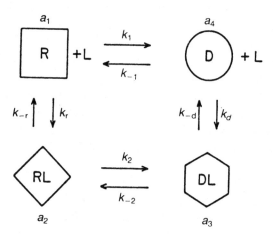

Fig. 8.6. Model for exact adaptation based on receptor desensitization. The receptor exists in the states R (active) and D (desensitized); the ligand, L, binds to these two forms with different affinities. Upon stimulation, each of the four receptor states (R, RL, D, DL) contributes to the response with a certain weight measured by the activity coefficient, $a_i$ $(i = 1, \ldots 4)$. The transitions between the more active forms R, RL and the desensitized forms D, DL are reversible and correspond either to a simple conformational change or to some covalent modification (Segel *et al.*, 1985, 1986).

forms, and that the contribution of each of them is weighted by an **activity coefficient** $a_i$ ($i = 1, \ldots 4$), as indicated in fig. 8.6. The activity $A(t)$ is thus defined by relation (8.1), where $R(t)$, $D(t)$, $X(t)$ and $Y(t)$ denote the concentrations of the four receptor forms in the course of time:

$$A(t) = a_1 R(t) + a_2 X(t) + a_3 Y(t) + a_4 D(t) \tag{8.1}$$

The time evolution of the receptor system is governed by the kinetic equations:

$$\frac{dR}{dt} = -k_1 R + k_{-1} D - k_r R \times L + k_{-r} X$$

$$\frac{dX}{dt} = -k_2 X + k_{-2} Y + k_r R \times L - k_{-r} X$$

$$\frac{dY}{dt} = k_2 X - k_{-2} Y + k_d D \times L - k_{-d} Y$$

$$\frac{dD}{dt} = k_1 R - k_{-1} D - k_d D \times L + k_{-d} Y \tag{8.2}$$

The sum of these equations leads to the conservation relation (8.3) for the total amount of receptor:

$$R + X + Y + D = R_T \tag{8.3}$$

Equations (8.2) admit a unique steady state, which is always stable. In the simple, general case where the binding and dissociation of the ligand are much faster than the processes of desensitization and resensitization of the receptor, eqns (8.2) and (8.3) reduce to a single differential equation governing the time evolution of the system. The integration of that equation yields an analytical expression for the four receptor forms in the course of time (Segel *et al.*, 1986).

In the absence of stimulation ($L = 0$), only the two forms R and D are present; their concentration at steady state is denoted $R_0$ and $D_0$, respectively. The basal activity $A_0$ is then given by eqn (8.4) (see Segel *et al.*, 1986, for further details) in which constant $K_1$ is equal to the ratio $k_{-1}/k_1$ of the resensitization and desensitization constants for the receptor, defined in fig. 8.6:

$$A_0 = a_1 R_0 + a_4 D_0 = R_0 (a_1 + a_4 K_1^{-1}) = \frac{R_T}{1 + K_1} (a_1 K_1 + a_4) \tag{8.4}$$

A stimulus corresponds to the increase of the ligand from an initial concentration $L_i$ to a final concentration $L_f$. For exact adaptation to occur, the activity, after an initial increase due to the redistribution of the receptor between the forms R, D, X and Y, which differ by their activity coefficient ($a_2$ having the largest value), must return to the basal value $A_0$ whatever the amplitude of the stimulation, i.e. regardless of the value of $L_f$. Mathematically, this translates into condition:

$$A(t) \rightarrow A_0, \quad \text{for } t \rightarrow \infty \tag{8.5}$$

Given that the steady-state concentrations of the four receptor forms are nonlinear functions of the amplitude of the stimulus, we would expect that different values of $L_f$ give rise to different redistributions of the receptor between the forms R, D, X and Y, so that, at steady state, relation (8.1) would give a value for the activity that would differ from the basal value, $A_0$. In conformity with intuition, this phenomenon indeed occurs in the general case and corresponds to partial adaptation (fig. 8.7, curves $b$ and $c$). It is, however, possible to show (Segel *et al.*,

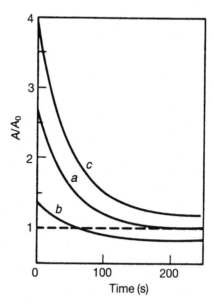

Fig. 8.7. Exact versus partial adaptation in the model based on receptor desensitization. The activity $A$ (eqn (8.1)) is shown for the stimulus $\gamma = 0 \rightarrow 10$. Whereas exact adaptation occurs when $a_2 = a_2{}^*$ (curve $a$), partial adaptation obtains for $a_2$ equal to $a_2{}^*/2$ (curve $b$ ) or 1.5 $a_2{}^*$ (curve $c$), where $a_2{}^*$ denotes the value of $a_2$ prescribed by eqn (8.6) for exact adaptation (Segel *et al.*, 1985, 1986; Goldbeter, 1987a).

1986) that, for a given value of the kinetic constants and of the activity coefficients $a_1$ and $a_4$ associated with the basal activity $A_0$, there exists a unique choice of the activity coefficients $a_2$ and $a_3$ such that exact adaptation occurs whatever the value of the stimulus. The values of the activity coefficients $a_2$ and $a_3$ which ensure exact adaptation are given by:

$$a_2 = \frac{a_1(k_2 + k_{-1}) - a_4(k_2 - k_1)}{k_{-1} + k_1}$$

$$a_3 = \frac{-a_1(k_{-2} - k_{-1}) + a_4(k_{-2} + k_1)}{k_{-1} + k_1} \tag{8.6}$$

In the case where the transitions between the active and desensitized states are purely conformational, the cyclical nature of the scheme in fig. 8.6 imposes the condition of detailed balance (8.7), where $K_2 = k_{-2}/k_2$ with $c = K_R/K_D$ and $K_R = k_{-r}/k_r$, $K_D = k_{-d}/k_d$:

$$K_1 = K_2 c \tag{8.7}$$

When desensitization is due to some covalent modification, the particular condition (8.7) does not necessarily hold, because the direct and reverse steps in the cyclical system correspond to the direct steps of distinct reactions catalysed by different enzymes (Segel *et al.*, 1986; Waltz & Caplan, 1987). Thus, in the case of receptor phosphorylation in *Dictyostelium*, the constants $k_1$, $k_2$ and $k_{-1}$, $k_{-2}$ relate, respectively, to the direct steps of the reactions catalysed by the protein kinase and by the phosphatase, reactions for which the reverse steps are being neglected. Conditions (8.6) then suffice to ensure exact adaptation for all stimuli, as indicated by fig. 8.7 and by fig. 8.8a where the changes in the activity are shown, in response to stimuli of increasing magnitude.

Figure 8.8 shows that the receptor system adapts exactly to successive stimuli corresponding to two sequential increments in ligand concentration. Such an adaptation is accompanied by an increase in the level of modified receptor. The latter returns to the nondesensitized state when the ligand is removed from the medium. The property of exact adaptation is useful in studying the effect of periodic stimulation as it allows us to define a reference, basal value for the activity.

Besides applying to cAMP or hormonal stimuli, the model for receptor adaptation was recently shown to apply to adaptation of the ryanodine-sensitive $Ca^{2+}$ channel (see chapter 9) in response to successive increments in cytosolic $[Ca^{2+}]$ (Sachs *et al.*, 1995).

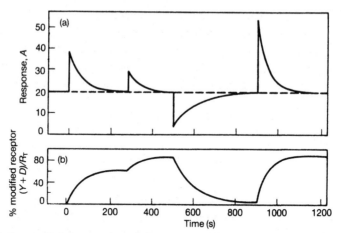

Fig. 8.8. Sequential changes in activity generated by the receptor in response to the successive stimuli in which the level of the ligand, $L/K_R$, passes from 0 to 1, 1 to 10, 10 to 0, and 0 to 10. The upper curve (a) for $A$ is determined according to eqn (8.1), when conditions (8.6) for exact adaptation are met, while the lower curve (b) shows the concomitant variation in the fraction of desensitized receptor (Segel *et al.*, 1986; Goldbeter, 1987a).

### Response to pulsatile stimulation

The simplest form of periodic stimulus, from the analytical point of view, is that of a square wave already considered in the numerical simulations carried out in the model for cAMP synthesis in *Dictyostelium* (fig. 8.1). Such a stimulation, which closely resembles the secretion of a GnRH pulse *in vivo* (Moenter *et al.*, 1992), consists in raising the normalized concentration of the ligand from an initial value $\gamma_0 \ (= L_0/K_R)$ up to the value $\gamma_1 \ (= L_1/K_R)$ for a time interval $\tau_1$, and then bringing the ligand back to the basal value $\gamma_0$ for an interval $\tau_0$, before repeating the operation (fig. 8.9). The period and amplitude of the square-wave stimulus are thus given by:

$$T = \tau_0 + \tau_1 \tag{8.8}$$

$$\delta\gamma = \gamma_1 - \gamma_0 \tag{8.9}$$

The concentrations of the different receptor forms and the activity $A(t)$ can be normalized, by dividing by the total quantity of receptor, $R_T$. The definition of the activity then takes the form of relation:

$$\alpha(\gamma, t) = a_1 \, r + a_2 \, x + a_3 \, y + a_4 \, d \tag{8.10}$$

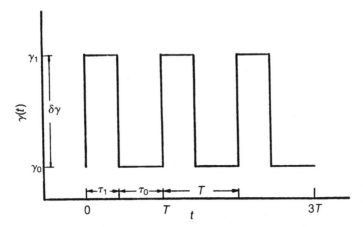

Fig. 8.9. Square-wave signal considered for the response to pulsatile stimulation in the general model based on receptor desensitization. The duration of stimulation and the interval between successive stimuli are denoted by $\tau_1$ and $\tau_0$, respectively (Li & Goldbeter, 1989b).

The response of the system remains governed by eqns (8.2). As in the case of the response to a single step increase in ligand (Segel *et al.*, 1986), it is possible to reduce the system of differential equations to a single variable, e.g. the total quantity of desensitized receptor, $Y + D$, and to express the concentration of each receptor form as a function of that quantity (Li & Goldbeter, 1989b). Integrating that equation yields an analytical expression for the evolution of the system in the transient phase during which it approaches the asymptotic state; the latter state is reached once the adaptation phase is completed.

As confirmed by fig. 8.10, the system responds to the square-wave stimulus (part c) by an activity change of similar period; after a transient phase during which its amplitude decreases, the activity reaches a constant level (part b). At the same time the amplitude of the periodic variation in the quantity of desensitized receptor stabilizes at a constant level (part a).

### *Optimal pattern of pulsatile stimulation maximizing target cell responsiveness*

To determine the effect of parameters $\tau_0$, $\tau_1$ and $\delta\gamma$, which characterize the periodic stimulus, on the response of the system, we can measure the latter by the integrated activity above the basal level $\alpha_0 = A_0/R_T$ over

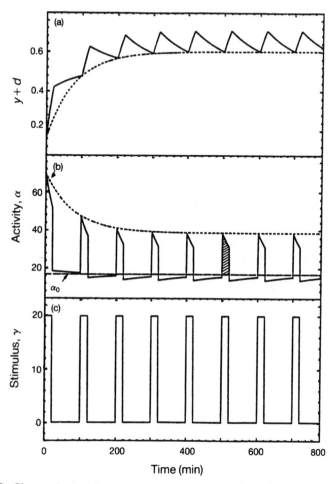

Fig. 8.10. Change in activity elicited by periodic, pulsatile stimulation in the model based on receptor desensitization. The periodic variation of the ligand, in the form of a square wave (c), gives rise to a periodic change in activity according to eqn (8.10) (b), which after several cycles stabilizes at a constant amplitude. This adaptation to the periodic stimulus is accompanied by periodic desensitization of the receptor (a), which also stabilizes at a constant amplitude higher than the initial level of desensitization prior to the onset of stimulation. The dashed area in the activity peak above the basal value $\alpha_0$ is the integrated activity, $\alpha_T$, adopted as measure of the receptor response to pulsatile stimulation after adaptation has occurred (Li & Goldbeter, 1989b).

a period $T$. This quantity, $\alpha_T$, represented by the dashed area in fig. 8.10b, is given by an analytical expression that contains the different parameters of the pulsatile signal (Li & Goldbeter, 1989b). Similar results are obtained when the response is measured by the maximum amplitude of the activity.

If we define $\beta$ as the ratio $\tau_1/\tau_0$ that characterizes the pattern of periodic stimulation, we can search for an optimum pattern maximizing the integrated activity $\alpha_T$, by solving:

$$\frac{\partial \alpha_T}{\partial \beta} = 0 \qquad (8.11)$$

The existence of an optimum value $\beta^*$ solution of eqn (8.11) (Li & Goldbeter, 1989b) is confirmed by the numerical study of the integrated activity as a function of $\beta$ (fig. 8.11). This result indicates that the relative duration of the stimulation phase with respect to the interval between two stimuli governs the amplitude of the activity change elicited by the signal.

In contrast, the derivative of the integrated activity with respect to the period $T$, for a given ratio $\tau_1/\tau_0$, is always positive (Li & Goldbeter, 1989b). The relation:

$$\frac{\partial \alpha_T}{\partial T} > 0 \qquad (8.12)$$

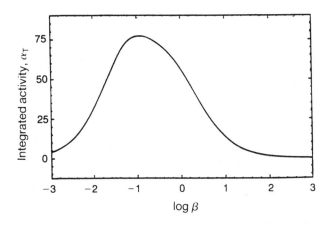

Fig. 8.11. Variation of the integrated activity $\alpha_T$ of the receptor, as a function of the ratio $\beta = \tau_1/\tau_0$ characterizing the periodic stimulus. The presence of a maximum denotes the existence of an optimal pattern of the signal at a given period $T = \tau_0 + \tau_1$ (Li & Goldbeter, 1989b).

thus implies that the activity does not go through a maximum when the period of the stimulus increases. This result is confirmed by numerical simulations indicating that the integrated activity of the receptor system rises and eventually reaches a constant level as a function of $T$, at a fixed value of $\beta$. As indicated by the curve shown in fig. 8.12, the same result holds when $T$ rises as a result of an increase in $\tau_0$, at a fixed duration $\tau_1$ of the stimulus. These curves, obtained for three increasing values of the stimulus amplitude, are similar to those usually obtained for a cellular response as a function of agonist concentration (Boeynaems & Dumont, 1980). Here, however, each curve is obtained as a function of the period of pulsatile stimulation, at a given amplitude. This result therefore establishes the possibility of a **frequency encoding** of the signal, such that the response of the target cell is modulated by the frequency rather than by the amplitude of the pulsatile stimulus (Goldbeter & Li, 1989).

The fact that the integrated activity rises and eventually reaches a constant level when the period of the stimulus increases, at a fixed value

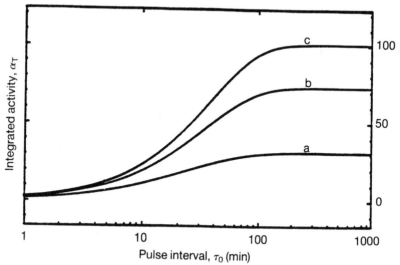

Fig. 8.12. Variation of the integrated activity of the receptor, $\alpha_T$, as a function of signal period when the latter increases owing to a rise in the interval $\tau_0$, at a given stimulus duration $\tau_1$ equal to 6 min. The curve is then obtained for three increasing values of the stimulus amplitude, $\delta\gamma = \gamma_1 - \gamma_0 = 0.1$, 1, 10, with $\gamma_0 = 0.001$. Here, in contrast to fig. 8.11, the activity does not go through a maximum as the period increases. The progressive rise in $\alpha_T$ with $\tau_0$ and, hence, with the period $T = \tau_1 + \tau_0$, provides the bases for the frequency encoding of the pulsatile stimulus (Goldbeter & Li, 1989b).

of $\tau_1$, originates from the receptor resensitizing more and more as the interval $\tau_0$ between two successive stimuli increases. The integrated activity of the receptor thus rises with $T$, until it reaches a maximum value.

If a receptor resensitizes significantly in 10 min, we can easily see that a stimulus applied at regular intervals of 2 to 3 h (or more) will elicit a maximum response. However, another view of cellular dynamics would rather take into account the capacity of the receptor to generate as many responses of significant amplitude as possible in a given amount of time. This idea is well expressed in a short poem by the French poet Eugène Guillevic (born in 1907) taken from a recent collection entitled *Le Chant* (Guillevic, 1990; © Editions Gallimard):

> Dans un minimum
> De temps
>
> Mettre un maximum
> De chant —

To characterize this property Li & Goldbeter (1989b) considered the **cellular responsiveness**, $\alpha_R$, defined by:

$$\alpha_R = \frac{\alpha_T}{\alpha_{Tstep}} \times \frac{\alpha_T}{T} \tag{8.13}$$

The quantity $\alpha_R$ is the product of two terms: the first, which measures the magnitude of the response, compares the activity $\alpha_T$ with the maximum integrated activity $\alpha_{Tstep}$ elicited by a unique stimulus of infinite duration of stimulation $\tau_1$; the second, which measures the amount of response per unit time, yields the mean integrated activity, i.e. $\alpha_T$ divided by the period of stimulation $T$. This definition of cellular responsiveness allows quantification of the number of responses of significant amplitude that the system can generate in a given amount of time; similar results would be obtained with other, related quantities defined in a similar manner.

Is there a frequency of stimulation that maximizes cell responsiveness? An analytical expression for $\alpha_R$ has been derived in the receptor model; studying the conditions for an extremum as a function of the durations $\tau_1$ and $\tau_0$ shows that there exists a pair of values of these parameters, $(\tau_1{}^*, \tau_0{}^*)$, that corresponds to a maximum in $\alpha_R$ (Li & Goldbeter, 1989b). This conclusion is illustrated in fig. 8.13, where cellular responsiveness is plotted as a function of $\tau_1$ and $\tau_0$: for fixed values of

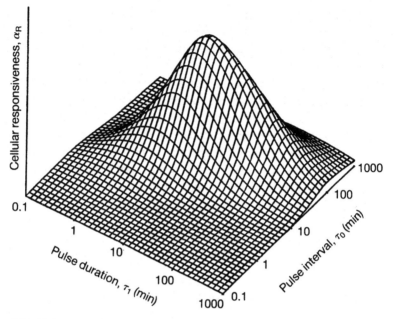

Fig. 8.13. Existence of an optimal periodic signal maximizing cellular responsiveness, $\alpha_R$. The quantity $\alpha_R$, defined by eqn (8.13), is determined numerically as a function of the duration of stimulation, $\tau_1$, and the interval between two stimuli, $\tau_0$. The optimum $(\tau_1{}^*, \tau_0{}^*)$ denotes the existence of a periodic signal maximizing the cellular responsiveness measured by $\alpha_R$. The figure is established for values of the kinetic parameters chosen so as to yield the values $(\tau_1{}^* = 6$ min, $\tau_0{}^* = 54$ min$)$ observed for the optimal, pulsatile pattern of GnRH in the experiments of Pohl *et al.* (1983) on the induction of LH and FSH secretion by the pituitary in response to exogenous GnRH stimulation in the rhesus monkey (Li & Goldbeter, 1989b).

the kinetic parameters characterizing desensitization and resensitization of the receptor, a particular pair of values of the duration of stimulation and of the interval between successive stimuli is seen to maximize the capability of the receptor to generate, in a given amount of time, the largest possible number of significant responses measured by the integrated activity above the basal level.

The optimum values $\tau_1{}^*$, $\tau_0{}^*$ depend on the parameters of the model. In particular, as shown by fig. 8.14, $\tau_1{}^*$ depends markedly on the desensitization rate constant $k_2$ (fig. 8.14a), while $\tau_0{}^*$ exhibits the strongest dependence on the resensitization constant $k_{-1}$ (fig. 8.14b). These results agree with intuition: the receptor needs less time to switch back into its active state when resensitization is faster; hence, the optimum interval

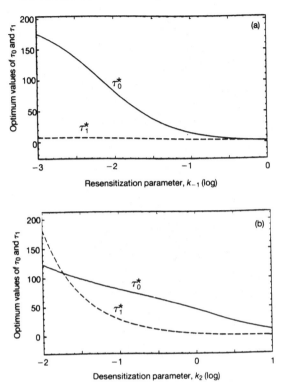

Fig. 8.14. Dependence of the optimal pattern of pulsatile stimulation on the kinetics of receptor desensitization and resensitization. Shown are the variation of the optimal duration $\tau_1^*$ and optimal interval $\tau_0^*$ as a function of the desensitization rate constant $k_2$ (b) and resensitization rate constant $k_{-1}$ (a) (Li & Goldbeter, 1989b).

$\tau_0^*$ should diminish as $k_{-1}$ increases, while $\tau_1^*$ remains largely unaffected by such parameter variation. Conversely, $\tau_1^*$ decreases, while $\tau_0^*$ remains practically unchanged, when desensitization proceeds at a faster pace.

The existence of an optimal pattern of pulsatile stimulation is independent of the amplitude of the stimulus. As shown in fig. 8.15, an optimum in $\alpha_R$ is observed at saturation (part a), half saturation (part b) or much below saturation by the ligand (part c). The position of the optimum – i.e. the particular pair of values ($\tau_1^*$, $\tau_0^*$) – as well as the absolute value of $\alpha_R$ vary, however, as a function of the magnitude of the pulsatile stimulus.

Similar results are obtained when a stimulus more realistic than a

(a)

(b)

(c)

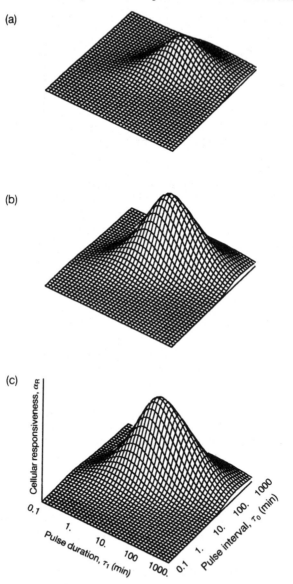

Fig. 8.15. Optimal pattern of cell responsiveness observed at three increasing magnitudes of stimulation in the general model based on receptor desensitization. The variation in cell responsiveness $\alpha_R$, defined by eqn (8.13), is determined numerically as a function of the duration of stimulation, $\tau_1$, and the interval between two stimuli, $\tau_0$, for $\gamma_1 = 0.1$ (a), 1 (b) and 10 (c); the basal value of the normalized ligand level, $\gamma_0$, is equal to 0.001. The optimum $(\tau_1^*, \tau_0^*)$ as well as the absolute value of $\alpha_R$ are seen to vary with the amplitude of the pulsatile stimulus (Li & Goldbeter, 1992).

square wave is chosen, such as the repetition, at regular intervals, of an instantaneous elevation of the ligand followed by its exponential degradation (Li & Goldbeter, 1989b). The lack of analytical expression for cell responsiveness then compels us to resort to the numerical simulations; the latter corroborate the findings of fig. 8.13 by again demonstrating the existence of an optimal pattern of pulsatile stimulation maximizing cellular responsiveness.

## 8.5 Optimal frequency of pulsatile signalling in intercellular communication

The study of a general model for the response of a receptor subjected to desensitization shows that continuous stimulation leads to a unique response, owing to the phenomenon of adaptation, while periodic stimulation can, in contrast, elicit a large number of significant responses provided that the characteristics of the pulsatile signal remain close to those that allow for maximum responsiveness in target cells.

This analysis applies to the two examples of intercellular communication by pulsatile signals discussed above, namely the response to cAMP signals in *Dictyostelium* amoebae, and the secretion of gonadotropic hormones by pituitary cells in response to pulsatile signals of GnRH. The model shows that in each case the characteristics of the desensitization process dictate the optimal periodic signal that maximizes cellular responsiveness. The results obtained with the general model developed in section 8.4 will be validated for the cAMP signalling system by resorting to the specific model for cAMP synthesis based on receptor desensitization. Whether these results might be of general applicability, beyond the pulsatile secretion of cAMP and GnRH, is discussed below for the cases of insulin and growth hormone signalling.

### *Optimal frequency of the pulsatile signal in* Dictyostelium

In *D. discoideum*, receptor desensitization results from reversible phosphorylation (Devreotes & Sherring, 1985; Klein, C. *et al.*, 1985). The study of a model based on this process (chapter 5) allows us to account for the oscillations and relay of cAMP signals observed in the course of aggregation. In conditions where the level of extracellular cAMP is controlled and varied periodically, simulations of the model indicate (fig. 8.1) that cAMP synthesis is more important when the interval between successive stimuli is 5 min rather than 1 min. Before examining the results of a more detailed study of the effect of periodic stimulation in that model, it is useful to turn to the simpler, general receptor model discussed in the preceding sections, to see what are its predictions concerning the efficiency of pulsatile signals of cAMP in *Dictyostelium*.

Although the model presented in section 8.4 does not contain any explicit coupling between cAMP binding to the receptor and the activation of adenylate cyclase, this model is formulated in a sufficiently general manner to allow us to determine how the response of the cAMP-synthesizing system depends on the frequency of stimulation by extracellular cAMP pulses. It is enough to assume that the activity of adenylate cyclase is proportional to the activity $A(t)$ generated by the binding of the ligand, cAMP, to the receptor; the activity coefficients associated with the four receptor forms could then be viewed as affinity constants for the formation of complexes between these receptor states and adenylate cyclase (Segel *et al.*, 1986). When the self-amplifying loop involving extracellular cAMP is included in this model, the properties of oscillations and relay of cAMP are observed (Barchilon & Segel, 1988), just as in the model by Martiel & Goldbeter (1987a), based in a more explicit manner on the detailed kinetics of cAMP synthesis.

When introducing in the general model for square-wave stimulation the values of the parameters for desensitization and resensitization of the cAMP receptor (table 5.3) already considered in the simulations of fig. 8.1 using the model by Martiel & Goldbeter (1987a), we obtain a surface similar to that shown in fig. 8.13 when cell responsiveness $\alpha_R$, related to the capacity of synthesizing cAMP, is determined as a function of the duration of stimulation $\tau_1$ and the interval between two cAMP pulses $\tau_0$. The periodic signal giving rise to maximum cell responsiveness corresponds to a stimulus duration of 3.8 min and to an interval of 5 min between two successive signals (Li & Goldbeter, 1989b). These values are close to those observed for the periodic signal emitted by centres in the course of aggregation. They are also close to the values which, in the model by Martiel & Goldbeter (1987a), produce a quasi-maximal response once the system has adapted to periodic stimulation (fig. 8.1a).

When applying these results to intercellular communication in cellular slime moulds, we need to remember that the amoebae relaying cAMP signals during aggregation are excitable. These cells therefore possess an absolute and a relative refractory period, but only the latter period characterizes the response of the system when self-amplification in cAMP synthesis is suppressed under conditions where the level of extracellular cAMP is artificially clamped (chapter 5). The results obtained here relate to the latter situation. The existence of an absolute refractory period should modify the characteristics of the optimal signal maximizing cell responsiveness. Thus, the optimal interval $\tau_0{}^*$ separating successive stimuli could be slightly larger in that case, which would agree with the periodicity observed for oscillations in suspensions of *D. discoideum* cells (fig. 5.12).

It is possible to check the results obtained in the general model of fig. 8.6 by carrying out a detailed study of the effect of pulsatile stimulation in the specific model for cAMP signalling based on receptor desensitization analysed in chapter 5. Such a study shows (Li & Goldbeter, 1990) that the results obtained for the simplified model analysed above hold, with only minor modifications, when cellular responsiveness is defined in terms of the quantity of cAMP produced in a given amount of time rather than in terms of the receptor-generated activity (that the latter is more vaguely defined is the price we have to pay for ensuring wide applicability of the general model...). Taking the synthesis of intracellular cAMP as a physiological response fits with the role of cAMP in the control of both aggregation and differentiation (Simon *et al.*, 1989; Harwood *et al.*, 1992) after starvation. Similar results would, however, be obtained for other intracellular messengers whose synthesis is triggered by the binding of extracellular cAMP to its receptor.

To determine quantitatively how the response of the signalling system varies with the pattern of periodic stimulation, it is useful to consider first the total amount of intracellular cAMP, $\beta_T$, synthesized above the basal level over a period once a constant amplitude has been reached in response to pulsatile stimulation (dashed area in fig. 8.16). How this quantity varies with the duration $\tau_1$ and interval $\tau_0$ is shown in fig. 8.17. At very large values of $\tau_0$ the response increases and later saturates as the duration of stimulation increases. At lower values of $\tau_0$, the response passes through a maximum as $\tau_1$ rises: the increase in response due to prolonged stimulation is indeed counterbalanced by enhanced receptor desensitization.

As pointed out above, however, it is likely that the relevant measure of cellular responsiveness is not the magnitude of a single response but the capability of generating a large number of significant responses in a given amount of time. To this effect we can introduce the cell responsiveness, $\beta_R$, which is now defined by eqn (8.14), by analogy with eqn (8.13):

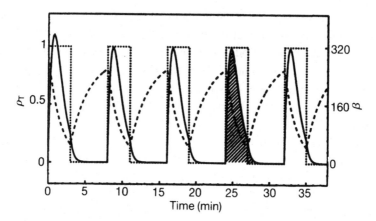

Fig. 8.16. Synthesis of intracellular cAMP ($\beta$, solid line) elicited by a periodic signal of extracellular cAMP ($\gamma$, dotted line) in the model for the cAMP signalling system of *Dictyostelium* based on receptor desensitization. Also indicated is the variation of the fraction of active (unphosphorylated) receptor, $\rho_T$ (dashed line). During the on-phase of stimulation ($\tau_1 = 3$ min) the normalized level of $\gamma$ equals 10; for a dissociation constant $K_R$ of $10^{-7}$ M for the cAMP receptor, this level of $\gamma$ corresponds to 1 $\mu$M extracellular cAMP. During the off-phase ($\tau_0 = 5$ min), the level of extracellular cAMP is nil. The curves are generated by integration of eqns (5.16) for the parameter values of table 5.3. The shaded area represents the integrated synthesis of cAMP, $\beta_T$, above the basal level obtained in the absence of stimulation. Actual levels of cAMP or of total cAMP synthesized over a period are obtained by multiplying $\beta$ or $\beta_T$ by $K_R$ (Li & Goldbeter, 1990) .

$$\beta_R = \frac{\beta_T}{\beta_{Tstep}} \times \frac{\beta_T}{T} \qquad (8.14)$$

Here, $\beta_{Tstep}$ is the amount of cAMP synthesized in response to a step increase in extracellular cAMP of infinite duration. While the term ($\beta_T/T$) measures the quantity of cAMP synthesized per unit time during one period of the signal, the first term of the product scales the response with that elicited by a single-step increase in cAMP. This scaling factor ensures that not only is the average cAMP level taken into account but also the magnitude of the cAMP peak.

Shown in fig. 8.18 are the results of numerical simulations in which cell responsiveness $\beta_R$ is determined, by numerical integration of eqns (5.16), as a function of the parameters $\tau_0$ and $\tau_1$ characterizing the square-wave pulse of extracellular cAMP. As in fig. 8.13 established for the more general model, fig. 8.18 demonstrates the existence of an opti-

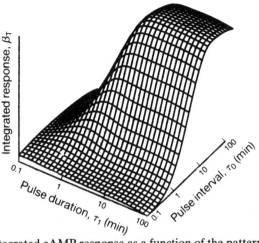

Fig. 8.17. Integrated cAMP response as a function of the pattern of square-wave stimulation by extracellular cAMP. The quantity of cAMP synthesized over a period ($\beta_T$, shaded area in fig. 8.16) is shown as a function of the durations $\tau_1$ of the on-phase and $\tau_0$ of the off-phase. Parameter values are as in fig. 8.16 (Li & Goldbeter, 1990).

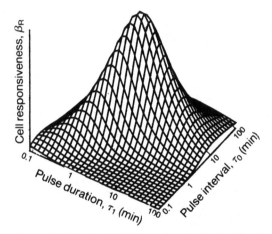

Fig. 8.18. Cell responsiveness ($\beta_R$) as a function of pulse duration and pulse interval. The function $\beta_R$ defined by eqn (8.14) reaches a maximum close to 63 for $\tau_1{}^* = 2.70$, $\tau_0{}^* = 4.75$ min; this pair of values represents the optimal pattern of periodic stimulation. The surface is determined for the set of parameter values of figs. 8.16 and 8.17 (Li & Goldbeter, 1990).

mal pattern of stimulation that corresponds to maximum target cell responsiveness. For parameter values found in experiments on the cAMP signalling system (see table 5.3), the optimum is given by the values $\tau_1{}^* = 2.70$ min, $\tau_0{}^* = 4.75$ min when the value of $\gamma$ during stimulation is equal to 10. This value corresponds to a cAMP level of $10^{-6}$ M, which is close to receptor saturation. The response illustrated in fig. 8.16 was obtained for conditions close to this particular pattern of periodic stimulation. For a $10^{-7}$ M stimulus ($\gamma = 1$ during stimulation), the optimal pattern in the same conditions corresponds to the values $\tau_1^* = 4.21$ min, $\tau_0^* = 5.21$ min (Li & Goldbeter, 1990).

The values of $\tau_0^*$ and $\tau_1^*$ that define the optimal pattern are markedly influenced by the parameters that govern receptor desensitization and resensitization. In the model, these processes depend primarily on the rate constants $k_2$ and $k_{-1}$, respectively. Table 8.1 indicates how a change in either one or both of these constants affects the values of $\tau_0^*$ and $\tau_1^*$. A comparison of rows and columns in table 8.1 shows that a five-fold increase (decrease) in $k_2$ leads to a decrease (increase) in $\tau_1^*$ by a roughly similar factor, while a similar result holds for $\tau_0^*$ as a function of $k_{-1}$.

Thus the data in fig. 8.16 corroborate the results obtained in the simpler receptor model: as in fig. 8.13, there exists a pattern of periodic stimulation that maximizes cell responsiveness measured here by the capability of generating the largest amount of intracellular cAMP in a given amount of time, in response to pulsatile stimulation. The values $\tau_1^* = 2.70$ min, $\tau_0^* = 4.75$ min obtained for the optimal pattern of periodic stimulation match those observed in the experiments and are close to those obtained in fig. 8.13 in the general model for similar parameter values. Very similar results are obtained when other definitions of cellular responsiveness are chosen, or when the excitable properties of *D. discoideum* cells are taken into account (Li & Goldbeter, 1990).

### *Optimal frequency of pulsatile stimulation by GnRH for secretion of the gonadotropic hormones LH and FSH*

As mentioned in section 8.4, the results of fig. 8.13 apply generally to the case of periodic hormone secretion and, in particular, to the pulsatile release of GnRH by the hypothalamus. In that case the activity coefficients $a_i$ of the different forms of the GnRH receptor could represent an ionic conductance for $Ca^{2+}$ whose influx into target cells in the pituitary is associated with the secretion of gonadotropic hormones (Conn *et al.*, 1986).

The studies of Knobil and coworkers (Knobil, 1980; Pohl *et al.*, 1983) have shown that target cells in the pituitary respond to the GnRH sig-

Table 8.1. *Dependence of the optimum pattern of stimulation* ($\tau_1^*$, $\tau_2^*$) *(in min) on the rate constants of receptor desensitization* ($k_2$) *and resensitization* ($k_{-1}$)

| Resensitization rate constant $k_{-1}$ (min$^{-1}$) | Desensitization rate constant, $k_2$ (min$^{-1}$) | | |
|---|---|---|---|
| | 0.1332 | 0.666 | 3.33 |
| 0.072 | $\tau_1^* = 13.50$ | 3.32 | 0.78 |
| | $\tau_0^* = 23.65$ | 20.3 | 19.2 |
| 0.36 | $\tau_1^* = 11.3$ | 2.70 | 0.66 |
| | $\tau_0^* = 6.18$ | 4.75 | 4.06 |
| 1.8 | $\tau_1^* = 9.5$ | 2.28 | 0.53 |
| | $\tau_0^* = 1.6$ | 1.24 | 0.93 |

Data are obtained as described in the legend to fig. 8.16 for a stimulus of $10^{-6}$ M cAMP ($\gamma = 10$ during stimulation), at constant values of the ratios $(k_{-1}/k_1) = 10$ and $(k_{-2}/k_2) = 0.005$. The data at the centre of the table correspond to the optimum in fig. 8.18. Other parameter values are as in fig. 8.16.

nals only when the latter possess the appropriate frequency observed in physiological conditions. The hormone GnRH is secreted in the rhesus monkey and in humans at a frequency of one 6 min pulse per hour (Dierschke *et al.*, 1970; see also fig. 8.2). The theoretical results shown in fig. 8.13 were obtained for parameter values for the kinetics of desensitization and resensitization such that the optimum cell responsiveness occurs for the periodic signal ($\tau_1^* = 6$ min, $\tau_0^* = 54$ min), analogous to the signal observed for GnRH.

The phenomenon of desensitization of target cells in the pituitary in conditions of constant stimulation by GnRH is well known (Smith & Vale, 1981; Adams, Cumming & Adams, 1986; Conn *et al.*, 1986, 1987), but its molecular bases are not yet fully clarified. It is probable that a number of distinct processes contribute to this desensitization. One of these is the removal of GnRH-binding sites from the membrane through receptor internalization; this process of down regulation is rather slow as it occurs over several hours (Zilberstein, Zakut & Naor, 1983). In principle, desensitization could also partly be due to the depletion of the intracellular pool of gonadotropic hormones (Naor *et al.*, 1982), but the phenomenon still occurs in the absence of secretion of LH and FSH (Jinnah & Conn, 1986), so that it must involve the uncoupling of the receptor from the secretion of gonadotropic hormones, much as desensitization uncouples cAMP binding to the receptor from cAMP synthesis in *Dictyostelium* (Theibert & Devreotes, 1983).

Desensitization could also be due to some modification of the receptor, as in bacteria or in the slime mould, or to a mechanism acting beyond the receptor (Smith, Perrin & Vale, 1983). A likely candidate for the latter type of mechanism could be provided by the inactivation of calcium channels (Stojilkovic *et al.*, 1989).

The results shown in fig. 8.13 were obtained for parameter values that are in a certain measure arbitrary, given the lack of available experimental data. The choice retained for the kinetic parameters possesses, however, a predictive value as it indicates that the optimal pattern of pulsatile stimulation observed in the experiments results from a process of reversible desensitization that is faster than that of receptor down regulation through the internalization mechanism mentioned above. Indeed, to obtain in the model an optimal signal of 6 min duration with an interval of 54 min between successive stimuli, it is necessary that desensitization proceeds with a characteristic time of 1 to 2 min, while resensitization upon removal of the ligand must proceed with a characteristic time of the order of 30 min (Li & Goldbeter, 1989b). Experiments performed in cultures of pituitary cells (Naor *et al.*, 1982; Zilberstein *et al.*, 1983) indicate that the phase of decrease of the response to a constant signal of GnRH is biphasic (fig. 8.19). The first phase, which is over in a few minutes, could correspond to the fast

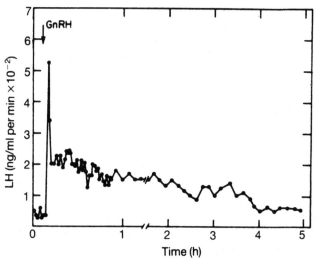

Fig. 8.19. Desensitization of the GnRH receptor in pituitary cells upon prolonged incubation with GnRH (Zilberstein *et al.*, 1983).

desensitization process predicted by the model, while the second, slower phase might be associated with receptor down regulation; the first process would be responsible primarily for fixing the optimal pattern of pulsatile stimulation by GnRH.

Whatever the details of the molecular mechanism of desensitization, the results on the existence of an optimal pattern of stimulation by GnRH pulses is confirmed not only by *in vivo* observations in the rhesus monkey (see above), but also by the systematic experiments of McIntosh & McIntosh (1986) and Liu & Jackson (1984) on the effect of different patterns of pulsatile stimulation by GnRH on the secretion of gonadotropic hormones in cultures of pituitary cells. The response of pituitary cells as a function of the duration of the stimulus and of the interval between successive stimuli (fig. 8.20) – which parameters correspond to the quantities $\tau_1$ and $\tau_0$ in the simulations presented in fig. 8.13

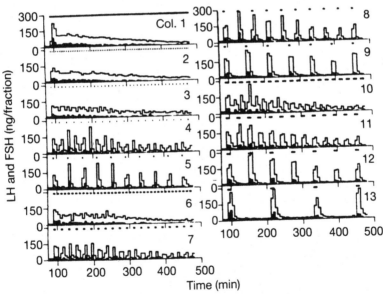

Fig. 8.20. Secretion of LH (open areas) and FSH (filled areas) in cultures of sheep pituitary cells subjected to different patterns of square-wave stimulation by GnRH (McIntosh & McIntosh, 1986). The cells were stimulated in each experiment with one of the following patterns of GnRH pulses indicated by horizontal bars: column 1, continuous GnRH; col. 2, 2 min every 5 min; col. 3, 2 min every 10 min; col. 4, 2 min every 20 min; col. 5, 2 min every 40 min; col. 6, 5 min every 10 min; col. 7, 5 min every 20 min; col. 8, 5 min every 40 min; col. 9, 5 min every 60 min; col. 10, 10 min every 20 min; col. 11, 10 min every 30 min; col. 12, 10 min every 60 min; col. 13, 10 min every 120 min.

– demonstrate the existence of a periodic signal that generates maximal secretion of LH and FSH (fig. 8.21). The characteristics of that signal are close to those of the autonomous, pulsatile signal of GnRH observed under physiological conditions and also match the optimal frequency determined *in vivo* in the experiments of Pohl *et al.* (1983) mentioned in section 8.3.

### Application to other cases of pulsatile hormonal signalling: insulin and growth hormone

Beyond the cases of cAMP and GnRH discussed above, it is likely that frequency encoding of pulsatile signals in intercellular communication is also encountered in many other instances, for hormones in particular, and, possibly, for growth factors (Brewitt & Clarke, 1988). Two additional cases of pulsatile hormonal signalling where frequency encoding might be effective are considered here. Examined in turn are growth hormone (GH), and the hormones involved in glucose homeostasis (insulin and glucagon).

The case of GH is of particular interest as the frequency of pulsatile GH secretion appears to be sex-dependent in many species including rat, chicken and human (Tannenbaum & Martin, 1976; Eden, 1979;

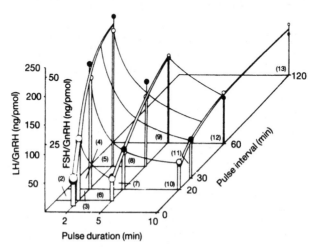

Fig. 8.21. Demonstration of an optimal pattern of pulsatile stimulation by GnRH inducing maximum secretion of LH and FSH in cultures of pituitary cells, based on the data of fig. 8.20; numbers in parentheses refer to the various conditions considered in that figure (McIntosh & McIntosh, 1986).

Millard *et al.*, 1987; Johnson, 1988; Asplin *et al.*, 1989; Winer, Shaw & Baumann, 1990). Thus, in the rat (fig. 8.22), while the higher frequency in GH release in the female results in quasi-constant levels of the hormone, well-defined secretory episodes occur in the male, with a frequency of one pulse every 3–4 h; the pulses are separated by intervals of 2–2.5 h during which GH drops down to barely detectable levels (Waxman *et al.*, 1991). The ultradian rhythmic secretion of GH appears to result from the antagonistic effects of pulsatile stimulation by GH-releasing factor (GHRH; Gelato & Merriam, 1986) and inhibition by somatostatin (Tanenbaum & Ling, 1984; Martha *et al.*, 1988).

Interestingly, the fact that the temporal profiles of GH are sexually differentiated is responsible for significant developmental differences in male and female, with respect to both growth and liver function (Norstedt & Palmiter, 1984; Morgan, MacGeoch & Gustafsson, 1985; Noshiro & Negishi, 1986; MacLeod & Shapiro, 1989; Waxman *et al.*, 1991). Thus, growth is usually slower in the female. That this results from the existence of a quasi-continuous pattern of GH is shown by experiments in which male or female hypophysectomized rats (unable to secrete GH) are administered exogenous GH; when the hormone is given in the form of six to seven pulses per day, both male and female rats achieve the body weight gain specific to males. In contrast, when GH is given continuously, the weight gain is reduced and growth of

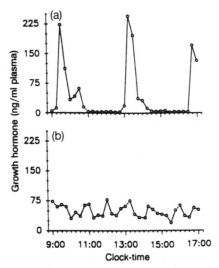

Fig. 8.22. Pulsatile secretion of growth hormone (GH) in (a) the male and (b) the female rat (Waxman *et al.*, 1991).

male and female animals acquires the female-specific characteristics (Waxman *et al.*, 1991, 1995).

A similar effect of the temporal pattern of GH is also seen in the expression of several hepatic enzymes including cytochrome P450-linked steroid hydroxylases and drug-metabolizing enzymes (Norstedt & Palmiter, 1984; Morgan *et al.*, 1985; Noshiro & Negishi, 1986; MacLeod & Shapiro, 1989). In an effort to determine what feature of the pulsatile GH stimulus is predominant in exerting this differential effect – i.e. pulse amplitude, pulse duration, or interpulse interval – Waxman *et al.* (1991) carried out a detailed study in the rat, using a pump for modulating the exogenous GH signal. Continuous infusion of the hormone was shown to stimulate expression of several female-pre-dominant hepatic enzymes (see also Mode *et al.*, 1981). In contrast, pul-satile administration of GH at a frequency of six pulses per day caused the expression of the male-specific hepatic enzyme designated CYP2C11. While the frequency of seven pulses per day proved as effec-tive as that of six pulses per day in promoting growth, however, the higher frequency was ineffective in inducing the male-specific liver response. This shows that different temporal patterns of GH can be rec-ognized by different target tissues (Waxman *et al.*, 1991).

The fact that the seven pulse per day frequency of GH stimulation is ineffective for the male-specific expression of the liver enzyme indicates that a critical time interval is needed for recovery in the hepatic tissue (Waxman *et al.*, 1991). This observation fits well with the finding that hepatocyte GH receptors in the adult rat internalize after a GH pulse and return to the membrane after a minimum of 3 h (Bick, Youdim & Hochberg, 1989a,b). Increased efficiency of pulsatile versus continuous GH infusion is also observed at the cellular level, with respect to growth factor mRNA synthesis (Isgaard *et al.*, 1988) and control of a regulator of male-specific gene expression (Waxman *et al.*, 1995).

As for GnRH, it is likely that the frequency encoding of GH pulses will lead to clinical applications. Of particular significance in this respect is the finding that patterned GH administration can produce enhanced growth in the hypophysectomized rat (Jansson *et al.*, 1982; Clark *et al.*, 1985; Hindmarsh *et al.*, 1990). Thus, pulsatile administration of 12 or 36 mU/day of GH proves much more effective than continuous infusion of corresponding amounts of the hormone; moreover, the growth-promoting effect increases with the frequency of pulsatile GH delivery and appears to saturate near nine pulses per day at the dose of 18 mU/day (fig. 8.23). Related studies show increased efficiency of pul-

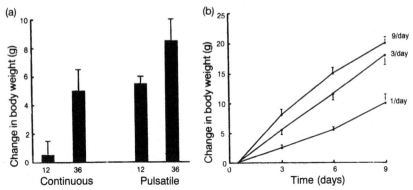

Fig. 8.23. Pulsatile administration of 12 or 36 mU GH/day promotes the growth of hypophysectomized rats more effectively than the continuous administration of the same doses of the hormone (a). As shown in (b), the effect on growth of the pulsatile administration of a given amount of GH (18 mU/day) by intravenous injection depends on the frequency of GH pulses (Hindmarsh *et al.*, 1990; adapted from Clark *et al.*, 1985).

satile versus continuous GHRH administration in GH-deficient children, the 3-hourly pulsatile treatment reaching similar effects with much lower total doses of GHRH (Hindmarsh *et al.*, 1990). Given that the role of GH is not limited to its growth-promoting effects, it is likely that appropriately patterned GH administration will prove to be of clinical interest for the increasing number of therapeutic uses of the hormone, one of which is the treatment of the 'wasting' syndrome in the elderly. It will be interesting to determine whether the increased efficiency of pulsatile delivery extends to a GH-releasing synthetic hexapeptide (Bowers, Alster & Frentz, 1992) and to a related, recently characterized nonpeptidyl inducer of GH secretion (Smith *et al.*, 1993).

With the increased effectiveness of pulsatile GH signalling on growth and hepatic steroid metabolism, it appears that the frequency encoding of the hormone can be accounted for by the general model based on receptor desensitization. The existence of an optimal pulse interval of the order of 2–3 h would be related to the time required for the GH receptor to recover from desensitization or internalization, i.e. down regulation.

Another important example of pulsatile signalling pertains to the hormones controlling glucose homeostasis in the blood. Since the finding that, in fasting monkeys, the blood glucose level oscillates with a period close to 15 min (Goodner *et al.*, 1977), similar oscillations have been observed in humans where the phenomenon is accompanied by synchronous oscillations of insulin and glucagon (Lang *et al.*, 1979, 1982;

for reviews, see Lefèbvre *et al.*, 1987; Weigle, 1987), two hormones that, respectively, lower and raise the level of glucose in the blood. Oscillations in insulin and glucose levels of longer period, of the order of 120 min, have also been observed in humans; a computer model suggests that these slower oscillations result from the glucose–insulin feedback loop (Sturis *et al.*, 1991).

The question again arises of the function of the high-frequency oscillations in insulin and glucagon. Animal studies showed that the pulsatile administration of insulin at the physiological frequency has greater hypoglycaemic effects than continuous infusion of the hormone (Matthews *et al.*, 1983), while pulsatile stimulation by glucagon likewise proves more effect than constant stimulation in enhancing hepatic glucose production *in vitro* (Weigle, Koerker & Goodner, 1984; Komjati *et al.*, 1986; Weigle, 1987). Here again the frequency of glucagon pulses appears to be a primary factor in optimizing hepatocyte responsiveness (Weigle & Goodner, 1986).

Insights into the possible physiological function of insulin oscillations with a period close to 13 min come from studies of the dynamics of the insulin receptor after a hormone pulse. Studies with perifused rat hepatocytes have shown (Goodner, Sweet & Harrison, 1988) that in these cells the receptor is rapidly internalized after a brief pulse of insulin, and returns to the membrane in a typical time of the order of 13 min once the pulse is over (fig. 8.24). There thus exists a close link between the situation observed for cAMP in *Dictyostelium* and that for GnRH and GH. In contrast to the latter two hormones, whose physiological periodicity is longer (in keeping with the longer time required for receptor recovery), here the optimum frequency is shorter, in accordance with the faster recovery of the insulin receptor.

With respect to clinical applications, the above studies have been extended to humans. It was found that pulsatile insulin and glucagon delivery have greater metabolic effects than continuous hormone administration (Lefèbvre *et al.*, 1987; Paolisso *et al.*, 1989). Frequency encoding is again observed, as pulsatile delivery of insulin at the frequency of one pulse per 13 min proves more effective than the lower frequency of one pulse per 26 min (Paolisso *et al.*, 1991). It is tempting to relate these observations to those of Goodner *et al.* (1988) on the dynamics of the insulin receptor after a pulse (fig. 8.24). It remains open whether there are possible therapeutic applications of these results to the treatment of some physiological disorders of glucose homeostasis. Of particular interest in this regard is the finding that impaired pulsatile

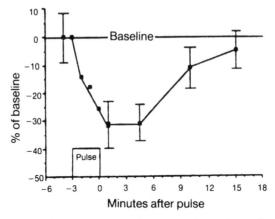

Fig. 8.24. Rapid internalization of the insulin receptor in rat hepatocytes, followed by return to the membrane, after a brief insulin pulse (Goodner *et al.*, 1988).

secretion of insulin might be related to some noninsulin-dependent cases of diabetes (O'Rahilly, Turner & Matthews, 1988).

## 8.6 Superior efficacy of periodic versus chaotic or stochastic signalling in intercellular communication

So far the pulsatile, periodic signal has been compared only with a continuous signal. The former proved more effective than the latter in eliciting sustained responses in target cells subjected to reversible desensitization in the continuous presence of the ligand. Moreover, the general model described in section 8.4 as well as the model for cAMP signalling showed that there exists an optimal frequency of the periodic, pulsatile stimulus that optimizes target cell responsiveness. The question arises as to how the periodic stimulus compares with a pulsatile stimulus whose duration and/or pulse interval vary in a stochastic or chaotic manner. The question of the physiological significance of chaos has been raised (Pool, 1989). The analysis of the effect of pulsatile signalling in intercellular communication provides an opportunity for assessing the efficiency of chaos versus that of periodic behaviour.

Li & Goldbeter (1992) carried out a detailed analysis of the general model based on receptor desensitization (see section 8.4), in which the duration $\tau_1$ and/or the pulse interval $\tau_0$ of the square-wave stimulus were varied in a random manner or chaotically. The latter, aperiodic variations were generated by means of the logistic map. In each case cell responsiveness, $\alpha_R$, defined by eqn (8.13) was determined and compared with the responsiveness obtained for the optimum pattern $(\tau_1^*, \tau_0^*)$ of periodic stimulation (see fig. 8.13). The results of that study indicate that the optimum, periodic stimulus is always more effective than the random or stochastic pulsatile signal in maximizing target cell responsiveness. Moreover, random or chaotic variations in the interval $\tau_0$ appear to be more detrimental than corresponding variations in the duration $\tau_1$ of the square-wave pulse; the largest loss in responsiveness is observed when both $\tau_0$ and $\tau_1$ vary in a stochastic or chaotic manner as compared with the optimal, periodic case (fig. 8.25).

That the random pulsatile signal is less effective than the periodic one accounts for the observations of Nanjundiah (1988), who studied the effect of random cAMP signals in *Dictyostelium* by varying the interpulse interval from 0 to 10 min in a stochastic manner. These experiments indicated that, like constant cAMP stimuli or cAMP pulses delivered at 2 min intervals (Wurster, 1982), the random cAMP signal fails to induce differentiation in *Dictyostelium* cells, in contrast to cAMP pulses delivered at 5 min intervals (Gerisch *et al.*, 1975).

As regards pulsatile hormone secretion, in addition to the examples of GnRH, GH, insulin and glucagon discussed above, there exist a number of hormones for which the pulsatile pattern of secretion appears to be extremely noisy. For these hormones, it might well be that such an irregular pulsatile pattern, although less effective than a truly periodic one, proves nevertheless more effective than constant stimulation.

### 8.7 Generation of the pulsatile signal

Whereas continuous stimulation gives rise to adaptation, pulsatile signals confer enhanced sensitivity by allowing target cells to repeatedly generate a response whose amplitude does not significantly decrease in the course of time. The most straightforward and robust manner for generating a pulsatile signal is to produce it rhythmically. What is of course needed is that the period of stimulation be adjusted to the characteristics of the receptor system in target cells; such an adjustment is achieved in the course of evolution.

The analysis of the models shows that the value of the optimal frequency of stimulation is dictated primarily by the rate of receptor resensitization in the absence of ligand. For *Dictyostelium* amoebae, the adjustment between the period of stimulation and the refractory period is ensured by the fact that both originate from the same signalling

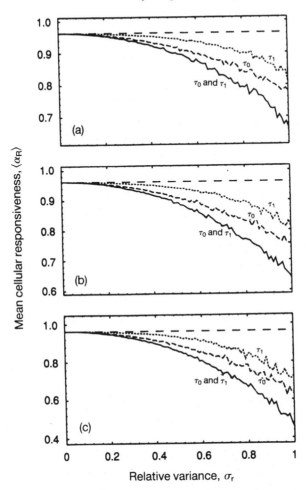

Fig. 8.25. The optimal pattern of periodic, pulsatile stimulation is more effective than random or chaotic signals. Shown is the mean responsiveness, $\langle\alpha_R\rangle$, as a function of the relative variance $\sigma_r = \sigma/\tau^*$ characterizing the stochastic or chaotic variations in the pulsatile signal ($\tau^*$ denotes either $\tau_1^*$ or $\tau_0^*$, depending on the quantity that is varied). Considered are random variations obeying (a) the truncated Gaussian distribution, (b) the uniform, white noise distribution, and (c) the chaotic distribution generated by the logistic map. The value of $\langle\alpha_R\rangle$ associated with stochastic or chaotic variations only in $\tau_1$ (dotted curve), only in $\tau_0$ (dashed curve) or in both $\tau_1$ and $\tau_0$ (solid curve) is plotted in each panel. The upper, dashed horizontal line indicates the optimal value $\alpha_R^*$ of cell responsiveness corresponding to the optimal periodic signal. For each value of $\sigma_r$, the value of $\langle\alpha_R\rangle$ is obtained by averaging over a series of 600 successive pulses (Li & Goldbeter, 1992).

mechanism. The molecular mechanism of cAMP oscillations is indeed based precisely on desensitization of the receptor, which is itself involved in the optimal response to periodic cAMP pulses.

In the case of pulsatile hormone secretion, the situation is somewhat different as the mechanism that produces the rhythm is largely independent of the mode of action of the hormone on target cells. When the hormone is secreted by the hypothalamus, as is the case for GnRH or GHRH, the pulse generator is of neuronal nature (see below). Insulin is a special case in that the hormone is secreted by pancreatic β-cells. The mechanism underlying the pulsatile secretion of insulin and glucagon remains unclear. Much work has been devoted to the bursting oscillations observed in the membrane potential of β-cells stimulated by glucose (Perez-Armendariz *et al.*, 1985). These oscillations generally have a period of the order of 15 s, much smaller than that characterizing insulin secretion. Recent observations indicate, however, that besides the existence of glucose-induced oscillations of intracellular $Ca^{2+}$ resembling this bursting electrical activity there exist slower oscillations in $Ca^{2+}$, with a period of the order of 2 min (Valdeolmillos *et al.*, 1989).

Longo *et al.* (1991) reported similar oscillations in intracellular $Ca^{2+}$, of slightly longer period, which occur in rat pancreatic islets in synchrony with oscillations in oxygen consumption and insulin secretion; these oscillations would be driven metabolically. More specifically, it is the glycolytic oscillator (see chapter 2) that would be responsible for the periodic release of insulin by β-cells. Such a hypothesis, also put forward by Lipkin *et al.* (1983), who observed glycolytic oscillations in isolated perifused rat fat cells, has been elaborated in further detail by Tornheim and co-workers who studied mainly glycolytic oscillations in skeletal muscle extracts (Andrés *et al.*, 1990; Tornheim *et al.*, 1991) before extending their studies to rat pancreatic islets (Longo *et al.*, 1991). According to these authors, glycolytic oscillations would be involved in the stimulus-secretion coupling for insulin release by pancreatic β-cells by causing periodic elevations in intracellular $Ca^{2+}$; oscillations in the ATP/ADP ratio occurring in the course of glycolytic oscillations would cause opening and closing of ATP-sensitive $K^+$ channels (see fig. 2.35). Closure of these channels would cause depolarization and the resultant influx of $Ca^{2+}$ through voltage-dependent channels (Andrés *et al.*, 1990). This very attractive possibility, supported by the observation in the pancreatic islet of $Ca^{2+}$ oscillations and metabolic oscillations concomitant with periodic insulin release (Longo *et al.*, 1991) and by recently reported evidence for glycolytic oscillations in β-cells (Chou *et al.*, 1992), provides a link between the glucose stimulus and the modulation of pulsatile insulin release through variation of the ATP/ADP ratio in the course of glycolytic oscillations.

Of a different, neuronal nature is the GnRH pulse generator located in the arcuate nuclei in the hypothalamus (Wilson *et al.*, 1984; Lincoln *et al.*, 1985). Elucidating the mechanism underlying the pulsatile release of GnRH with a periodicity of the order of 1h is one of the most fascinating problems that remains to be solved in the study of cellular rhythms. Much progress should be made thanks to the recent demonstration (Krsmanovic *et al.*, 1992; Wetsel *et al.*, 1992) of intrinsic pulsatile GnRH secretion by immortalized neuronal lines; hormone secretion is dependent on voltage-sensitive $Ca^{2+}$ channels for its maintenance (Krsmanovic *et al.*, 1992). These observations show that GnRH neurons already possess the pulse generator underlying the rhythmic secretion of the hormone; that neuronal rhythms of even longer period exist at the cellular level has also been recently demonstrated for circadian rhythms in *Bulla* eyes (Michel *et al.*, 1993).

The mechanism of neuronal oscillations has been much clarified by the analysis of models based on equations of the type proposed by Hodgkin & Huxley (1952). Thus, such models have been used to account for simple periodic oscillations in the squid axon (Huxley, 1959; Aihara & Matsumoto, 1982), or bursting in neurons or β-cells of the pancreas (Plant & Kim, 1976; Rinzel & Lee, 1986; Rinzel, 1987; Adams & Benson, 1989; Av-Ron *et al.*, 1991). Neuronal oscillators, however, have a period that generally ranges from hundredths of second to a few tens of seconds, and the question arises as to how such oscillators could generate long-term periodicities of the type responsible for the hourly, pulsatile secretion of GnRH. Such periodicities may well rely on a transcriptional oscillator involving protein synthesis and genetic regulatory mechanisms, as observed for the pulsatile secretory mechanism of GH (Zeitler *et al.*, 1991) or the circadian clock (Hardin *et al.*, 1992; Khalsa, Whitmore & Block, 1992; Takahashi, 1992, 1993; Takahashi *et al.*, 1993; Young, 1993; see also chapter 11).

The mechanism of the GnRH pulse generator might well be based, alternatively, on an interplay of voltage-dependent conductances, as in traditional neuronal oscillators; what is needed, however, is the coupling with some process that would slow down the oscillations and bring the period from tens of seconds to an hour. One mechanism, presented here as a working hypothesis (Li, 1992; Y.X. Li & A. Goldbeter, unpublished results), is that such slow neuronal oscillations could involve the activation of a $K^+$ channel through phosphorylation by a $Ca^{2+}$-activated protein kinase (fig. 8.26). The build up of intracellular $Ca^{2+}$ during the bursting phase would progressively activate a $K^+$ conductance through phosphorylation; the subsequent efflux of $K^+$ would terminate the plateau and lead to repolarization. The lengthening of the period up to the order of 1h could originate from the slow action of a phosphatase that would control the rate of depolarization of the membrane, leading to a new phase of bursting. An example of numerical simulations of such a model where a purely electrical neuronal oscillator (Rinzel & Lee, 1986) is coupled to the slow phosphorylation–dephosphorylation of a $Ca^{2+}$-dependent $K^+$ channel is shown in fig. 8.27. A mirror mechanism yielding similar results would involve the activation of the $K^+$ channel through dephosphorylation by a $Ca^{2+}$-dependent phosphatase such as calcineurin (Armstrong, 1989), and its slow inactivation by a protein kinase.

The above, putative mechanism is given here to illustrate how long-period neuronal bursting might in principle originate from the coupling of a purely electrical mechanism based on the interplay of voltage-dependent ionic conductances with a slow biochemical controlling process. Such a mixed electrical–biochemical mechanism, which

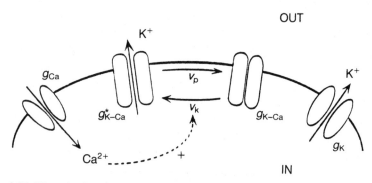

Fig. 8.26. Hypothetical model for the generation of long-period neuronal bursting of the type seen in GnRH-secreting cells. Besides the $Ca^{2+}$ ($g_{Ca}$) and $K^+$ ($g_K$) conductances, it is assumed that intracellular $Ca^{2+}$ activates another $K^+$ channel by inducing its phosphorylation through a $Ca^{2+}$-activated protein kinase whose maximum rate is denoted $v_k$; this channel is inactivated through dephosphorylation by a phosphatase of maximum rate $v_p$. The active and inactive forms of the $Ca^{2+}$-dependent $K^+$ conductance are denoted by $g^*_{K-Ca}$ and $g_{K-Ca}$, respectively. The increase in the duration of the active and silent phases of bursting would arise from the slowness of the action of the kinase and phosphatase modulated by parameters $v_k$ and $v_p$ (Li, 1992; Y.X. Li & A. Goldbeter, unpublished results).

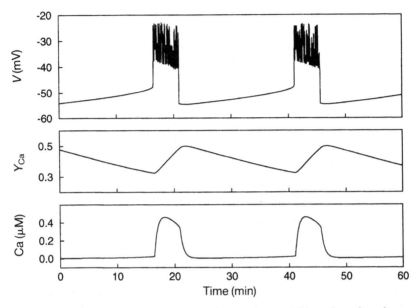

Fig. 8.27. Typical bursting pattern generated by the model based on the scheme shown in fig. 8.26. The curves are obtained in a model for neuronal bursting based on the equations of Sherman *et al.* (1988) in which the $Ca^{2+}$-dependent $K^+$ conductance, rather than being activated directly upon binding of $Ca^{2+}$ to the channel, is regulated through reversible phosphorylation by a $Ca^{2+}$-activated kinase. Shown from top to bottom are the membrane potential, the fraction of open (i.e. phosphorylated) $Ca^{2+}$-dependent $K^+$ channels, and the concentration of cytosolic $Ca^{2+}$. The long period of the bursting pattern is dictated by the low values of the maximum rates of phosphorylation ($v_k$) and dephosphorylation ($v_p$) (Y.X. Li & A. Goldbeter, unpublished results).

remains to be validated by experimental studies, resorts to plausible components. Control of ionic channels through reversible phosphorylation is indeed well documented (Levitan, 1985, 1994; Esguerra *et al.*, 1994; Lieberman & Mody, 1994). In addition, $Ca^{2+}$-influx through a voltage-dependent channel is part of the GnRH pulse generator mechanism (Krsmanovic *et al.*, 1992). It would be of interest to determine whether phosphorylation–dephosphorylation plays a role in the termination of the plateau phase and in setting the length of the silent phase of the pulse generator and, if so, whether the protein kinase or the phosphatase involved is controlled by $Ca^{2+}$.

Oscillatory release of GnRH could also rely on an autocrine loop involving negative (Bourguignon *et al.*, 1994) or positive (Krsmanovic *et al.*, 1993) feedback. The possibility of positive feedback in GnRH secretion is suggested by the presence of GnRH receptors on the surface of immortalized GnRH-secreting hypothalamic neurons (Krsmanovic *et al.*, 1993). The analysis of a model based on such autocatalytic regulation confirms that it may give rise to sustained oscillations in GnRH release (H. Schepers and A. Goldbeter, unpublished results), in a manner reminiscent of the mechanism generating cAMP pulses in *Dictyostelium* cells (see chapter 5).

The mechanism of pulsatile secretion might well differ from hormone to hormone. Thus, for insulin, the periodic change in the membrane potential of pancreatic β-cells leading to $Ca^{2+}$ influx and insulin secretion might well be passively driven by a continuous biochemical oscillator. As discussed in section 2.9 (see also fig. 2.35), glycolytic oscillations could be triggered by an increase in glucose; the resulting change in the ATP/ADP ratio would in turn lead to periodic changes in the membrane potential via the modulation of an ATP-dependent $K^+$ conductance (Corkey *et al.*, 1988). If this conjecture were supported by further experimental observations, the insulin pulse generator would be primarily of biochemical nature, whereas in the putative mechanism discussed for the GnRH pulse generator the mechanism would be of a mixed electrical–biochemical nature.

# Part V

## Calcium oscillations

# 9

# Oscillations and waves of intracellular calcium

## 9.1 Experimental observations on cytosolic Ca$^{2+}$ oscillations

Ca$^{2+}$ oscillations are among the most significant findings of the last decade in the field of intracellular signalling. Together with the mitotic oscillator, which underlies the eukaryotic cell division cycle (examined in chapter 10), Ca$^{2+}$ oscillations are also one of the most important periodic phenomena uncovered in recent years in the field of biochemical and cellular oscillators. Ca$^{2+}$ oscillations are of interest for a variety of reasons. First, they occur in a large number of cell types, either spontaneously or after stimulation by hormones or neurotransmitters. Second, it is by now clear that they represent the oscillatory phenomenon that is the most widespread at the cellular level, besides the rhythms encountered in electrically excitable cells. Third, Ca$^{2+}$ oscillations are often associated with the propagation of Ca$^{2+}$ waves within the cytosol, and sometimes between adjacent cells; even though its physiological significance remains to be determined, this phenomenon has become one of the most important examples of spatiotemporal organization at the cellular level.

Since their first direct observation in fertilized mouse oocytes (Cuthbertson & Cobbold, 1985) and hormone-stimulated hepatocytes (Woods *et al.*, 1986, 1987), which followed their earlier, theoretical prediction (Rapp & Berridge, 1977; Kuba & Takeshita, 1981) and indirect characterization (Rapp & Berridge, 1981), the number of experimental reports on Ca$^{2+}$ oscillations has mushroomed at an increasing pace in recent years; these experimental results have been examined in several reviews (Berridge & Galione, 1988; Berridge, Cobbold & Cuthbertson, 1988; Berridge, 1989, 1990; Cuthbertson, 1989; Rink & Jacob, 1989; Cobbold & Cuthbertson, 1990; Jacob, 1990a; Petersen & Wakui, 1990; Tsien & Tsien, 1990; Meyer & Stryer, 1991; Tsunoda, 1991; Fewtrell,

1993) and in a special issue of *Cell Calcium* (Cuthbertson & Cobbold, 1991). The related, spatial aspects of $Ca^{2+}$ signalling have also been the subject of several reviews (Amundson & Clapham, 1993; Berridge & Dupont, 1994) and of another special issue of *Cell Calcium* (Dissing, 1993). Oscillations and waves of $Ca^{2+}$ were recently covered in a CIBA Foundation Symposium (Berridge, 1995). Although theoretical studies have been fewer, a number of models accounting for the periodic generation of $Ca^{2+}$ spikes have been proposed, some of which have also been extended to the spatial propagation of $Ca^{2+}$ signals (for reviews, see Dupont & Goldbeter, 1992b; Fewtrell, 1993; Stucki & Somogyi, 1994). The purpose of this chapter is to examine the experimental aspects and the theoretical modelling of $Ca^{2+}$ waves and oscillations.

Drawing an exhaustive list of experimental studies on $Ca^{2+}$ oscillations is beyond the scope of this chapter; the above-mentioned reviews should be consulted to that end. Here, only those references that pertain directly to the elaboration of the models or allow the comparison of theoretical predictions with experimental data are retained. Before we address theoretical aspects, it is useful to briefly recall the main properties of $Ca^{2+}$ oscillations as determined from a large number of experimental studies pertaining to a variety of different cell types. The spatial aspects of $Ca^{2+}$ signalling are considered further below, in section 9.6.

Cytosolic $Ca^{2+}$ oscillations arise either spontaneously (Holl *et al.*, 1988; Malgaroli, Fesce & Meldolesi, 1990) or in response to stimulation by extracellular signals, with periods ranging from nearly 1 s to tens of minutes, depending on the cell type. Among the most studied cells with regard to $Ca^{2+}$ oscillations are cardiac cells (fig. 9.1), oocytes, hepatocytes (fig. 9.2), endothelial cells (fig. 9.3), fibroblasts, pancreatic acinar cells and pituitary cells (see the above-mentioned reviews, and Stojilkovic & Catt (1992) for a specific review of $Ca^{2+}$ oscillations in pituitary cells). The shape of the oscillations is highly variable (Berridge *et al.*, 1988; Fewtrell, 1993); while in some cases they are quasi-sinusoidal, sometimes, as in endothelial cells (see fig. 9.3), they take the form of abrupt spikes, which are often preceded by a gradual increase reminiscent of the pacemaker, depolarizing potential seen in oscillatory neurons or cardiac cells (DiFrancesco, 1993).

It has been repeatedly observed (Berridge & Galione, 1988; Berridge *et al.*, 1988; Jacob *et al.*, 1988; Berridge, 1989, 1990; Cuthbertson, 1989; Jacob, 1990a; Petersen & Wakui, 1990; Tsien & Tsien, 1990; Tsunoda, 1991; Fewtrell, 1993) that oscillations occur only in a certain range of

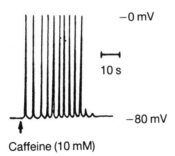

Caffeine (10 mM)

Fig. 9.1. Intracellular Ca$^{2+}$ oscillations in rat myocytes stimulated by caffeine. Oscillations are recorded as a train of action potentials caused by periodic changes in cytosolic Ca$^{2+}$, which are triggered by caffeine via calcium-induced calcium release (Capogrossi *et al.*, 1987).

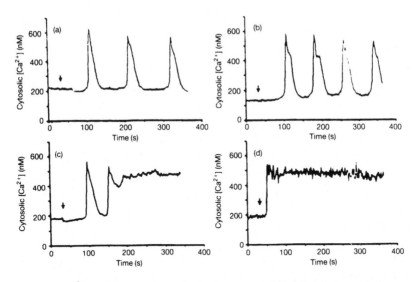

Fig. 9.2. Ca$^{2+}$ oscillations in a single hepatocyte stimulated (arrow) by vasopressin, at increasing levels of hormone stimulation. The concentrations of vasopressin used (in nM) were (a) 0.5, (b) 1, (c) 3, and (d) 5 (Rooney *et al.*, 1989).

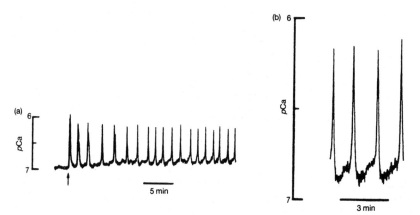

Fig. 9.3. (a) $Ca^{2+}$ oscillations in a single endothelial cell stimulated (arrow) by
0.3 μM histamine; an expanded section of the trace is shown in (b) (Jacob *et al.*,
1988).

stimulation (see fig. 9.3), and that the frequency of $Ca^{2+}$ spikes increases
with the intensity of the stimulus (see fig. 9.2). Besides the induction of
oscillations by external signals, it is often possible to elicit a train of $Ca^{2+}$
spikes by increasing the level of extracellular or intracellular $Ca^{2+}$, or
the level of inositol 1,4,5-trisphosphate ($IP_3$; see Osipchuk *et al.*, 1990).
The latter messenger is synthesized in response to external signals and
is known to raise the level of cytosolic $Ca^{2+}$ through mobilization from
intracellular stores (Berridge & Irvine, 1989; Berridge, 1993).

As regards the physiological significance of $Ca^{2+}$ oscillations and
waves, it is conceivable that the rapid spatial propagation of $Ca^{2+}$ signals
provides a useful communication mechanism between distinct parts of
the cell or between different, adjacent cells in a tissue (see, e.g.,
Sanderson, Charles & Dirksen, 1990; Charles *et al.*, 1992). Moreover,
although the cellular roles of $Ca^{2+}$ oscillations have not yet been
explored in detail, they appear to play an important role in secretory
processes (Rapp & Berridge, 1981; Law, Pachter & Danies, 1989;
Malgaroli & Meldolesi, 1991; Stojilkovic *et al.*, 1993; Tse *et al.*, 1993). A
recent study (Pralong, Spät & Wollheim, 1994) provides the experimental demonstration that $Ca^{2+}$ oscillations may drive the periodic evolution of biochemical reactions.

## 9.2   Models for $Ca^{2+}$ oscillations requiring periodic variation in $IP_3$

Given that $Ca^{2+}$ mobilization is triggered by $IP_3$, whose synthesis follows
external stimulation (Berridge & Irvine, 1989; Berridge, 1993), the

mechanism of oscillations may in principle result from regulation of $IP_3$ production, or from a process beyond $IP_3$ synthesis, such as the exchange of $Ca^{2+}$ between the cytosol and intracellular stores, or from a combination of both. Thus, a significant distinction between the various models is that some require the periodic variation of $IP_3$ concomitantly with $Ca^{2+}$ oscillations. Experimentally, this aspect is still difficult to test, since $IP_3$, in contrast to cytosolic $Ca^{2+}$, cannot yet be measured continuously within cells. The occurrence of simultaneous oscillations of $Ca^{2+}$ and $IP_3$, over a brief period of time, has nevertheless been reported for mouse pancreatic β-cells (Barker *et al.*, 1994).

The first model for $Ca^{2+}$ oscillations that incorporates the role of $IP_3$ is due to Meyer & Stryer (1988). In this model, schematized in fig. 9.4a, the external stimulus triggers the synthesis of $IP_3$, which mobilizes $Ca^{2+}$ from an intracellular store, thus leading to a rise in cytosolic $Ca^{2+}$. A positive feedback loop arises from the assumed activation of $IP_3$ synthesis by $Ca^{2+}$. This cross-regulation gives rise to sustained oscillations in cytosolic $Ca^{2+}$, accompanied by a periodic variation in $IP_3$. The activation of phosphoinositidase C by $Ca^{2+}$ has been observed in some cell types (Eberhard & Holz, 1988; Harootunian *et al.*, 1991; Meyer & Stryer, 1991).

In its early version (Meyer & Stryer, 1988), this model did not account for the increase in the mean level of cytosolic $Ca^{2+}$ with the external stimulus nor for the relationship between the level of the stimulus and the time needed to reach the first $Ca^{2+}$ spike, which time interval is known as **latency**. These shortcomings are obviated in a subsequent version of the model (Meyer & Stryer, 1991), which incorporates the inhibition of $Ca^{2+}$ release from the intracellular store at high levels of cytosolic $Ca^{2+}$; such a mechanism for the termination of a $Ca^{2+}$ spike is supported by some experimental studies (Parker & Ivorra, 1990; Zholos *et al.*, 1994). A related version of that model based on a more detailed mechanism for $Ca^{2+}$ inhibition of $IP_3$-stimulated $Ca^{2+}$ release has been considered (Keizer & De Young, 1992).

Another model based on the necessary periodic variation of $IP_3$ has been proposed by Cobbold and Cuthbertson (see Woods *et al.* 1987; Berridge *et al.*, 1988; Sanchez-Bueno *et al.*, 1990) and studied mathematically by Cuthbertson & Chay (1991). In this model, schematized in fig. 9.4b, the stimulus elicits the activation of phosphoinositidase C, through activation of G-proteins, thus leading to the production of $IP_3$ and diacylglycerol (DAG). Here again, $IP_3$ liberates $Ca^{2+}$ from an intracellular store, but the regulatory effect exerted by cytosolic $Ca^{2+}$ is to

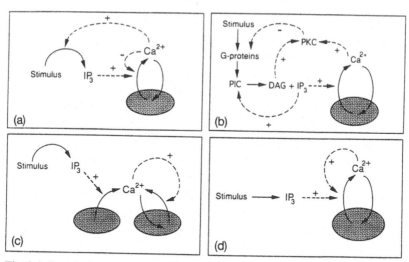

Fig. 9.4. Regulatory mechanisms at the core of the various theoretical models considered for $Ca^{2+}$ oscillations (Dupont & Goldbeter, 1992b). Sketched for each model are the intracellular processes triggered by the external stimulus; the shaded compartments represent $Ca^{2+}$ pools. (a) Model based on the cross-activation of $Ca^{2+}$ release and $IP_3$ synthesis; the three variables in the initial version of this model (Meyer & Stryer, 1988) are $IP_3$, cytosolic $Ca^{2+}$, and $Ca^{2+}$ in the $IP_3$-sensitive store, while the latter variable is replaced by the degree of inhibition of $Ca^{2+}$ release in a more recent version (Meyer & Stryer, 1991). (b) Model based on the regulation of $IP_3$ synthesis by $Ca^{2+}$ via the negative control of the receptor or a G-protein (Cuthbertson & Chay, 1991); also important for oscillations is the assumed activation of phosphoinositidase C (PIC) by diacylglycerol (DAG) or $IP_3$. The three variables here are the fraction of active G-protein, the level of DAG or $IP_3$, and cytosolic $Ca^{2+}$. PKC, protein kinase C. (c) Model based on $Ca^{2+}$-induced $Ca^{2+}$ release (CICR) with two distinct $Ca^{2+}$ pools sensitive to $IP_3$ and $Ca^{2+}$, respectively (Dupont & Goldbeter, 1989; Goldbeter *et al.*, 1990; Dupont *et al.*, 1991); the two variables are the concentration of $Ca^{2+}$ in the cytosol and in the $Ca^{2+}$-sensitive $Ca^{2+}$ store. (d) Model based on CICR with a single pool sensitive to both $Ca^{2+}$ and $IP_3$; the variables are the same as in the model illustrated in (c) (Dupont & Goldbeter, 1993; see also Somogyi & Stucki, 1991). Extensions of this model incorporate inhibition by high cytosolic $Ca^{2+}$ levels of $IP_3$-mediated $Ca^{2+}$ release (De Young & Keizer, 1992; Atri *et al.*, 1993; Keizer & De Young, 1994; Li & Rinzel, 1994). See text for details, and the original references for a listing of the kinetic equations describing each model.

activate protein kinase C (PKC), in concert with DAG. PKC, in turn, inhibits the G-protein or the receptor through reversible phosphorylation. Although the authors claim that the resulting negative feedback loop plays a primary role in the mechanism of oscillations, it appears that these occur only when an additional, positive feedback is incorporated into the model: $IP_3$ synthesis has to be activated by either DAG or

$IP_3$. Such a putative regulation has not yet been corroborated by experimental observations.

A conspicuous property of the above models is the requirement for a periodic variation in $IP_3$ in the course of $Ca^{2+}$ oscillations. Although observations in fibroblasts tend to support the occurrence of cross-activation between $Ca^{2+}$ and $IP_3$ (Harootunian *et al.*, 1991), sustained $IP_3$ oscillations have been neither observed nor ruled out directly in any experiment; an exception is the recent observation, over a limited period of time, of simultaneous oscillations of $Ca^{2+}$ and $IP_3$ in mouse pancreatic β-cells (Barker *et al.*, 1994). There is indirect evidence, however, that $IP_3$ variations need not accompany $Ca^{2+}$ spiking. In this respect, the most compelling finding is that $Ca^{2+}$ spikes can be elicited by a nonmetabolizable analogue of $IP_3$ (Wakui, Potter & Petersen, 1989). Moreover, $Ca^{2+}$ oscillations have been observed in some experimental preparations lacking the $IP_3$ signalling pathway, such as skinned cardiac cells (Fabiato & Fabiato, 1975), or in the presence of inhibitors of $IP_3$ synthesis (Malgaroli *et al.*, 1992).

## 9.3 Models for $Ca^{2+}$ oscillations based on $Ca^{2+}$-induced $Ca^{2+}$ release

That $Ca^{2+}$ spiking may arise in the absence of $IP_3$ oscillations is corroborated by the analysis of another class of models based on a different feedback mechanism. According to that mechanism proposed by Berridge (Berridge & Galione, 1988; Berridge & Irvine, 1989), $IP_3$ elicits the mobilization of $Ca^{2+}$ from an intracellular store while cytosolic $Ca^{2+}$ is transported into an $IP_3$-insensitive store from which it is released in a process activated by $Ca^{2+}$. The latter phenomenon, known as $Ca^{2+}$-induced $Ca^{2+}$ release (CICR), has long been demonstrated experimentally in muscle (Endo, Tanaka & Ogawa, 1970) and cardiac (Fabiato & Fabiato, 1975) cells, but recent evidence points to its occurrence in other cell types such as oocytes (Busa *et al.*, 1985), chromaffin cells (Malgaroli *et al.*, 1992), and pancreatic acinar cells (Wakui, Osipchuk & Petersen, 1990).

The first model based on CICR was proposed by Kuba & Takeshita (1981) for $Ca^{2+}$ oscillations in sympathetic neurons (see Friel & Tsien, 1992, for a recent study of $Ca^{2+}$ oscillations in this type of cell); the role of $IP_3$, unknown at that time, was not taken into account. Dupont and Goldbeter, in collaboration with Berridge, later considered the model shown in fig. 9.4c, of a more general nature, where CICR is initiated by the $IP_3$-mediated build up of cytosolic $Ca^{2+}$ (Dupont & Goldbeter, 1989;

Dupont, Berridge & Goldbeter, 1990, 1991; Goldbeter, Dupont & Berridge, 1990). As shown below, this model suffices to account for a large number of observations on signal-induced $Ca^{2+}$ spikes and demonstrates that $Ca^{2+}$ oscillations can occur in the absence of a periodic variation of $IP_3$. One insight provided by the theoretical analysis was the demonstration that for oscillations to occur, $Ca^{2+}$ release must be activated by cytosolic rather than intraluminal $Ca^{2+}$ as was initially envisaged (Berridge & Galione, 1988; Berridge, 1989). The existence of two separate pools of $Ca^{2+}$ is not required for oscillations; as shown in section 9.5, similar results are indeed obtained when we assume the existence of a single $Ca^{2+}$ pool sensitive to both $Ca^{2+}$ and $IP_3$ (fig. 9.4d).

Other related models have been proposed for $Ca^{2+}$ oscillations, one of which invokes the inhibition by intravesicular $Ca^{2+}$ of its transport into the cytosol (Swillens & Mercan, 1990); since total cell $Ca^{2+}$ is assumed to be constant, however, such regulation is tantamount to CICR. In that model, oscillations of $IP_3$ occur due to the assumed activation by $Ca^{2+}$ of the metabolic transformation of $IP_3$. Another model, directly based on CICR, has been proposed (Somogyi & Stucki, 1991) for $Ca^{2+}$ oscillations in hepatocytes. Fundamentally, that model does not differ significantly from the one described above (fig. 9.4c); what differentiates it is the recourse to polynomial kinetics, similar to that considered in the Brusselator model (Lefever & Nicolis, 1971), to describe the underlying biochemical processes, and the consideration of a single pool of $Ca^{2+}$ sensitive to $IP_3$ as well as $Ca^{2+}$. In contrast to experimental observations, the steady-state level of cytosolic $Ca^{2+}$ in that model remains independent of external stimulation.

More recently, Jafri *et al.* (1992) proposed an electrochemical model for $Ca^{2+}$ oscillations in *Xenopus* oocytes based on the exchanges of $Ca^{2+}$ between the cytosol and the endoplasmic reticulum; the regulation considered is again based on CICR, as the $Ca^{2+}$ conductance in that model is activated in a nonlinear, cooperative manner by cytosolic $Ca^{2+}$. Another single pool model for $Ca^{2+}$ oscillations and waves in *Xenopus* oocytes, due to Atri *et al.* (1993), incorporates both CICR and $Ca^{2+}$ inhibition of $Ca^{2+}$ release at higher cytosolic $Ca^{2+}$ levels.

The bell-shaped dependence of $IP_3$-induced $Ca^{2+}$ release as a function of cytosolic $Ca^{2+}$ observed in certain cells (Finch *et al.*, 1991; Bezprozvanny *et al.*, 1991) is also at the core of a single-pool model for $Ca^{2+}$ oscillations based on a dual action of $Ca^{2+}$ on the $IP_3$ receptor (De Young & Keizer; 1992). This model, which originally contained nine variables, was later reduced to two variables only (Keizer & De Young,

1994; Li & Rinzel, 1994). The latter reductions therefore yield a minimal model for $Ca^{2+}$ oscillations, like the earlier, two-pool minimal model considered below, which takes into account only CICR and not the inhibition of $Ca^{2+}$ release at high levels of cytosolic $Ca^{2+}$. A one-pool version of this model in which $Ca^{2+}$ and $IP_3$ behave as co-agonists for $Ca^{2+}$ release is presented in section 9.4. A model based on the bell-shaped calcium dependence of the ryanodine-sensitive calcium channel was recently proposed for calcium dynamics in cardiac myocytes (Tang & Othmer, 1994b).

Finally, a still different mechanism for $Ca^{2+}$ spiking in T lymphocytes has been studied in a theoretical model (Dolmetsch & Lewis, 1994) in which sustained oscillations result from an interplay between the depletion of $Ca^{2+}$ intracellular stores and the activation of $Ca^{2+}$ entry into the cell via depletion-activated $Ca^{2+}$ channels.

### *Minimal model based on $Ca^{2+}$-induced $Ca^{2+}$ release*

The model considered (Dupont & Goldbeter, 1989; Dupont et al., 1990, 1991; Goldbeter et al., 1990), schematized in fig.9.5 (see also fig. 9.4c), relies on the hypothesis (Berridge, 1988, 1990, 1993; Berridge & Galione, 1988; Berridge & Irvine, 1989) that an external stimulus triggers the synthesis of a certain amount of $IP_3$ that induces the release of $Ca^{2+}$ from an $IP_3$-sensitive pool; the amount of $Ca^{2+}$ thus released is controlled by the level of the stimulus through modulation of the level of $IP_3$. The amount of $IP_3$ is measured by a parameter, $\beta$, that can be viewed as the saturation function of the $IP_3$ receptor; $\beta$ therefore varies between 0 and 1 according to the degree of stimulation. It is assumed that the $Ca^{2+}$ concentration in the $IP_3$-sensitive pool remains constant, owing to fast replenishment that could involve activation of the uptake of external $Ca^{2+}$, as proposed in the capacitative model of $Ca^{2+}$ entry (Putney, 1986; Berridge, 1990). Cytosolic $Ca^{2+}$ is pumped into an $IP_3$-insensitive compartment; $Ca^{2+}$ in this compartment is released into the cytosol in a process activated by cytosolic $Ca^{2+}$. The binding of $Ca^{2+}$ to buffers is viewed as reaching rapid equilibration, so that only changes in free $Ca^{2+}$ are considered in the model; the role of $Ca^{2+}$ buffers in oscillations and waves of cytosolic $Ca^{2+}$ has recently been investigated theoretically by Wagner & Keizer (1994).

An important simplification in the above model is that the level of stimulus-induced, $IP_3$-mediated $Ca^{2+}$ release is treated as an adjustable parameter. As a consequence, the model is minimal as it contains only

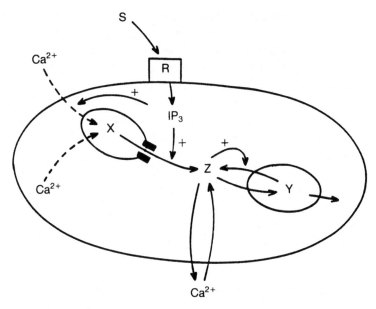

Fig. 9.5. Scheme of the model considered for signal-induced, intracellular $Ca^{2+}$ oscillations based on $Ca^{2+}$-induced $Ca^{2+}$ release (CICR). The stimulus (S) acting on a cell surface receptor (R) triggers the synthesis of $IP_3$; the latter intracellular messenger elicits the release of $Ca^{2+}$ from an $IP_3$-sensitive store (X, concentration $X$) at a rate proportional to the saturation function ($\beta$) of the $IP_3$ receptor. Fast replenishment of X could involve activation of $Ca^{2+}$ uptake from the extracellular medium into the $IP_3$-sensitive store. Cytosolic $Ca^{2+}$ (Z, concentration $Z$) is pumped into an $IP_3$-insensitive intracellular store; $Ca^{2+}$ in the latter store (Y, concentration $Y$) is released into the cytosol in a process activated by cytosolic $Ca^{2+}$. This feedback, known as CICR, plays a primary role in the origin of $Ca^{2+}$ oscillations. In its minimal version, the model contains only two variables, i.e. $Y$ and $Z$, as $X$ is treated as constant owing to fast replenishment of the $IP_3$-sensitive $Ca^{2+}$ pool (Dupont et al., 1991).

two variables whose time evolution is governed by the following kinetic equations (Dupont & Goldbeter, 1989; Dupont et al., 1990; Goldbeter et al., 1990):

$$\frac{dZ}{dt} = v_0 + v_1\beta - v_2 + v_3 + k_f Y - kZ$$

$$\frac{dY}{dt} = v_2 - v_3 - k_f Y \qquad (9.1)$$

with:

$$v_2 = V_{M2} \frac{Z^n}{K_2^n + Z^n}$$

$$v_3 = V_{M3} \frac{Y^m}{K_R^m + Y^m} \times \frac{Z^p}{K_A^p + Z^p} \qquad (9.2)$$

In these equations, $Z$ and $Y$ denote the concentration of free $Ca^{2+}$ in the cytosol and in the $IP_3$-insensitive pool; $v_0$ refers to a constant input of $Ca^{2+}$ from the extracellular medium; $v_1\beta$ denotes the $IP_3$-modulated release of $Ca^{2+}$ from the $IP_3$-sensitive store. The rates $v_2$ and $v_3$ refer, respectively, to the pumping of $Ca^{2+}$ into the $IP_3$-insensitive store, and to the release of $Ca^{2+}$ from that store into the cytosol in a process activated by cytosolic $Ca^{2+}$; $V_{M2}$ and $V_{M3}$ denote the maximum values of these rates. Parameters $K_2$, $K_R$ and $K_A$ are threshold constants for pumping, release and activation; $k_f$ is a rate constant measuring the passive, linear leak of $Y$ into $Z$; $k$ relates to the assumed linear transport of cytosolic $Ca^{2+}$ into the extracellular medium. Equations (9.2) allow for cooperativity in pumping, release and activation; $n$, $m$ and $p$ denote the Hill coefficients characterizing these processes. While pumping is known to be characterized by a cooperativity index of the order of 2 (Carafoli & Crompton, 1978), higher degrees of positive cooperativity have been reported for the activation of $Ca^{2+}$ release by cytosolic $Ca^{2+}$ (see, e.g., Abramson *et al.*, 1993).

In the above equations, all parameters and concentrations are defined with respect to the total cell volume; thus, to obtain the actual value of $Y$ and $K_R$ in the intracellular store, we have to multiply these values by the ratio of the cellular to total storage volumes. Given that latency and period can be viewed as local properties of the oscillations in any particular region of the cell, the spatial aspects of $Ca^{2+}$ signalling can initially be disregarded; these aspects are considered in section 9.4.

The steady-state concentration of cytosolic $Ca^{2+}$ predicted by eqns (9.1) is given by:

$$Z_0 = (v_0 + v_1\beta)/k \qquad (9.3)$$

This relation indicates that the external stimulation produces a rise in cytosolic $Ca^{2+}$ through the increased level of $IP_3$.

In writing eqns (9.2), provision was made that these equations can in principle generate periodic behaviour. The question is indeed how the external signal induces not just a rise in cytosolic $Ca^{2+}$, but also the oscillations that are observed in a large variety of cells. The usefulness of two-variable models is that they are amenable to phase plane analysis. In particular, there exists a powerful criterion due to Poincaré and Bendixson (see Minorsky, 1962), which allows us to rule out the occurrence of sustained oscillations in two-variable systems. Although negative in nature, this criterion is particularly useful in cases such as the present one, as it permits us to conclude, *a priori*, whether or not sustained oscillations are at all possible in a given two-variable system (Minorsky, 1962; Nicolis & Prigogine, 1977).

Applied to system (9.2), the Poincaré–Bendixson criterion states that sustained oscillations will never occur as long as the quantity $B$ defined by eqn (9.4) cannot change sign:

$$B = \partial \dot{Z}/\partial Z + \partial \dot{Y}/\partial Y \qquad (9.4)$$

In this equation $\dot{Z}$ and $\dot{Y}$ denote the time variation of $Z$ and $Y$ given by the kinetic eqns (9.1). Substituting the latter equations into (9.4), and taking into account the constancy of the terms $v_0$ and $v_1\beta$, leads to:

$$B = -\partial v_2/\partial Z + \partial v_3/\partial Z - k + \partial v_2/\partial Y - \partial v_3/\partial Y - k_f \qquad (9.5)$$

If the rate of release of $Ca^{2+}$ from the $Y$-containing store depends only on $Y$ and not on cytosolic $Ca^{2+}$, $Z$, the second term in (9.5) is nil, while the fifth one will be negative, given that the rate of release increases with $Y$. If we assume that the rate of pumping of $Z$ into $Y$ depends only on the level of cytosolic $Ca^{2+}$, the derivative of $v_2$ with respect to $Y$ will vanish while the derivative of $v_2$ with respect to $Z$ will be positive. Then, quantity $B$ will always be negative, thus ruling out sustained oscillations. We see, therefore, that in the absence of additional time delays CICR from intracellular stores cannot induce oscillations in the case where this process solely relies on the filling up of the store and on its discharge when a threshold level of intravesicular $Ca^{2+}$ is reached.

Given the negative nature of the first and last terms in (9.5), the quantity $B$ will be capable of changing sign only if either one of the second and third derivatives is positive, i.e. if $v_3$ increases with $Z$ and/or if $v_2$ increases with $Y$. The latter condition would correspond to the activation of $Ca^{2+}$ pumping from the cytosol into the store by $Ca^{2+}$ already sequestered in the latter compartment. Conversely, the former condition would imply the activation of the release process by $Ca^{2+}$ present in the cytosol. This activation is one possible form (Fabiato, 1985) of the phenomenon of CICR demonstrated long ago in skeletal muscle and cardiac cells (Endo *et al.*, 1970; Fabiato & Fabiato, 1975) and, more recently, in several other cell types (Busa *et al.*, 1985; Malgaroli *et al.*, 1990; Wakui *et al.*, 1990). While initially the accent was placed on the activation of the release by $Ca^{2+}$ present in the intracellular store, a view taken up in some recent experiments (Missiaen *et al.*, 1992; see Bezprozvanny & Ehrlich, 1994, for a recent analysis of this issue), other recent studies appear to move the emphasis towards an activating role of cytosolic $Ca^{2+}$ in this process (Fabiato, 1985; Bezprozvanny, Watras & Ehrlich, 1991; Finch, Turner & Goldin, 1991; Iino & Endo, 1992; Miyazaki *et al.*, 1992a). As this form of the CICR mechanism, which is observed experimentally, provides a possible source of oscillatory behaviour according to the above analysis based on the Poincaré–Bendixson criterion, eqns (9.2) are based on the assumption that the rate $v_3$ of $Ca^{2+}$ release from the endoplasmic reticulum is activated by cytosolic $Ca^{2+}$. The other possible source of oscillations, which relies on the activation of the pumping rate $v_2$ from the cytosol by $Ca^{2+}$ sequestered in the endoplasmic reticulum, has not been documented by experimental observations and is therefore not considered in this minimal model.

### *Sustained Ca²⁺ oscillations*

The steady-state concentration of sequestered calcium, $Y_0$, can readily be obtained by setting $(dY/dt)$ equal to zero and solving the resulting algebraic equation, while taking into account eqns (9.2) and (9.3). The stability properties of the unique steady state can be determined by linear stability analysis, as outlined in chapter 2. Figure 9.6 shows some typical results of this analysis, as a function of parameters $K_R$ and $(v_0 + v_1\beta)$. For sufficiently large values of $K_R$, the steady state is stable for $\beta = 0$ or at low, finite $\beta$ values and corresponds then to a low $Ca^{2+}$ level (see fig. 9.10, below); a stable steady state corresponding to a high, constant level of cytosolic calcium is established at large values of $\beta$. It is for intermediate values of $\beta$ that the steady state becomes unstable and sustained oscillations in $Z$ and $Y$ occur (fig. 9.7).

The waveform of the oscillations predicted by the model for cytosolic $Ca^{2+}$ (fig. 9.7) resembles that of the spikes observed for a number of cells stimulated by external signals. In particular, the rise in cytosolic $Ca^{2+}$ is preceded by a rapid acceleration that starts from the basal level; although it originates from a different, nonelectrical mechanism, this pattern, which is reminiscent of the pacemaker potential that triggers autonomous spiking in nerve and cardiac cells (DiFrancesco, 1993), has been observed (Jacob *et al.*, 1988) in epithelial cells stimulated by histamine (see fig. 9.3). As in the model by Meyer & Stryer (1988), the oscillations of $Ca^{2+}$ in the intracellular store have a saw-tooth appearance (see the dashed curve in fig. 9.7). Here, however, the phenomenon does

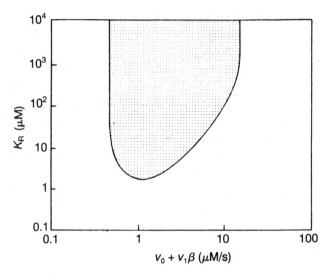

Fig. 9.6. Stability diagram established as a function of the threshold constant for release, $K_R$, and of the total (basal plus signal-triggered) influx of $Ca^{2+}$ into the cytosol ($v_0 + v_1\beta$). The diagram is obtained by linear stability analysis of eqns (9.1)–(9.2) around the unique steady-state solution admitted by these equations. Parameter values are: $V_{M2} = 100\ \mu M/s$, $V_{M3} = 1\ mM/s$, $m = n = p = 2$, $K_2 = 1\ \mu M$, $K_A = 2.5\ \mu M$, $k = 2\ s^{-1}$, $k_f = 0$. The steady state is unstable in the dotted domain; sustained oscillations of $Ca^{2+}$ occur under these conditions (Dupont & Goldbeter, 1989).

not require the periodic variation of $IP_3$ whose constant level is reflected in parameter $\beta$.

The oscillations in fig. 9.7 have a period of the order of a few seconds, as observed, for example, in cardiac cells (Fabiato & Fabiato, 1975) and pituitary gonadotropes (Shangold, Murphy & Miller, 1988; Stojilkovic *et al.*, 1993; Tse *et al.*, 1993). Such period duration results from the choice of parameter values (see legend to fig. 9.7). It is not uncommon to observe variations of more than two orders of magnitude between different cell types for parameters such as the pumping rate of calcium into the sarcoplasmic or endoplasmic reticulum (Carafoli & Crompton, 1978), with particularly high values being found in cardiac cells. As demonstrated in fig. 9.8, lower values can readily produce periods of the order of 1 min, as observed for $Ca^{2+}$ oscillations in hepatocytes (Woods *et al.*, 1986, 1987) and endothelial cells (Jacob *et al.*, 1988).

In the phase plane formed by concentrations $Y$ and $Z$, sustained oscillatory behaviour corresponds to the evolution towards a limit cycle (fig. 9.9). Also indicated in fig. 9.9 are the nullclines $(dZ/dt) = 0$ and

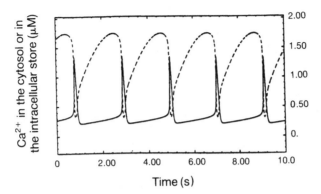

Fig. 9.7. Oscillations in cytosolic $Ca^{2+}$ ($Z$, solid line) brought about by an increase in $\beta$ up to 30.1% triggered by external stimulation. The concentration of $Ca^{2+}$ in the $IP_3$-insensitive intracellular store ($Y$, dashed line) has a concomitant saw-tooth variation; for the sake of clarity, the actual level of $Y$ (defined with respect to the total cell volume) has been decreased by 0.35 $\mu$M. The curves are obtained by integration of eqns (9.1)–(9.2) for the following parameter values, which are in a physiological range (Carafoli & Crompton, 1978): $v_0 = 1$ $\mu$M/s, $k = 10$ s$^{-1}$, $k_f = 1$ s$^{-1}$, $v_1 = 7.3$ $\mu$M/s, $V_{M2} = 65$ $\mu$M/s, $V_{M3} = 500$ $\mu$M/s, $K_2 = 1$ $\mu$M, $K_R = 2$ $\mu$M (defined with respect to the total cell volume), $K_A = 0.9$ $\mu$M, $m = n = 2$, $p = 4$. Oscillations also occur for $p = 2$, with a slightly larger period (Goldbeter *et al.*, 1990).

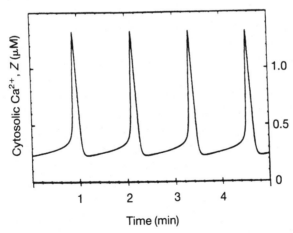

Fig. 9.8. Oscillations in cytosolic $Ca^{2+}$ obtained as in fig. 9.7 for $v_0 = 1$ $\mu$M/min, $k = 10$ min$^{-1}$, $k_f = 1$ min$^{-1}$, $v_1\beta = 2.7$ $\mu$M/min, $V_{M2} = 65$ $\mu$M/min, $V_{M3} = 500$ $\mu$M/min; other parameters are as in fig. 9.7 (Dupont & Goldbeter, 1992a).

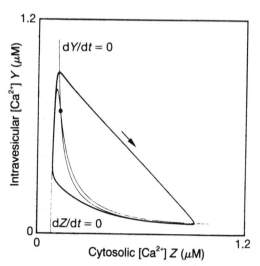

Fig. 9.9. Evolution towards a limit cycle in the phase plane $(Z, Y)$. The $Y$ and $Z$ nullclines are indicated; their intersection (large dot) defines the (unstable) steady state. The limit cycle trajectory is obtained by integration of eqns (9.1)–(9.2) for the following parameter values: $k = 9.5 \text{ s}^{-1}$, $m = n = 2$, $p = 4$, $v_0 = 0.5 \ \mu\text{M/s}$, $v_1\beta = 0.6 \ \mu\text{M/s}$, $V_{M2} = 80 \ \mu\text{M/s}$, $V_{M3} = 5 \times 10^3 \ \mu\text{M/s}$, $K_R = 0.5 \ \mu\text{M}$, $K_A = 0.9 \ \mu\text{M}$; other parameter values are as in fig. 9.6 (Dupont & Goldbeter, 1989).

$(dY/dt) = 0$, whose intersection defines the steady state. As in the model for glycolytic oscillations (see chapter 2), it is possible to show that the steady state is unstable when the two nullclines intersect in a region of sufficiently negative slope $(dY/dZ)$ on the $Z$ nullcline, as in the case of fig. 9.9. As indicated by eqn (9.3), the abscissa of the steady state in the phase plane moves from left to right when $\beta$ (i.e. the stimulation) increases. This explains why a sufficiently low value of $\beta$ will corre-spond (as in fig. 9.10a) to a stable steady state, located in the region of positive slope on the $Z$ nullcline, while an increase in $\beta$ will induce oscillations as soon as the slope becomes sufficiently negative (see fig. 9.10b,c). A further rise in stimulation will produce a stable steady state, corresponding to a high $Z$ value (as in fig. 9.10d), if the abscissa value given by eqn (9.3) is that of a point located on the $Z$ nullcline in the region where the slope $(dY/dZ)$ is either not sufficiently negative, or positive.

The results shown in fig. 9.7 and 9.8 were obtained in the case where $Ca^{2+}$ pumping into the intracellular store, $Ca^{2+}$ release from this com-partment, and activation of this release by cytosolic $Ca^{2+}$ are all charac-

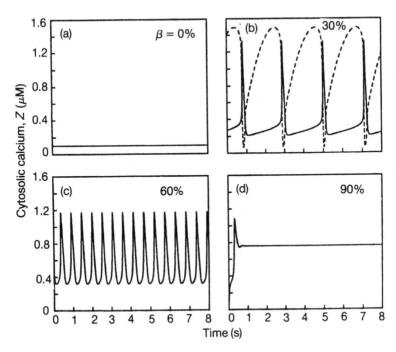

Fig. 9.10. Dynamics of cytosolic $Ca^{2+}$ in response to external stimulation in the minimal model for $Ca^{2+}$ oscillations based on $Ca^{2+}$-induced $Ca^{2+}$ release. In the absence of stimulation ($\beta = 0\%$), a stable, low steady-state level of cytosolic $Ca^{2+}$ ($Z$, solid line) is established. Upon increasing the stimulation, oscillations develop with a frequency that rises with the stimulus level (compare graphs with $\beta = 30\%$ and $\beta = 60\%$). Above a critical level of stimulation, oscillations disappear and a stable, high steady-state level of cytosolic $Ca^{2+}$ is established ($\beta = 90\%$). The curves were generated by numerical integration of eqns (9.1) with $v_0 = 1\ \mu M/s$, $v_1 = 7.3\ \mu M/s$, $V_{M2} = 65\ \mu M/s$, $V_{M3} = 500\ \mu M/s$, $K_R = 2\ \mu M$, $K_A = 0.9\ \mu M$, $K_2 = 1\ \mu M$, $k_f = 1\ s^{-1}$, $k = 10\ s^{-1}$, $n = m = 2$, $p = 4$. The oscillatory range of $\beta$ (condition (9.7)) is bounded by the two critical values $\beta_{c1} = 0.2945$ and $\beta_{c2} = 0.7671$. In the upper, right part, the dashed line represents the variation of $[Ca^{2+}]$ ($Y$) in the $Ca^{2+}$-sensitive store (the value shown should be increased by 0.5 $\mu M$); oscillations with higher values of $Y$ are obtained at larger values of the threshold constant for release, $K_R$ (Dupont *et al.*, 1990). Oscillations can also occur, for different parameter values, with $n = m = p = 1$ (Dupont & Goldbeter, 1992a).

terized by positive cooperativity. Although such cooperativity is observed experimentally (Carafoli & Crompton, 1978; Abramson *et al.*, 1993), we may wonder whether it is required for oscillatory behaviour. The analysis of the model indicates (Goldbeter & Dupont, 1990) that this is not the case: oscillations can indeed occur for $m = n = p = 1$, i.e. when each of the three mechanisms of pumping, release and activation is described by a purely Michaelian process. The characteristics of $Ca^{2+}$ oscillations are somewhat different from those obtained under the conditions of fig. 9.7 or 9.8: the period and the amplitude both increase significantly when the three Hill coefficients are equal to unity, while the waveform is much smoother. A similar increase in the amplitude is seen in the allosteric model for glycolytic oscillations when decreasing or suppressing cooperativity (see chapter 2).

The fact that oscillations occur when the kinetic expressions for pumping, release and activation are of the Michaelian type raises the possibility that periodic behaviour might also occur when each of these expressions becomes linear, since Michaelian functions reduce to linear ones in the domain of first-order kinetics. To test this possibility, let us consider the case where the expressions (9.2) for $v_2$ and $v_3$ transform into expressions:

$$v_2 = k_2 Z, \quad v_3 = k_3 YZ \qquad (9.6)$$

Under these conditions system (9.1) still admits a unique steady state, but linear stability analysis shows that the latter is always stable (Goldbeter & Dupont, 1990); this rules out the occurrence of sustained oscillations around a nonequilibrium unstable steady state. This result holds with previous studies of two-variable systems governed by polynomial kinetics; these studies indicated that a nonlinearity higher than quadratic is needed for limit cycle oscillations in such systems (Tyson, 1973; Nicolis & Prigogine, 1977). Thus, in system (9.1), it is essential for the development of $Ca^{2+}$ oscillations that the kinetics of pumping or activation be at least of the Michaelian type. Experimental data in fact indicate that these processes are characterized by positive cooperativity associated with values of the respective Hill coefficients well above unity, thus favouring the occurrence of oscillatory behaviour.

### Control of $Ca^{2+}$ oscillations by the stimulus and by extracellular $Ca^{2+}$

One of the most salient results of the model based on the CICR mechanism is represented in fig. 9.10. There, the temporal evolution of cytosolic $Ca^{2+}$ is shown at different values of $\beta$ associated with increasing levels of stimulation. In the absence of stimulation, a stable steady state is established, corresponding to a low level of cytosolic $Ca^{2+}$ (fig. 9.10a). Upon increasing the value of $\beta$, the steady state becomes unstable and oscillations appear, with a frequency that increases with the degree of stimulation (fig. 9.10b and c). A further increase in stimulation produces low-amplitude oscillations around a higher mean level of cytosolic $Ca^{2+}$. Finally, above a critical degree of stimulation, oscillations disappear and the system evolves towards a stable steady state corresponding to a high level of cytosolic $Ca^{2+}$ (fig. 9.10d). Similar results are obtained upon raising the level of extracellular $Ca^{2+}$ (see below). The sequence of dynamic behaviour predicted by the model in response to increasing stimulation is observed in many cells (Woods *et al.*, 1987; Jacob *et al.*, 1988; Harootunian, Kao & Tsien, 1988; Meyer & Stryer, 1991; Malgaroli *et al.*, 1992; see fig. 9.2). The results shown in fig. 9.10 also indicate that the same mechanism can produce oscillations of very different waveforms, depending on the value of a single parameter such as that measuring the level of $IP_3$; this may contribute to the variety of waveforms produced in a given type of cell by diverse agonists (Fewtrell, 1993), which may differentially affect the synthesis of $IP_3$.

The analysis of eqns (9.1) thus shows that for a given set of parameter values, $Ca^{2+}$ oscillations occur whenever parameter $\beta$, which rises with the level of the external stimulus and measures the saturation of the $IP_3$ receptor, lies in a range bounded by two critical values, i.e. when condition (9.7) holds :

$$\beta_{c1} < \beta < \beta_{c2} \tag{9.7}$$

The critical values $\beta_{c1}$ and $\beta_{c2}$ depend on other parameters and, in particular, on $v_0$ and $v_1$.

In fact, what actually controls the oscillations is the sum $v_0 + v_1\beta$: oscillations can thus result either from an increase in stimulation leading to enhanced $Ca^{2+}$ release from the $IP_3$-sensitive store ($v_1\beta$), or from an increase in the influx of extracellular $Ca^{2+}$ ($v_0$) (Goldbeter *et al.*, 1990; Dupont *et al.*, 1991). As expected, and illustrated in fig. 9.11, the two situations, which might be encountered simultaneously upon stimulation, result in a similar dependence of the frequency on the control parameter. Besides showing the existence of a range of stimulation giving rise to repetitive $Ca^{2+}$ spiking, the data in fig. 9.11 also indicate that oscillations can occur in the absence of stimulation when the influx of extracellular $Ca^{2+}$ is in an appropriate range. The latter property provides an explanation for the occurrence of spontaneous $Ca^{2+}$ spiking in cardiac cells (Tsien, Kass & Weingart, 1979; Kort, Capogrossi & Lakatta, 1985; Stern, Capogrossi & Lakatta, 1988) and pituitary somatotropes (Holl *et al.*, 1988). It may also account for the $K^+$-induced $Ca^{2+}$ oscillations described in sympathetic neurons (Lipscombe *et al.*, 1988). The

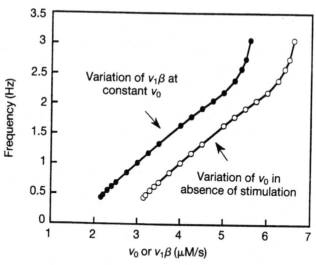

Fig. 9.11. Dependence of frequency of $Ca^{2+}$ oscillations on the magnitude of external stimulation and on extracellular $Ca^{2+}$ in the minimal model for signal-induced $Ca^{2+}$ oscillations based on $Ca^{2+}$-induced $Ca^{2+}$ release. The curve on the left shows the effect of an increase in stimulation, measured by $\beta$, at a fixed value of the extracellular influx of $Ca^{2+}$, i.e. $v_0 = 1$ μM/s, for $v_1 = 7.3$ μM/s. The curve on the right shows the effect of a variation in the influx of extracellular $Ca^{2+}$ ($v_0$) in the absence of stimulation ($\beta = 0$). The points were obtained by numerical integration of eqns (9.1) for different values of $v_0$ or $\beta$. For the para-meter values considered, which are those of fig. 9.10, oscillations occur when the sum ($v_0 + v_1\beta$) is in the range 3.15–6.6 μM/s (Dupont *et al.*, 1991).

following results focus mainly on the induction of oscillations by a sig-nal-induced rise in $IP_3$ corresponding to an increase in $\beta$.

### Response to a transient increase in IP₃

The existence of a finite range of $IP_3$ levels producing oscillations is also illustrated by numerical experiments demonstrating that the effect of an instantaneous increase in $IP_3$, followed by an exponential decrease, results in a train of damped oscillations (figs. 9.12a and b). Such behav-iour was observed experimentally in *Xenopus* oocytes following injec-tion with $IP_3$ (fig. 9.12c and d). The model accounts for these observations provided that the $IP_3$ pulse initially brings parameter $\beta$ above the critical value $\beta_{c2}$; the progressive decline in $\beta$ caused by $IP_3$ decay would carry the system through the oscillatory domain defined by condition (9.7), before a stable steady state is re-established when $\beta$ drops below $\beta_{c1}$.

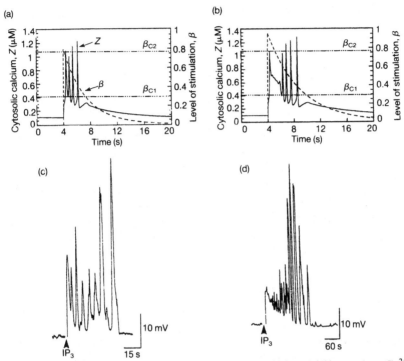

Fig. 9.12. Theoretical ((a) and (b)) and experimental ((c) and (d)) transient $Ca^{2+}$ oscillations elicited by a pulse of $IP_3$ applied intracellularly to a resting, nonoscillatory cell. (a) and (b) The pulse of $IP_3$ selected is such that the system crosses the oscillatory domain as $\beta$ rises initially above the critical value $\beta_{c2} = 76.71\%$ and then decreases exponentially below the lower critical value $\beta_{c1} = 29.45\%$. Parameter values are as in fig. 9.10. For (a), $\beta$ decays from the initial value $\beta_i = 85\%$ according to the equation $\beta = \beta_i \exp[-0.3(t - t_p)]$, while for (b), $\beta = \beta_i \exp[-0.2(t - t_p)]$ with $\beta_i = 96\%$; for both curves, the time of instantaneous increase in $\beta$, $t_p$, is equal to 4 s. (c) and (d) In the experimental system (*Xenopus* oocytes), the cells were injected with either a low (c) or a high (d) dose of $IP_3$. Since oocytes have a $Ca^{2+}$-sensitive $Cl^-$ conductance, the membrane potential oscillations reflect an underlying $Ca^{2+}$ oscillation. The amount of $IP_3$ injected (arrows) was varied by adjusting the duration of the iontophoretic current (–20 nA), which was 1 s in (c) and 10 s in (d) (Dupont *et al.*, 1991).

When $\beta$ is initially brought just above $\beta_{c2}$, in the model (fig. 9.12a) as in the experiments (fig. 9.12c), the amplitude of the transients is at first high, then declines and later may rise again while the interval between successive $Ca^{2+}$ peaks increases towards the end of the oscillations. When the initial value of $\beta$ is well above $\beta_{c2}$, the model predicts that a phase of monotonous decrease of $Ca^{2+}$ follows the initial spike before the beginning of transient oscillations (fig. 9.12b); such behaviour is observed, accordingly, when oocytes are injected with larger amounts of $IP_3$ (fig. 9.12d).

## *Correlation of latency with period of $Ca^{2+}$ oscillations*

In hepatocytes the period of $Ca^{2+}$ oscillations correlates with the time required for observing the first peak in $Ca^{2+}$ after the onset of stimulation (Rooney, Sass & Thomas, 1989). Specifically, this time interval, called latency, increases roughly linearly with the period of $Ca^{2+}$ oscillations as stimulation decreases. Given that this observation brings further insight into the mechanism of signal-induced $Ca^{2+}$ mobilization and provides an additional test for any theoretical explanation of the oscillatory phenomenon, it is worth examining the relationship between period and latency of $Ca^{2+}$ transients in the model based on CICR. The model shows that the existence of an approximately linear correlation between these two quantities is a natural consequence of the two-pool mechanism of $Ca^{2+}$ oscillations (Dupont et al., 1990).

In the model, latency $L$ can be defined as the time needed to reach the first $Ca^{2+}$ spike after the onset of an instantaneous increase in $\beta$. Simulations indicate that a progressive increase in stimulation, measured by the final value $\beta_f$, results in a decrease in latency towards a plateau value (Dupont et al., 1990). This result agrees with experimental observations showing that latency declines towards a constant minimum value as agonist concentration increases in blood platelets (Sage & Rink, 1987), blowfly salivary gland (Berridge et al., 1988), adrenal glomerulosa cells (Quinn, Williams & Tillotson, 1988), and hepatocytes (Rooney et al., 1989).

To compare the predictions of the model with the experimental results of Rooney et al. (see fig. 9.13b), period and latency were determined for different final values of $\beta$ in the cases where $\beta$ increases in a stepwise manner (fig. 9.13a). The results are in good agreement with the experimental observations since period correlates with latency in a manner that is, to a good approximation, linear. The model predicts that such a correlation is obtained only at values of $v_0$ for which the level of $Ca^{2+}$ in the cytosol and, hence, in the $IP_3$-insensitive store, is relatively low prior to stimulation (Dupont et al., 1990).

The model based on CICR provides an explanation for the similar dependence of period and latency on external stimulation. The effect of the stimulus is to elicit a rise in $\beta$ that produces an increase in $Y$ and $Z$ which both evolve to new (unstable) steady-state values. The accumula-

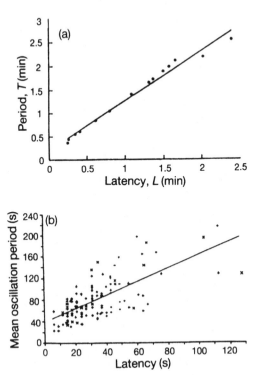

Fig. 9.13. Correlation of period and latency of Ca²⁺ oscillations at different levels of stimulation. (a) The theoretical curve is obtained by integration of eqns (9.1) in response to step increases in $\beta$ from zero to final values ranging from $\beta_{c1}$ = 0.286 to $\beta_{c2}$ = 0.846; $v_0$ = 1.4 μM/min, $v_1$ = 6 μM/min, $V_{M2}$ = 65 μM/min, $V_{M3}$ = 500 μM/min, $k_f$ = 1 min⁻¹, $k$ = 10 min⁻¹; other parameter values are those of fig. 9.10 (Dupont *et al.*, 1991). (b) Experimental results obtained in hepatocytes by Rooney *et al.* (1989).

tion of $Y$ and $Z$ is more rapid at the larger value of $\beta$ considered, because the accumulation of $Z$ in that phase is primarily governed by the influx $v_1\beta$ from the IP₃-sensitive store. Hence, the slope of the rise in $Z$ before the spike is steeper, so that the threshold for self-amplified release is reached more rapidly and latency is diminished; for the same reason, the period of the oscillations is reduced when the value of $\beta$ increases (compare the two curves shown in fig. 9.14 obtained at different values of $\beta$). It is therefore not surprising to find in this model a strong correlation between period and latency.

The data in figs. 9.13a and 9.14 were obtained for parameter values yielding oscillations whose period is of the order of minutes. Similar relationships between period and latency are obtained when parameter

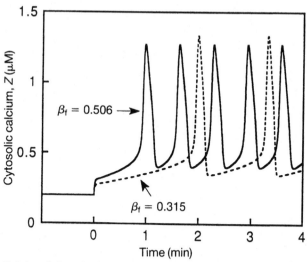

Fig. 9.14. Origin of the correlation between period and latency of $Ca^{2+}$ oscillations. The correlation is made clear by comparing oscillations in cytosolic $Ca^{2+}$ resulting from an instantaneous increase in $\beta$ at time zero, from $\beta = 0$ up to the values 0.315 and 0.506. The curves, generated by integration of eqns (9.1) for the parameter values of fig. 9.13, show the shortening of both period and latency at higher stimulation, due to the faster accumulation of $Ca^{2+}$ between spikes (Dupont *et al.*, 1991).

values are used for which the period of $Ca^{2+}$ spikes is of the order of seconds, as in fig. 9.7.

### *Phase-shift and transient suppression of $Ca^{2+}$ oscillations by $Ca^{2+}$ pulses*

If CICR release plays a prominent role in the mechanism of $Ca^{2+}$ oscillations, then the latter should be particularly sensitive to perturbations in cytosolic $Ca^{2+}$. The effect of pulses of cytosolic $Ca^{2+}$ has been determined by numerical experiments, the results of which are shown in fig. 9.15. In part a, an instantaneous, moderate increase in $Z$ by some 0.2 μM results in a delay of the next high-amplitude peak in cytosolic $Ca^{2+}$. This delay occurs when the perturbation is applied shortly after the minimum of $Ca^{2+}$ oscillations. When the same $Ca^{2+}$ pulse is given a little later, as shown in part b, a phase advance is observed. The dependence of the phase shift on the phase $\phi$ at which the perturbation occurs is illustrated by the phase response curve shown in fig. 9.16. This curve, established at a fixed value of the perturbation equal to 0.18 μM, indi-

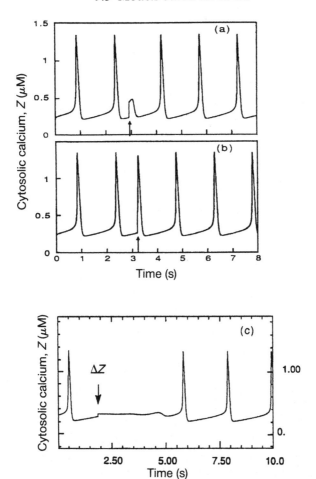

Fig. 9.15. Phase shift and transient suppression of $Ca^{2+}$ oscillations by $Ca^{2+}$ pulses. The curves show the response of the minimal model to a pulse of cytosolic $Ca^{2+}$ resulting in (a) a delay or (b) an advance of the next high-amplitude peak in cytosolic $Ca^{2+}$. In (c), a tiny $Ca^{2+}$ pulse of precise magnitude applied at the appropriate phase after the minimum of $Ca^{2+}$ oscillations, as the system passes in the immediate vicinity of the steady state, results in their transient suppression. Parameter values are those of fig. 9.10 with $\beta = 0.3$. The pulse is given as an instantaneous increase in $Z$ by of 0.2045 µM for (a) and (b); for (c), a pulse of 0.040967 µM is given 1.432 s after the maximum in $Z$ (Dupont *et al.*, 1991).

cates that during and immediately after a peak of $Ca^{2+}$ there is an absolute refractory period during which no phase shift occurs. Later, a delay such as that shown in fig. 9.15a, corresponding to a negative value of $\Delta\phi$, is observed but only over a small range of $\phi$. A sharp discontinuity occurs thereafter as the delay transforms into a phase advance

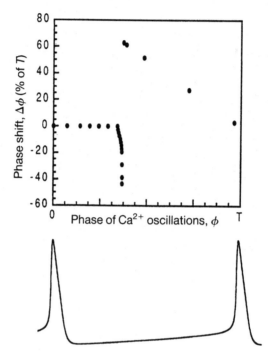

Fig. 9.16. Phase response curve predicted by the minimal model for signal-induced $Ca^{2+}$ oscillations based on $Ca^{2+}$ induced $Ca^{2+}$ release. Shown is the phase shift of $Ca^{2+}$ oscillations induced by a pulse of 0.18 μM applied at different phases $\phi$ of the oscillations; $\phi = 0$ corresponds to the peak in cytosolic $Ca^{2+}$ ($Z$) that is represented in the lower panel. A phase advance ($\Delta\phi > 0$) occurs when the $Ca^{2+}$ peak resulting from perturbation exceeds 1 μM. A phase delay ($\Delta\phi < 0$) occurs when the resulting peak is less than that value (see also fig. 9.15a and b). The phase shift is expressed as a fraction of the period $T$. Parameter values are those of fig. 9.15; for these parameter values, $T = 2.07$ s. The phase response curve remains qualitatively unchanged when the period is of the order of minutes (Dupont *et al.*, 1991).

corresponding to a positive value of $\Delta\phi$. Phase delays or advances can also be obtained by varying the amplitude of the pulse at a given phase of the oscillations.

Phase advances have been observed in experiments where perturbations of cytosolic $Ca^{2+}$ were achieved by the photorelease of caged $Ca^{2+}$ (Harootunian *et al.*, 1988). The occurrence of phase delays has not yet been observed with $Ca^{2+}$ pulses, but the model predicts that such delays should occur over a much narrower range of phases of the oscillations.

Another phenomenon predicted by the model is the transient suppression of oscillations by a $Ca^{2+}$ pulse of precise magnitude delivered at the right phase (fig. 9.15c). This behaviour occurs when the pulse brings the system into the immediate vicinity of the steady state (Winfree, 1980). Because the latter state is an unstable focus, oscillations of increasing magnitude will slowly develop. The simulations indicate that the duration of transient suppression of the oscillations increases as the pulse brings the system closer and closer to that steady state.

## 9.4 Extending the model for $Ca^{2+}$ oscillations based on CICR

*One-pool model involving $Ca^{2+}$ and $IP_3$ as co-agonists for $Ca^{2+}$ release*

Recent experimental investigations have also focused on the intracellular $Ca^{2+}$ pools involved in the generation of repetitive $Ca^{2+}$ spikes. In its initial version described above, the CICR model (see fig. 9.5) assumed the existence of two types of $Ca^{2+}$ pool: one sensitive to $IP_3$ and one insensitive to $IP_3$ but sensitive to cytosolic $Ca^{2+}$. The channels present in the membrane of the latter $Ca^{2+}$ store are responsible for CICR; these channels are characterized experimentally by their sensitivity to ryanodine and caffeine (Tsien & Tsien, 1990). $Ca^{2+}$ channels involved in CICR may, however, be more widespread than was previously considered, given the recent characterization of receptors/$Ca^{2+}$ channels sensitive to ryanodine but not to caffeine (Giannini *et al.*, 1992). The distinction between $IP_3$-sensitive and caffeine-sensitive stores has been made clear in a variety of cell types such as adrenal chromaffin cells (Robinson & Burgoyne, 1991), Purkinje neurons (Walton *et al.*, 1991), acinar cells (Foskett & Wong, 1991), and smooth muscle cells (Tribe, Borin & Blaustein, 1994). The various stores, characterized by different $Ca^{2+}$ release properties, have generally distinct cellular locations, though they are both thought to be part of the endoplasmic or sarcoplasmic reticulum.

In other cells, however, the existence of a unique type of nonmitochondrial $Ca^{2+}$ pool has been demonstrated. Thus, in the neurosecretory cell line PC12, $Ca^{2+}$ release evoked by caffeine/ryanodine or $IP_3$ originates from the same $Ca^{2+}$ pool (Zacchetti *et al.*, 1991). Rapid release of $Ca^{2+}$ by purified $IP_3$ receptors isolated from mammalian brain and reconstituted into vesicles requires cytosolic $Ca^{2+}$ as well as $IP_3$ (Finch *et al.*, 1991); there, $IP_3$ and $Ca^{2+}$ behave as co-agonists for $Ca^{2+}$ release (for further discussion of these results, see Combettes & Champeil, 1994; Finch & Goldin, 1994). The same property characterizes the $IP_3$

receptor from Purkinje cells of canine cerebellum (Bezprozvanny *et al.*, 1991) and hamster egg (Miyazaki et al., 1992a). When $Ca^{2+}$ and $IP_3$ behave as co-agonists for the induction of $Ca^{2+}$ release, the regulatory process may be viewed either as $Ca^{2+}$-sensitized, $IP_3$-induced $Ca^{2+}$ release (IICR) (Miyazaki *et al.*, 1992a) or as $IP_3$-sensitized CICR (Dupont & Goldbeter, 1993). With regard to the mechanism of $Ca^{2+}$ oscillations, referring to the action of $Ca^{2+}$ on IICR as a form of CICR emphasizes the prominent role of the positive feedback exerted by cytosolic $Ca^{2+}$ on its release from intracellular stores. Therefore, in this chapter devoted to the modelling of $Ca^{2+}$ oscillations, the ($IP_3$-independent) CICR and $Ca^{2+}$-sensitized IICR will be referred to as $IP_3$-insensitive and $IP_3$-sensitive forms of CICR, respectively.

The question arises as to the possibility of $Ca^{2+}$ oscillations in a one-pool model based on CICR. We may argue, indeed, that in a model with a single pool sensitive to both $IP_3$ and $Ca^{2+}$, the rise in $IP_3$ after stimulation could prevent oscillations by inducing the depletion of the $Ca^{2+}$ pool, which would annihilate the destabilizing effect of CICR. To address this issue, Dupont & Goldbeter (1993) examined the possibility of sustained $Ca^{2+}$ oscillations in a modified version of the original CICR model containing a single $Ca^{2+}$ pool. Related studies of oscillations in a one-pool model based on $IP_3$-sensitive CICR were carried out independently by De Young & Keizer (1992; see also Keizer & De Young, 1994; Li & Rinzel, 1994) and Atri *et al.* (1993), who took into account the inhibition of $IP_3$-induced $Ca^{2+}$ release observed at high levels of cytosolic $Ca^{2+}$ (Finch *et al.*, 1991; Bezprozvanny *et al.*, 1991).

In the version of the one-pool model considered below, the same $Ca^{2+}$ channel is assumed to be sensitive to both $IP_3$ and $Ca^{2+}$ behaving as co-agonists (Finch *et al.*, 1991). An alternative possibility is that two distinct $Ca^{2+}$ channels, sensitive to $Ca^{2+}$ or $IP_3$, coexist in the same $Ca^{2+}$ pool. However, only the first version of the one-pool model readily gives rise to $Ca^{2+}$ oscillations (Dupont & Goldbeter, 1993); the predictions of this model are compared below with those of the two-pool model based on CICR. Besides a number of common properties, the one- and two-pool models based, respectively, on $IP_3$-sensitive and $IP_3$-insensitive CICR lead to distinctive predictions that might provide an explanation for differences in $Ca^{2+}$ oscillations observed in various cell types.

When we consider one type of pool possessing channels activated by both $IP_3$ and $Ca^{2+}$ (fig. 9.17b), the $Ca^{2+}$ exchange processes are still globally represented by eqns (9.1) but the detailed nature of some of the

(a)

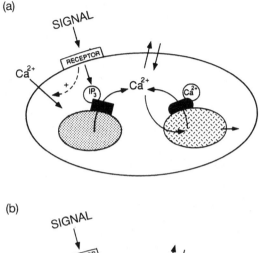

(b)

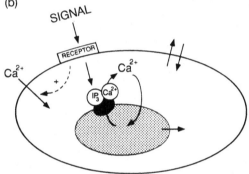

Fig. 9.17. Schematic representation of (a) the two-pool and (b) the one-pool versions of the model for signal-induced $Ca^{2+}$ oscillations based on $Ca^{2+}$-induced $Ca^{2+}$ release. In the one-pool model it is assumed that a single type of $Ca^{2+}$ channel is activated both by $IP_3$ and $Ca^{2+}$ behaving as co-agonists (Dupont & Goldbeter, 1993).

processes has to be modified (compare fig. 9.17a and b). Pumping into the unique $IP_3$- and $Ca^{2+}$-sensitive store is still given by eqn (9.2) while the release of $Ca^{2+}$ into the cytosol now takes the form:

$$v_3 = \beta V_{M3} \frac{Y^m}{K_R^m + Y^m} \times \frac{Z^p}{K_A^p + Z^p} \tag{9.8}$$

where $\beta$ represents the degree of saturation by $IP_3$ of this 'bi-activated' receptor and $V_{M3}$ the maximum rate of release. As in the two-pool version (see eqn (9.2)) the last factor reflects the assumption that $Ca^{2+}$ release is activated by cytosolic $Ca^{2+}$. Though the activation of the $IP_3$ receptor by cytosolic $Ca^{2+}$ has in some cases been shown to be followed by an inhibition of the same receptor at higher cytosolic $Ca^{2+}$ concentration (Finch *et al.*, 1991; Bezprozvanny *et al.*, 1991), the model predicts that, even if present, this inhibition is not required for the generation of sustained $Ca^{2+}$ oscillations.

In the two-pool model, the term $v_1\beta$ in eqn (9.1) denotes the constant influx of $Ca^{2+}$ from the $IP_3$-sensitive pool. If such a term is suppressed in the one-pool model – because the effect of $IP_3$ on $Ca^{2+}$ release is then expressed by eqn (9.8) – and if we consider only a constant $Ca^{2+}$ influx, $v_0$, from the extracellular medium, we lose an important property of the CICR model, namely that the mean cytosolic $Ca^{2+}$ concentration rises with the stimulation level. In most cell types, indeed, a low (high) concentration of agonist generates a constant low (high) level of cytosolic $Ca^{2+}$, when the stimulus is outside the range leading to sustained oscillations (see above). One way to obviate this shortcoming is to assume that the stimulation, besides inducing $IP_3$ synthesis, also leads to a direct activation of $Ca^{2+}$ entry from the extracellular medium into the cytosol (see fig. 9.17b). Such a stimulus-activated $Ca^{2+}$ entry has been reported in some cell types and could be triggered by depletion of the intracellular stores (Putney, 1991; Dolor *et al.*, 1992; Hoth & Penner, 1992; Törnquist, 1992). The replenishment of intracellular stores through depletion-activated $Ca^{2+}$ channels plays a central role in a model for $Ca^{2+}$ oscillations in T lymphocytes (Dolmetsch & Lewis, 1994).

To take into account stimulus-activated $Ca^{2+}$ entry, the simplest assumption has been retained, namely that the influx from the extracellular medium triggered by external stimulation is proportional to parameter $\beta$, as in the $IP_3$-regulated release from the $Ca^{2+}$ pool (Dupont & Goldbeter, 1993). The influx $v_{in}$ from the extracellular medium takes thus the form of eqn (9.9) which is similar to that of eqn (9.1):

$$v_{in} = v_0 + v_1\beta \qquad (9.9)$$

where $v_0$ still denotes the constant rate of $Ca^{2+}$ influx in the absence of stimulation while $v_1$ is now the maximum rate of stimulus-induced influx from the extracellular medium into the cytosol.

Although based on distinct assumptions, the model where $Ca^{2+}$ and $IP_3$ behave as co-agonists is mathematically similar to the two-pool model, the only difference being that $V_{M3}$ in eqn (9.2) is replaced by $\beta V_{M3}$ in eqn (9.8). Oscillations of $Ca^{2+}$ therefore readily arise in this one-pool model, as shown in fig. 9.18, where cytosolic $Ca^{2+}$ oscillations obtained by numerical simulations are represented together with the variation of the $Ca^{2+}$ content of the pool sensitive to $IP_3$ and $Ca^{2+}$.

Compared in fig. 9.19 are the steady-state level $(Z_0)$ and the envelope of the oscillations of cytosolic $Ca^{2+}$ in the one- and two-pool versions of

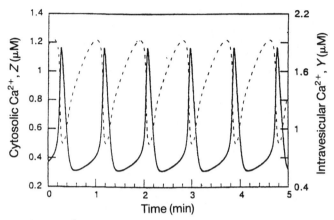

Fig. 9.18. Typical $Ca^{2+}$ oscillations generated by the one-pool model based on $IP_3$-sensitive $Ca^{2+}$-induced $Ca^{2+}$ release schematized in fig. 9.17b. The solid and dashed lines represent the evolution of cytosolic and intravesicular $Ca^{2+}$, respectively. Curves are obtained by numerical integration of eqns (9.1) in which the rate $V_3$ is expressed by eqn (9.8), with $\beta = 0.4$, $v_0 = v_1 = 3.4\ \mu M/min$, $V_{M2} = 50\ \mu M/min$, $V_{M3} = 650\ \mu M/min$, $K_2 = 1\ \mu M$, $K_R = 2\ \mu M$, $K_A = 0.9\ \mu M$, $k = 10\ min^{-1}$, $k_f = 1\ min^{-1}$, $n = m = 2$ and $p = 4$. Initial conditions are $Y = 1.8739\ \mu M$, $Z = 0.3682\ \mu M$ (Dupont & Goldbeter, 1993).

the model. The solid line indicates the steady-state level of cytosolic $Ca^{2+}$ and, for intermediate stimuli for which sustained oscillations occur, the maximum and minimum values reached by the $Ca^{2+}$ concentration in the cytosol; the dashed line represents the unstable steady state. In both cases the model exhibits the property that cytosolic $Ca^{2+}$ progressively rises with the level of stimulation. While the amplitude of the oscillations in the two-pool model decreases only slightly as the system passes through the oscillatory domain, it decreases more significantly in the one-pool model.

Common to the one- and two pool models analysed here is the role of intravesicular $Ca^{2+}$ as a counterbalance to the increase in cytosolic $Ca^{2+}$ in the course of oscillations. The rise in cytosolic $Ca^{2+}$ is indeed accompanied by a concomitant decrease in the $Ca^{2+}$ level in the $Ca^{2+}$-sensitive pool (see dashed line in fig. 9.18 for oscillations of intravesicular $Ca^{2+}$ in the one-pool model, and fig. 9.7 for the corresponding curve in the two-pool model). The level of cytosolic $Ca^{2+}$ drops thereafter as a result of the decreased rate of release from the pool and of the extrusion of $Ca^{2+}$ from the cell. The level of $Ca^{2+}$ in the store begins to rise again as soon as the rate of pumping from the cytosol exceeds the rate of $IP_3$-sensitive or $IP_3$-insensitive CICR. Since the level of intravesicular $Ca^{2+}$ begins to

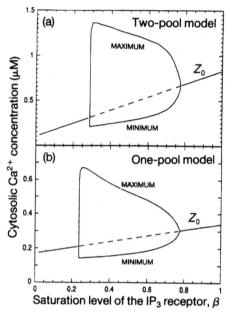

Fig. 9.19. Concentration of cytosolic $Ca^{2+}$ as a function of the stimulation level ($\beta$) in the one- and two-pool models based, respectively, on the $IP_3$-sensitive and $IP_3$-insensitive $Ca^{2+}$-induced $Ca^{2+}$ release. The solid lines represent the stable level of cytosolic $Ca^{2+}$ or the maximum and minimum cytosolic $Ca^{2+}$ concentration reached during oscillations; the dashed line indicates the steady-state level of cytosolic $Ca^{2+}$ in the domain of $\beta$ values where this state is unstable and oscillations occur. Parameter values are $k = 10$ $min^{-1}$, $k_f = 1$ $min^{-1}$, $n = m = 2$ and $p = 4$. Moreover, for the upper (lower) panel, $v_0 = 1$ (1.7) $\mu$M/min, $v_1 = 7.3$ (1.7) $\mu$M/min, $V_{M2} = 65$ (25) $\mu$M/min, $V_{M3} = 500$ (325) $\mu$M/min, $K_2 = 1$ (0.5) $\mu$M, $K_R = 2$ (1) $\mu$M, $K_A = 0.9$ (0.45) $\mu$M. The lower values considered for some parameters in the one-pool model have been adjusted so as to limit the amplitude of the first $Ca^{2+}$ spike to the 1–2 $\mu$M range. The concentrations of intravesicular and cytosolic $Ca^{2+}$ are defined with respect to the total cell volume; the actual intravesicular $Ca^{2+}$ concentration is therefore larger than on the given scale. The curves are established by linear stability analysis and numerical integration of eqns (9.1); the expression of $V_3$ in the two-pool and one-pool versions of the model is given by eqns (9.2) and (9.8), respectively (Dupont & Goldbeter, 1993).

rise well before complete depletion of the pool, this pool is never empty.

A noticeable difference between the one- and two-pool models based on $IP_3$-sensitive and $IP_3$-insensitive CICR pertains to the concentration of intravesicular $Ca^{2+}$ as a function of the stimulation level for various agonist concentrations (see fig. 9.20). In the two-pool model, the steady-

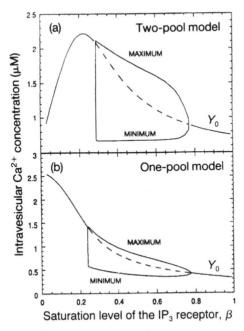

Fig. 9.20. Concentration of intravesicular Ca$^{2+}$ as a function of the stimulation level in the one- and two-pool models based on the two versions of Ca$^{2+}$-induced Ca$^{2+}$ release. The curves are established as described in fig. 9.19, where solid and dashed lines have been defined (Dupont & Goldbeter, 1993).

state level of intravesicular Ca$^{2+}$ ($Y_0$) first rises and then decreases when stimulation increases; this behaviour reflects the fact that cytosolic Ca$^{2+}$ – whose level increases with stimulation – at first replenishes the Ca$^{2+}$-sensitive pool but later favours its depletion once CICR becomes significant. In contrast, in the one-pool model, $Y_0$ decreases over the whole range of stimulation, because the predominant Ca$^{2+}$ efflux from the single pool is directly proportional to the stimulation level reflected by $\beta$ (see eqn (9.8)).

The fact that $Y_0$ has a large value in the absence of stimulation in the one-pool model, in contrast to the two-pool model has consequences that might help to distinguish experimentally between the two situations. When increasing $\beta$ in a stepwise manner from zero up to a finite value in the oscillatory range, the first Ca$^{2+}$ spike in the one-pool model occurs immediately regardless of the final value of $\beta$ (fig. 9.21); the time between the stimulus and the first Ca$^{2+}$ spike – i.e. the latency – is therefore very short compared with the period of oscillations. The reason is that the pool, filled to capacity before stimulation, discharges its content

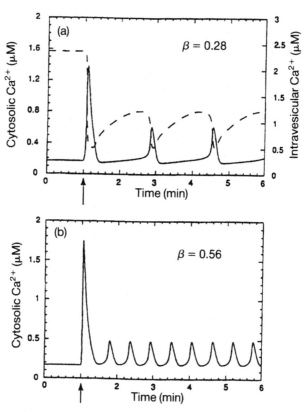

Fig. 9.21. Oscillations in cytosolic $Ca^{2+}$ ($Z$, solid line) triggered by two stimuli of increasing magnitude in the one-pool model schematized in fig. 9.17b, where $Ca^{2+}$ and $IP_3$ behave as co-agonists. After 1 min (arrow), the value of parameter $\beta$ measuring stimulation is increased instantaneously from zero up to 0.28 (a) or 0.56 (b). Notice that the first $Ca^{2+}$ spike each time is much larger than the following spikes and occurs with negligible latency. The variation of intravesicular $Ca^{2+}$ ($Y$, dashed line) is also indicated in (a). The curves are obtained by integration of eqns (9.1), with $V_3$ given by eqn (9.8), for the parameter values of figs. 9.19b and 9.20b, with $k_f = 1.1$ min$^{-1}$. Initial conditions correspond to the stable steady state in the absence of stimulation ($\beta = 0$): $Y = 2.355$ μM and $Z = 0.17$ μM (Dupont & Goldbeter, 1993).

as soon as $\beta$ rises, as expected from eqn (9.8); latency nevertheless slightly decreases as the value of $\beta$ increases (compare fig. 9.21a and b). A roughly linear correlation between period and latency therefore holds in the one-pool model, but the time scales for period and latency are widely different (Dupont & Goldbeter, 1993), in contrast with the predictions of the two-pool model (see figs. 9.13a and 9.14) and with experimental observations in hepatocytes (fig. 9.13b). Comparing the period–latency relationship in the one- and two-pool models based, respectively, on IP$_3$-sensitive and IP$_3$-insensitive CICR shows (fig. 9.22) that only the two-pool version yields satisfactory agreement with data obtained for hepatocytes.

Another difference between the one- and two-pool models based on CICR pertains to the magnitude of the first Ca$^{2+}$ spike that follows stimulation. As shown in fig. 9.21, the first spike triggered by the increase in $\beta$ is significantly larger than the following spikes, whose amplitude settles to a reduced level. In contrast, the first spike in the two-pool model has the same magnitude as the following ones (see fig. 9.14). The difference in behaviour again originates from the fact that, in the two-pool model, the Ca$^{2+}$ level in the Ca$^{2+}$ store is initially low and progressively increases after stimulation; the level it reaches just before the first discharge is dictated by the threshold for CICR in the cytosol and is therefore the same as for the next spikes. In the one-pool model, in contrast, the analysis of the model shows that regardless of parameter values, the Ca$^{2+}$ store is initially filled to capacity; because the store is so much charged and begins to release its content immediately as a result of the rise in IP$_3$ that follows stimulation, the threshold for CICR in the cytosol is exceeded rapidly and a large spike occurs with negligible latency. For the next spikes, when the threshold for CICR is reached in the cytosol, the level of Ca$^{2+}$ in the store is below its maximum capacity so that the spikes will have a reduced magnitude (fig. 9.21).

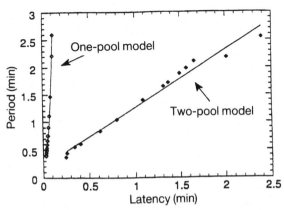

Fig. 9.22. Period versus latency relationship predicted by the one- and two-pool models based, respectively, on the IP$_3$-sensitive and IP$_3$-insensitive Ca$^{2+}$-induced Ca$^{2+}$ release. Parameter values are those of figs. 9.21 and 9.19 (upper panel) for the one- and two-pool models, respectively (Dupont & Goldbeter, 1993).

From an experimental point of view, it is of interest to note that both situations appear to be encountered, depending on the cell type. Thus, while the first $Ca^{2+}$ spike has often the same magnitude as the successive ones and occurs after a finite latency in hepatocytes (see Rooney *et al.*, 1989, and also Fig. 3 in Woods *et al.*, 1987), as predicted by the two-pool model, the first spike is larger and occurs immediately upon stimulation in fibroblasts (Harootunian *et al.*, 1988, 1991), in accordance with the predictions of the one-pool model based on CICR. In the latter cells, however, experimental observations (Harootunian *et al.*, 1991) can also be accounted for in terms of a mechanism involving the cross-activation of $Ca^{2+}$ and $IP_3$ (Meyer & Stryer, 1988).

The response of hepatocytes to high levels of certain agonists some-times appears to be more complex: a high initial spike is followed by a brief silent phase and by a train of $Ca^{2+}$ spikes of smaller magnitude (see Fig. 10 in Woods *et al.*, 1987). Such a behaviour, however, does not con-tradict the predictions of the two-pool model based on CICR. On the contrary, a similar time course, also observed in oocytes (Dupont *et al.*, 1991) can be accounted for by the model when the initial stimulation is so high that it brings the system transiently above the oscillatory range of parameter $\beta$ (see fig. 9.12).

While the above theoretical results and some experimental studies (Rooney *et al.*, 1989) point to CICR as being involved in the mechanism of $Ca^{2+}$ oscillations in hepatocytes, other experimental observations can apparently not be reconciled with this hypothesis. Thus, the observation that ryanodine fails to suppress $Ca^{2+}$ oscillations in these cells does not hold with a mechanism involving CICR through the ryanodine-sensitive $Ca^{2+}$ channel (Cobbold, Sanchez-Bueno & Dixon, 1991), but it does not rule out the existence of a putative ryanodine-insensitive $Ca^{2+}$ channel involved in CICR. The finding (Noel *et al.*, 1992) that $Ca^{2+}$ oscillations may occur in hepatocytes in conditions where the synthesis of $IP_3$ is being prevented nevertheless supports the involvement of CICR in the mechanism of $Ca^{2+}$ oscillations in these cells.

In some cell types the distinction between $IP_3$- or $Ca^{2+}$-sensitive pools is not clear-cut; sometimes, indeed, the $Ca^{2+}$ pools appear to be sensitive to the two messengers to various degrees (Shoshan-Barnatz *et al.*, 1990; Lytton & Nigam, 1992). Modelling such a complex network of pools and regulations would lead to intermediate behaviours in comparison with the extreme cases described here. The models analysed above thus demonstrate that CICR, sensitized or not by $IP_3$, keeps providing a plausible, robust mechanism to account for experimental observations about $Ca^{2+}$ oscillations in a number of cell types regardless of whether the cell possesses a single or two distinct types of $Ca^{2+}$ pool.

### Model incorporating desensitization of the IP₃ receptor and IP₃ variations

The dynamics of cytosolic $Ca^{2+}$ does not always possess a simple periodic nature. Thus, upon stimulation, some cells exhibit complex transients that resemble bursting oscillations (Woods *et al.*, 1987; Berridge *et al.*, 1988; Cuthbertson, 1989). It is not possible to generate such complex behaviour when the models schematized in fig. 9.17 contain only two variables, namely $Y$ and $Z$.

A most plausible, additional variable is $IP_3$. The fact that $IP_3$ variations do not accompany $Ca^{2+}$ oscillations in some cells does not mean that $IP_3$ oscillations could not occur in other cells, concomitantly with $Ca^{2+}$ transients. To allow for $IP_3$ oscillations we have to couple the variation of $IP_3$ with that of $Ca^{2+}$. As evidenced in chapters 4 and 6, complex oscillations can originate from the interplay between two (or more) endogenous oscillatory mechanisms. Such an interplay should occur when incorporating into the CICR-based model the activation of $IP_3$ synthesis by cytosolic $Ca^{2+}$ – which process is at the core of the oscillatory mechanism proposed by Meyer & Stryer (1988) – and desensitization of the $IP_3$ receptor (see Meyer & Stryer, 1990; Pietri, Hilly & Mauger, 1990; Payne & Potter, 1991; Hajnoczky & Thomas, 1994). The variables in this extended model can be as many as five and are listed in table 9.1, together with the variables of the minimal model considered above.

A plausible conjecture, supported by preliminary observations (Dupont, 1993; G. Dupont & A. Goldbeter, unpublished results) is that complex oscillations would result from the existence of two oscillatory processes characterized by different time scales: (i) a slow $Ca^{2+}$ variation arising from the reversible desensitization of the $IP_3$ receptor, and (ii) fast $Ca^{2+}$ transients occurring on top of that periodic variation as a result of the mechanism based on CICR. However, the latter mechanism can still readily generate simple periodic oscillations in $Ca^{2+}$ in such an extended model; as illustrated by the results shown in fig. 9.23, these oscillations are passively accompanied by a simple, periodic variation in $IP_3$ and in the fraction of nondesensitized $IP_3$ receptors.

Complex oscillations of $Ca^{2+}$, including chaos, can originate from various mechanisms, as shown by the analysis of several modifications of the one- and two-pool models discussed above (J. Borghans, G. Dupont

Table 9.1. *Variables considered in the minimal and extended models for Ca$^{2+}$ oscillations based on Ca$^{2+}$-induced Ca$^{2+}$ release*

| Minimal model | Extended model |
|---|---|
| Cytosolic Ca$^{2+}$ | Cytosolic Ca$^{2+}$ |
| Ca$^{2+}$ in the Ca$^{2+}$-sensitive pool | Ca$^{2+}$ in the Ca$^{2+}$-sensitive pool |
| | Ca$^{2+}$ in the IP$_3$-sensitive pool |
| | IP$_3$ |
| | Fraction of active IP$_3$ receptor |

As regards regulatory steps, besides the IP$_3$-mediated release of Ca$^{2+}$ and CICR from the IP$_3$-insensitive store, which are considered in the minimal model, the extended model incorporates activation of IP$_3$ synthesis by Ca$^{2+}$ and desensitization of the IP$_3$ receptor.

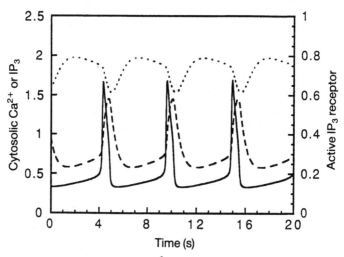

Fig. 9.23. Oscillations of cytosolic Ca$^{2+}$ (in μM, solid line) following sustained stimulation by an external signal in the extended version of the model based on Ca$^{2+}$-induced Ca$^{2+}$ release (CICR). Also represented are the normalized concentration of IP$_3$ (dashed line) and the fraction of active, nondesensitized IP$_3$ receptors (dotted line), which are both driven here passively by oscillations based on CICR. In the extended model, which contains the five variables listed in table 9.1, IP$_3$-mediated release of Ca$^{2+}$ and CICR are supplemented with the following additional regulatory processes: activation of IP$_3$ synthesis by Ca$^{2+}$, and desensitization of the IP$_3$ receptor induced by IP$_3$ (as in the present simulations) or cytosolic Ca$^{2+}$ (Dupont *et al.*, 1991).

& A. Goldbeter, unpublished results). Chaos has recently been found in a closely related model for $Ca^{2+}$ oscillations (Shen & Larter, 1995).

### Comparing various models for $Ca^{2+}$ oscillations

A variety of models for $Ca^{2+}$ oscillations have ben proposed, some of which have already been mentioned (see fig. 9.4 and, for reviews, Dupont & Goldbeter, 1992b; Fewtrell, 1993). The main distinction between the various models for $Ca^{2+}$ oscillations lies in whether a periodic variation in $IP_3$ necessarily accompanies the periodic variation in $Ca^{2+}$. So far, the only mechanism where $IP_3$ oscillations are not required is that based on CICR. However, as described above, if $IP_3$ synthesis and/or its metabolic transformation are controlled in any manner by $Ca^{2+}$, oscillations in $IP_3$ would follow passively those of cytosolic $Ca^{2+}$ produced by that mechanism (Swillens & Mercan, 1990; Dupont *et al.*, 1991). Moreover, the fact that CICR might underlie $Ca^{2+}$ oscillations in a variety of cells does not rule out the possibility that other mechanisms (such as those depicted in fig. 9.4a or b) might in some cells replace, or act in conjunction with CICR to produce oscillations.

Several models based on CICR and incorporating the triggering role of $IP_3$ have been analysed; these models differ mainly in the kinetics considered for the control of the $Ca^{2+}$ channel by $Ca^{2+}$ and/or $IP_3$. Thus, the one-pool models proposed by De Young & Keizer (1992) and Atri *et al.* (1993) incorporate CICR as well as inhibition of $Ca^{2+}$ release by high levels of cytosolic $Ca^{2+}$; the number of variables in the former model has been reduced to only two (Keizer & De Young, 1994; Li & Rinzel, 1994). A related model, recently proposed (Li *et al.*, 1994), provides a detailed account of $Ca^{2+}$ oscillations in pituitary gonadotrophs. The predictions of these theoretical models compare well with a number of experimental observations.

The model based on CICR presented in this chapter, which also contains only two variables, has been subjected to an extensive comparison with experimental data. As indicated above, agreement was reached with respect to the effect of external stimulation, extracellular $Ca^{2+}$, or a transient rise in $IP_3$. Furthermore, the model provides a molecular explanation for the quasi-linear correlation between period and latency observed at different levels of stimulation. That model also accounts for the immediate resetting of $Ca^{2+}$ oscillations by $Ca^{2+}$ pulses of sufficient magnitude but indicates that small-amplitude pulses should induce a delay rather than a phase advance when applied near the minimum of

$Ca^{2+}$ oscillations; moreover, the simulations predict that there exists a critical phase where a pulse of precise magnitude can transiently suppress $Ca^{2+}$ oscillations. Such a transient suppression of the oscillations by a critical perturbation is a general property of chemical and biological oscillators (Winfree, 1980).

The model shows how a given biochemical mechanism can produce very different patterns of oscillations depending on the value of a single parameter such as that ($\beta$) measuring the extent of stimulation (fig. 9.10), the rate constant for $Ca^{2+}$ extrusion from the cell (Goldbeter *et al.*, 1990), or the input of $Ca^{2+}$ from the extracellular medium, whose effect is similar to that of parameter $\beta$. Such a variability in the waveform of the oscillations provides a plausible explanation for the differences in $Ca^{2+}$ transients seen in a given cell with different agonists (Woods *et al.*, 1987; Berridge *et al.*, 1988; Thomas, Renard & Rooney, 1991; Fewtrell, 1993) or between individual cells responding to the same stimulus (Prentki *et al.*, 1988). Individual cells probably differ by the number of receptors for the agonist, the activity of enzymes involved in $IP_3$ metabolism, the velocity of $Ca^{2+}$ pumps, etc. The theoretical study of models for $Ca^{2+}$ signalling shows that such a heterogeneity in biochemical parameters could readily account for the observed diversity in oscillatory behaviour.

The experimental variation of one or other parameter controlling $Ca^{2+}$ signalling provides useful information on the mechanism of $Ca^{2+}$ oscillations as well as additional tests for theoretical models. Of particular interest, in this respect, are the results obtained recently by Camacho & Leichleiter (1993), who succeeded in manipulating the $Ca^{2+}$ pump located in the endoplasmic reticulum membrane in *Xenopus* oocytes (the rate of this pump corresponds to parameter $V_{M2}$ in the minimal model based on CICR). Increasing the pumping rate into the intravesicular pool results in a decrease in the period of $Ca^{2+}$ waves (which propagate at an unchanged rate) and, hence, in the period of $Ca^{2+}$ oscillations. It is not obvious how we can account for such a result using the model based on CICR, since an increase in $V_{M2}$ often produces the opposite effect, i.e. period lengthening, because the time required for the building of cytosolic $Ca^{2+}$ up to the threshold for CICR generally rises owing to faster pumping into the intravesicular store. For other parameter values, however, an increase in $V_{M2}$ in that model at least does not cause a lengthening of the period. This happens, for example, when the release process is favoured by smaller values of parameters $K_A$ (cytosolic threshold constant for CICR) or $K_R$ (intravesicular threshold constant for release) and larger values of the maximum rate of release, $V_{M3}$ (G. Dupont & A. Goldbeter, unpublished results). Recent numerical simulations indicate (Dupont & Goldbeter, 1994) that the results of Camacho & Leichleiter (1993) can be explained once the effect of $Ca^{2+}$ diffusion is taken into account (see also section 9.6).

Questions of current debate pertain to the nature of the 'quantal release' of $Ca^{2+}$ triggered by $IP_3$ (Tregear, Dawson & Irvine, 1991; see Swillens, Combettes & Champeil (1994) for a theoretical analysis of this issue), and to the role of a new messenger, cyclic ADP-ribose, which appears to control the release of $Ca^{2+}$ from intracellular stores in some cells (Lee, 1993; Galione, 1992, 1993) – a process that may itself be mediated by $Ca^{2+}$, via calmodulin (Lee, H.C., *et al.*, 1994). Also debated is the existence of two distinct pools of $Ca^{2+}$ sensitive to $IP_3$ and $Ca^{2+}$, respectively. In the models for $Ca^{2+}$ spiking that involve $IP_3$ oscillations (fig. 9.4a and b) a single pool of $Ca^{2+}$ is considered. The model based on CICR also indicates that the existence of two distinct pools is not a necessary prerequisite for oscillatory behaviour (see also Somogyi & Stucki, 1991; De Young & Keizer, 1992; Jafri *et al.*, 1992; Atri *et al.*, 1993; Li *et al.*, 1994). The above results indicated that $Ca^{2+}$ spikes can indeed occur in the case where $IP_3$ and $Ca^{2+}$, acting on distinct receptors or behaving as co-agonists, elicit the release of $Ca^{2+}$ from a single pool (Dupont & Goldbeter, 1993). Such a model, schematized in fig. 9.4d, yields results similar to those obtained with the two-pool model (fig. 9.4c). To retain the observed dependence of the mean level of cytosolic $Ca^{2+}$ on the stimulus we need to assume, however, that both the level of $IP_3$ and the influx of $Ca^{2+}$ from the extracellular medium into the cytosol increase as a result of external stimulation. When such an assumption is not made, the resting level of cytosolic $Ca^{2+}$ remains independent of the stimulus (Somogyi & Stucki, 1991).

Although the possibility exists that redundant mechanisms may operate in certain cells (Galione *et al.*, 1993), of possible help in distinguishing between different experimental situations is the result that the the one- and two-pool models based, respectively, on $IP_3$-sensitive and $IP_3$-insensitive CICR, lead to different predictions with respect to the magnitude of the first spike (figs. 9.14 and 9.21) and the relationship between the latency and period of $Ca^{2+}$ oscillations obtained at different levels of stimulation (fig. 9.22). While in the two-pool model period and latency are often of the same order of magnitude and correlate in a roughly linear manner, as generally observed in hepatocytes, the one-pool model predicts that latency should always be negligible with respect to the period of $Ca^{2+}$ oscillations. Furthermore, the theoretical analysis shows that the magnitude of the first spike and its latency are closely linked, as observed, for example, in fibroblasts and hepatocytes. Thus, we obtain either a large initial spike appearing without latency (one-pool model), or an initial spike of the same magnitude as the following ones, appearing after a time lag (two-pool model). Given that the first situation occurs in fibroblasts (Harootunian *et al.*, 1988) whereas the second is encountered in hepatocytes at moderate stimulation level (Woods *et al.*, 1987; Rooney *et al.*, 1989), $Ca^{2+}$ oscillations in these cell types could originate from a mechanism involving either a single pool sensitive to both $Ca^{2+}$ and $IP_3$ behaving as co-agonists, or two pools sensitive to $IP_3$ or $Ca^{2+}$, respectively.

## 9.5 Frequency encoding of $Ca^{2+}$ oscillations

Although a recent experimental study (Pralong *et al.*, 1994) showed that $Ca^{2+}$ oscillations may drive the periodic operation of some biochemical reactions, most studies so far concentrate on the mechanism of $Ca^{2+}$ oscillations and on their various properties, rather than on their physiological mode of action within the cell. In some instances, e.g. secretory processes (Rapp & Berridge, 1981; Law *et al.*, 1989; Malgaroli & Meldolesi, 1991; Stojilkovic *et al.*, 1993; Tse *et al.*, 1993), it appears that the cellular response is correlated with the frequency of $Ca^{2+}$ spikes; the latter frequency, in turn, increases with the level of stimulation. The idea that the external signal might be encoded in terms of the frequency of $Ca^{2+}$ oscillations (Berridge & Galione, 1988; Berridge *et al.*, 1988; Jacob *et al.*, 1988; Jacob, 1990a; Petersen & Wakui, 1990; Tsien & Tsien, 1990; Meyer & Stryer, 1991; Tsunoda, 1991) raises the question of what transduction mechanism might link the cellular response to the frequency of $Ca^{2+}$ spikes. One simple possibility, implying the existence of a spike counter (Meyer & Stryer, 1991), is that the response is triggered each time the level of cytosolic $Ca^{2+}$ exceeds a threshold.

Another likely way of encoding the external signal in terms of the frequency of $Ca^{2+}$ oscillations is through protein phosphorylation (Berridge *et al.*, 1988; Goldbeter *et al.*, 1990; Goldbeter & Dupont, 1991; Dupont & Goldbeter, 1992a). Consider a kinase activated in a Michaelian manner by cytosolic $Ca^{2+}$ and a phosphatase, that both act on a protein substrate, W, whose fraction in the phosphorylated form is denoted $W^*$ (fig. 9.24). The time variation of $W^*$ associated with the oscillations of cytosolic $Ca^{2+}$ (Z) is given by eqn (9.10) (see Goldbeter &

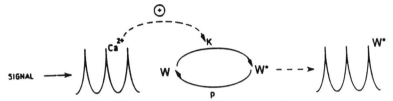

Fig. 9.24. Protein phosphorylation driven by $Ca^{2+}$ oscillations. Schematized is a protein W phosphorylated by a kinase K into the form W*; the latter form is dephosphorylated by a phosphatase P. If kinase K is activated by $Ca^{2+}$, oscillations in cytosolic $Ca^{2+}$ would be accompanied by periodic variations in the level of phosphorylated substrate (Dupont & Goldbeter, 1992a).

Koshland, 1981, and legend to fig. 9.25):

$$\frac{dW^*}{dt} = (v_P/W_T)\left[(v_K/v_P)\frac{1-W^*}{K_1+1-W^*} - \frac{W^*}{K_2+W^*}\right] \qquad (9.10)$$

with:

$$v_K = V_{MK}\frac{Z}{K_a+Z}$$

In these equations, $v_P$ and $v_K$ denotes the effective, maximum rates of the kinase and phosphatase at a given value of $Z$; $K_1$ and $K_2$ denote the Michaelian constants of these enzymes normalized by dividing by the total amount of protein substrate, $W_T$; $V_{MK}$ is the maximum rate of the kinase when it is fully activated by $Ca^{2+}$, with an activation constant $K_a$.

Owing to the periodic variation in cytosolic $Ca^{2+}$ ($Z$) and to the subsequent variation in kinase activity, the value of $W^*$ periodically rises and decreases with $Z$ (fig. 9.25). A comparison of curves $a$ and $b$ in fig. 9.25 shows that the mean level of $W^*$ increases with the frequency of $Ca^{2+}$ oscillations (i.e. with the level of stimulation, measured by $\beta$); this effect is due partly to a change in the mean value of $Z$ which is higher at the larger value of $\beta$ considered. Equally significant, however, is the fact that when the period of oscillations is longer, the protein undergoes significant dephosphorylation from one $Ca^{2+}$ peak to the next; much less dephosphorylation occurs between successive $Ca^{2+}$ spikes at the higher frequency of oscillations, resulting in the maintenance of a larger fraction of protein phosphorylated. Similar results are obtained when the frequency of $Ca^{2+}$ oscillations is increased at constant $\beta$ through a decrease in the threshold constant $K_R$ (Dupont & Goldbeter, 1992a); this procedure allows us to hold the mean level of $Z$ unchanged.

Efficient frequency encoding through protein phosphorylation

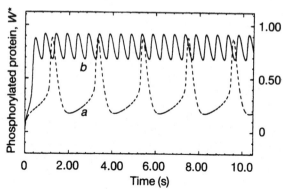

Fig. 9.25. Oscillations in the level of a protein phosphorylated by a kinase acti-vated by cytosolic $Ca^{2+}$. (Curve *a*) Time evolution of the fraction of phosphory-lated protein in the course of the $Ca^{2+}$ oscillations shown in fig. 9.7, with $\beta = 30.1\%$; (curve *b*) evolution of protein phosphorylation at the higher level of stimulation, $\beta = 64.4\%$. The mean fraction of phosphorylated protein over a period, $\langle W^* \rangle$, is equal to 0.32 and 0.80 for curves *a* and *b*, respectively. The curves are obtained by integration of eqns (9.1)–(9.2) and (9.10), with $W^* = 0$ at time zero, for $v_P = 5\ \mu M/s$, $V_{MK} = 40\ \mu M/s$, $K_a = 2.5\ \mu M$, $W_T = 1\ \mu M$, $K_1 = K_2 = 0.1$; these values are in the physiological range for protein phosphory-lation systems (Goldbeter & Koshland, 1987). For these parameter values, the phosphorylation curve for $W^*$ possesses a sharp threshold, where $W^* = 0.5$, in $v_K/v_P = 1$ (see chapter 10 and Goldbeter & Koshland, 1981). In eqn (9.10), $v_K$ and $v_P$ denote the maximum rates of kinase and phosphatase at a given value of $Z$; $V_{MK}$ is the maximum rate of the kinase at saturation by $Z$; $K_a$ denotes the constant of activation of the kinase by cytosolic $Ca^{2+}$; $K_1 = K_{m1}/W_T$, $K_2 = K_{m2}/W_T$ where $K_{m1}$ and $K_{m2}$ denote the Michaelis constants of kinase and phosphatase, while $W_T$ is the total amount of protein substrate (Goldbeter *et al.*, 1990).

occurs, however, only under precise kinetic conditions. The analysis indicates that saturation of the converter enzymes – which favours the occurrence of 'zero-order ultrasensitivity' (Goldbeter & Koshland, 1981) – allows a better encoding of $Ca^{2+}$ oscillations as the mean value of the fraction of phosphorylated protein varies over a much larger range (0.2 to 0.95) with the frequency of $Ca^{2+}$ oscillations than it does when the kinase and the phosphatase are not saturated by their sub-strate (compare curves *a* and *b* obtained for the same oscillations in fig. 9.26).

The analysis further shows that to obtain a significant variation in phosphorylated protein, cytosolic $Ca^{2+}$ would have to oscillate in such a manner that the ratio $v_K/v_P$ goes above a threshold, here equal to unity, when $Z$ reaches its maximum, and below it when $Z$ decreases to its min-imum (Goldbeter *et al.*, 1990; Dupont & Goldbeter, 1992a). Such fine tuning clearly depends on the absolute and relative values of the maxi-

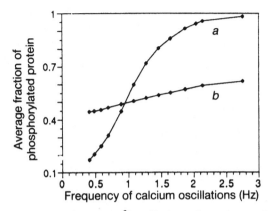

Fig. 9.26. Frequency encoding of $Ca^{2+}$ oscillations through protein phosphorylation. The curves show the average value of the fraction of phosphorylated pro-
tein, $\langle W^* \rangle$, over a period in the course of the oscillations in $Ca^{2+}$ in the range of
frequencies obtained as a function of $\beta$. Data (points) are obtained as in
fig. 9.26, for the same parameter values, with $K_1 = K_2$ equal to 0.01 (curve $a$) or
10 (curve $b$) (Goldbeter *et al.*, 1990).

mum rates of the kinase and phosphatase, of their Michaelis constants,
and on the value of the activation constant $K_a$, which corresponds to the
$Ca^{2+}$ concentration that yields the half-maximum kinase activity.

Positive cooperativity in the activation of the kinase by $Ca^{2+}$ yields
similar results; however, by allowing for changes of larger amplitude in
the amount of phosphorylated substrate during $Ca^{2+}$ oscillations, such
cooperativity permits the system to be sensitive to fast variations in the
cytosolic $Ca^{2+}$ level at small or intermediate velocities of the converter
enzymes (Dupont & Goldbeter, 1992a; for a detailed analysis of
calmodulin kinase activation by $Ca^{2+}$ pulses, see Michelson &
Schulman, 1994). The above results also hold in the mirror image situation where $Ca^{2+}$ would activate the phosphatase rather than the kinase;
this situation might arise, for example, in the case of calcineurin
(Armstrong, 1989).

## 9.6 Intracellular Ca²⁺ waves

### *Experimental observations on Ca²⁺ waves*

The spatial propagation of $Ca^{2+}$ waves has long been observed in a variety of egg types after fertilization (Gilkey *et al.*, 1978; Jaffe, 1983, 1991,
1993; Busa & Nuccitelli, 1985). In these cells, waves of $Ca^{2+}$ propagate
over the cortex, from the site of fertilization. More recently, the wavelike propagation of $Ca^{2+}$ signals has been observed in other cells in

which $Ca^{2+}$ oscillations were previously characterized (for recent reviews, see Amundson & Clapham, 1993, and Berridge & Dupont, 1994). Thus propagating $Ca^{2+}$ waves have been observed within the cytoplasm of cardiac cells (Kort *et al.*, 1985; Takamatsu & Wier, 1990), endothelial cells (Jacob, 1990b), hepatocytes (Rooney *et al.*, 1990; Thomas *et al.*, 1991), astrocytes (Cornell-Bell *et al.*, 1990; Yagodin *et al.*, 1994; Roth *et al.*, 1995), and pancreatic acinar cells (Nathanson *et al.*, 1992, 1994). $Ca^{2+}$ oscillations and waves thus appear to be closely related phenomena (Berridge & Irvine, 1989). The velocity of $Ca^{2+}$ waves varies in different cells; the wave propagates at a rate of the order of 10 μm/s on the surface of oocytes (Jaffe, 1983, 1993) and of 30 μm/s in hepatocytes (Thomas *et al.*, 1991), and at a rate close to 100 μm/s in the cytoplasm of cardiac cells (Kort *et al.*, 1985; Takamatsu & Wier, 1990). The frequency of the waves has been shown to increase with calcium influx in *Xenopus* oocytes (Girard & Clapham, 1993).

There appears to be two main types of $Ca^{2+}$ wave, designated as types 1 and 2, although the distinction is somewhat arbitrary as a continuum should exist between the two sorts of wave (Dupont & Goldbeter, 1992b, 1994). For type 1, exemplified by $Ca^{2+}$ waves in cardiac cells (Kort *et al.*, 1985; Takamatsu & Wier, 1990), the $Ca^{2+}$ signal propagates along the cell as a sharp band; two successive bands sometimes progress concomitantly (fig. 9.27). In contrast, for waves of type 2, the level of $Ca^{2+}$ progressively rises along the entire cell before returning to basal levels in a quasi-homogeneous manner; such behaviour is observed in hepatocytes (Thomas *et al.*, 1991) as well as in small eggs (Jaffe, 1993) and endothelial cells (Jacob, 1990b) (see fig. 9.28). Waves of this type have also been referred to as tides (Tsien & Tsien, 1990). A similar distinction betwen $Ca^{2+}$ waves in myocytes and oocytes was underlined by Meyer (1991).

A third type of $Ca^{2+}$ wave was demonstrated in *Xenopus* oocytes expressing acetylcholine receptors coupled to $IP_3$ production (Lechleiter *et al.*, 1991). While the majority of waves were planar or concentric, the latter patterns sometimes transformed into spiral waves. Although it is difficult to judge whether the waves are of type 1 or 2 in that study, because of the subtraction of successive images, waves of type 2 are clearly seen in experiments performed on the same cells under similar conditions (Brooker *et al.*, 1990). Spiral waves of $Ca^{2+}$ have also been observed (Lipp & Niggli, 1993) in single cardiac cells.

The spatiotemporal patterns of cytosolic $Ca^{2+}$ bear much resemblance to the concentric and spiral waves of cAMP that underlie the aggrega-

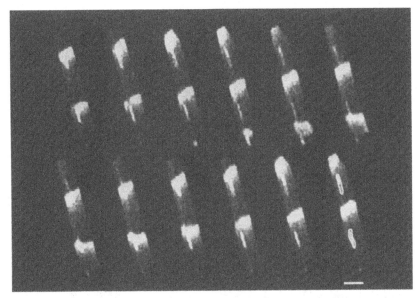

Fig. 9.27. Intracellular $Ca^{2+}$ waves in a guinea-pig cardiac myocyte (Lipp & Niggli, 1993). The grey levels represent ratios of the two fluorescent $Ca^{2+}$ indicators Fluo-3 and Fura-Red. The propagation velocity of the waves is 65 μm/s and the scale bar 20 μm. (Picture kindly provided by Dr E. Niggli.)

tion of *Dictyostelium discoideum* amoebae after starvation (see figs. 5.3 (p. 167) and 5.6 (p. 169)). As in the case of $Ca^{2+}$, waves of cAMP are associated with temporal oscillations (fig. 5.12, p. 174). Whereas cAMP waves occur at the supracellular level, the above-described waves of $Ca^{2+}$ propagate within the cytosol of single cells. Other experiments indicate, however, that in some cell types such as epithelial (Sanderson *et al.*, 1990) or glial cells (Charles *et al.*, 1991, 1992) $Ca^{2+}$ waves triggered by mechanical stimulation may propagate from cell to cell. This intercellular propagation appears to be mediated by the passage of $Ca^{2+}$ or $IP_3$ through gap junctions (Boitano, Dirksen & Sanderson, 1992).

### Empirical models for Ca²⁺ waves

The experimental study of $Ca^{2+}$ waves in fertilized eggs (Gilkey *et al.*, 1978; Jaffe, 1983; Busa & Nuccitelli, 1985) preceded the observation of temporal oscillations of $Ca^{2+}$ in these and other cells. Likewise, the theoretical study of spatiotemporal $Ca^{2+}$ patterns was initially uncoupled from the study of models for $Ca^{2+}$ oscillations. Thus, empirical models were proposed to account for the propagation of $Ca^{2+}$ waves in amphib-

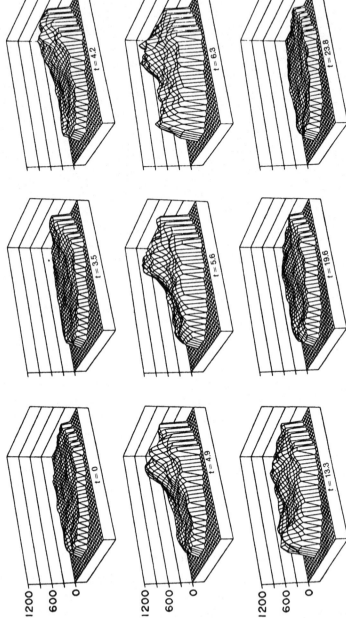

Cytosolic free Ca²⁺ (nM)

ian eggs after fertilization (Cheer *et al.*, 1987) and in cardiac cells (Backx *et al.*, 1989). While the analytical results in the former study were based on a continuous, *ad hoc* description of the CICR process by a cubic function – which description was also used (see Murray, J.D., 1989) to predict the occurrence of stationary, spatial Ca²⁺ patterns – the latter, numerical study relied on a discontinuous representation of CICR.

The propagation of concentric and spiral waves of Ca²⁺ in *Xenopus* oocytes has also been studied (Lechleiter *et al.*, 1991) by means of simulations based on cellular automata (Gerhard, Schuster & Tyson, 1990; Markus & Hess, 1990): such a mathematical representation of excitable systems in terms of a set of rules simulated on a computer considers the existence of a finite number of cell states (excitable, excited, refractory). The comparison of numerical simulations with experiments suggested (Lechleiter *et al.*, 1991) that the species responsible for the spatial propagation of the wave is cytosolic Ca²⁺ rather than IP₃. Moreover, the characteristics of the phenomenon fit with the view (Berridge & Irvine, 1989) that CICR is the primary mechanism underlying Ca²⁺ wave propagation.

## *Incorporation of Ca²⁺ or IP₃ diffusion into models for Ca²⁺ oscillations*

Besides the empirical studies described above, another approach consists in the incorporation of diffusion terms into molecular models accounting for Ca²⁺ oscillations. Thus, incorporation of IP₃ and Ca²⁺ diffusion along a single spatial dimension into the model based on cross-activation of IP₃ synthesis and Ca²⁺ release (fig. 9.4a) leads to the spatial propagation of a Ca²⁺ front corresponding to a type 1 wave (Meyer & Stryer, 1991). According to that study, the role of cytosolic Ca²⁺ in the spatial propagation of Ca²⁺ signals would be only a secondary one. This conclusion was based on the hypothesis that the diffusion coefficient of IP₃ is larger than that of Ca²⁺ by more than one order of magnitude. In support of this view, but in contrast with the results of earlier studies (see, e.g., Lechleiter *et al.*, 1991), experiments (Allbritton, Meyer & Stryer, 1992) indicate that IP₃ diffuses much faster than Ca²⁺. The smaller value for the Ca²⁺ diffusion coefficient might, however, be due to the

Fig. 9.28. Cytosolic Ca²⁺ waves in hepatocytes (Rooney *et al.*, 1991). In response to phenylephrine, a wave of Ca²⁺ is seen to propagate from the right to the left extremity of a single cell; the time *t* (in s) is indicated for successive snapshots of the intracellular spatial Ca²⁺ distribution.

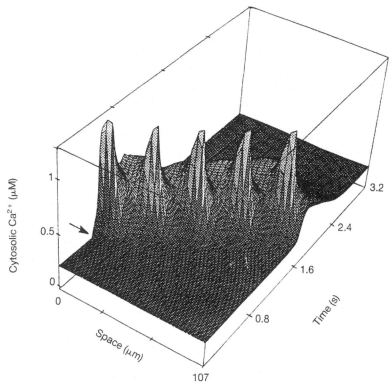

Fig. 9.29. One-dimensional propagation of a $Ca^{2+}$ wave following transient stimulation. The concentration of cytosolic $Ca^{2+}$ is shown as a function of time and space, under conditions where five equidistant $Ca^{2+}$-sensitive $Ca^{2+}$ pools are distributed along the single spatial dimension, which is equivalent to a longitudinal section of a cell. The cell length is 107 μm, and the diffusion coefficient of $Ca^{2+}$ is 400 μm²/s. The wave shown in the figure propagates at a rate close to 80 μm/s, which is of the order of that observed in cardiac cells of similar dimensions (Kort *et al.*, 1985; Takamatsu & Wier, 1990). The data are obtained by dividing space into a mesh of 30 points (which can be thought of as communicating boxes); each $Ca^{2+}$-sensitive $Ca^{2+}$ pool occupies three points of that mesh. Diffusion is represented by means of finite differences with zero-flux boundary conditions. The intravesicular $Ca^{2+}$ dynamics is generated according to eqn (9.11), while the evolution of $Z$ is given by eqn (9.12); thus it is considered that cytosolic $Ca^{2+}$ diffuses over all points of the spatial mesh while in points where a $Ca^{2+}$-sensitive $Ca^{2+}$ pool is located, the diffusion of $Z$ is coupled to eqns (9.1) in which only the terms related to exchanges between $Z$ and $Y$ are retained; both the influx of $Ca^{2+}$ from the extracellular medium ($v_0$) and efflux from the cytosol ($-kZ$) occur at the two extremities of the mesh. At time 0.5 s, the left extremity of the mesh is stimulated (arrow) by increasing $\beta$ from zero to 0.8 for 1 s; alternatively, for cells stimulated by an increase in extracellular $Ca^{2+}$, the stimulation can be viewed as an increase in $v_0$ from 2 to 8 μM/s for 1 s. Parameter values are $v_0 = 2$ μM/s, $v_1 = 7.5$ μM/s, $K_A = 0.85$ μM; other parameter values are as in fig. 9.10. Before the onset of stimulation, the values of $Y$ and $Z$ along the mesh are the homogeneous, steady-state values obtained for $\beta = 0$, i.e. $Z = 0.2$ μM and $Y = 1.81$ μM (Dupont *et al.*, 1991).

conditions in which the experimental measurement was performed; diffusion was indeed assayed after depletion of $Ca^{2+}$ from the medium, i.e. in the absence of any initial $Ca^{2+}$ buffering.

To determine whether the mechanism can underlie the propagation of the two types of wave, it is useful to consider the model based on CICR, governed by eqns (9.1), and to incorporate the diffusion of intracellular $Ca^{2+}$ into these equations. In view of the data obtained on the relative rate of diffusion of $Ca^{2+}$ and $IP_3$ within the cytosol (Allbritton *et al.*, 1992), it will be necessary to incorporate the diffusion of $IP_3$ as well. It is reasonable, however, to consider the effect of cytosolic $Ca^{2+}$ diffusion as being predominant in a model based on CICR. When assuming that the $Ca^{2+}$-sensitive $Ca^{2+}$ pool is homogeneously distributed within the cell, in the case of two-dimensional diffusion, eqns (9.1) become:

$$\frac{\partial Z}{\partial t} = v_0 + v_1\beta - v_2 + v_3 + k_f Y - kZ + D_Z\left(\frac{\partial^2 Z}{\partial x^2} + \frac{\partial^2 Z}{\partial y^2}\right)$$

$$\frac{dY}{dt} = v_2 - v_3 - k_f Y \tag{9.11}$$

In the above equations, $D_Z$ denotes the diffusion coefficient of $Ca^{2+}$ within the cytosol, along the two spatial dimensions $x$ and $y$. The various terms appearing in these equations remain defined by eqns (9.2).

Before turning to the effect of two-dimensional diffusion, it is useful to consider the simpler situation of a single spatial dimension, a case for which the effect of discrete $Ca^{2+}$-sensitive pools of $Ca^{2+}$ has been determined (Dupont *et al.*, 1991). Then, the equation for $Z$ in (9.11) is replaced by:

$$\frac{\partial Z}{\partial t} = v_0 + v_1\beta - v_2 + v_3 + k_f Y - kZ + D_Z\frac{\partial^2 Z}{\partial x^2} \tag{9.12}$$

Considered in fig. 9.29 is the situation where five pools are distributed along the cell at equal intervals. The left extremity of the cell is stimulated at time zero by a transient increase in $IP_3$ ($\beta$); a sharp front of $Ca^{2+}$ subsequently propagates towards the right at a rate close to 100 μm/s. This pattern of wave propagation resembles that observed in cardiac cells. Simulations indicate that the wave propagates more slowly, and eventually stops when the distance between the pools increases (see also Meyer & Stryer, 1991).

Shown in figs. 9.30 and 9.31 are the two-dimensional patterns of wavelike propagation of a $Ca^{2+}$ signal in the model based on CICR

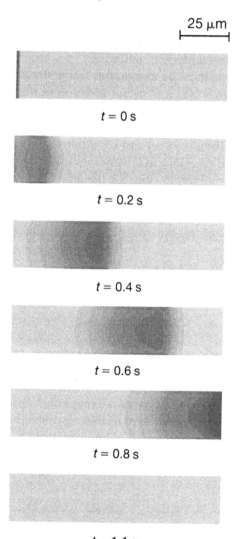

25 μm
⊢————————⊣

$t = 0$ s

$t = 0.2$ s

$t = 0.4$ s

$t = 0.6$ s

$t = 0.8$ s

$t = 1.1$ s

Fig. 9.30. Spatial propagation of a sharp $Ca^{2+}$ front of the type seen in cardiac cells (type 1 wave). Shown are six successive stages of the transient pattern obtained by numerical integration of eqns.(9.11) of the model based on CICR, from which the term $v_1\beta$ related to stimulation has been removed and to which the diffusion of cytosolic $Ca^{2+}$ has been added. In these simulations, the $Ca^{2+}$-sensitive $Ca^{2+}$ pool is assumed to be distributed homogeneously within the cell. The latter is represented as a two-dimensional mesh of $20 \times 60$ points and diffusion is approximated by finite differences; boundary conditions are of the zero-flux type. The terms related to $Ca^{2+}$ influx from ($v_0$) and into ($kZ$) the extracellular medium only appear in the points located on the borders of the mesh. The diffusion coefficient of $Ca^{2+}$ is equal to 400 μm²/s; other parameter

when a continuous, homogeneous distribution of $Ca^{2+}$-sensitive pools is assumed (Dupont & Goldbeter, 1992b, 1994). In fig. 9.30, the wave propagation rate is close to 100 µm/s; the model therefore predicts the correct magnitude of the wave propagation rate observed in cardiac myocytes (Kort *et al.*, 1985; Takamatsu & Wier, 1990) when the period of the oscillations is of the order of 1 s, as in these cells. In line with other studies (Cheer *et al.*, 1987; Backx *et al.*, 1989; Lechleiter *et al.*, 1991), a value of 400 µm²/s is taken for the $Ca^{2+}$ diffusion coefficient. As discussed above, such a value is significantly higher than the value of recently reported by Allbritton *et al.* (1992).

While the spatial propagation shown in fig. 9.30 is of type 1, as in simulations carried out along a single spatial dimension (Dupont *et al.*, 1991), for other parameter values the model based on CICR can also generate waves of type 2, i.e. $Ca^{2+}$ tides, as shown in fig. 9.31. Whereas in fig. 9.30 parameter values are such that the half-width of a $Ca^{2+}$ spike in the course of spatially homogeneous oscillations is of the order of 0.1 s, as in cardiac cells (Fabiato & Fabiato, 1975), for the parameter values considered in fig. 9.31 the period of the oscillations is longer and the peaks of cytosolic $Ca^{2+}$ are wider, of the order of 10–20 s, as in hepatocytes (Woods *et al.*, 1987; Thomas *et al.*, 1991). As a result, in the presence of diffusion, the $Ca^{2+}$ signal progressively rises along the cell and drops thereafter in a quasi-homogeneous manner; the rate of wave propagation in these conditions is close to 30 µm/s. These properties match the experimental observations on wave propagation in hepatocytes (Rooney *et al.*, 1991; Thomas *et al.*, 1991). Similar results have been reported by Thomas *et al.* (1991) who used the model based on CICR to simulate along one spatial dimension the results of their experiments on liver cells. That CICR is the mechanism underlying $Ca^{2+}$ wave propagation in these cells is further supported by the observation (Rooney *et al.*, 1991) that $Ca^{2+}$ waves can be initiated by *t*-butyl hydroperoxide (TBHP), a compound that bypasses the effect of $IP_3$ and increases the level of cytosolic $Ca^{2+}$ by inhibiting its pumping into $Ca^{2+}$ pools.

---

values are listed in the legend to fig. 9.29, with $v_0 = 1$ µM/s, $V_{M2} = 90$ µM/s, $K_2 = 0.9$ µM, $k = 8$ s⁻¹. The black bar in the top part indicates that initially the model cell is stimulated on the left by a transient rise in cytosolic $Ca^{2+}$ up to 1.3 µM, while in the rest of the cell cytosolic $Ca^{2+}$ is at a stable resting level of 0.1 µM (the second variable, intravesicular $Ca^{2+}$, is initially at its stable steady-state level in all points of the mesh); further $Ca^{2+}$ fronts propagate when the stimulus is maintained. The scale of $Ca^{2+}$ concentration ranges from 0 (white) to 1.3 µM (black) (Dupont & Goldbeter, 1992b, 1994).

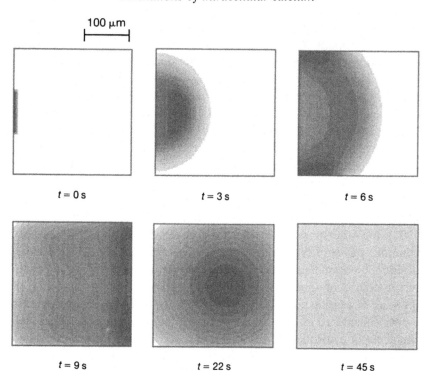

Fig. 9.31. Spatial propagation of a $Ca^{2+}$ tide resembling the waves seen in hepa-tocytes, oocytes, or endothelial cells (type 2 wave). The transient pattern is obtained as in fig. 9.30 for parameter values yielding oscillations of a period of the order of 1 min: $v_0 = 1.68$ μM/min, $V_{M2} = 93$ μM/min, $V_{M3} = 500$ μM/min, $K_A = 0.66$ μM, $K_2 = 1$ μM, $k = 16.8$ min$^{-1}$, $k_f = 1$ min$^{-1}$; other parameter values are as in fig. 9.10. The spatial mesh contains 30 × 30 points (similar results are obtained with a mesh of 60 × 60 points). The black bar in the upper, left part denotes the initial, transient stimulation, which consists in raising locally the level of cytosolic $Ca^{2+}$ to 1.5 μM at the left extremity while the rest of the cell is in the resting level of 0.1 μM; the scale of $Ca^{2+}$ concentration extends from 0 (white) to 1.5 μM (black) (Dupont & Goldbeter, 1992b, 1994).

The data in figs. 9.30 and 9.31 were obtained under conditions where the stimulation is only transient, so that a single front is observed. If the stimulus is maintained at an appropriate value, the spatiotemporal phe-nomenon becomes sustained (Dupont & Goldbeter, 1994); successive waves then propagate along the cell from the stimulation site as has been described in hepatocytes (Rooney *et al.*, 1991; Thomas *et al.*, 1991; see fig. 9.28).

The effect of $Ca^{2+}$ diffusion has also been investigated in the one-pool model based on CICR, described in section 9.4. Except for the effect on the amplitude and latency of the first spike, altready observed for oscillations (Dupont and Goldbeter, 1993), and for the role of $IP_3$, no major difference was found between the behaviours of the one- and two-pool versions of the model based on CICR with respect to $Ca^{2+}$ wave propagation. The primary role played by the autocatalytic CICR regulation in the occurrence of $Ca^{2+}$ oscillations is thus recovered for the propagation of intracellular waves. The role of $IP_3$ in wave propagation nevertheless differs in the one- and two-pool versions of the CICR model. In the two-pool model, oscillations and waves can occur in the absence of $IP_3$, provided that the $Ca^{2+}$ influx ($v_0$) from the external medium is sufficiently high. In contrast, for $Ca^{2+}$ oscillations and waves to occur in the one-pool model in the presence of $IP_3$ diffusion, $IP_3$ at the site of stimulation has to exceed a critical level (Dupont & Goldbeter, 1994). The requirement for a sufficient level of $IP_3$ in $Ca^{2+}$ wave propagation has been observed in *Xenopus* oocytes (DeLisle & Welsh, 1992; Lechleiter & Clapham, 1992), and demonstrated by the use of antibodies to the $IP_3$ receptor (Miyazaki *et al.*, 1992b).

Among the questions of interest that could be investigated by means of the model based on CICR are the dependence of $Ca^{2+}$ waves on the spatial distribution of $Ca^{2+}$-sensitive $Ca^{2+}$ pools, as well as the relation between the propagation rate, the geometry of the cell and the period and half-width of $Ca^{2+}$ spikes. An analytical study of wave propagation, taking advantage of the existence of only two variables, has been performed in this model (Sneyd, Girard & Clapham, 1993). Two-dimensional simulations of $Ca^{2+}$ wave propagation, also using models based on CICR, have been presented to account specifically for a variety of wavelike patterns observed in oocytes (Girard *et al.*, 1992; Atri *et al.*, 1993). As regards the propagation of intercellular $Ca^{2+}$ waves, a recent theoretical study raises the possibility (Sneyd, Charles & Sandersen, 1994) that in addition to CICR, which underlies intracellular propagation of the waves, a process of $IP_3$ regeneration is necessary to account for their passage from cell to cell.

In conclusion, the study of models for $Ca^{2+}$ signalling indicates that the phenomena of $Ca^{2+}$ oscillations and waves are closely intertwined. The spatiotemporal patterns correspond to the propagation of a $Ca^{2+}$ front in a biochemically excitable or oscillatory medium, at a rate much higher than that associated with simple diffusion. Such a property could also underlie a possible role of $Ca^{2+}$ waves in intercellular communication (Charles *et al.*, 1991). The results presented in figs. 9.30 and 9.31 show that a unique mechanism, based on CICR, can account for the two types of wave seen, on the one hand, in cardiac cells and *Xenopus* oocytes and, on the other hand, in hepatocytes, smaller eggs, and endothelial cells. It is likely that similar results on the two types of $Ca^{2+}$ wave could also be generated, for appropriate values of the parameters, by other models such as that based on the cross-coupling between $Ca^{2+}$ and $IP_3$ (Meyer & Stryer, 1988, 1991). A similar remark holds for the models proposed by De Young & Keizer (1992; see also Keizer & De Young, 1994; Roth *et al.*, 1995) and Atri *et al.* (1993). The simulations confirm that the differences between the waves originate from the characteristics of the underlying oscillations in $Ca^{2+}$; the form of the wave is closely related to the period and half-width of the $Ca^{2+}$ spikes observed under spatially homogeneous conditions. Moreover, the geometry of the cell also plays a role as elongated forms tend to favour waves of type 1 (see fig. 9.30). It is clear, however, that a continuum should exist between the situations considered in figs. 9.30 and 9.31: at intermediate periods and half-widths of $Ca^{2+}$ spikes, waves of intermediate appearance should be observed.

The above discussion underscores the dual role of models whose function is to provide, on the one hand, a unifying explanation of experimental observations as well as a clearer understanding of the role played by each component in a highly complex regulated system and, on the other hand, to lead to testable predictions. On both counts, models for $Ca^{2+}$ signalling provide an example of how added insights may be gained from a synergy between experiments and theory in studying the dynamics of cellular processes.

# Part VI
## The mitotic oscillator

# 10

# Modelling the mitotic oscillator driving the cell division cycle

## 10.1 The eukaryotic cell cycle is driven by a biochemical oscillator

Few cellular processes are as crucial as that governing cell division. The tight control of cell division indeed plays a prominent role in development and differentiation, while unrestrained proliferation is associated with the cancerous state. The control of the eukaryotic cell division cycle therefore represents a central issue in cell biology (for reviews of recent experimental advances, see the special issues of *Science*, **246**:603–640 (1989) and of the *Journal of Cell Science*, Suppl. 12 (1989), as well as volume LVI of *Cold Spring Harbor Symposium on Quantitative Biology* (1991), and Cross *et al.*, 1989; Hunter, 1992; Norbury & Nurse, 1992; Murray & Hunt, 1993; Heichman & Roberts, 1994; King, Jackson & Kirschner, 1994; Nurse, 1994; Peter & Herskowitz, 1994; Sherr, 1994). The cell cycle is classically portrayed as a sequence of phases (fig. 10.1): following mitosis (M) come, successively, the $G_1$ phase leading to the S phase of DNA replication, and the $G_2$ phase, which separates the latter from the next M phase; sometimes cells stay in a quiescent phase, $G_0$, prior to their entry into $G_1$. The fact that in dividing cells mitosis recurs at regular intervals (whose duration, varying with the cell type, ranges from 10 min up to 24 h or even more) has for a long time raised the possibility that the cell cycle is driven by a continuous biochemical oscillator.

The hypothesis that a biochemical oscillator of the limit cycle type controls the onset of mitosis has been proposed for long (Sel'kov, 1970; Gilbert, 1974, 1978; Winfree, 1980, 1984; for an earlier discussion of cell division in terms of a chemical oscillatory process, see Rashevsky, 1948) and was put forward, in particular detail, on the basis of experiments performed on the slime mould *Physarum* (Kauffman, 1974; Kauffman

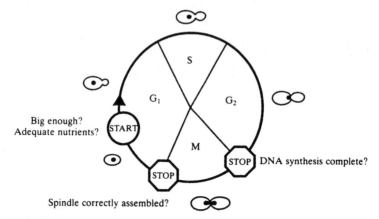

Fig. 10.1. The somatic cell cycle in budding yeast, illustrating the successive phases of the cycle as well as checkpoints (Murray, 1989b).

& Wille, 1975). The division cycle occurs in these cells with a periodicity close to 12 h. Mixing experiments relying on the fusion of plasmodia taken at different times over the cycle showed phase advances or delays that would be typical of the behaviour expected if mitosis were driven by a continuous oscillator of a moderate relaxation nature (Kauffman, 1974; Kauffman & Wille, 1975; Wille, Scheffey & Kauffman, 1977). Implicit in the limit cycle description is the assumption that one of the variables of the oscillator behaves as a mitogenic factor: once this variable exceeds a certain threshold, mitosis would ensue. A specific prediction of the limit cycle model of mitosis is that finely tuned perturbations may transiently suppress oscillations. In this case, mitosis would eventually resume, with undefined phase, possibly after a delay corresponding to a few cycles in which the mitogenic factor oscillates below its threshold level. As the trajectory followed by the oscillator eventually approaches the asymptotic limit cycle, mitosis would occur when the threshold is again exceeded.

An alternative view holds that cell division is brought about by the accumulation of a mitogenic factor that, above some threshold, triggers mitosis; the latter, discontinuous event brings this factor back to a low value and a new cycle starts as the accumulation of the mitogenic factor resumes. Models of such a discontinuous nature (Fantes *et al.*, 1975) differ in several respects from the continuous biochemical oscillatory mechanism; for a comparative discussion of the two classes of mechanism, see Tyson & Sachsenmaier (1978) and Winfree (1980, 1984).

At the time when the limit cycle model for mitosis in *Physarum* was proposed, no data on the biochemical nature of the oscillatory variables were available and an abstract mathematical model was used to simulate the results from plasmodial fusion experiments (Kauffman, 1974; Kauffman & Wille, 1975). In recent years, however, significant progress has been made on the biochemical characterization of the control mechanisms that trigger mitosis. These experimental advances, which rely both on genetic evidence obtained from yeast and on biochemical studies performed on sea urchin and amphibian eggs, support the existence of a continuous biochemical oscillator driving the cell division cycle. Early observations pointing to the existence of such a cytoplasmic oscillator were made in sea urchin embryos (Mano, 1970) and *Xenopus* eggs (Hara, Tydeman & Kirschner, 1980). More recent work indicates that the oscillator is based on a universal mechanism involving the periodic activation of a protein kinase, product of the gene *cdc2* in the fission yeast *Schizosaccharomyces pombe* (the gene is referred to as *CDC28* in the budding yeast *Saccharomyces cerevisiae*) or of its homologues in other eukaryotes (Cross *et al.*, 1989; Nurse, 1990; Hunter, 1992; Norbury & Nurse, 1992; Murray & Hunt, 1993; Nasmyth, 1993; King *et al.*, 1994; Nurse, 1994; see also *Cold Spring Harbor Symposium on Quantitative Biology*, vol. LVI, 1991).

These results have opened the way to the construction of more realistic models for the mitotic oscillator. The purpose of this chapter is briefly to present these models and to classify them according to the type of regulation responsible for oscillatory behaviour. The way sustained oscillations are generated is examined in detail in a minimal model based on the cascade of phosphorylation–dephosphorylation cycles that controls the onset of mitosis in embryonic cells. Extensions of the cascade model taking into account additional, recently uncovered phosphorylation–dephosphorylation cycles are considered. Ways of arresting the cell division cycle in that model and the control of the mitotic oscillator by growth factors are also discussed.

## 10.2 Brief summary of experimental advances

In amphibian eggs (fig. 10.2), the cell cycle has a simple form as it consists in the periodic alternation of mitosis and interphase. Mitosis is associated with the activation of a factor known as maturation (or mitosis) promoting factor (MPF). Activation of MPF is triggered by the building up to a threshold of cyclin, a protein whose name reflects its

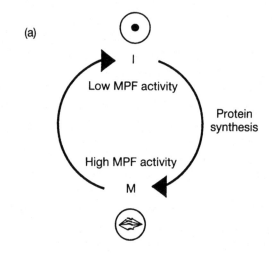

(a)

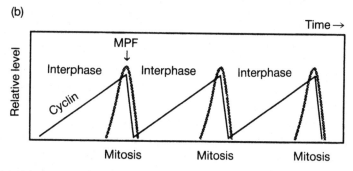

(b)

Fig. 10.2. (a) The early amphibian embryonic cell cycle. (b) Role of maturation (or mitosis)-promoting factor (MPF) and cyclin in the periodic alternation of interphase and mitosis in that cycle. ((a) reproduced from Murray, 1989b; (b) reproduced from Minshull *et al.*, 1989).

periodic variation during the cell cycle (Evans *et al.*, 1983). There exists a family of different cyclins, among which cyclin B is associated with the $G_2/M$ transition. It was later found that MPF consists of a complex between cyclin B and a protein kinase, known as cdc2 kinase; inactivation of MPF after mitosis is brought about by cyclin degradation (Draetta & Beach, 1989; Labbé *et al.*, 1989; Minshull *et al.*, 1989; Gautier *et al.*, 1990).

The interaction of cdc2 kinase with cyclin is in fact central to the mechanism of the mitotic oscillator in all eukaryotic cells (Nurse, 1990). The situation encountered in the early stages of amphibian development represents the simplest form of mitotic trigger mechanism

(Murray & Kirschner, 1989a): in such rapidly dividing embryonic cells, where the period of the mitotic oscillator is of the order of 30 min, the accumulation of cyclin alone suffices to drive the cell cycle (Murray & Kirschner, 1989b), which may be viewed as a cdc2 cycle (Murray, 1989a). Oscillations in cdc2 kinase activity have been demonstrated in *Xenopus* egg extracts (Félix *et al.*, 1989).

In yeast (Hartwell & Weinert, 1989; Nurse, 1990; Murray, 1992; Murray & Hunt, 1993; Moreno & Nurse, 1994; Peter & Herskowitz, 1994; Schwob *et al.*, 1994) and somatic cells (Cross, Roberts & Weintraub, 1989; Murray, 1992; Norbury & Nurse, 1992; Heichman & Roberts, 1994) the cell cycle is subjected to additional controls, e.g. before the initiation of DNA replication where the checkpoint has been designated as 'start' (see fig. 10.1). Here also, however, the transition between the $G_2$ and M phases is brought about by the periodic activation of cdc2 kinase, while the transition between the $G_1$ and S phases appears to be controlled by $G_1$-specific cyclins and, in higher eukaryotes, by another cyclin-dependent protein kinase, cdk2, closely related to cdc2 (Dulic, Lees & Reed, 1992; Elledge *et al.*, 1992; De Bondt *et al.*, 1993). In somatic cells, besides cdc2 and cdk2, other members of a family of cyclin-dependent kinases have been found, which play a role, not yet fully clarified, in different points of the cell cycle (see, e.g. Hunter & Pines, 1994). The following discussion focuses on the simpler situation that prevails in the early stages of amphibian embryogenesis, although it is likely that key features of the mitotic control system operating in embryonic cells remains at the core of more complex cell cycle mechanisms. Although the detailed mechanism of cdk2 activation remains to be established, experimental evidence shows that cdck2 kinase also undergoes a periodic activation–inactivation cycle that can run independently of the cdc2 cycle (Gabrielli *et al.*, 1992). Recent observations indicate, however, that cyclin-dependent kinases active at the $G_1$/S transition, in yeast as in somatic cells, are controlled not only by post-translational modification but also by protein inhibitors (Gu *et al.*, 1993; Xiong *et al.*, 1993; Hengst *et al.*, 1994; Moreno & Nurse, 1994; Peter & Herskowitz, 1994; Schwob *et al.*, 1994; Morgan, 1995).

Advances made over the last six years have permitted the characterization of the biochemical mechanism through which cdc2 kinase becomes activated; for reviews, see Draetta (1990), Maller (1990), King *et al.* (1994) and Morgan (1995). Such a process relies primarily (fig. 10.3) on the dephosphorylation of cdc2 kinase on tyrosine (Tyr) 15 by a tyrosine phosphatase (Draetta & Beach, 1989; Gautier *et al.*, 1989;

*Modelling the mitotic oscillator*

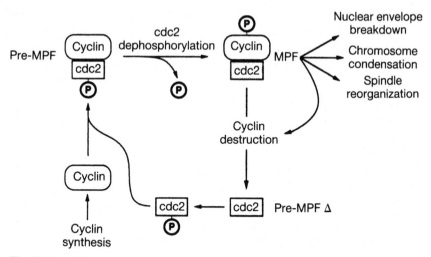

Fig. 10.3. Schematic representation of the control of MPF by phosphorylation–dephosphorylation of the cdc2 subunit in embryonic cells (adapted from Minshull *et al.*, 1989). Also shown is the phosphorylation of cyclin, which is not considered in the model of fig. 10.4, which is based solely on the regulation of cdc2 kinase by reversible dephosphorylation.

Gould & Nurse, 1989; Morla *et al.*, 1989). The latter enzyme has been identified as the product of the *cdc25* gene in fission yeast (Russell & Nurse, 1986; Kumagai & Dunphy, 1991; Strausfeld *et al.*, 1991). The cdc25 phosphatase is itself regulated through reversible phosphorylation (Lorca *et al.*, 1991a; Kumagai & Dunphy, 1992). Some experimental observations suggest (Newport & Kirschner, 1984; Cyert & Kirschner, 1988; Dorée, Lorca & Picard, 1991; King *et al.*, 1994) that the activation of cdc2 may directly or indirectly possess an autocatalytic nature. This view is supported by the observation that cdc25 is activated by cdc2 in human cells (Hoffmann *et al.*, 1993; Strausfeld *et al.*, 1994); a similar self-amplification loop involves cdk2 and its associated phosphatase cdc25A at the $G_1/S$ transition in these cells (Hoffmann, Draetta & Karsenti, 1994). As discussed below, autocatalysis by cdc2 plays a prominent role in several models proposed for the mitotic oscillator.

The inhibitory effect of the phosphorylation of cdc2 kinase on Tyr15 stems from the fact that this residue is located in the ATP-binding domain of the enzyme (Gould & Nurse, 1989). In vertebrate cells, but apparently not in yeast, phosphorylation of threonine (Thr) 14, a second residue located in that site, also inhibits cdc2 kinase (Krek & Nigg, 1991; Norbury, Blow & Nurse, 1991). In fission yeast, the kinase phosphorylating Tyr15 is the product of the gene *wee1* (Russell & Nurse, 1987; Parker, Atherton-Fessler & Piwnica-Worms, 1992). In vertebrate cells this kinase appears to differ from the one that phosphorylates Thr14 (Parker & Piwnica-Worms, 1992), so that two kinases and two phosphatases acting on cdc2 kinase might well be needed at the $G_2/M$ transition (Krek & Nigg, 1991; Parker & Piwnica-Worms, 1992). However, it appears that cdc2 kinase is fully active only when a third residue, Thr161 in vertebrate cells (Solomon, Lee & Kirschner, 1992) or Thr167 in fission yeast (Gould *et al.*, 1991), is phosphorylated. Inactivation of cdc2 kinase after mitosis requires dephosphorylation of that residue (Lorca *et al.*, 1992). The kinase and phosphatase associated with the phosphorylation and dephosphorylation of Thr161 have been designated CAK (Solomon, Lee & Kirschner, 1992) and INH (Lee *et al.*, 1991; Solomon *et al.*, 1992), respectively. CAK is associated with a specific cyclin and is also known as cdk7 or MO15 (Fisher & Morgan, 1994; Mäkela *et al.*, 1994); this observation indicates the existence of a cascade in which cyclin-dependent kinases are controlled by other members of the cdk family. While some studies (Lee *et al.*, 1991; Solomon *et al.*, 1992) point to a direct activation of cdc2 kinase by phosphorylation of Thr161, other results suggest (Ducommun *et al.*, 1991) that this modification merely represents a necessary step for the proper binding of cyclin to cdc2 kinase.

Because of the primary role of cyclin in the activation of cdc2 kinase, it is clear that cyclin degradation plays an essential role in the control of mitosis (Minshull *et al.*, 1989; Murray, Solomon & Kirschner, 1989; King *et al.*, 1994). Important in this respect is the observation that cdc2 kinase activates cyclin degradation (Félix *et al.*, 1990), via the ubiquitin pathway (Glotzer, Murray & Kirschner, 1991; Hershko *et al.*, 1994). The mechanism of that activation appears to rely on a phosphorylation cascade containing at least one phosphorylation–dephosphorylation cycle, the product of which is the active form of a cyclin protease. According to Murray & Kirschner (1989a) and to Félix *et al.* (1990), the activation of cdc2 kinase by cyclin and the subsequent activation of cyclin protease by cdc2 kinase provides a negative feedback loop on which a minimal cell cycle oscillator might be based.

A conspicuous feature of cyclin and cdc2 interactions is the occurrence of sharp thresholds in the dependence of cdc2 kinase activation by cyclin, and in the dependence of cyclin degradation on active cdc2 kinase (Minshull *et al.*, 1989; Murray & Kirschner, 1989b; Félix *et al.*, 1990; Karsenti, Verde & Félix, 1991; King *et al.*, 1994). Besides the

nature of the mechanism of the mitotic oscillator, the question therefore arises as to what is the origin of these threshold phenomena.

Triggered by the above-described developments, the experimental
study of the cell cycle has rapidly accelerated over the last years, particularly for yeast and somatic cells. Among the main, closely intertwined
topics currently explored are the elucidation of the role of various
cyclin-dependent kinases in progression along the cell cycle, the ordering of the M and S phases (Amon, Irniger & Nasmyth, 1994; Hayles *et
al.*, 1994; Nurse, 1994), the role and regulation of DNA replication complexes (Heichman & Roberts, 1994), the operation of checkpoints
(Murray, 1992; Minshull *et al.*, 1994), the mechanisms by which external
signals such as growth factors trigger cell cycling (Nourse *et al.*, 1994),
the control of cell cycle progression by cdk inhibitors (Xiong *et al.*, 1993;
Peter & Herskowitz, 1994) and the potential role of these inhibitors and
cyclins in cancer (Hunter & Pines, 1991, 1994). Currently investigated
are also the developmental control of the cell cycle (Edgar & O'Farrell,
1989; O'Farrell *et al.*, 1989; Edgar *et al.*, 1994) and the spatial patterns of
cell division observed as mitotic domains in the embryos of *Drosophila*
(Foe, 1989; Foe, Odell & Edgar, 1993) and zebrafish (Kane, Warga &
Kimmel, 1992).

## 10.3 Models based on positive feedback by cdc2 kinase

In their early theoretical studies of the mitotic oscillator, Kauffman *et
al.* (Kauffman, 1975; Kauffman & Wille, 1975; Tyson & Kauffman,
1975) resorted to the abstract, Brusselator model (Lefever & Nicolis,
1971) for their simulations of mixing experiments in which *Physarum*
plasmodia taken at different phases of the cell cycle were fused. Like
most models proposed for limit cycle behaviour, the Brusselator relies
on an autocatalytic step for producing the instability leading to oscillations; an advantage of this simple model is that the temporal evolution
is governed by two polynomial, nonlinear kinetic equations (Lefever &
Nicolis, 1971).

The first models for the mitotic oscillator specifically based on the
interaction between cyclin and cdc2 kinase also relied on positive feedback. The experimental basis for autocatalytic regulation of cdc2 kinase
stems primarily from observations showing that catalytic amounts of
active MPF promote the transition from inactive to active MPF, which
consists of a complex between cyclin and the active form of cdc2 kinase;
spontaneous activation of this transition, however, does not normally

occur *in vivo* (Cyert & Kirschner, 1988). Recent experiments in human cells have shown (Hoffmann *et al.*, 1993; Strausfeld *et al.*, 1994) that cdc2 kinase can phosphorylate, and thereby activate, the cdc25 phosphatase that itself activates cdc2; such regulation creates a positive feedback loop in the control of MPF (see section 10.5).

In the model proposed by Hyver & Le Guyader (1990), two systems of equations are considered. In the first version, inactive p34 (i.e. cdc2 kinase) transforms into active p34, either spontaneously or in an autocatalytic manner; active p34 then combines with cyclin to yield active MPF. This situation is described by four differential equations, of a polynomial nature, in which the highest nonlinearities are of the quadratic type. In a second version of this model, governed by three kinetic equations of a similar form, the authors consider the effect of an activation of MPF by MPF itself as well as cyclin, and show that oscillations develop when the degradation of cyclin is brought about by the formation of a complex between cyclin and MPF. That study was the first to show the occurrence of sustained oscillations in a model based on the interactions between cyclin and cdc2 kinase. The type of kinetics considered for these interactions remained, however, remote from the actual kinetics of phosphorylation–dephosphorylation cycles.

A second, simpler model, due to Norel & Agur (1991), contains only two variables, cyclin and active MPF. The model relies on an autocatalytic reaction that is very similar to the one considered in the Brusselator model, i.e. it originates from a trimolecular step; degradation of cyclin occurs at a rate proportional to the concentration of MPF. This model readily produces oscillations in cyclin and MPF. Here again, the kinetics considered for MPF self-activation and cyclin disappearance are not based on the type of kinetics encountered in phosphorylation–dephosphorylation cascades. Although rather crude, these phenomenological equations should be regarded as being among the simplest capable of producing oscillations on the basis of the observed qualitative interactions between cyclin and MPF.

Building on similar ideas but starting from a more detailed reaction scheme, Tyson (1991) proposed a model for the mitotic oscillator based on the formation of a complex between cyclin and cdc2 kinase, followed by the activation of this complex. Essential to the oscillatory mechanism is the assumption that the active complex, i.e. MPF, promotes its own activation in a nonlinear manner. The kinetic equations, of a polynomial form, reduce under some simplifying assumptions to the equations of the two-variable Brusselator model. Inactivation of MPF is not considered as being regulated by MPF. A recent, more extended version of that model, applied to yeast (Novak & Tyson, 1993a, 1995) and to the embryonic cell cycle (Novak & Tyson, 1993b), incorporates the latter feature as well as the effect of the second regulatory phosphorylation of cdc2 kinase on Thr161/167 (see above). In yeast, the model suggests a mechanism for the control exerted by cellular size on the mitotic oscillator. The models studied by Novak and Tyson are the most detailed ones proposed so far. Other theoretical models for the mitotic oscillator were recently presented by Obeyesekere, Tucker & Zimmerman (1994) and Thron (1994).

Common to most of these models is the primary role played by autocatalysis in the origin of sustained oscillations in the mitotic control system. In the absence of self-activation by cdc2 kinase, oscillations would not occur in this class of model. Let us consider next a model that bypasses the absolute requirement for such a positive feedback. In contrast to the above-described models which rely on polynomial kinetic equations, the model examined in the following section is based on enzyme kinetics of the Michaelis–Menten type, closer to the phosphorylation–dephosphorylation nature of the reactions that control the activation of cdc2 kinase.

## 10.4 A phosphorylation–dephosphorylation cascade model for the mitotic oscillator in embryonic cells

That cyclin degradation represents a key element of the dynamics of the mitotic control system was demonstrated in amphibian embryonic cells. Several authors had suggested (Hunt, 1989; Murray & Kirschner, 1989a,b) that MPF, besides inducing events leading to mitosis such as chromosome condensation and the breakdown of the nuclear envelope, also triggers cyclin proteolysis, leading to MPF inactivation. Félix *et al.* (1990) corroborated this by showing that cdc2 kinase promotes the activation of a protease that degrades cyclin. The activation of cyclin protease is the end step of a phosphorylation–dephosphorylation cascade comprising at least one cycle in which phosphorylation is catalysed by cdc2 kinase (see also Hershko *et al.*, 1994).

The fact that cyclin activates cdc2 kinase whereas cdc2 kinase promotes cyclin degradation suggested that this negative feedback regulation could provide a mechanism on which a minimum cell cycle oscillator might be based (Murray & Kirschner, 1989a; Félix *et al.*, 1990). To produce oscillations, however, negative feedback must be coupled to thresholds and time delays whose existence has repeatedly been emphasized in experimental studies of cyclin–cdc2 kinase interactions (Murray & Kirschner, 1989a; Félix *et al.*, 1990). Provided that the mitotic control system possesses these additional features, oscillations could in principle result from the delayed negative feedback present in the cyclin–cdc2 kinase interactions. Félix *et al.* (1990) suggested that, in such conditions, 'the system cannot reach equilibrium: it keeps oscillating, because threshold levels of both cyclin protein and cdc2 kinase activities trigger post-translational reactions with built-in time delays, and the destruction of cyclin causes the loss of cdc2 kinase activity'.

### Minimal cascade model involving cyclin and cdc2 kinase

On the basis of this hypothesis, Goldbeter (1991a) proposed a minimal cascade model for the mitotic oscillator. So as to deal with the simplest structure of the mitotic oscillator, the analysis focused on the cell cycle in early amphibian embryos, where the two main factors are cyclin B (referred to below as cyclin) and cdc2 kinase (Murray & Kirschner, 1989a,b). The basic assumption (see fig. 10.4) is that cyclin is synthesized at a constant rate and triggers the activation of cdc2 kinase. To keep the model simple and to allow for the straightforward generation of thresholds (see below), the formation of a complex between cyclin and cdc2 kinase is not taken into account; instead, it is assumed that cyclin drives cdc2 activation by enhancing the velocity of an 'activase' (Murray & Kirschner, 1989a,b), which, in the light of the above discussion, represents the cdc25 phosphatase (Kumagai & Dunphy, 1991; Strausfeld *et al.*, 1991). Evidence for such a direct activation of the cdc25 phosphatase by cyclin has been reported (Galaktionov & Beach, 1991; Zheng & Ruderman, 1993). Moreover, the existence of an active monomeric form of cdc2 kinase is supported by some experimental observations (Brizuela, Draetta & Beach, 1989; Lorca *et al.*, 1992) and can also be inferred from the fact that the activity of cdc2 kinase persists for some time before falling after cyclin has been degraded (Moreno, Hayles & Nurse, 1989). The fact that the complex between cyclin and cdc2 is not considered explicitly represents a weakness of this relatively simple model, which nevertheless yields useful insights into the mechanism of mitotic oscillations.

Recent experiments indicate that cdc25 is itself activated by phosphorylation (Lorca *et al.*, 1991a; Kumagai & Dunphy, 1992; Millar & Russell, 1992), probably by cdc2 kinase (Hoffmann *et al.*, 1993;

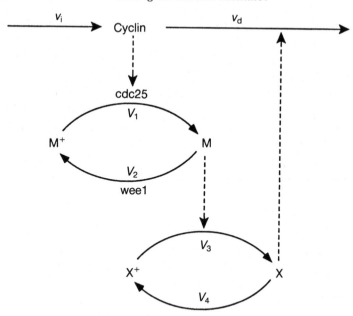

Fig. 10.4. Minimal cascade model for the mitotic oscillator. The cascade incorporates cyclin synthesis and degradation, activation of the phosphorylated form $M^+$ of cdc2 kinase through dephosphorylation into the form M by phosphatase cdc25 ($E_1$), which is itself activated by cyclin, inactivation of active cdc2 kinase M into $M^+$ by the kinase wee1 ($E_2$), phosphorylation of inactive cyclin protease $X^+$ into the active form X by cdc2 kinase ($E_3$), and inactivation of X into $X^+$ by the phosphatase $E_4$ (see text for details).

Strausfeld *et al.*, 1994); this situation is considered in section 10.5, where an extension of the minimal cascade model is envisaged (this extension also permits us to assess the effect of self-activation by cdc2 kinase). A further assumption is that the maximum activity of the kinase wee1 (Parker *et al.*, 1992) that inactivates cdc2 – the cdc2 'inactivase' (Murray & Kirschner, 1989a,b) – remains constant throughout the cell cycle.

In line with the observation that the kinase activity of the cdc2 protein promotes cyclin degradation (Félix *et al.*, 1990), it is assumed that cdc2 kinase activates a cyclin protease, designated X, by reversible phosphorylation (fig. 10.4); the maximum activity of the phosphatase inactivating that protease is taken as constant throughout the cycle. There is evidence that the pathway of cyclin degradation is itself a bicyclic phosphorylation cascade, the first step of which would be controlled by cdc2 kinase (Félix *et al.*, 1990; Hershko *et al.*, 1994).

Consideration of a multicyclic rather than monocyclic cascade leading to the activation of the protease by cdc2 kinase would, however, not significantly affect the results of the analysis.

Thus, the three variables of the minimal model described in fig. 10.4 are cyclin, the active (i.e. dephosphorylated) form of cdc2 kinase, and the active (i.e. phosphorylated) form of cyclin protease. The dynamics of the bicyclic cascade of post-translational modification is governed by the following system of kinetic equations (Goldbeter, 1991a):

$$\frac{dC}{dt} = v_i - v_d X \frac{C}{K_d + C} - k_d C \tag{10.1a}$$

$$\frac{dM}{dt} = V_1 \frac{(1 - M)}{K_1 + (1 - M)} - V_2 \frac{M}{K_2 + M} \tag{10.1b}$$

$$\frac{dX}{dt} = V_3 \frac{(1 - X)}{K_3 + (1 - X)} - V_4 \frac{X}{K_4 + X} \tag{10.1c}$$

with:

$$V_1 = \frac{C}{K_c + C} V_{M1}, \quad V_3 = M V_{M3} \tag{10.2a,b}$$

In the above equations, $C$ denotes the cyclin concentration, while $M$ and $X$ represent the fraction of active cdc2 kinase and the fraction of active cyclin protease, respectively; $(1 - M)$ thus represents the fraction of inactive (i.e. phosphorylated) cdc2 kinase, while $(1 - X)$ represents the fraction of inactive (i.e. dephosphorylated) cyclin protease. As to parameters, $v_i$ and $v_d$ denote, respectively, the constant rate of cyclin synthesis and the maximum rate of cyclin degradation by protease X reached for $X = 1$; $K_d$ and $K_c$ denote the Michaelis constants for cyclin degradation and for cyclin activation of the phosphatase acting on the phosphorylated form of cdc2 kinase; $k_d$ represents an apparent first-order rate constant related to nonspecific degradation of cyclin (this facultative reaction, whose contribution is much smaller than that of cyclin degradation by protease X, is not needed for oscillations; its sole effect is to prevent the boundless increase of cyclin under conditions where the specific protease would be inhibited, as is discussed further below).

The remaining parameters $V_i$ and $K_i$ ($i = 1, \ldots 4$) denote the effective maximum rate and the Michaelis constant for each of the enzymes $E_i$ ($i = 1, \ldots 4$) involved in the two cycles of post-translational modification, namely, on one hand, the phosphatase cdc25 ($E_1$) and the kinase wee1 ($E_2$) acting on the cdc2 molecule, and on the other hand, the kinase cdc2 ($E_3$) and the phosphatase ($E_4$) acting on the cyclin protease (see fig. 10.4). For each converter enzyme, the two parameters $V_i$ and $K_i$ are divided by the total amount of relevant target protein, i.e. $M_T$ (total amount of cdc2 kinase) for enzymes $E_1$ and $E_2$, and $X_T$ (total amount of cyclin protease) for enzymes $E_3$ and $E_4$; both $M_T$ and $X_T$ are considered below as constant throughout the cell cycle (although the synthesis of cdc2 varies during the cell cycle, compensatory changes in its degradation rate result in the maintenance of a practically constant level of cdc2 (Welch & Wang, 1992)).

The expressions for the effective, maximum rates $V_1$ and $V_3$ are given by eqns (10.2a,b). Expression (10.2a) reflects the assumption that cyclin activates phosphatase $E_1$ in a Michaelian manner; $V_{M1}$ denotes the maximum rate of that enzyme reached at saturating levels of cyclin. Equation (10.2b) expresses the proportionality of the effective maximum rate of cdc2 kinase to the fraction $M$ of active enzyme; $V_{M3}$ denotes the maximum velocity of the kinase reached for $M = 1$.

All nonlinearities in the model are of the Michaelian type. In other words, no form of positive cooperativity is assumed in the proteolysis of cyclin, in the activation by cyclin of the cdc25 phosphatase acting on cdc2, or in any of the reactions of covalent modification. The equations acquire a polynomial nature, much as in the other models considered for the mitotic oscillator (Hyver & Le Guyader, 1990; Norel & Agur, 1991; Tyson, 1991; Novak & Tyson, 1993a,b), when the converter enzymes of the cascade operate in the domain of first-order kinetics; however, no threshold occurs under such conditions and it is much more difficult, if not impossible, to observe oscillations. The self-amplification effect due to the possible (indirect) activation of cdc2 kinase by the active form of the enzyme (Cyert & Kirschner, 1988; Murray & Kirschner, 1989a; Hoffmann *et al.*, 1993; Strausfeld *et al.*, 1994) is not considered at this stage (see below). One goal of the analysis of this minimal model is indeed to determine whether oscillations can arise solely as a result of the negative feedback provided by cdc2-induced cyclin degradation and of the thresholds and time delays built into the cyclin–cdc2 cascade of covalent modification.

### Thresholds in cyclin and cdc2 kinase action

Before showing how the cascade model accounts for oscillations, it is useful to see how it provides an explanation for the origin of thresholds in the control of cdc2 kinase by cyclin and in the activation of cyclin protease by cdc2 kinase. The thresholds can be demonstrated by determining at steady state the dependence of the fraction of active cdc2 kinase, $M$, on cyclin concentration, $C$, and the dependence of the fraction of active cyclin protease, $X$, on $M$. This can readily be done, thanks to the modular structure of the cascade model that allows separation of variables in eqns (10.1), which at steady state reduce to algebraic equations. The problem of finding the steady-state dependence of $M$ on $C$ and of $X$ on $M$ can therefore be solved by studying each of the two cycles separately (see fig. 10.4). In the first covalent modification cycle, involving cdc2 kinase, $M$ is expressed at steady state from eqn (10.1b) as a function of the ratio of maximum modification rates $V_1$ and $V_2$. In the second cycle, involving cyclin protease, the steady state value of $X$ is obtained from eqn (10.1c) as a function of the maximum rates $V_3$ and $V_4$.

The general problem of how the steady-state fraction of protein modified in a single covalent modification cycle depends on the maximum rates of the converter enzymes has been addressed previously (Goldbeter & Koshland, 1981, 1982a, 1984). The expression obtained for that fraction is the solution of a second-degree equation. Applying this expression (Goldbeter & Koshland, 1981, 1982a) to the dependence of $M$ on $V_1$ and $V_2$ yields the relation:

$$M = \frac{[M]}{M_T} = \frac{\phi + \left[ \phi^2 + 4K_2\left(\frac{V_1}{V_2} - 1\right)\left(\frac{V_1}{V_2}\right) \right]^{1/2}}{2\left(\frac{V_1}{V_2} - 1\right)} \tag{10.3}$$

with:

$$\phi = \left(\frac{V_1}{V_2} - 1\right) - K_2\left(\frac{K_1}{K_2} + \frac{V_1}{V_2}\right)$$

where $V_1$ is given by eqn (10.2a).

Likewise, the application of this analysis to the second cycle of the cascade yields, at steady state, the following relation for $X$ as a function of $V_3$ and $V_4$:

$$X = \frac{[X]}{X_T} = \frac{\phi' + \left[ \phi'^2 + 4K_4 \left( \frac{V_3}{V_4} - 1 \right) \left( \frac{V_3}{V_4} \right) \right]^{1/2}}{2 \left( \frac{V_3}{V_4} - 1 \right)} \tag{10.4}$$

with:

$$\phi' = \left( \frac{V_3}{V_4} - 1 \right) - K_4 \left( \frac{K_3}{K_4} + \frac{V_3}{V_4} \right)$$

where $V_3$ is given by eqn (10.2b).

In a covalent modification cycle five parameters govern the amount of protein modified at steady state (Goldbeter & Koshland, 1981, 1982a): the maximum rates of the converter enzymes ($V_1$ and $V_2$ in the case of the cycle acting on M), the Michaelis constants of these enzymes ($K_{m1}$ and $K_{m2}$), and the total amount of modified protein ($M_T$). To discuss the behaviour of the modification cycle it is useful to introduce the reduced (dimensionless) Michaelis constants $K_1 = K_{m1}/M_T$ and $K_2 = K_{m2}/M_T$, as done in eqns (10.1)–(10.3). The main result of the kinetic analysis (Goldbeter & Koshland, 1981, 1982a) is that the fraction of modified protein at steady state, $M$, varies in markedly different manners as a function of the ratio of maximum modification rates $V_1/V_2$, depending on the values of the reduced Michaelis constants $K_1$ and $K_2$. When $K_1, K_2 \gg 1$ (or $K_{m1}, K_{m2} \gg M_T$), i.e. when the modifying enzymes mostly function in the first-order kinetic domain, $M$ progressively rises from zero to unity in a smooth, Michaelian manner as the ratio $V_1/V_2$ increases. In contrast, when $K_1, K_2 \ll 1$ (or $K_{m1}, K_{m2} \ll M_T$), i.e. when the converter enzymes are saturated by their respective substrates and function in the zero-order kinetic domain, $M$ varies with the ratio $V_1/V_2$ in a sigmoidal manner characterized by a threshold that becomes steeper and steeper as $K_1$ and $K_2$ decrease below unity. The possibility of generating steep thresholds in this way was referred to as **zero-order ultrasensitivity** (Goldbeter & Koshland, 1981). Experimental support for this theoretically predicted behaviour is furnished by several examples of enzymes regulated by phosphorylation–dephosphorylation (LaPorte & Koshland, 1983; Meinke, Bishop & Edstrom, 1986).

These results, applied to the steady-state behaviour of the first modification cycle of the cascade model of fig. 10.4, are shown in fig. 10.5a where the fraction of active, i.e. dephosphorylated cdc2 kinase is plotted not as a function of the ratio $V_1/V_2$ but as a function of cyclin

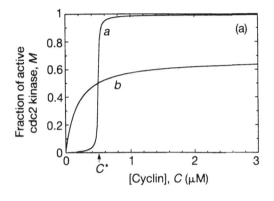

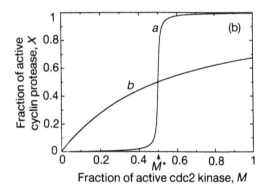

Fig. 10.5. Dependence of (a) the fraction of active cdc2 kinase ($M$) on cyclin and (b) the fraction of active cyclin protease ($X$) on $M$. Curves in (a) are generated by means of eqn (10.3), yielding the steady-state value of $M$ as a function of the ratio of maximum modification rates $V_1/V_2$, which is itself a function of cyclin concentration, $C$, according to eqn (10.2a). Curves in (b) are generated with eqn (10.4) yielding the steady-state value of $X$ as a function of the ratio $V_3/V_4$, which is proportional to $M$, as indicated by eqn (10.2b). $C^*$ and $M^*$ refer to the thresholds apparent in curves $a$ of (a) and (b), respectively. The curves are established for the following parameter values (in min$^{-1}$): $V_{M1} = 3$, $V_2 = 1.5$, $V_{M3} = 1$, $V_4 = 0.5$; moreover, $K_c = 0.5\ \mu M$. In both (a) and (b), curves $a$ are obtained for $K_i = 0.005$ while curves $b$ are obtained for $K_i = 10$ ($i = 1, \ldots 4$). The actual values of the maximum rates and Michaelis constants of the converter enzymes $E_1$ and $E_2$ are obtained by multiplying $V_{M1}$, $V_2$ and $K_1$, $K_2$ by $M_T = 4\ \mu M$ (Labbé *et al.*, 1989; Minshull *et al.*, 1989); the resulting value close to $10^{-5}$ M/min for the maximum rate of cdc2 kinase matches that observed experimentally (see, e.g. Félix *et al.*, 1989); the actual values of the corresponding parameters for enzymes $E_3$ and $E_4$ in the second modification cycle are obtained by multiplying $V_{M3}$, $V_4$ and $K_3$, $K_4$ by $X_T$, for which a value of $4\ \mu M$ is taken. The value of $K_c$ is in the range of experimentally determined cyclin concentrations (Minshull *et al.*, 1989) (Goldbeter, 1991a).

concentration, which governs the ratio of maximum rates of the phosphatase ($E_1$) and kinase ($E_2$) acting on cdc2 kinase; indeed the maximum rate $V_1$ depends on cyclin concentration according to eqn (10.2a). Curve *a* in fig. 10.5a corresponds to the case of zero-order kinetics ($K_1 = K_2 = 0.005$). The sharp threshold that characterizes this curve is associated with a threshold cyclin concentration, $C^*$ (see fig. 10.5a and eqn (10.7) below). This threshold disappears when the values of $K_1$ and $K_2$ exceed unity; the curve then becomes Michaelian (curve *b*).

A similar discussion in the case of the second cycle of the mitotic cascade indicates that the fraction of active cyclin protease, $X$, abruptly increases beyond a threshold value $M^*$ of the fraction of active cdc2 kinase ($M$) to which the maximum rate $V_3$ is proportional, as shown by eqn (10.2b). Such a threshold, however, is only observed (fig. 10.5b) when the modifying enzymes – i.e. cdc2 kinase and the phosphatase that reverts its action – are saturated by their respective substrates; this occurs when $K_3$, $K_4 \ll 1$, i.e. $K_{m3}$, $K_{m4} \ll X_T$, e.g. when $K_3 = K_4 = 0.005$ (curve *a*) but not when $K_3 = K_4 = 10$ (curve *b*).

The position of the thresholds $C^*$ and $M^*$ can be determined by means of the expressions obtained in the kinetic analysis of covalent modification cycles (Goldbeter & Koshland, 1981, 1982a). Thus, for the first cycle, this analysis indicates that the threshold of the transition curve in $M$ as a function of the ratio of modification rates ($V_1/V_2$) occurs for the value given by eqn (10.5) at which $M = 0.5$:

$$\left(\frac{V_1}{V_2}\right)^* = \frac{1 + 2K_1}{1 + 2K_2} \tag{10.5}$$

Because $V_1$ is a function of $C$, the threshold value $C^*$ of cyclin concentration corresponding to the value $(V_1/V_2)^*$ can be obtained from eqns (10.2a) and (10.5), yielding successively:

$$\left(\frac{V_{M1}}{V_2}\right)\left(\frac{C^*}{K_c + C^*}\right) = \frac{1 + 2K_1}{1 + 2K_2} \tag{10.6}$$

or:

$$C^* = K_c\left(\frac{1 + 2K_1}{1 + 2K_2}\right)\bigg/\left[\left(\frac{V_{M1}}{V_2}\right) - \frac{1 + 2K_1}{1 + 2K_2}\right] \tag{10.7}$$

In the particular case considered in fig. 10.5a, $K_1 = K_2$ and $(V_{M1}/V_2) = 2$ so that $C^* = K_c = 0.5$ μM.

A similar expression for the threshold $M^*$ of the transition curve for cyclin protease X as a function of the ratio $V_3/V_4$ can be obtained from eqn (10.2b) and from eqn (10.8):

$$\left(\frac{V_3}{V_4}\right)^* = \frac{1 + 2K_3}{1 + 2K_4} \tag{10.8}$$

$$M^* = \left(\frac{1 + 2K_3}{1 + 2K_4}\right)\Big/\left(\frac{V_{M3}}{V_4}\right). \tag{10.9}$$

In the case considered in fig. 10.5b, $K_3 = K_4$ and $V_{M3} = 2V_4$, so that $M^* = 0.5$.

### *Role of thresholds and time delays in the mechanism of oscillations*

The model not only provides a plausible explanation for the origin of the threshold levels observed in the action of cyclin and cdc2 kinase during the mitotic cycle, but it also shows how these thresholds are necessarily linked to the time delays that play a primary role in the onset of oscillations (Murray & Kirschner, 1989a; Félix *et al.*, 1990).

The system governed by eqns (10.1a–c) admits a single steady state, which, for appropriate values of the parameters, can become unstable. The range of parameter values corresponding to instability can be determined by linear stability analysis performed around the steady state. While linear stability analysis leads to a third-degree polynomial equation, obtaining the steady state requires solving a ninth-degree equation, which, in the simpler case where the term related to non-specific degradation of cyclin is omitted in eqns (10.1a–c) (i.e. $k_d = 0$), reduces to a fifth-degree equation. Instability occurs as soon as at least one root of the characteristic equation acquires a positive real part. Then the system cannot reach the unstable steady state and evolves to sustained oscillations in the course of time (fig. 10.6).

The mechanism of oscillations in the mitotic cascade can best be comprehended by looking at the transition curves for $M$ and $X$ in the case where these curves admit a sharp threshold, as in fig. 10.5a and b. Resorting to these steady-state curves is justified by the fact that $M$ varies much faster than $C$, while $X$ closely follows the variation in $M$ (fig. 10.5); in first approximation, $C$ can thus be treated as a slowly varying parameter. A similar approach has been successfully used in the discussion of bursting oscillations in electrically excitable cells (Rinzel,

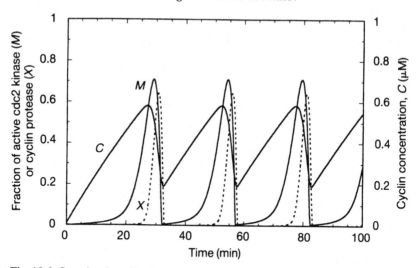

Fig. 10.6. Sustained oscillations in the minimal cascade model involving cyclin and cdc2 kinase (see fig. 10.4). The time evolution of the cyclin concentration ($C$), the fraction of active cdc2 kinase ($M$) and the fraction of active cyclin protease ($X$) is obtained by numerical integration of eqns (10.1) under the conditions of curves $a$ in fig. 10.5, where a threshold exists in the dependence of $M$ on $C$ and of $X$ on $M$, with $v_i = 0.025$ μM/min, $v_d = 0.25$ μM/min, $K_d = 0.02$ μM, and $k_d = 0.01$ min$^{-1}$. Initial conditions are $C = 0.01$ μM, $M = X = 0.01$ (Goldbeter, 1991a).

1987) and in biochemical systems such as the multiply regulated enzyme model considered in chapter 4 (see section 4.6, and Decroly & Goldbeter, 1987) or the cAMP signalling system in *Dictyostelium* (see section 6.3, and Martiel & Goldbeter, 1987b).

Starting from a low value of cyclin, as in fig. 10.6, we see that C at first accumulates at a nearly constant rate; this results from the fact that the variation of cyclin is dominated by the constant input term $v_i$, which largely exceeds the nonspecific degradation term ($-k_d C$) as well as the action of the cyclin protease X, since the latter is then predominantly in its inactive state, given that $M < M^*$ because $C < C^*$. The cyclin level continues to rise until it reaches the threshold value $C^*$ at which cdc2 kinase abruptly becomes activated, as predicted by the transition curve in curve $a$ in fig. 10.5a. Very rapidly the fraction of active cdc2 kinase, $M$, increases until it reaches the threshold value $M^*$ beyond which the second cycle of the cascade switches on, producing a steep transition in the fraction of active cyclin protease $X$ (curve $a$ in fig. 10.5b).

The abrupt activation of the cyclin protease signals the onset of the second phase of the see-saw movement: the cyclin level begins to drop precipitously as the rate of cyclin degradation by X exceeds the constant input $v_i$. As C drops below the threshold $C^*$ of the transition curve for M, cdc2 kinase becomes inactivated in the first cycle of the cascade; as a consequence, M drops below the threshold $M^*$ of the transition for X so that the cyclin protease becomes inactivated in the second cycle. Both M and X having returned to low values, the level of C may rise again as a new cycle of oscillation begins.

That sustained oscillations in the mitotic cascade correspond to the evolution toward a limit cycle in the (C, M, X) space is demonstrated in fig. 10.7. Indeed, for a given set of parameter values the system always reaches the same oscillatory regime characterized by a fixed amplitude and frequency, regardless of initial conditions.

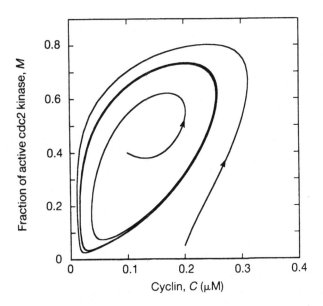

Fig. 10.7. Limit cycle behaviour of the cascade model for the mitotic oscillator. The curves are obtained by projecting the trajectory of the three-variable system governed by eqns (10.1) onto the cyclin-cdc2 kinase (C, M) plane. Two sets of initial conditions are considered, one inside and the other outside the limit cycle; arrows indicate the direction of the time evolution. Parameter values are $K_i = 0.1$ ($i = 1, \ldots 4$), $V_{M1} = 0.5$ min$^{-1}$, $V_2 = 0.167$ min$^{-1}$, $V_{M3} = 0.2$ min$^{-1}$, $V_4 = 0.1$ min$^{-1}$, $v_i = 0.023$ $\mu$M/min, $v_d = 0.1$ $\mu$M/min, $K_c = 0.3$ $\mu$M, $K_d = 0.02$ $\mu$M, $k_d = 3.33 \times 10^{-3}$ min$^{-1}$; for these values, the period of the oscillations is equal to 36 min. The initial value of X for the two trajectories shown is 0.01 (Goldbeter, 1991a).

The model thus shows how thresholds in the phosphorylation–dephosphorylation cascade controlling cdc2 kinase play a primary role in the mechanism of mitotic oscillations. The model further shows how these thresholds are necessarily associated with time delays whose role in the onset of periodic behaviour is no less important. The first delay indeed originates from the slow accumulation of cyclin up to the threshold value $C^*$ beyond which the fraction of active cdc2 kinase abruptly increases up to a value close to unity. The second delay comes from the time required for $M$ to reach the threshold $M^*$ beyond which the cyclin protease is switched on. Moreover, the transitions in $M$ and $X$ do not occur instantaneously once $C$ and $M$ reach the threshold values predicted by the steady-state curves; the time lag in each of the two modification processes contributes to the delay that separates the rise in $C$ from the increase in $M$, and the latter increase from the rise in $X$. The fact that the cyclin protease is not directly inactivated when the level of cyclin drops below $C^*$ prolongs the phase of cyclin degradation, with the consequence that M and X will both become inactivated to a further degree as $C$ drops well below $C^*$.

The above discussion explains why the period of the oscillations is largely determined by the rate of accumulation of cyclin up to the threshold beyond which it activates cdc2 kinase. This phase of accumulation, which corresponds to the interphase of the mitotic cycle in amphibian embryonic cells, is indeed longer than the mitotic phase, which corresponds to the peak in cdc2 kinase activity. Such a result holds with the observation (Murray & Kirschner, 1989b) that an increase in the rate of cyclin accumulation shortens the period of the cell cycle.

The question arises as to whether thresholds are needed in the two cycles of the cascade of fig. 10.4, or if a single threshold in either one of these cycles suffices to allow for sustained oscillatory behaviour. Although the occurrence of thresholds in the transition curves for both M and X definitely favours oscillations, stability diagrams established as a function of the various parameters of the model indicate that periodic

---

Fig. 10.8. Stability diagram established as a function of the reduced Michaelis constants $(K_1, K_2)$ of the first cycle of the minimal cascade model of fig. 10.4, versus the reduced Michaelis constants $(K_3, K_4)$ of the second cycle. The domain of oscillations corresponds to the domain of instability of the unique steady state admitted by eqns (10.1). The stability properties of the steady state are determined by linear stability analysis. The diagrams are established for (a) equal or (b) unequal values of $(K_1, K_2)$ on the one hand, and $(K_3, K_4)$ on the other. Parameter values are as in fig. 10.6 (Guilmot & Goldbeter, 1995).

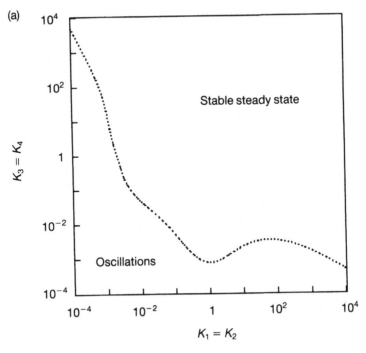

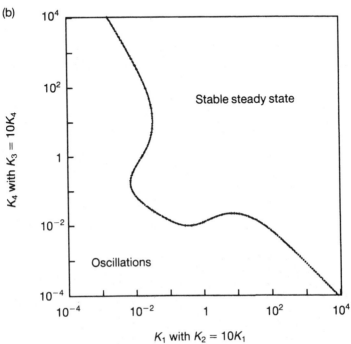

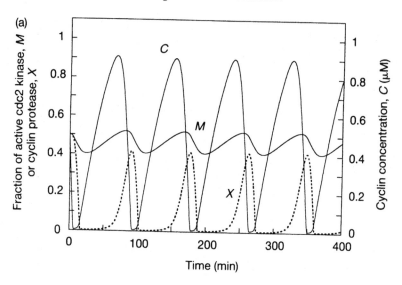

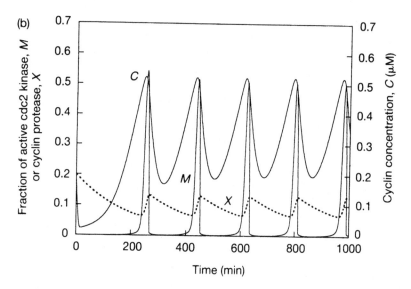

Fig. 10.9. Sustained oscillations in the minimal cascade model involving cyclin and cdc2 kinase, in conditions where a single threshold exists in (a) the second or (b) the first cycle of the cascade (Guilmot & Goldbeter, 1995). Figure 10.6, in contrast, was established for the case where a threshold occurs in both cycles. Parameter values are those of Fig. 10.8a. Initial conditions are $C$ (in $\mu M$) $= M = X = 0.5$ in (a) and 0.2 in (b). Moreover, in (a) $K_1 = K_2 = 100$ and $K_3 = K_4 = 0.001$, while in (b) $K_1 = K_2 = 0.001$ and $K_3 = K_4 = 100$.

behaviour may nevertheless occur when only one of the cycles of the cascade functions in a switch-like manner (Romond, Guilmot & Goldbeter, 1994; Guilmot & Goldbeter, 1995). This is shown by the stability diagram established in fig. 10.8 as a function of the reduced Michaelis constants $K_1$, $K_2$ versus $K_3$, $K_4$. Whereas the region where the four constants are much less than unity constitutes the core of the oscillatory domain, the latter extends a narrow tongue along the whole range of $K_1$ and $K_2$ values, from well below to well above unity, when $K_3$ and $K_4$ are sufficiently low. This result is obtained, with slight differences, both with equal (fig. 10.8a) and unequal (fig. 10.8b) pairs of ($K_1$, $K_2$) and ($K_3$, $K_4$) values.

The results of fig. 10.8 therefore indicate that a single threshold in either one of the two cycles of the minimal cascade shown in fig. 10.4 suffices to produce sustained oscillatory behaviour. If the threshold is in the $M$ vs $C$ curve, the huge variation in $M$ will produce significant changes in $X$ even if the curve showing the dependence of $X$ on $M$ is rather smooth. Conversely, if the threshold is in the $X$ vs $M$ dependence, the large-amplitude variation in $X$ will produce significant changes in $C$ and, hence, in $M$. The oscillations obtained when the threshold is only in the first or second cycle are shown in parts a and b, respectively, of fig. 10.9. What characterizes these curves is the fact that the amplitude of the oscillations is larger for $M$ and smaller for $X$ when the threshold lies in the cycle acting on M, while the reverse is true when the threshold lies in the second cycle, acting on X.

The existence of two thresholds clearly favours the occurrence of sustained oscillations of significant amplitude in both $M$ and $X$ (see fig. 10.6), even if a single threshold suffices for oscillatory behaviour. The domain of parameter values for which oscillations occur is, however, much reduced when only one cycle of the bicyclic cascade possesses a sharp threshold. It remains possible that the requirement for at least one threshold depends on the number of cycles comprised in the cascade. Extending the model to include additional phosphorylation–dephosphorylation cycles allows us to address this issue as well as the influence of self-amplification in cdc2 kinase activation.

## 10.5 Extending the cascade model for the mitotic oscillator and testing the effect of autocatalysis

Recent experimental evidence indicates that the cdc25 phosphatase that activates cdc2 kinase is itself regulated by reversible phosphorylation;

thus, cdc25 appears to be activated by phosphorylation and inactivated by dephosphorylation (Izumi, Walker & Maller, 1992; Kumagai & Dunphy, 1992). This additional cycle of covalent modification can be included in an extended cascade model (Goldbeter, 1993a). Of particular significance for the mitotic oscillatory mechanism is the fact that the kinase that activates cdc25 phosphatase might be cdc2 kinase itself (Lorca *et al.*, 1991a; Kumagai & Dunphy, 1992; Millar & Russell, 1992; Hoffmann *et al.*, 1993; Strausfeld *et al.*, 1994). Such a regulation would result in the indirect, autocatalytic activation of cdc2 kinase.

Besides the possible activation of cdc25 by cdc2, there are in principle at least two other ways by which autocatalysis in the activation of cdc2 could occur: the wee1 kinase, which inactivates cdc2, is itself inhibited by phosphorylation; indirect autocatalysis in cdc2 activation would result if cdc2 were the kinase that inactivates wee1 (Smythe & Newport, 1992). Recent studies (Parker *et al.*, 1993; Wu & Russell, 1993) have shown, however, that the wee1 kinase is inactivated through phosphorylation by nim1 kinase. This seems to rule out out the form of positive feedback that would result from a direct inactivation of wee1 kinase by cdc2 kinase itself. Alternatively, autophosphorylation of cdc2 on Thr161/167 could result in self-amplification in cdc2 kinase activity (Krek & Nigg, 1991); this possibility also seems to be excluded since no intramolecular or intermolecular phosphorylation of cdc2 on this residue has been observed (Solomon *et al.*, 1992). Phosphorylation of Thr161/167 is catalysed by a specific kinase variously known as CAK, MO15 of cdk7 (Solomon *et al.*, 1992; Fisher & Morgan, 1994; Mäkela *et al.*, 1994).

An extended cascade model incorporating the activation of the cdc25 phosphatase through reversible phosphorylation is schematized in fig. 10.10 – a further extension of the model, along similar lines, should include the reversible phosphorylation of wee1 kinase (Smythe & Newport, 1992). For simplicity and definiteness, it is assumed that cyclin promotes the phosphorylation of the inactive cdc25 phosphatase $P^+$ into the active, phosphorylated form $P$ by enhancing the activity of the kinase $E_5$, which may or may not be identical with the active form of cdc2 kinase, M. The cdc25 phosphatase is inactivated by another phosphatase, $E_6$. A recent study (Grieco, Avvedimento & Gottesman, 1994) shows that in *Xenopus* egg extracts this phosphatase can itself be activated by the cAMP-dependent protein kinase; this explains how signals acting through cAMP may control cell cycle progression by blocking the activation of MPF. The equations describing the time evolution of

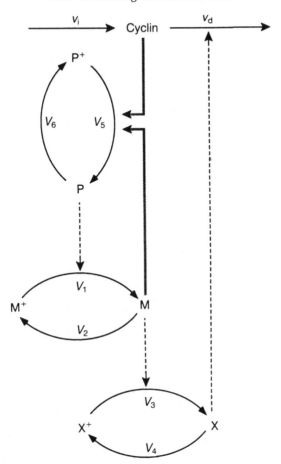

Fig. 10.10. Extended cascade model for the mitotic oscillator (Goldbeter, 1993a). The model shown in fig. 10.4 is extended by taking into account the reversible phosphorylation of the cdc25 phosphatase acting on cdc2 kinase; the inactive form $P^+$ of cdc25 phosphatase is activated through phosphorylation into the form P by a kinase $E_5$, which is probably identical with the active form of cdc2 kinase, M (Hoffmann *et al.*, 1993; Strausfeld *et al.*, 1994); also indicated is the putative activation of kinase $E_5$ by cyclin (if $E_5$ is identical with M, the activation follows from the binding of cyclin to cdc2 kinase). The active form P of cdc25 is inactivated through dephosphorylation catalysed by the phosphatase $E_6$ (see text for details). The regulation of wee1 kinase ($E_2$) through phosphorylation–dephosphorylation (Smythe & Newport, 1992) is not considered.

this tricyclic cascade model now include a kinetic equation for the evolution of the fraction of active cdc25 phosphatase, P (Goldbeter, 1993a):

$$\frac{dP}{dt} = V_5 \frac{(1-P)}{K_5 + (1-P)} - V_6 \frac{P}{K_6 + P} \tag{10.10}$$

The time evolution of $C$, $M$ and $X$ remains governed by eqns (10.1a–c), but the effective maximum rate of dephosphorylation of cdc2 kinase is now proportional to the fraction of active cdc25 phosphatase, $P$:

$$V_1 = V_{M1}P \qquad (10.11)$$

Two situations are now considered, depending on how the phosphorylation of $P^+$ to $P$ takes place. If this reaction is catalysed by a kinase $E_5$ other than cdc2 kinase and if we assume, for simplicity, that the enzyme is merely controlled by cyclin behaving as an allosteric activator, the rate of $E_5$ will be given by:

$$V_5 = \frac{C}{K_c + C}\, V_{M5} \qquad (10.12)$$

Oscillations readily occur in these conditions, as shown in fig. 10.11. These oscillations are very similar to those obtained in the bicyclic cascade model (fig. 10.6), but now the active phosphatase $P$, which closely follows the waveform of cyclin $C$, is the trigger that elicits the sharp rise

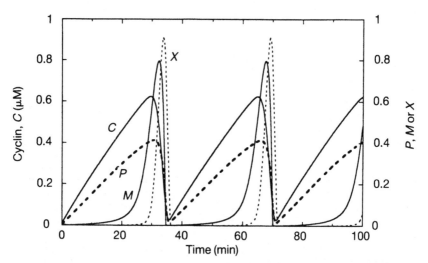

Fig. 10.11. Sustained oscillations in the extended cascade model of fig. 10.10 for the mitotic oscillator, in the absence of autocatalysis by cdc2 kinase. Shown are the fraction of active cdc25 phosphatase ($P$), the fraction of active cdc2 kinase ($M$), the cyclin concentration ($C$), and the fraction of active cyclin protease ($X$). The curves are obtained by numerical integration of eqns (10.1a–c), and (10.10)–(10.12). Parameter values are (in min$^{-1}$) $V_{M1} = 4$, $V_2 = 1.5$, $V_{M3} = 1$, $V_4 = 0.5$, $V_{M5} = 1.25$, $V_6 = 1$; moreover, $K_c = 0.2$ μM; other parameter values are as in fig. 10.6. Initial conditions are: $C = P = M = X = 0.01$ (Goldbeter, 1993a).

in $M$ and $X$, successively. The increase in $X$ leads to a decrease in $C$, followed, sequentially, by a decrease in $P$, $M$ and $X$.

What could be the role of autocatalysis in cdc2 kinase activation? The extended cascade model of fig. 10.10 allows us to incorporate such an effect by assuming that cdc2 kinase directly phosphorylates, and thereby activates, the cdc25 phosphatase. That such a situation prevails is supported by recent experimental observations (Hoffmann *et al.*, 1993; Strausfeld *et al.*, 1994). When considering that $E_5$ coincides with M, and keeping the role of cyclin as an allosteric effector of the kinase – the latter assumption agrees with the fact that cyclin B behaves as a regulatory subunit of cdc2 kinase – the kinetic equation for P remains identical with eqn (10.10) but the effective maximum rate of $E_5$ is now given by:

$$V_5 = (M + \alpha) \left( \frac{C}{K_c + C} V_{M5} \right) \tag{10.13}$$

The general form of eqn (10.13) is only one of the possible phenomenological expressions that incorporates the activation of cdc2 kinase by M as well as the possible direct activation of phosphatase $E_5$ by cyclin; the latter effect is measured by parameter $\alpha$, which has to remain smaller than unity for self-activation by M to be significant. The oscillatory behaviour of the mitotic cascade model in these conditions is represented in fig. 10.12. What distinguishes these oscillations from those observed in the absence of autocatalysis by cdc2 kinase (fig. 10.11) is the activation curve of cdc25 phosphatase: here, the increase in $P$ is clearly biphasic, reflecting the dual, parallel mode of control assumed in eqn (10.13). In the first, linear phase, $P$ accumulates at a nearly constant rate owing to the progressive rise in $C$. As soon as $M$ reaches a sufficient level, comparable to $\alpha$, a second phase is observed in which the rise in $P$ becomes autocatalytic. Furthermore, $P$ reaches a plateau at a value close to unity, much as $M$ and $X$.

Thus, oscillations become much more abrupt in the presence of autocatalysis by cdc2 kinase, but it appears that such an autocatalysis is not required for sustained oscillations in the mitotic cascade. In the presence of self-activation by cdc2 kinase, however, oscillations can occur in the absence of zero-order ultrasensitivity in the cdc2 kinase phosphorylation cycle. Thresholds due to zero-order ultrasensitivity in any cycle of the cascade nevertheless continue to favour oscillatory behaviour. The threshold due to self-amplification in cdc2 kinase activation (Hoffmann *et al.*, 1993) may, however, substitute for the threshold due to zero-order ultrasensitivity in the first cycle of the cascade. This is what might

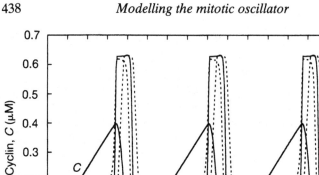

Fig. 10.12. Sustained oscillations in the extended cascade model of fig. 10.10 for the mitotic oscillator, in the presence of autocatalysis by cdc2 kinase. Shown are the fraction of active cdc25 phosphatase ($P$), the fraction of active cdc2 kinase ($M$), the cyclin concentration ($C$), and the fraction of active cyclin protease ($X$). The curves are obtained by numerical integration of eqns (10.1a–c), (10.10), (10.11) and (10.13). Initial conditions and parameter values are as in fig. 10.11 with $\alpha = 0.25$ and $V_{M5} = 5$ min$^{-1}$ (Goldbeter, 1993a).

happen if the formation of a complex between cdc2 and cyclin B was explicitly considered in the model. Oscillations would then originate from the coupling between autocatalysis by cdc2 and the negative feedback resulting from cdc2-induced cyclin degradation. For the essential part, the preceding results and the discussion in the following section remain valid, regardless of the origin of the threshold in cdc2 activation.

## 10.6 Arresting the mitotic oscillator

Any model for the biochemical oscillator controlling the onset of mitosis should account for the fact that, in some cells or at certain stages of development, mitosis can be prevented and cells cease to divide. In dynamic terms, such a transient or permanent suppression of mitotic activity can be viewed as a switch of the mitotic control system from an oscillatory into a nonoscillatory regime corresponding to a stable steady state. Models for the mitotic oscillator allow the discussion of possible mechanisms whereby the mitotic clock might be arrested.

How can oscillations be suppressed in the minimal cascade model for the mitotic oscillator involving cyclin and cdc2 kinase? In principle, any parameter of the model possesses a range in which the steady state is unstable and oscillations occur. One may, however, classify the parameters within two categories, according to whether they relate directly to cyclin synthesis or degradation, or to the kinetics of the converter enzymes acting on cdc2 kinase and cyclin protease.

Parameters of the first kind are the rate of cyclin synthesis ($v_i$), the maximum rate of cyclin degradation by the specific protease X ($v_d$), the Michaelis constant of the latter enzyme ($K_d$), and the apparent first-order rate constant for nonspecific cyclin degradation ($k_d$). Clearly the nonspecific degradation of cyclin should remain negligible, since (at least in the absence of positive feedback) no oscillations would occur if such a process predominated over the specific action of protease X. This is confirmed by the construction of stability diagrams showing that the domain of oscillations in parameter space shrinks as the value of $k_d$ increases, until a value of that rate constant is reached beyond which oscillations disappear (Romond *et al.*, 1994).

The analysis of the model indicates that what is important for oscillations is an appropriate balance between cyclin synthesis and degradation (J.M. Guilmot & A. Goldbeter, unpublished results): cyclin has to accumulate up to the critical level $C^*$, of the order of $K_c$, in order to activate the first cycle of the cascade, but has to decrease below that threshold once the protease is turned on. A significant perturbation in the balance between cyclin synthesis and degradation may thus lead to the suppression of oscillatory behaviour: at given values of the cyclin specific and nonspecific degradation rates, oscillations occur only in a range bounded by two critical values of the rate of cyclin synthesis, $v_i$, as is well shown by the stability diagram established as a function of $v_i$ and $v_d$ (fig. 10.13). This diagram also indicates that, at a given value of $v_i$, oscillations occur in a range bounded by two critical values of $v_d$.

An alternative way to halt the mitotic oscillator relies on changes in the parameters that govern the kinetics of the phosphorylation–dephosphorylation cycles in the cascade. Of particular importance for oscillations are the thresholds in the dependence of $M$ on $C$ (curve *a*, fig. 10.5a) and of $X$ on $M$ (curve *a*, fig. 10.5b). An effective way to stop the oscillations is to suppress the two thresholds altogether by increasing $K_1$, $K_2$, $K_3$, $K_4$ above unity. As is clear from fig. 10.8, switching the values of $K_i$ ($i = 1, \ldots 4$) from $5 \times 10^{-3}$ – as in the case considered in fig. 10.6 – to 10 will indeed suppress the oscillations.

Another way to stop the oscillations is to prevent the system from

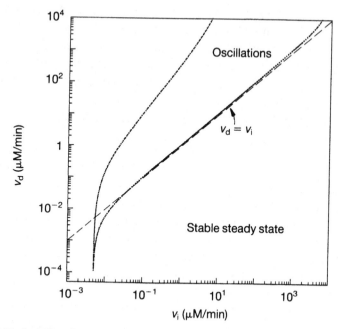

Fig. 10.13. Stability diagram established as a function of the rates of cyclin synthesis, $v_i$, and degradation, $v_d$, in the minimal cascade model of fig. 10.4. The domain of oscillations is determined as in fig. 10.8; parameter values are as in fig. 10.6. A narrow region of hard excitation (not shown) in which a stable limit cycle coexists with a stable steady state is observed just above part of the upper boundary of the instability domain (J.M. Guilmot & A. Goldbeter, unpublished results).

passing across the thresholds. For oscillations to occur, the system must indeed be capable of going back and forth through the thresholds. Therefore, the ratios $(V_1/V_2)$ and $(V_3/V_4)$ must be greater than their threshold values given by eqns (10.5) and (10.8), but not too large. This is confirmed by the stability diagram of fig. 10.14 established as a function of $V_{M1}$ vs $V_2$, under conditions where a threshold exists in the two cycles of the minimal cascade. As $V_1$ depends on $C$ according to eqn (10.2a), we see that $(V_1/V_2)$ tends to the ratio $(V_{M1}/V_2)$ as $C$ increases above $K_c$. In the conditions of fig. 10.14 where $K_1 = K_2$, the threshold is in $(V_1/V_2) = 1$. Accordingly, the domain of instability extends just above the line corresponding to the value $(V_{M1}/V_2) = 1$. The steady state recovers its stability, however, when the ratio $(V_{M1}/V_2)$ becomes larger than a critical value. Figure 10.14 thus indicates that there are four different ways to quit the oscillatory domain when we tinker only with the maxi-

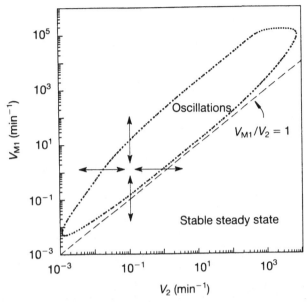

Fig. 10.14. Stability diagram established as a function of the maximum rates of cdc25 phosphatase ($V_{M1}$) and wee1 kinase ($V_2$), in the minimal cascade model of Fig. 10.4. The domain of oscillations is determined as in fig. 10.8; parameter values are as in fig. 10.6. In these conditions, a sharp threshold exists both in the first and second cycles of the cascade (see curves *a* in fig. 10.5a and b). The dashed line indicates the locus of the threshold value of ($V_{M1}/V_2$), which is equal here to unity. Arrows denote ways to quit (or enter) the oscillatory domain by manipulating the rates $V_{M1}$ or $V_2$ (Goldbeter & Guilmot, 1995).

mum rates of enzymes $E_1$ and $E_2$. As indicated by the arrows in the stability diagram of fig. 10.14, we may thus arrest the mitotic oscillator by increasing or decreasing the maximum rate of the cdc25 phosphatase, $V_{M1}$, or that of wee1 kinase, $V_2$ (Goldbeter & Guilmot, 1995).

To illustrate the important notion that there exists a window of the ratio ($V_1/V_2$) producing oscillations, let us divide $V_{M1}$ by 2 in the case of fig. 10.6. We see from eqns (10.5)–(10.6) that when the maximum rate of enzyme $E_1$ in the minimal cascade model of fig. 10.4 is so halved, the threshold ($V_1/V_2$)* in the first cycle cannot be reached at any finite value of the cyclin concentration. As a consequence, $M$ will never increase abruptly and will reach a low, steady state value; the protease X will also remain at a low level because cdc2 kinase is not sufficiently activated ($M$ remains below the threshold $M$*), and cyclin will increase up to a high level determined by the rate of its nonspecific degradation

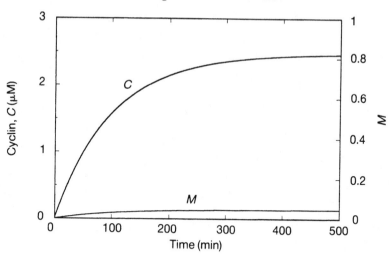

Fig. 10.15. Arrest of the mitotic oscillator through inhibiting the cdc25 phosphatase that activates cdc2 kinase. The evolution of the fraction of active cdc2 kinase ($M$) in the minimal cascade model of fig. 10.4 is shown, together with the cyclin concentration ($C$) under the conditions of fig. 10.6, with $K_1 = K_2 = K_3 = K_4 = 0.01$, and $V_{M1} = 1.5$ instead of 3 min⁻¹. This halving of the maximum rate of cdc25 ($E_1$) results in the arrest of the mitotic oscillator in a state characterized by a high level of cyclin and low levels of active cdc2 kinase and cyclin protease (the value of the fraction of active cyclin protease $X$, not shown, remains lower than 1%) (Goldbeter, 1993a).

(fig. 10.15). If, in contrast, the rate of $E_1$ was increased or the rate of $E_2$ decreased so that the ratio $(V_{M1}/V_2)$ goes above the domain of oscillations, the system would reach a stable steady state characterized by large levels of active cdc2 kinase and cyclin protease, and a small level of cyclin.

Preventing the system from passing the second threshold $M^*$ characterizing the dependence of the cyclin protease on cdc2 kinase is as effective in arresting the mitotic oscillator. This is confirmed by the stability diagram established as a function of $V_{M3}$ vs $V_4$ (fig. 10.16), which has the same appearance as the corresponding diagram established for the first cycle of the cascade in fig. 10.14 (Goldbeter & Guilmot, 1995). Again there are four ways to quit the oscillatory domain by manipulating the ratio $(V_3/V_4)$. Shown in fig. 10.17 is the result of dividing by 2 the maximum rate of cdc2 kinase, i.e. halving the value of $V_{M3}$ considered in fig. 10.6, in the minimal model governed by eqns (10.1a–c). The increase in $C$ can still bring $M$ beyond the threshold of the $M$ vs $C$ curve (curve $a$, fig. 10.5a) but, as predicted from eqn (10.9), the increase in $M$ is not

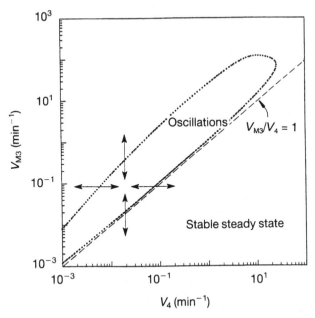

Fig. 10.16. Stability diagram established as a function of the maximum rates of cdc2 kinase ($V_{M3}$) and of the phosphatase that inactivates cyclin protease ($V_4$), in the minimal cascade model of fig. 10.4. The domain of oscillations is determined as in fig. 10.8; parameter values are as in fig. 10.6. As for fig. 10.14, a sharp threshold exists both in the first and second cycles of the cascade. The dashed line indicates the locus of the threshold value of ($V_{M3}/V_4$), which is equal here to unity. Arrows denote ways to quit (or enter) the oscillatory domain by manipulating the rates $V_{M3}$ or $V_4$ (Goldbeter & Guilmot, 1995).

sufficient to bring the protease beyond the threshold of the $X$ vs $M$ curve (curve $a$, fig. 10.5b). As a result, the system evolves toward a stable steady state characterized by high levels of cyclin and active cdc2 kinase, and a low level of active cyclin protease. The latter stable steady state can be related to the cytostatic factor (CSF)-mediated metaphase arrest of unfertilized vertebrate eggs (Murray, 1989b; Murray *et al.*, 1989; Murray & Hunt, 1993), which is also characterized by high levels of both cyclin and active MPF. A similar association of such a stable steady state with metaphase arrest was made in other models for the mitotic oscillator (Hyver & Le Guyader, 1990; Norel & Agur, 1991; Tyson, 1991; Novak & Tyson, 1993b).

Exit from meiosis and from metaphase arrest is known to result from an increase in cytosolic $Ca^{2+}$. The cytostatic factor appears to be associated with the activity of p39$^{mos}$ (also known as Mos), the product of the

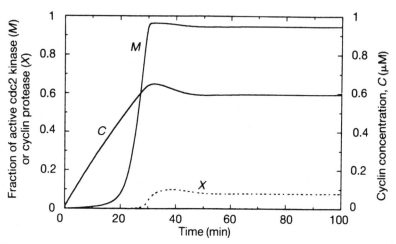

Fig. 10.17. Arrest of the mitotic oscillator through inhibiting the activation of cyclin protease (Goldbeter, 1993a). The evolution of the fractions of active cdc2 kinase ($M$) and cyclin protease ($X$) in the minimal cascade model of fig. 10.4 are shown, together with the cyclin concentration ($C$) under the conditions of fig. 10.6, with $V_{M3} = 0.5$ instead of 1 min$^{-1}$. This halving of the maximum rate of cdc2 kinase ($E_3$) results in the arrest of the mitotic oscillator in a state characterized by high levels of active cdc2 kinase and cyclin, similar to the state of metaphase arrest in vertebrate eggs (see Murray, 1989b; Murray & Kirschner, 1989b; Murray *et al.*, 1989).

proto-oncogene c-*mos* (Sagata *et al.*, 1989); it was suggested that a phosphorylation event catalysed by p39$^{mos}$ was preventing cyclin degradation and the exit from meiotic phase (more recent data link c-*mos* to oocyte maturation; see Hunt, 1992; Yew, Mellini & Vande Woude, 1992). Evidence was previously presented in favour of a role for the $Ca^{2+}$-activated protease calpain in inducing the proteolysis of CSF (Watanabe *et al.*, 1989). The degradation of the proto-oncogene product p39$^{mos}$ is not necessary for cyclin proteolysis and exit from meiotic phase: instead, the increase of cytosolic $Ca^{2+}$ upon fertilization would trigger a phosphorylation event involving calmodulin (Lorca *et al.*, 1991a, 1993). Such a phosphorylation would release the cyclin degradation pathway from inhibition in metaphase-arrested eggs. In terms of the cascade shown in fig. 10.4, the effect of the $Ca^{2+}$ signal would be analogous to increasing the rate of an enzyme activating the cyclin protease. Thus, the passage from the metaphase-arrested stage (fig. 10.17) to oscillatory behaviour (fig. 10.6) occurs in the model upon doubling the maximum rate of cdc2 kinase, which activates the cyclin protease through phosphorylation; a similar effect is obtained when inhibiting the phosphatase $E_4$, which reverses this activation step. Recent observations indicate (Kosako, Gotoh & Nishida, 1994) that the metaphase arrest induced by Mos is mediated by the mitogen-activated protein kinase kinase, designated as MAPKK; this links cell cycle control to the mitogen-activated protein (MAP) kinase, which is involved in a major signal transduction pathway. MAP kinase is also involved, with Mos, in a spindle assembly checkpoint that prevents premature MPF inactivation through cyclin B degradation in *Xenopus* egg extracts (Minshull *et al.*, 1994).

The above discussion has shown how the mitotic oscillator can be arrested in a variety of ways, by inhibiting the activity of any of the kinases and phosphatases involved in the cascade controlling the activation of cdc2 at the $G_2/M$ transition. Recent observations indicate that cell cycling in yeast and somatic cells is regulated in such a way, by means of inhibitors of cyclin-dependent kinases at the $G_1/S$ transition (Gu *et al.*, 1993; Xiong *et al.*, 1993; Hengst *et al.*, 1994; Moreno & Nurse, 1994; Peter & Herskowitz, 1994; Schwob *et al.*, 1994; Morgan, 1995). Terminal cell cycle arrest and differentiation have been shown to be governed by changes in the levels of such protein inhibitors (Halevy *et al.*, 1995). Furthermore, some growth factors, such as interleukin-2 in T lymphocytes (Nourse *et al.*, 1994), promote the exit of cells from the quiescent state, and induce cycling, by causing the elimination of a cyclin-dependent kinase inhibitor.

The actual cascade of phosphorylation–dephosphorylation cycles controlling the onset of mitosis is much more complicated than the simple model represented in fig. 10.4. Therefore, the number of enzymes that may stop or trigger cell cycling is much greater than is suggested by the results obtained in the minimal model. This model, however, illustrates well the different ways in which modulating the activity of the converter enzymes may stop the mitotic oscillator. Multiplying the opportunities of controlling the arrest – or the induction – of cell division may be one of the main reasons for the existence of the large number of cycles that form the cascade timing the onset of mitosis.

## 10.7 Further extensions and concluding remarks

### *The mitotic oscillator: a role for positive and negative feedback*

The onset of mitosis in eukaryotic cells is controlled by a phosphorylation–dephosphorylation cascade behaving as a continuous biochemical oscillator. The outcome of the cascade is the periodic activation of cdc2 kinase. Activation of cdc2 kinase is brought about by the progressive increase in cyclin and occurs primarily through dephosphorylation by the cdc25 phosphatase.

A number of models for the mitotic oscillator are based on the positive feedback exerted by cdc2 kinase on its own activation (Hyver & Le Guyader, 1990; Norel & Agur, 1991; Tyson, 1991; Novak & Tyson, 1993a). Alternatively, the fact that cyclin degradation is triggered by cdc2 kinase suggested (Hunt, 1989; Murray & Kirschner, 1989a; Félix *et al.*, 1990) that a cell cycle oscillator might be based on the resulting negative feedback loop. The analysis of the minimal cascade model involving cyclin and cdc2 kinase presented above shows that sustained oscillations may indeed originate from such a negative feedback, provided that thresholds exist in the dependence of cdc2 kinase activation on cyclin, and in the activation of cyclin protease by cdc2 kinase. Such thresholds arise naturally as a result of zero-order ultrasensitivity (Goldbeter & Koshland, 1981, 1982a), a switch-like property that is inherent to all phosphorylation–dephosphorylation cycles. The latter manifest this specific property when the converter enzymes operate in or near the zero-order kinetic domain.

Time delays also play a significant role in the onset of oscillations in the cascade. Such time delays result here in a natural way from the thresholds in the activation curves of cdc2 kinase and cyclin protease, and are by no means inserted in an *ad hoc* manner into the kinetic equations (10.1), as is often done for systems governed by time-delay differential equations.

The minimal cascade model for the mitotic oscillator shown in fig. 10.4 belongs to the class of biochemical models in which oscillations are based on negative feedback; such models have been studied mostly for the cooperative end-product inhibition of the first enzyme in a metabolic pathway (Walter, 1970) or for control by repression at the genetic level (Goodwin, 1965). To this class belongs the model analysed in chapter 11 for circadian oscillations in the *Drosophila period* protein (PER). The majority of models for metabolic oscillations are nevertheless based on positive feedback. Thus, such a regulatory mechanism is involved in the best-known examples of biochemical oscillatory behaviour, including glycolytic oscillations in yeast and muscle (chapter 2), the periodic generation of cyclic AMP signals in the slime mould *Dictyostelium discoideum* (chapter 5), and $Ca^{2+}$ oscillations in cells stimulated by hormones or neurotransmitters (chapter 9).

Although the cascade model shows that the mitotic oscillator might be based solely on negative feedback associated with thresholds and resulting time delays, positive feedback could also underlie or contribute to the periodic activation of cdc2 kinase. Thus, other models for the mitotic oscillator are based on the self-activation of cdc2 kinase (Hyver & Le Guyader, 1990; Norel & Agur, 1991; Tyson, 1991; Novak & Tyson, 1993a,b). Such an autocatalysis most probably arises in an indirect manner through the activation of the cdc25 phosphatase by cdc2 kinase (Lorca *et al.*, 1991a; Kumagai & Dunphy, 1992; Millar & Russell, 1992; Hoffmann *et al.*, 1993; Strausfeld *et al.*, 1994). The results described in section 10.5 show how the indirect self-activation of cdc2 kinase can be naturally incorporated into the model for the phosphorylation–dephosphorylation cascade controlling the onset of mitosis in amphibian embryonic cells (fig. 10.10). Numerical simulations then indicate that positive feedback renders the oscillations more abrupt and confirm the results obtained for the minimal cascade model in showing that autocatalysis is not required for sustained oscillatory behaviour. In fact, the mitotic oscillator might well rely on two, parallel, overlapping mechanisms capable of producing the periodic activation of cdc2 kinase; such a dual feedback design might represent a safeguard ensuring that oscillations occur over a wide range of physiological conditions.

Another possibility is that, in the absence of autocatalysis, no sharp threshold for activation of cdc2 kinase exists in the first cycle of the cascade, in which the wee1 kinase and the cdc25 phosphatase act on cdc2. Self-amplification of cdc2 kinase through activation of the cdc25 phosphatase (Hoffmann *et al.*, 1993) would then replace zero-order ultrasensitivity as a source for the threshold that favours the occurrence of sustained oscillations. The mitotic oscillator would thus rely on the conjunction of positive and negative feedback, the latter arising from cdc2-induced inactivation of cdc2, through the cyclin degradation triggered by cdc2 kinase.

That the mechanism of the mitotic oscillator may rest on both positive and negative feedback has also been pointed out in a recent, detailed theoretical study of the periodic activation of cdc2 kinase in embryonic cells and extracts (Novak & Tyson, 1993b). According to Novak & Tyson, however, the mechanism of cdc2 kinase oscillations would rely on positive feedback in oocyte extracts whereas in intact embryos the mechanism would rest primarily on the negative feedback loop involving cdc2-induced cyclin degradation. In yeast, according to Tyson (1991), the cell cycle would function more like an excitable system than a true oscillator, as in embryonic cells; cell growth above a threshold would kick the cyclin–cdc2 kinase cascade into an excitable loop that would induce mitosis, while the latter would return the mitotic control system to the stable, excitable state. Global oscillatory dynamics can, however, be recovered when the variations in the biochemical parameters which change during growth are incorporated into a more exhaustive model (Novak & Tyson, 1993a).

### Extension of the cascade favours oscillations

Incorporation of the activation of cdc25 through phosphorylation illustrates how the minimal cascade model for the mitotic oscillator can be extended. The modular structure of this model is such that additional cycles of phosphorylation–dephosphorylation may readily be incorporated into the theoretical description if and when such cycles are identified in the experiments. Each additional cycle will increase by one the number of variables considered in the model. Thus, a necessary extension will consist in incorporating the recently determined control of wee1 kinase through reversible phosphorylation (Smythe & Newport, 1992; Tang, Coleman & Dunphy, 1993). Besides introducing new potential sites for the control of the cascade – and thus new ways of arresting the mitotic oscillator – one effect of the additional phosphorylation–dephosphorylation cycles is to relax in a certain measure the requirements for sharp thresholds in the dependence of cdc2 kinase on cyclin and of cyclin protease on cdc2 kinase. Even in the absence of autocatalysis by cdc2 kinase and of a zero-order ultrasensitivity threshold in the cycle acting on cdc25, the oscillatory domain in the $K_1 = K_2$ vs $K_3 = K_4$ parameter space is indeed larger in the three-cycle extended model studied in section 10.5, compared to that obtained (see fig. 10.8a) for the minimal two-cycle cascade model (Romond *et al.*, 1994). This is analogous to the result obtained in studies of metabolic oscillations in chains of enzyme reactions controlled by end-product inhibition, where the degree of cooperativity in feedback inhibition required for sustained oscillations decreases as the number of reaction steps in the chain increases (Walter, 1970; Hunding, 1974; Tyson & Othmer, 1978).

### The somatic cell cycle as a double, cdc2–cdk2 oscillator

The cell cycle in yeast and in somatic cells appears to be more complex than in embryonic cells as it is subjected to additional controls linking, for example, the onset of mitosis to the successful completion of DNA replication or to the reaching of a critical cellular size (Cross *et al.*, 1989;

Hartwell & Weinert, 1989; Murray & Kirschner, 1989a; Murray, 1992; Norbury & Nurse, 1992; Murray & Hunt, 1993; Nasmyth, 1993; King *et al.*, 1994; Nurse, 1994). It appears, however, that the periodic activation of cdc2 kinase remains at the core of the mechanism controlling the onset of mitosis in these cells (Cross *et al.*, 1989; Nurse, 1990; Draetta, 1990; Norbury & Nurse, 1992). A second oscillator involving $G_1$-specific cyclins and cdc2 kinase, or cdk2 kinase in higher eukaryotes (Dulic, Lees & Reed, 1992; Elledge *et al.*, 1992), could control, on a similar basis, the periodic onset of DNA replication (Murray & Kirschner, 1989b). In support of this conjecture, evidence for oscillations in cdk2 independent of mitotic cyclins and of cdc2 kinase has been presented (Gabrielli *et al.*, 1992). The regulation of cdk2 bears much resemblance to that of cdc2; moreover, the self-amplification arising from the activation of cdc25 by cdc2 (Hoffmann *et al.*, 1993) is also observed for cdk2 (Hoffmann *et al.*, 1994). Differences in the mechanism of cdc2 and cdk2 oscillations stem, however, not simply from the nature of the cyclins involved in activating the kinases but also from the fact that cyclin-dependent kinase inhibitors appear to control the $G_1/S$ transition (Xiong *et al.*, 1993; Hengst *et al.*, 1994; Moreno & Nurse, 1994; Peter & Herskowitz, 1994; Schwob *et al.*, 1994). A higher level of complexity in the oscillatory mechanism may also arise in somatic cell cycles, owing to the existence of a positive feedback exerted by cdc2 kinase and $G_1$-cyclins on the synthesis of the latter proteins (Cross & Tinkelenberg, 1991; Dirick & Nasmyth, 1991).

A key problem in the yeast and somatic cell cycles is the ordering of the M and S phases (Nurse, 1994), since mitosis should not occur before DNA replication is completed, while the latter process should not start before cell division, preceded by chromosome segregation, has successfully ended. The proper ordering of the M and S phases is the subject of current investigations (Amon *et al.*, 1994; Hayles *et al.*, 1994; Nurse, 1994), which point to the existence of mutual inhibitory interactions between the reactions controlling the $G_1/S$ and $G_2/M$ transitions. The role of these interactions is to prevent the operation of the cyclin-dependent kinase active in one transition when the other transition is in progress.

From a dynamic point of view, we may thus see the cell cycle as driven by two self-sustained, limit cycle oscillators, one involving cdc2, and the other cdk2. The mechanism of oscillations in each case would be similar to the one described above, and would involve a cyclin and its associated cyclin-dependent kinase (cdk) and cyclin protease, as well as

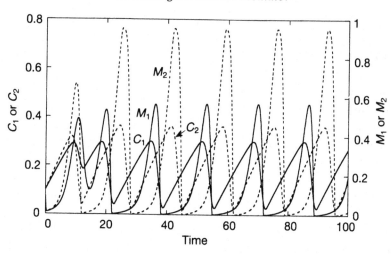

Fig. 10.18. Oscillations in a double cdc2–cdk2 oscillator model. The model assumes the existence of two oscillators active in the somatic cell cycle, each of which is governed by kinetic equations similar to eqns (10.1a–c). The three variables for the second oscillator (which is the one studied so far in this chapter) are a $G_2$ cyclin, i.e. cyclin B ($C_2$), cdc2 kinase ($M_2$), and a cyclin protease ($X_2$) specific for cyclin B; the corresponding variables for the first oscillator are a $G_1$ cyclin ($C_1$), the kinase cdk2 ($M_1$), and a cyclin protease ($X_1$) specific for the $G_1$ cyclin. Mutual inhibition of the two oscillators is assumed to occur through direct activation of $X_1$ by $M_2$ and of $X_2$ by $M_1$: the degradation of each of the two cyclins is thus activated by the other cyclin-dependent kinase. Similar results would be obtained if the synthesis of $C_1$ were inhibited by $C_2$ or $M_2$, and the synthesis of $C_2$ was inhibited by $C_1$ or $M_1$. Referring to the parameters in eqns (10.1a–c) that govern cdc2 and cdk2 oscillations, parameter values for the first and second oscillator (values for the latter are given in parentheses) are: $K_c$ = 0.5 (0.5) µM, $K_d$ = 0.02 (0.04) µM, $v_i$ = 0.025 (0.03) µM/min, $v_d$ = 0.25 (0.25) µM/min, $k_d$ = 0.01 (0.01) min$^{-1}$, $K_1 = K_2 = K_3 = K_4$ = 0.01 (0.01), and (in min$^{-1}$) $V_{M1}$ = 3 (3), $V_2$ = 1 (1), $V_{M3}$ = 2 (3), $V_4$ = 0.5 (1). Moreover, to include mutual inhibition of the two oscillators, $v_d$ is multiplied, for the first oscillator, by the term $M_2/(K_{i2} + M_2)$, and for the second oscillator by the term $M_1/(K_{i1} + M_1)$, with $K_{i1}$ = 0.1 and $K_{i2}$ = 0.11. Initial conditions are $C_1 = C_2 = 0.1$ µM, $M_1 = M_2 = 0.1$, $X_1 = X_2 = 0.01$. For the sake of clarity, the variations of $X_1$ and $X_2$ are not shown on the graph. The simulation aims only at yielding a qualitative picture of the interaction of the two oscillators; the time scale chosen (in minutes) is much shorter than the duration of the somatic cell cycle (A. Goldbeter, unpublished results).

the specific cdc25 phosphatase and wee1 kinase acting on the cdk. The proper timing of the peaks in each of the two periodically activated kinases would result from the inhibition exerted by each of them on the activation of the other. Such a minimal, double oscillator model contains six variables (three for each oscillator) and would function with the couples (cyclin 1 ≡ G$_2$ cyclin, cdc2 ≡ cdk1) and (cyclin 2 ≡ G$_1$ cyclin, cdk2). Moreover, the synthesis of cyclin 1 is inhibited by cdk2 whereas synthesis of cyclin 2 is inhibited by cdk1; alternatively, the degradation of cyclin 1 is activated by cdk2, whereas cyclin 2 degradation is activated by cdc2. Numerical simulations of this model show (fig. 10.18) that the two oscillators, characterized by different parameter values, synchronize at a common period; peaks of cdc2 and cdk2 follow at regular intervals, in antiphase – cdc2 reaching its maximum when cdk2 is largely inactive, and vice versa – as observed in the experiments. In the absence of additional regulations, the two oscillators are coupled but largely independent, as one could function periodically when the other is at a stop. Under physiological conditions, the existence of checkpoints ensures that such a situation does not occur and slows down the cell cycle clock.

### *Growth-factor control of the cell cycle oscillator(s)*

While some cytokines control cell cycle arrest (Kimchi, 1992), the mirror image situation, i.e. the passage from the quiescent, G$_0$ phase to the entry into the cell cycle at phase G$_1$, often requires the presence of growth factors (Pardee, 1989). Unrestrained cell proliferation may occur, however, in cells that begin to secrete a growth factor for which they possess a specific membrane receptor; such autocrine stimulation has been associated with the development of several types of malignant transformation (Sporn & Todaro, 1980). One way by which cell proliferation might be controlled by the presence of growth factors is via the synthesis of cyclins. Thus, the control exerted by growth factors on the cell cycle could be introduced in the double oscillator model mentioned above or, for simplicity, in the three-variable single oscillator version, by making the synthesis of cyclin depend on the presence of a growth factor.

When the prolonged incubation of the growth factor with its receptor does not cause desensitization, the rate of cyclin synthesis in eqn (10.1a) simply becomes proportional to the saturation function of the receptor by the growth factor, i.e.:

$$v_i = v_{i0} + v_{Mi}\left(\frac{F}{1+F}\right) \tag{10.14}$$

where $F$ denotes the concentration of growth factor normalized by division through its dissociation constant for the receptor; $v_{i0}$ and $v_{Mi}$ are, respectively, the basal rate of cyclin synthesis in the absence of growth factor and the maximum increase in $v_i$ above the basal level, obtained at saturating levels of F. The analysis of eqns (10.1a–c) showed that no oscillations will take place as long as $v_i$ is smaller than a critical value (see fig. 10.13). Now, however, the onset of sustained periodic behaviour of the mitotic cascade becomes subservient to the continued presence of adequate amounts of growth factor. The model shows how the presence of a sharp threshold in the activation of cdc2 kinase by cyclin can result in the requirement for a critical duration of stimulation by the growth factor for mitosis to ensue. Indeed, starting with a low initial value of cyclin, C has to accumulate up to the threshold $C^*$ for a peak in cdc2 kinase to occur. Because the rate of cyclin synthesis $v_i$ is now a function of the growth factor presence according to eqn (10.14), mitosis will not occur if the growth factor is withdrawn before C has accumulated up to $C^*$. On the other hand, if the growth factor is maintained until C exceeds $C^*$ and removed thereafter, a single round of cell division will occur. If the growth factor is permanently present so that $v_i$ (but not the basal rate $v_{i0}$) remains in the oscillatory range, sustained oscillations in cdc2 kinase will be observed.

If the growth factor induces desensitization or down regulation of its receptor, the results obtained in chapter 5 for the cAMP signalling system in *Dictyostelium* indicate that periodic behaviour will disappear if the growth factor is maintained at a constant level. However, successive rounds of cell division can be elicited when the growth factor is given in a pulsatile manner. Such a result may be related to the observation (Brewitt & Clarke, 1988) that growth and differentiation of the lens, an epithelial tissue, occurs in vitro only when platelet-derived growth factor (PDGF) is given in pulses, every 4 h, whereas continuous application of this growth factor remains ineffective.

The response to pulsatile stimulation of the mitotic oscillator by a growth factor may in fact be complex. The dynamic behaviour of the cascade indeed results from the relative weights of a large number of parameters such as the maximum rates of cyclin synthesis and degradation, the rates of receptor desensitization and resensitization (or down regulation and return to the membrane), and the interval between successive pulses. If the frequency of stimulation and the rate of cyclin synthesis are sufficiently high, cdc2 kinase may remain quasi-permanently in its active state; in its effects on the cell cycle, this situation would be equivalent to that of a stable steady state characterized by high $M$ and

$X$ values (see fig. 10.17). However, irregular oscillations of cdc2 kinase have been observed in the model, at least in the transient phase. The possibility of chaotic dynamics resulting from the periodic stimulation of the mitotic oscillator by pulses of growth factor remains very hypothetical. The occurrence of chaos in relation to the cell cycle has been discussed by Mackey (1985), and by Lloyd, Lloyd & Olsen (1992), who used an abstract model of the mitotic oscillator subjected to forcing by a sinusoidal input with much shorter period.

### *Minimal, core models versus extended models*

Most models proposed for the mitotic oscillator aim at reducing, by means of suitable kinetic assumptions, the dynamics of the mitotic cascade to a two- or three-variable core involving cyclin and cdc2 kinase. These models were distinguished according to the form of the kinetic equations and to the type of feedback considered. Early models based on positive feedback were described by simple polynomial kinetics; in these models, autocatalysis was represented as a direct activation of cdc2 kinase in the form of a nonlinear (e.g. quadratic) term in the evolution equation for this variable. In the extended cascade model presented in section 10.5, autocatalysis – which is not required for the occurrence of oscillations – was represented in a more appropriate manner, through the activation of the cdc25 phosphatase through phosphorylation by cdc2 kinase (Hoffmann *et al.*, 1993; Strausfeld *et al.*, 1994). While the converter enzymes in this and other cycles of the cascade are described by more realistic, Michaelian, kinetic laws, the role of cyclin in driving the cascade (Murray & Kirschner, 1989b) remains to be represented in a more faithful manner. Thus, to be taken into account explicitly in this extended model or in its minimal version is the formation of a complex between cyclin and cdc2 kinase. Future extensions of this cascade model should also take into account the other dephosphorylation (on Thr14) and phosphorylation (on Thr161/167) involved in the activation of cdc2 kinase (see above). Only the dephosphorylation of Tyr15 by cdc25 phosphatase, which represents the main step in the activation process, has been considered so far in this model. The phosphorylation of Thr161/167 has been incorporated into a model based on positive feedback (Novak & Tyson, 1993a,b, 1995).

Contrasting with the studies that focus on the core of the oscillatory mechanism, alternative, more detailed, studies aim at incorporating all known steps of the reaction mechanism and therefore consider a much larger number of biochemical variables. Such an approach, which is currently being pursued for the *Drosophila* (G. Odell, unpublished results; see Fig. 4.22 in Murray & Hunt, 1993) and *Xenopus* (Novak & Tyson, 1993b) embryonic cell cycles, as well as for yeast (Novak & Tyson, 1993a, 1995), takes more faithfully into account the full complexity of biochemical observations but often lacks the advantages provided by a well-defined core mechanism. As new features of the complex network of biochemical reactions controlling the cell cycle are continually being discovered, detailed modelling of this system represents a rather difficult, if not frustrating, task. Because of their relative simplicity, core models that capture the essence of the network may yield insights that cannot be reached so readily in more elaborate models, which possess a huge number of variables and of independent parameters. The same choice between the two strategies repeatedly arises in most modelling studies in biology. The distinction, however, is not always clear-cut: thus, although it contains only three variables, the minimal cascade model discussed above represents an intermediate level of complexity between the two extreme situations just described.

### Developmental control of the mitotic oscillator

Although the periodic activation of cdc2 kinase in *Xenopus* egg extracts is driven by the synthesis of cyclin (Murray & Kirschner, 1989b), mitotic oscillations could also be induced in other ways. Thus, in yeast, mitosis appears to be regulated by the cyclic accumulation of cdc25, the phosphatase that activates cdc2 (Moreno, Nurse & Russell, 1990). In *Drosophila* embryos also, experimental evidence (Edgar & O'Farrell, 1989; O'Farrell *et al.*, 1989; Edgar *et al.*, 1994) indicates that periodic transcription of *string*, the gene encoding the homologue of the cdc25 phosphatase, plays a primary role in driving cell cycles 14–16 and in establishing mitotic domains (Foe, 1989) after the first 13 rapid divisions. Distinct molecular mechanisms would control the latter cycles, which require only maternal proteins and RNA: the first seven divisions occur in the absence of any detectable periodic variation in cdc2 activity, while cdc2 oscillations due to cyclic phosphorylation–dephosphorylation on Thr161 are observed during cycles 8–13 (Edgar *et al.*, 1994). The interest of the *Drosophila* embryo lies in its providing the possibility of studying, within a single organism, various mechanisms for the control of the cell cycle in the course of development.

### Arresting the mitotic oscillator and the control of cell proliferation

With respect to future developments, a most important aspect of experimental and theoretical studies pertains to the arrest of the cell division cycle. One role of the cascade structure of the mitotic control system might well be to multiply the possibilities of suppressing – or inducing – oscillations. Thus, in the extended (but still simplified) model described in fig. 10.10 for the *Xenopus* embryonic cell cycle – which may serve as template for the mechanism underlying cdk2 oscillations in the somatic cell cycle – no less than four converter enzymes are involved in determining the level of cdc2 kinase activity, and an even larger number of enzymes directly or indirectly control cyclin degradation. One way to arrest the oscillations is by altering the balance between cyclin synthesis and degradation. Another way is by interfering with thresholds. As shown above, oscillations can be induced or suppressed by increasing or decreasing the maximum rates of the converter enzymes cdc25 relative to wee1, or of cdc2 kinase relative to the phosphatase inactivating the cyclin protease (figs. 10.14–10.17). Manipulating the maximum rates of the converter enzymes of the cascade by phosphorylation–dephosphorylation is probably much more efficient than controlling these enzymes by allosteric effectors, which often affect Michaelis constants rather than maximum rates. The efficient, foolproof control of enzymes such as cdc25 phosphatase, wee1 kinase and cyclin protease appears to play a primary role in the periodic operation of the mitotic control system.

The mirror image of cell cycle arrest is unrestrained cell proliferation. Rapid experimental developments are taking place in investigations into the relation between cancer and the mitotic cascade controlling cell proliferation. Of particular significance are the link between cyclins and malignant transformation (Hunter & Pines, 1991, 1994; Motokura *et al.*, 1991; Nicholas, Cameron & Honess, 1992; Hinds *et al.*, 1994) and the interaction of oncoproteins (Wang, 1992) or tumour-suppressing gene products such as the retinoblastoma protein RB with cdc2 kinase (Lees *et al.*, 1991; Lin *et al.*, 1991; Dalton, 1992). Also important is the recent characterization of the protein inhibitors of cyclin-dependent kinases in mammalian cells (Xiong *et al.*, 1993; Morgan, 1995), whose absence may lead to cell proliferation (Hunter & Pines, 1994). Possible future goals of theoretical models could be to address the dynamics of the somatic cell cycle and to test how mitogenic factors and the products of oncogenes or tumour-suppressing genes might affect the dynamics of the mitotic oscillator.

### *Egg fertilization as link between $Ca^{2+}$ spiking and the mitotic oscillator*

Finally, let us return to the topic of $Ca^{2+}$ oscillations dealt with in chapter 9. $Ca^{2+}$ oscillations and the mitotic oscillator are linked at the first stage of development, i.e. fertilization. The first direct observations of $Ca^{2+}$ oscillations (Cuthbertson & Cobbold, 1985) were related to activation of mouse oocytes by sperm. Such oscillations occur in all types of mammalian egg (see Miyazaki *et al.*, 1986; Miyazaki, 1988; Swann & Ozil, 1994; Ozil & Swann, 1995). Waves of intracellular $Ca^{2+}$ were also first observed in fertilized eggs (see section 9.6).

These observations are important for understanding the mechanism by which fertilization triggers the onset of development, and relevant to the successful fertilization of human oocytes in cases of assisted conception, particularly since the introduction of intracytoplasmic sperm injection (ICSI). Whereas normal sperm – oocyte fusion in humans is always followed by a train of periodic $Ca^{2+}$ spikes, ICSI only sometimes elicits similar $Ca^{2+}$ oscillations (Tesarik, Sousa & Testart, 1994). Much attention is focused on the function of periodic $Ca^{2+}$ spikes in successful fertilization by ICSI (Taylor, 1994; Tesarik, 1994). In some species, such as sea urchin, a single $Ca^{2+}$ spike at fertilization induces zygote development, while a larger number of spikes may be needed in mammalian species. The pattern of $Ca^{2+}$ spiking associated with successful fertilization may thus vary from species to species.

As discussed in section 10.6, the increase in $Ca^{2+}$ at fertilization appears to induce some phosphorylation event(s) via calmodulin-activated protein kinase (Lorca *et al.*, 1993). One result of this is the activation of the cyclin degradation pathway, which releases the oocyte from metaphase arrest. A recent study in rabbit oocytes (Collas *et al.*, 1995) was aimed at measuring MPF-associated kinase activity as a function of the number of $Ca^{2+}$ pulses applied to the oocytes. The results suggest that, whereas a single $Ca^{2+}$ pulse elicits only a transient decrease in MPF activity, a prolonged series of $Ca^{2+}$ spikes is needed to sustain the MPF inactivation associated with exit from the metaphase arrested state.

# Part VII
## Circadian rhythms

# 11

# Towards a model for circadian oscillations in the *Drosophila period* protein (PER)

## 11.1 Circadian rhythms as ubiquitous biological clocks

Most living organisms exhibit circadian rhythms which allow them to adapt to an environment that varies with a periodicity of 24 h; such rhythms are ubiquitous and govern so many key physiological functions that they have become synonymous with **biological clocks** (Chovnik, 1960; Halberg, 1960; Pittendrigh, 1960, 1961; Menaker, 1971; Bünning, 1973; Aschoff, 1981; Moore-Ede *et al.*, 1982; Edmunds, 1988). Likewise, for many, **chronobiology** refers specifically to the study of 24 h rhythms. In no way do we detract from the important role of circadian rhythms by stressing that such periodic phenomena comprise only one among many types of biological rhythm, besides those that have much shorter periods (see table 1.1).

Yet, circadian rhythms are of special interest because of both their prevalence among all eukaryotic organisms, unicellular or multicellular, and their immediate relevance to our human experience (Moore-Ede *et al.*, 1982): the sleep–wake and nutrition cycles remind us every day of the fundamental, physiological role of circadian rhythms in allowing us to cope with our periodically changing environment.

Another reason why circadian rhythms remain so fascinating is that, although these rhythms have been known for so long, their mechanism remains largely unknown. The purpose of this introductory section is to present a brief overview of circadian oscillations and of the current understanding of their molecular bases, which at present is undergoing rapid progress, and to indicate how theoretical models have been used to investigate the properties of these rhythms.

## An overview of circadian rhythms

Circadian rhythms have long been studied in plants (Sweeney, 1969; Bünning, 1973), mammals including rodents and humans (Moore-Ede *et al.*, 1982), molluscs, insects such as the fly *Drosophila* (Pittendrigh, 1960, 1961; see section 11.2), and unicellular organisms (Edmunds, 1988). Among the unicellular systems most studied for circadian rhythmicity – see Table 2.1 in the book by Edmunds (1988) for an exhaustive list – are algae such as *Gonyaulax* (Hastings & Sweeney, 1958), *Euglena* (Edmunds & Laval-Martin, 1984), and *Acetabularia* (Schweiger & Schweiger, 1977; Vanden Driessche, 1980), and fungi such as *Neurospora* (Feldman & Dunlap, 1983).

   Like other biochemical oscillations discussed in this book, circadian rhythms possess an endogenous nature. True circadian rhythms thus persist under constant environmental conditions, for example in constant light or darkness, with a **free-running period** close to 24 h. A second property that characterizes circadian rhythms is their sensitivity to light. These rhythms are readily **phase-shifted** by pulses of light applied at the appropriate time over the period, or **entrained** by light–dark cycles (see, e.g. Hastings & Sweeney, 1958; Pittendrigh, 1965). A third property shared by circadian rhythms in different organisms is their relative independence of temperature; this property is referred to as **temperature compensation**. The last two properties make sense if the organism is to adapt to the natural light–dark cycle, regardless of the temperature. That circadian rhythms are largely independent of temperature (Pittendrigh, 1954; Sweeney & Hastings, 1960) makes them nearly unique among other biological rhythms, whose frequency generally increases with temperature. How to account for such a property is one of the many challenges posed by circadian rhythms to experimental and theoretical investigators.

   Besides the unicellular organisms already mentioned, the goal of studies of circadian rhythms in higher organisms such as insects, molluscs, or mammals is to identify the cells responsible for generating the rhythm. These cells behave as a circadian pacemaker. The question remains as to whether a single pacemaker or multiple pacemakers govern various circadian rhythms within a given organism. Data supporting the latter view, as well as early developments in circadian rhythm research, have recently been reviewed by Pittendrigh (1993).

   Among the most investigated neuronal pacemakers are those located in the eyes of the molluscs *Aplysia* and *Bulla* (Jacklet, 1977, 1989b;

Block & Wallace, 1982). An important study (Michel *et al.*, 1993) recently showed that circadian rhythmicity occurs at the level of a single neuron in *Bulla*. This study indicated that the circadian variation of electrical activity involves the periodic variation of potassium conductance. Thanks to these advances, the *Bulla* circadian pacemaker has become a most promising experimental model for unravelling the molecular bases of circadian rhythmicity. Circadian pacemakers have also been identified in insects such as the cockroach (Page, 1982).

In mammals, the identification of circadian pacemakers was first accomplished in the rat. These studies (Moore & Eichler, 1972; Stephan & Zucker, 1972) showed that circadian rhythms in corticosterone levels or drinking and locomotor activity originate from a group of neurons in the hypothalamus, forming the **suprachiasmatic nucleus** (SCN). That the SCN drives the circadian clock was further shown by experiments in which circadian rhythms in rat SCN electrical activity persisted when the connections between the SCN and other parts of the brain were severed (Inouye & Kawamura, 1979). Numerous studies have shown, since, that the SCN represents the most important – but perhaps not unique (Moore-Ede, 1983) – circadian pacemaker in mammals (Rusak & Zucker, 1979; Moore-Ede *et al.*, 1982; Moore, 1983; Turek, 1985; Meijer & Rietveld, 1989; Ralph *et al.*, 1990). What highlighted the SCN as a possible circadian pacemaker was the existence of a direct neuronal connection, the retinohypothalamic tract, between the SCN and the retina. That connection permits the control of the circadian clock by light. Recent studies (Ding *et al.*, 1994) clarified the pharmacological basis of this control, which appears to be mediated by glutamate, $N$-methyl-D-aspartate (NMDA) and nitric oxide (NO); brief treatment of rat SCN *in vitro* with these compounds generated light-like phase shifts of circadian rhythms. How SCN neurons are capable of generating a 24 h periodicity remains an open, fundamental question in chronobiology.

Another system that may contain a pacemaker is the avian pineal gland, which generates a circadian rhythm in melatonin; the rhythm has also been observed in pineal cell cultures (see fig. 11.1) (Binkley, Riebman & Reilly, 1978; Takahashi, Hamm & Menaker, 1980; Takahashi *et al.*, 1989). Melatonin is involved in the control of the sleep–wake cycle, and so analogues of the hormone are currently being developed for the prevention of jet lag. A circadian rhythm of activity has been demonstrated for the enzyme $N$-acetyltransferase, which catalyses an early step in melatonin synthesis (Binkley, 1983).

Because they occur in so many organisms, unicellular or multicellular, it may seem unlikely that circadian rhythms all originate from a single molecular mechanism. Some biochemical processes, nevertheless, are probably shared by many of these rhythms. Thus, protein synthesis appears to play a central role in the mechanism of circadian periodicity. Experiments in organisms as varied as *Gonyaulax* (Karakashian & Hastings, 1963; Taylor *et al.*, 1982a,b), *Euglena* (Feldman, 1967), *Neurospora* (Dunlap & Feldman, 1988), or molluscs (Jacklet, 1977, 1989b; Eskin, Yeung & Klass, 1984) show indeed that inhibitors of protein synthesis suppress circadian oscillations, modify their period or produce significant phase shifts (see figs. 11.1 and 11.2). Similar results were obtained by means of an inhibitor of transcription (Raju *et al.*, 1991). The role of protein synthesis in circadian oscillations is considered in further detail below.

The study of circadian rhythms has seen a clear acceleration in recent years owing to the combined insights provided by genetics and molecular biology (Feldman, 1982; Hall & Rosbash, 1988; Takahashi, 1992, 1993; Dunlap, 1993; Takahashi *et al.*, 1993; Young, 1993). Mutants of circadian rhythms were obtained in *Drosophila* as early as 1971 (see section 11.2). The use of circadian mutants, recently reviewed by Dunlap (1993), has also proved fruitful in *Neurospora* (see below) and rodents. Clock mutants have thus been obtained in the hamster (Ralph & Menaker, 1988), and the mouse (Vitaterna *et al.*, 1994).

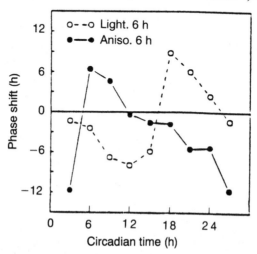

Fig. 11.1. Phase response curves obtained in chick pineal cell cultures for 6 h pulses of light and anisomycin (Aniso.), an inhibitor of protein synthesis (Takahashi *et al.*, 1989).

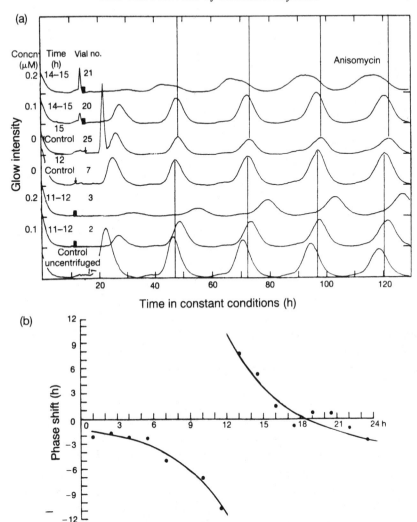

Fig. 11.2. Circadian rhythm of bioluminescence in *Gonyaulax polyedra* (Taylor *et al.*, 1989b). The curves in (a) show phase-shifts of the rhythm by pulses of anisomycin, an inhibitor of protein synthesis. The concentration of anisomycin used is indicated along the vertical axis, as well as the time at which the drug is administered. Experiments shown indicate phase-shifts produced by 1 h pulses of 0.1 μM (vials 2 and 20) and 0.2 μM (vials 3 and 21) anisomycin. Vertical lines indicate positions of control peaks (vials 7 and 25). Drug pulses given between hours 11 and 12 (vials 2 and 3) resulted in phase delays as compared to control (vial 7); pulses given from hours 14 to 15 resulted in phase advances compared to control (vial 25). (b) A phase response curve for 1 h pulses of 0.3 μM anisomycin. Conditions are as in (a). Time on the abscissa denotes beginning of the pulses, in hours since cells were transferred to constant conditions. A positive phase shift denotes a phase advance.

Once a mutant is obtained, the focus of investigation turns to the identification of the protein encoded by the mutated gene. The most promising results, so far, pertain to the *per* gene in *Drosophila* (discussed in section 11.2) and the *frq* gene in *Neurospora* (Feldman, 1982; Dunlap *et al.*, 1993; Loros *et al.*, 1993; Aronson *et al.*, 1994). A variety of genes whose expression follows a circadian pattern have recently been identified in plants (Giuliano *et al.*, 1988; Carter *et al.*, 1991; Kay & Millar, 1993; McKlung, 1993). All these studies have inaugurated a new era for research on the molecular bases of circadian rhythms (Hall & Rosbach, 1988; Takahashi, 1992, 1993; Takahashi *et al.*, 1993). Along similar lines, good progress is also being made on the origin of circadian rhythms in the pineal gland (Stehle *et al.*, 1993) and in the daily rhythm of bioluminescence in *Gonyaulax*, which involves a daily variation in the amounts of luciferase and of a luciferin-binding protein (Morse, Fritz & Hastings, 1990).

In humans, circadian rhythms governed by the SCN affect a large number of physiological functions, including the sleep–wake cycle and nutrition (Moore-Ede *et al.*, 1982; Touitou & Haus, 1992). Moreover, a variety of hormone levels (Van Cauter & Aschoff, 1989) or enzyme activities display circadian patterns. Circadian rhythms, accordingly, play important roles in both health and disease (Moore-Ede, Czeisler & Richardson, 1983). Clinical applications of circadian rhythm research include the use of light to treat maladaptation to night work (Czeisler *et al.*, 1990), jet lag, or seasonal affective disorder (Lewy *et al.*, 1987).

Besides the use of light to treat rhythm-related disorders, clinical work aims at taking into account the therapeutic implications of circadian rhythms. Thus, determining the appropriate time at which types of medication should be administered to maximize their efficacy, while minimizing unwanted side-effects, is the main goal of the rapidly developing field of **chronopharmacology** (Lemmer, 1989; Zhang & Diasio, 1994).

A case in point is provided by the chronotherapy of cancer (Hrushesky, 1994). 5-Fluorouridine (5-FU) is a widely used anticancer drug that interferes with DNA synthesis. Biochemical studies indicate that enzymes involved in the degradation and utilization of 5-FU follow inverse circadian activity patterns (Harris *et al.*, 1990; Daher *et al.*, 1991; Zhang *et al.*, 1993). These results are being incorporated into treatments in which the administration of 5-FU and other anticancer drugs by programmable pumps is chronomodulated in a circadian manner that aims at optimizing therapeutic efficacy while minimizing deleterious side-effects (Levi *et al.*, 1994).

### Models for circadian rhythms

As indicated by the above overview, the study of the molecular bases of circadian rhythms has accelerated in recent years, thanks to the combined approaches of biochemistry, molecular biology, and genetics. These advances pave the way for the construction of realistic models based on experimentally determined mechanisms, as attempted in section 11.3, below. However, even though detailed information on the underlying mechanisms was lacking at the time, theoretical models have been used since the 1960s to investigate the properties of circadian rhythms (Wever, 1965; Winfree, 1970, 1980; Pavlidis, 1971, 1973; Kronauer, 1984; Friesen, Block & Hocker, 1993). These models, of an abstract mathematical nature, were not based directly on biologically identifiable mechanisms or variables. A favourite type of model used in this context was the Van der Pol oscillator, which provides a prototype for limit cycle oscillations of the relaxation type in electrical systems (see, e.g. Minorsky, 1962; Andronov *et al.*, 1966).

Such theoretical studies proved useful in accounting in a qualitative manner for observations on the entrainment of circadian rhythms by a periodic **zeitgeber** such as light–dark cycles (Wever, 1965, 1966, 1972). They also permitted investigation of the consequences of an interaction between two circadian pacemakers (Daan & Berde, 1978; Kronauer *et al.*, 1982; Wever, 1987; Gander *et al.*, 1984a,b). One such consequence, also studied by means of models for coupled neuronal oscillators (Kawato & Suzuki, 1980; Carpenter & Grossberg, 1985), is the splitting of the free-running, circadian rhythm into two components.

Among circadian rhythms, one of the most studied by means of theoretical models is the sleep–wake cycle (Winfree, 1982; Daan, Beersma & Borbely, 1984; Moore-Ede & Czeisler, 1984; Strogatz, 1986, 1987). A related topic of current interest in view of its therapeutical applications, which has also been approached theoretically (Kronauer, 1990; Kronauer & Czeisler, 1993), pertains to the use of light to control the circadian pacemaker in humans. The property of temperature compensation, so characteristic of circadian rhythms, has also been the subject of theoretical investigations (Pavlidis & Kauzmann, 1969; Lakin-Thomas, Brody & Coté, 1991).

Finally, mathematical models have proved useful in accounting for the phase advances or delays induced by transient perturbations of the circadian clock (Winfree, 1970, 1980; Drescher, Cornelius & Rensing, 1982). These studies are of particular interest, given that phase response curves represent an important tool in the study of circadian rhythms (Turek, 1987). Theoretical studies further permitted Winfree (1980) to define different types of phase response curves, referred to as **type 0** or **type 1**, depending on the magnitude of the perturbation used; type 0 refers to strong phase resetting (see Czeisler *et al.*, 1989, for an application of this concept to the resetting of the human circadian pacemaker by light). Winfree (1980) also predicted the possibility of suppressing the circadian rhythm by a critical pulse of appropriate magnitude delivered at the appropriate phase over the period. The effect of such perturbation is to abolish the rhythm, at least transiently, by bringing the oscillatory system back to the vicinity of its singularity, i.e. the steady state. Such a prediction was subsequently corroborated by experimental observations, by means of pulses of protein synthesis inhibitor (Taylor *et al.*, 1982b) or light (Jewett, Kronauer & Czeisler, 1991).

This book focuses on theoretical models for rhythmic phenomena, related as closely as possible to underlying molecular mechanisms. The remarkable progress recently accomplished in clarifying the molecular bases of circadian rhythmicity will soon allow the construction of such theoretical models, built on the basis of available experimental data. Because the *Drosophila* circadian system, centred on the role of the *period* (*per*) gene, is presently the best understood in molecular terms, we focus on that system in the remaining part of this chapter. After briefly reviewing salient experimental results in section 11.2, section 11.3 presents the preliminary results obtained in the study of a theoretical model for circadian oscillations in the *Drosophila period* protein (PER).

## 11.2 The *period* (*per*) gene and the circadian clock in *Drosophila*

Some of the most remarkable advances in elucidating the molecular basis of circadian rhythms have been made in mutants of the fly *Drosophila*. Two types of circadian rhythm are known in that organism. The first affects the rest–activity cycle, and the second underlies the daily eclosion peaks of pupae. Both types of rhythm persist in constant light or temperature conditions (Pittendrigh, 1960, 1961).

By selection of strains in which flies emerged early or late in the daily eclosion peaks, Pittendrigh (1967) succeeded in obtaining strains whose periodic eclosion behaviour differed by about 4 h after 50 generations of selection. Another method, based on chemical mutagenesis, permitted Konopka & Benzer (1971) to obtain clock mutants in *Drosophila melanogaster.* This classic work yielded flies altered in their circadian system, owing to single mutations in a single gene that was called *per* (for *period*). Four phenotypes were characterized: the wild type (*per*⁺) has a free-running period of activity and eclosion close to 24 h; short-period mutants (*per*ˢ) have a period close to 19 h; in long-period mutants (*per*ʲ), the periodicity increases up to 29 h; finally, arrhythmic mutants (*per*⁰) have lost the circadian pattern of eclosion or activity (Konopka & Benzer, 1971; Konopka, 1979). The behaviour of the different types of circadian clock mutant is shown in fig. 11.3 for locomotor activity rhythm.

Interestingly, whereas the wild type exhibits the property of temperature compensation that is common to all circadian rhythms, the mutants *per*ʲ and *per*ˢ have lost this property (Konopka, Pittendrigth & Orr, 1989); in contrast to the wild type, the period of their activity rhythm respectively increases and decreases with temperature (fig. 11.4). A possible molecular basis for temperature compensation of circadian rhythms in *Drosophila* has been proposed in terms of inter- and intramolecular interactions of the *per* gene product (Huang, Curtin & Rosbash, 1995).

Thanks to genetic engineering, remarkable progress has been achieved over the last decade in elucidating the role of *per* in circadian rhythmicity. Thus, rhythmic behaviour has been restored in arrhythmic mutants of *D. melanogaster* by gene transfer, i.e. insertion of a piece of DNA carrying the normal *per* gene (Bargiello, Jackson & Young, 1984;

Fig. 11.3. Circadian rhythm of locomotor activity in the fly *Drosophila* (Konopka & Benzer, 1971). Shown, from top to bottom, are the rhythms for the normal fly, the arrhythmic mutant, and the *per*ˢ and *per*ʲ mutants. The rhythms are monitored in constant infrared light by an event recorder. Records read from left to right; each successive interval is replotted to the right of the preceding one.

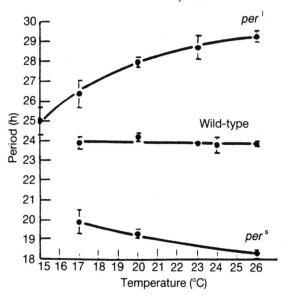

Fig. 11.4. Temperature dependence of the rhythm of locomotor activity in *Drosophila* (Konopka *et al.*, 1989). Shown are the temperature-compensated curve for the wild type, and the noncompensated variation for the *per*[l] and *per*[s] mutants.

Konopka, 1982; Coté & Brody, 1986; Baylies *et al.*, 1987): a decrease in gene dosage lengthens the period, while an increased dosage shortens it.

Surprisingly, the mutations in the *per* gene affect not only the circadian clock but also a high-frequency rhythm in *Drosophila*. Besides the circadian periodicity in eclosion and activity, *Drosophila* displays a rhythm of much shorter period associated with the courtship song of the male (von Schilcher, 1976; Hall, 1986, 1994). A male of *D. melanogaster* courts a female by vibrating one of his wings; this produces a specific temporal pattern of auditory stimulation known as the courtship song. The pattern consists in trains of several spikes, separated by an inter-pulse interval (IPI). The IPI lasts some tens of milliseconds, but fluctuates rhythmically with a period close to 54 s in the wild type *D. melanogaster*. The song pattern appears to be species–specific (Kyriacou & Hall, 1986). Kyriacou & Hall (1980) made the startling observation that mutations of the *per* gene affect not only the period or even the existence of circadian rhythms, but also the much shorter-period rhythm of the courtship song. Moreover, the effect of the mutations keeps the same trend for both rhythms. Thus, the *per*[s] mutant displays a courtship rhythm whose period declines from 54 to 43 s, while the *per*[1] mutant has a courtship song whose period has risen from 54 to 96 s. Finally, the arrhythmic mutant *per*[0], which has lost circadian rhythmicity, has lost also the rhythmic change in the interpulse interval that characterizes the courtship song in the normal flies (Kyriacou & Hall, 1980; Kyriacou *et al.*, 1990, 1993; Hall, 1986, 1994).

A remarkable property of the *per* gene product therefore is that it affects 24 h rhythms as well as a 1 min rhythm. How can a protein play such a dual role, with the additional constraint that the circadian rhythms have the same period in different species while the period of the courtship song is necessarily variable as it serves as a code for recognition of males by females of a given species?

A major cue to the role of the *per* locus in these rhythms has come from the molecular characterization of the gene and of its protein product (PER). These studies (Baylies *et al.*, 1987; Yu *et al.*, 1987b) showed that the *per* gene, which codes for a proteoglycan (Jackson *et al.*, 1986; Reddy *et al.*, 1986), is transcribed into a 4.5 kb mRNA, and that the corresponding protein contains about 1220 amino acid residues (a certain polymorphism in length exists for PER in different *Drosophila* strains; see Baylies *et al.*, 1993). Molecular mapping of the mutations futhermore indicated that the short-period phenotype *per*[s] results from the substitution of a single amino acid in the sequence at position 589, where serine is replaced by asparagine (Baylies *et al.*, 1987; Yu *et al.*, 1987b). The long-period phenotype *per*[1] originates from another point mutation that results in the replacement of valine by aspartic acid (Baylies *et al.*, 1987). The arrhythmic phenotype *per*[0] (also referred to as *per*[01]) also arises from a single mutation, but in this case the change of a glutamate codon (CAG) to a translational stop codon (TAG) at position 464 results in a truncated protein containing only 463 amino acid residues (Yu *et al.*, 1987b; Baylies *et al.*, 1987). This altered protein presumably has lost its function, so that flies carrying the mutation become arrhythmic.

These results constitute a breakthrough not only for research on biological rhythms but also, more generally, for behavioural studies. Indeed, as pointed out by Yu *et al.* (1987b), the characterization of the point mutations leading to altered rhythmic behaviour is the first example of a behavioural mutation identified at the level of the amino acid sequence of a protein.

The point mutations described above govern, in a still unknown manner, the slowing down, acceleration or disappearance of the circadian rhythm and of the courtship song. They do not explain the variability of the latter rhythmic process in different species of *Drosophila*. Here also, the study of the protein sequence suggested a possible clue. The PER sequence contains a Thr-Gly repeat region whose length changes in different species (Jackson *et al.*, 1986; Yu *et al.*, 1987b). That this repeat sequence might govern the characteristics of the song rhythm was suggested by the results of transformation experiments (Yu *et al.*, 1987a). These experiments, based on the phenotype of transformed flies in which the *per* gene lacks the portion corresponding to the entire Thr-Gly repeat region, showed that the effects of the *per* gene on the circadian and courtship song rhythms can be fully dissociated. Such a truncated gene sequence indeed restored normal circadian rhythmicity in a *per*[01] mutant, but failed to restore the normal rhythmicity of the male courtship song (Yu *et al.*, 1987a). Furthermore, removing the Thr-Gly region of PER has a much greater impact on the period of the courtship song rhythm than on that of the circadian clock. More recent data indicate, however, that rather than the variable length in the repeat region, up to four amino acid residues, varying from species to species, located downstream from the repeat sequence, may be at the root of song rhythm variability (Baylies *et al.*, 1993).

Analysis of the temporal and spatial expression of the *per* gene message in developing embryos of the fruit fly (James *et al.*, 1986) indicates that the *per* mRNA is expressed in the developing central nervous system, primarily within the brain and ventral ganglia. These locations correspond to the putative sites of generation or control of the circadian and courtship song rhythms, respectively. In *Drosophila* adults, subsequent studies (Siwicki *et al.*, 1988; Ewer *et al.*, 1992) showed that the *per* gene is expressed in neuronal and glial cells; *per* expression in the former type of cell appears to be required for stronger rhythmicity. The gene is expressed in the eyes, but *per* expression in the central brain is sufficient for generating the rhythm in locomotor activity (Ewer *et al.*, 1992; Zeng, Hardin & Rosbach, 1994).

The key question pertains of course to the function of PER in the mechanism of the circadian and courtship song rhythms. Through its role(s) in the nervous system of the fly, this protein controls both processes even if these differ in period by some three orders of magnitude. An interesting observation made by Dowse, Hall & Ringo (1987) is that a large fraction of mutants that lose circadian rhythmicity still display ultradian oscillations. This raises the possibility that PER may be involved in the production of circadian rhythmicity through the synchronization of rhythms of shorter period (Dowse & Ringo, 1993).

Bargiello *et al.* (1987) suggested that PER may exert its effect by controlling electrical communications between cells. The analysis of the electrical properties of cells of salivary glands of $per^0$, $per^+$, and $per^1$ flies indeed showed differences in the conductance of gap junctions in the mutants, compared to those of the wild type. The authors therefore proposed that PER affects the behavioural rhythms by altering gap-junctional communication of cells in the nervous system. This idea had to be abandoned, however, as subsequent observations (Saez *et al.*, 1992) indicated that the function of PER is not related to gap junctions.

In what represents a breakthrough, recent studies showed that PER possesses a dimerization domain that is common to several transcription factors (Huang, Edery & Rosbash, 1993). Together with the nuclear localization of PER in the brain of adult flies (Ewer *et al.*, 1992; Liu *et al.*, 1992), which turns out to be required for circadian rhythmicity (Baylies *et al.*, 1993; Sehgal *et al.*, 1994; Vosshall *et al.*, 1994), this suggests that PER behaves as a regulator of transcription, possibly by interacting with transcription factors (Takahashi, 1992; Huang *et al.*, 1993), since it does not seem to possess itself a DNA-binding domain or DNA-binding activity (Baylies *et al.*, 1993; Huang *et al.*, 1993). In such a manner the circadian rhythm in PER could direct the circadian variation of other proteins (Hardin *et al.*, 1992; Huang *et al.*, 1993), through the control of circadian-clock responsive elements, referred to as CCREs (Takahashi, 1993).

Such a mechanism may represent a unified mechanism for the generation of circadian rhythmicity in a wide variety of experimental systems (see Takahashi, 1992; Sassone-Corsi, 1994), in much the same way as the control of mitotic oscillations by regulation of cdc2 kinase (see chapter 10) in different eukaryotic organisms. In support of this view, besides the evidence that involves protein synthesis in the control of the circadian clock in various systems (see above, and Takahashi *et al.*, 1993), recent experiments in hamsters indicate that, in the SCN, light controls the expression of Fos, a component of a transcriptional regulatory complex, in a manner that, like light-induced phase-shifts, depends on circadian time (Kornhauser *et al.*, 1992).

That the circadian rhythm in *Drosophila* is among the most promising systems for unravelling the molecular basis of circadian rhythms is further corroborated by the findings (Hardin *et al.*, 1990, 1992) that *per* mRNA is produced rhythmically, in a circadian manner (fig. 11.5). Moreover, this variation is accompanied by a circadian rhythm in the

(a)

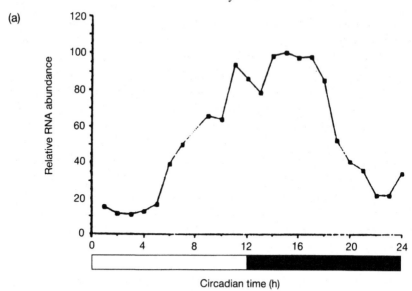

(b)

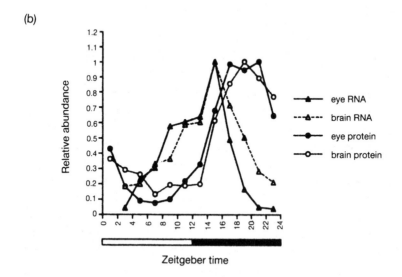

Fig. 11.5. (a) Circadian rhythm in *per* mRNA abundance in *Drosophila* (Hardin *et al.*, 1990). (b) The periodic variation of total PER protein is shown to lag by several h after the circadian change in *per* mRNA. The curves show the variation of *per* mRNA and PER in the eyes and brain of *Drosophila* flies (Zeng *et al.*, 1994).

degree of abundance (Zerr *et al.*, 1990) and of phosphorylation (Edery *et al.*, 1994) of PER; the peak in *per* mRNA precedes the peak in PER by about 4 h (Zeng *et al.*, 1994; see also fig. 11.5b). On the basis of this observation, Hardin *et al.* (1990, 1992) suggested that the *Drosophila* circadian rhythm results from a negative feedback exerted by the PER protein – or by some related product – on the synthesis of the *per* mRNA or on its translation; PER may act by binding to and removing a transcription activator (Zeng *et al.*, 1994). Recent results (Zeng *et al.*, 1994) support the view that this mechanism is capable of producing circadian rhythmicity at the cellular level. A similar negative feedback mechanism is also suggested to be at work for the *frq* gene product in *Neurospora* (Aronson *et al.*, 1994). It is noteworthy that repression of transcription has long been considered theoretically as a mechanism for the occurrence of oscillations in protein synthesis (Goodwin, 1963, 1965). A more specific model for PER oscillations based on these experimental observations is proposed below.

## 11.3 Model for circadian oscillations in the *Drosophila* PER protein

As summarized in the preceding section, much progress has recently been made in characterizing the molecular mechanism of circadian rhythms in *Drosophila*. First, *per* mRNA was shown to vary in a circadian manner. The fact that the amount of PER protein also varies with the same period but follows the mRNA rhythm by several hours suggested that the mechanism of circadian oscillations involves a negative feedback loop exerted by PER on the transcription of the *per* gene (Hardin *et al.*, 1990, 1992; Zeng *et al.*, 1994). This view is supported by the observation that nuclear localization of the protein is required for oscillations (Sehgal *et al.*, 1994; Vosshall *et al.*, 1994). Such an oscillatory mechanism is reminiscent of early theoretical studies based on the principles of genetic regulation laid down by Jacob & Monod (1961). These theoretical studies, initiated by Goodwin (1963, 1965), showed that oscillatory behaviour may originate from the repression of mRNA synthesis by the protein encoded by this mRNA. Negative feedback models of this sort were later studied with respect to oscillations in a chain of enzyme reactions controlled by end-product inhibition (see the discussion at the end of this chapter).

Post-translational modification of PER is also involved in the mechanism of its circadian oscillations (Zwiebel *et al.*, 1991). Thus, a recent study (Edery *et al.*, 1994) showed that PER becomes phosphorylated, probably at multiple residues. The phosphorylation status of PER could play a role in the circadian oscillatory mechanism, perhaps by controlling the nuclear localization of PER and/or its degradation. Recent observations indicate that the entry of PER into the nucleus is delayed (Curtin *et al.* 1995), possibly due to the requirement of PER phosphorylation.

Another important study (Zeng *et al.*, 1994) demonstrated that the overexpression of PER in the *Drosophila* eyes represses *per* transcription and suppresses circadian rhythmicity in these cells, without affecting circadian PER oscillations in other *per*-expressing cells in the brain, or the circadian rhythm in locomotor activity. This work also shows that the action of PER on transcription is intracellular, and suggests that 'each *per*-expressing cell contains an autonomous oscillator of which the *per* feedback loop is a component' (Zeng *et al.*, 1994).

The purpose of this section is to propose a theoretical model for circadian oscillations in the *Drosophila* PER protein and its mRNA (Goldbeter, 1995). The model is based on multiple phosphorylation of PER and on the inhibition of *per* transcription by a phosphorylated form of PER. Besides providing a plausible explanation for the delay between the circadian rhythms in mRNA and protein level, the theoretical model shows how variations in parameters such as the rate of degradation of PER or the rate of its translocation into the nucleus may change the period of these oscillations, or even suppress rhythmic behaviour. These results provide insights into ways by which modifications affecting the stability of PER may give rise to the $per^s$, $per^1$, or $per^0$ mutant phenotypes described in the preceding section. Investigating the effect of PER transport into the nucleus allows us to assess the role of the recently discovered *tim* (*timeless*) gene product (Sehgal *et al.*, 1994) in the mechanism of the *Drosophila* circadian clock.

An alternative, more detailed model for circadian PER oscillations in *Drosophila* has been developed independently be Abbot *et al.* (1995) (M. Rosbasch, personal communication; results of that study were first presented at the 4th conference of the Society for Research on Biological Rhythms held in May 1994 in Amelia Island, Florida). That model, which is also based on the negative feedback exerted by PER on *per* transcription, takes into account a larger number of phosphorylated residues and focuses on the role of PER phosphorylation in delaying the entry of the protein into the nucleus (Curtin *et al.* 1995).

### Model and kinetic equations

Based on the experimental observations summarized above, the model for the circadian variation in PER is schematized in fig. 11.6. We assume that *per* mRNA, whose cytosolic concentration is denoted by $M$, is synthesized in the nucleus and rapidly transfers to the cytosol, where it accumulates at a maximum rate $v_s$; there it is degraded enzymically, in a Michaelian manner, at a maximum rate $v_m$.

The rate of synthesis of PER, proportional to $M$, is characterized by an apparent first-order rate constant $k_s$. To take into account the fact that PER is multiply phosphorylated (Edery *et al.*, 1994), and to keep the model as simple as possible (the precise number of phosphorylated residues is still unknown), we consider only three states of the protein: unphosphorylated ($P_0$), monophosphorylated ($P_1$) and bisphosphorylated ($P_2$). The model could readily be extended to include a larger number of phosphorylated residues; such an extension would, however, unnecessarily complicate the model without altering significantly its dynamic behaviour.

The role of PER phosphorylation is still unclear. It has been suggested that phosphorylation may control nuclear localization and/or degradation of PER (Edery *et al.*, 1994). Here we assume that the fully phosphorylated form ($P_2$) is degraded in a Michaelian manner, at a maximum rate $v_d$, and also transported into the nucleus, at a rate characterized by the apparent first-order rate constant $k_1$. Transport of the

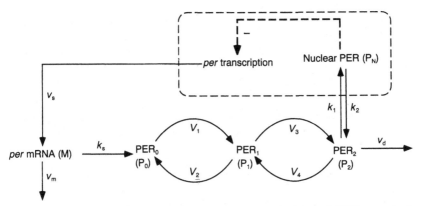

Fig. 11.6. Model for sustained oscillations in *per* mRNA and PER protein in *Drosophila* (Goldbeter, 1995). The model is based on the negative feedback exerted by nuclear PER on *per* transcription, and on the multiple phosphorylation of PER (see text for details).

nuclear, bisphosphorylated form of PER ($P_N$) into the cytosol is characterized by the apparent first-order rate constant $k_2$. In the absence of more detailed information, alternative assumptions could, at this stage, be retained. Thus, degradation of PER could also be directed at the nuclear bisphosphorylated form of the protein, or at the cytosolic unphosphorylated or monophosphorylated forms, which could also be transported into the nucleus. Probably, such changes would produce only minor modifications in dynamic behaviour, although delaying PER entry into the nucleus and degradation until the protein is fully phosphorylated – as considered here – should favour the occurrence of sustained oscillations. The assumption that only the fully phosphorylated form of PER enters the nucleus holds with the recent observations on delayed nuclear entry possibly due to PER phosphorylation (Curtin *et al.*, 1995)

Crucial to the mechanism of oscillations in the model is the negative feedback exerted by nuclear PER on the production of *per* mRNA. This negative feedback will be described by an equation of the Hill type in which $n$ denotes the degree of cooperativity, and $K_I$ the threshold repression constant. To simplify the model, we consider that $P_N$ behaves directly as a repressor; activation of a repressor upon binding of $P_N$ would not significantly alter the results (see Sinha & Ramaswamy, 1988).

If the concentrations of the corresponding PER species are denoted by $P_0$, $P_1$, $P_2$ and $P_N$, the time evolution of the five-variable model is governed by the following kinetic equations (Goldbeter, 1995):

$$\frac{dM}{dt} = v_s \frac{K_I^n}{K_I^n + P_N^n} - v_m \frac{M}{K_{m1} + M} \tag{11.1a}$$

$$\frac{dP_0}{dt} = k_s M - V_1 \frac{P_0}{K_1 + P_0} + V_2 \frac{P_1}{K_2 + P_1} \tag{11.1b}$$

$$\frac{dP_1}{dt} = V_1 \frac{P_0}{K_1 + P_0} - V_2 \frac{P_1}{K_2 + P_1} - V_3 \frac{P_1}{K_3 + P_1} + V_4 \frac{P_2}{K_4 + P_2} \tag{11.1c}$$

$$\frac{dP_2}{dt} = V_3 \frac{P_1}{K_3 + P_1} - V_4 \frac{P_2}{K_4 + P_2} - k_1 P_2 + k_2 P_N - v_d \frac{P_2}{K_d + P_2} \tag{11.1d}$$

$$\frac{dP_N}{dt} = k_1 P_2 - k_2 P_N \tag{11.1e}$$

The total (nonconserved) quantity of PER, $P_t$, is given by:

$$P_t = P_0 + P_1 + P_2 + P_N \qquad (11.2)$$

All parameters and concentrations in eqns (11.1)–(11.2) are defined with respect to the total cell volume. Besides the parameters already defined above, $V_i$ and $K_i$ ($i = 1, \ldots 4$) denote the maximum rate and Michaelis constant of the kinase(s) and phosphatase(s) involved in the reversible phosphorylation of $P_0$ into $P_1$, and of $P_1$ into $P_2$, respectively (see fig. 11.6).

### Sustained oscillations in PER protein and per mRNA

The linear stability analysis of eqns (11.1a–e) is currently being developed along the lines followed in preceding chapters. Pending the more detailed information to be gathered from such a study, numerical integration shows that for appropriate parameter values, instead of evolving toward a stable steady state, the system governed by eqns (11.1a–e) reaches a regime of sustained, periodic oscillations (fig. 11.7). Shown in fig. 11.7a is the temporal variation in the amount of *per* mRNA ($M$), together with the variation in the level of nuclear PER ($P_N$). Also shown is the periodic variation in the total amount of PER ($P_t$). The concomitant changes in the amounts of unphosphorylated ($P_0$) and phosphorylated cytosolic ($P_1$ and $P_2$) and nuclear ($P_N$) forms of PER are shown in fig. 11.7b, to illustrate the relative phase differences between these oscillating variables and those shown in fig. 11.7a.

Here again, sustained oscillations in PER protein and *per* mRNA correspond to the evolution towards a limit cycle in the phase plane. This is demonstrated in fig. 11.8 where the level of *per* mRNA, $M$, is plotted as a function of the total amount of PER protein, $P_t$, for two different initial conditions, one located inside and the other outside the limit cycle. In each case the system evolves towards the same, unique closed curve. Therefore sustained oscillations in PER and *per* mRNA are characterized by a unique amplitude and frequency for a given set of parameter values, regardless of initial conditions.

The sequence of events over one cycle of sustained oscillations in the model can be described as follows (see fig. 11.7). Starting from a low level of nuclear PER ($P_N$), *per* mRNA ($M$) is synthesized at a nearly constant, maximum rate, $v_s$, since the repression of transcription, exerted by $P_N$, is at a low. Subsequently *per* mRNA begins to accumulate in the cytosol; this leads to the synthesis of PER protein in its unphosphorylated state $P_0$, at a rate proportional to $M$. Sequential, reversible

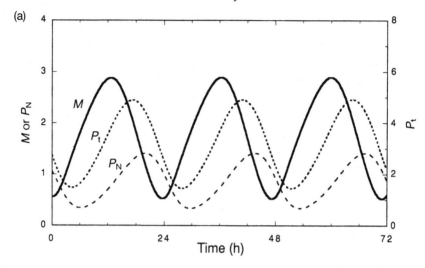

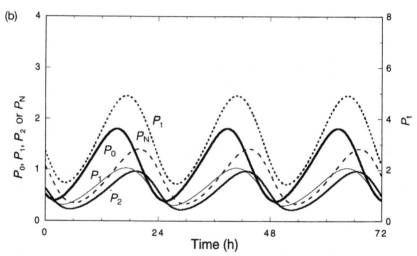

Fig. 11.7. Sustained oscillations in the model for regulation of *per* mRNA by the PER protein in *Drosophila* (Goldbeter, 1995). Shown in (a) is the temporal variation in *per* mRNA ($M$), together with the variation in nuclear PER ($P_N$) and in the total amount of PER protein ($P_t$). Concomitant changes in the unphosphorylated ($P_0$) and phosphorylated, cytosolic ($P_1$ and $P_2$) and nuclear forms of PER are shown in (b), together with the total, nonconserved amount of PER protein. The curves are obtained by numerical integration of eqns (11.1a–e); $P_t$ is given by eqn (11.2). Parameter values are: $v_s = 0.76$ μM/h, $v_m = 0.65$ μM/h, $k_s = 0.38$ h$^{-1}$, $v_d = 0.95$ μM/h, $k_1 = 1.9$ h$^{-1}$, $k_2 = 1.3$ h$^{-1}$, $K_I = 1$ μM, $K_{m1} = 0.5$ μM, $K_d = 0.2$ μM, $K_1 = K_2 = K_3 = K_4 = 2$ μM, $n = 4$, $V_1 = 3.2$ μM/h, $V_2 = 1.58$ μM/h, $V_3 = 5$ μM/h, $V_4 = 2.5$ μM/h. The tentative concentration scale, in μM, is given for clarity; this scale might be changed as more experimental information becomes available.

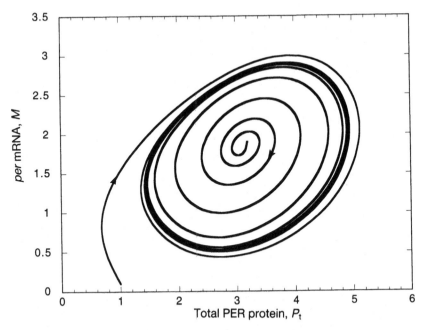

Fig. 11.8. Sustained oscillations in PER protein and *per* mRNA correspond to the evolution towards a limit cycle in the $(M, P_t)$ plane. Starting from two different sets of initial conditions, the system eventually reaches a unique, closed curve characterized by a period and amplitude that are fixed for the given set of parameter values. The initial conditions, located outside or inside the limit cycle, are, respectively, in μM (see remark about the concentration scale in the legend to fig. 11.7): $M = 0.1$, $P_0 = P_1 = P_2 = P_N = 0.25$ $(P_t = 1)$, and $M = 1.9$, $P_0 = P_1 = P_2 = P_N = 0.8$ $(P_t = 3.2)$. The trajectories are obtained as in fig. 11.7, for the same set of parameter values (Goldbeter, 1995).

phosphorylation of $P_0$ into $P_1$ and of $P_1$ into $P_2$ occurs thereafter. At that time, the synthesis of M still proceeds at a nearly maximum rate. As $P_2$ progressively accumulates, it is transported into the nucleus. It is only when the level of $P_N$ reaches a threshold that the transcription of the *per* gene becomes impaired; the rate of synthesis of M begins to drop. When its effective rate of synthesis becomes smaller than its rate of degradation, *per* mRNA starts to decrease. As a result of the drop in M, the levels of $P_0$, $P_1$, $P_2$ and $P_N$ begin to decrease, successively. These falls are compounded by the degradation of PER, directed here at the cytosolic, bisphosphorylated form $P_2$ (similar results would be obtained if the degradation would affect the nuclear form $P_N$). A phase of synthesis of M and a new cycle of the oscillations resume as soon as the level of $P_N$ drops below the threshold for repression of *per* transcription. Such a verbal explanation of the periodic sequence of events, on the basis of fig. 11.7, should not obscure the fact that sustained oscillations occur in the model only under specific conditions, in a restricted region of parameter space.

Although the precise role of multiple PER phosphorylation remains unclear, the above model assumes that full phosphorylation of PER (here, at only two residues, but the number of such residues could readily be increased) is required for both PER degradation, and entry into the nucleus. In this sense, the model suggests that, in dynamic terms, multiple phosphorylation of PER can be viewed as providing a series of delays that enhance the destabilizing effect of the negative feedback exerted by PER on the transcription of the *per* gene. As shown in many cases (see chapter 10 and, for additional examples, Landahl, 1969; May, 1973; an der Heiden, 1979; Mizraji, Acerenza & Hernandez, 1988), such time delays favour the occurrence of sustained oscillations in regulated biochemical systems.

Another factor that favours oscillations is the steepness of the function characterizing the repression of *per* transcription by nuclear PER. Such steepness depends on the degree of cooperativity, $n$, of the repression process, which is described in eqn (11.1a) by a Hill function. In agreement with such a destabilizing role of cooperativity, oscillations in fig. 11.7 were obtained for a value of $n = 4$; however, periodic behaviour can also occur in the model for $n = 2$ and for $n = 1$, but the domain in parameter space where sustained oscillations occur is smaller than for $n = 4$.

The simulations shown in fig. 11.7 are of a qualitative rather than a quantitative nature. Parameter values have indeed been chosen so as to yield a period of the order of 24 h. A scale in $\mu$M is given for the concentrations of *per* mRNA and PER protein; this scale is only tentative (a scale in nM or pM could be used, if necessary) since detailed biochemical information is still lacking on the kinetics of PER synthesis, degradation, reversible phosphorylation, and transport into and out of the nucleus, and on the intracellular levels of PER and *per* mRNA (see, e.g. Zeng *et al.*, 1994). What the data in fig. 11.7 show, however, is that if the period is close to 24 h, then the phase shift between the peak of *per* mRNA and the peak of total PER is of the order of 4 h. Such a result holds with the observation that the maximum in PER protein follows the peak in *per* mRNA by about 4 h (Zeng *et al.*, 1994; see fig. 11.5b). As shown in fig. 11.7a, the phase shift between nuclear PER and *per* mRNA is longer, of the order of 7 h, under the conditions of fig. 11.7.

### Altering the period, or suppressing rhythmicity, in the model for PER oscillations

Besides showing how the negative feedback exerted by PER can give rise to sustained oscillations in *per* gene expression, the model can be used to predict how changes in parameter values alter the period of oscillations or even suppress rhytmic behaviour. This is of particular value, given the existence of short- and long-period mutants of circadian rhythmicity in *Drosophila*. The model may help to pinpoint the type of mutation that may result in slowing down or accelerating circadian rhythmicity. All parameters of the model for PER oscillations can in principle control the period of the rhythmic phenomenon, to an extent that varies with the parameter considered. We focus here on a few control parameters that appear to play a prominent role in the mechanism of oscillatory behaviour.

The first parameter considered is $v_d$, which measures the maximum rate at which PER is degraded. When all other parameter values remain as in fig. 11.7, numerical simulations show that sustained oscillations occur in a window bounded by two critical values of this parameter, close to 0.45 and 2.6, respectively (in $\mu$M/h if a scale of $\mu$M is chosen for concentrations). In this interval of $v_d$ values, the period of the oscillations increases from a value close to 19.7 h up to a value close to 64 h. The period range in the window of $v_d$ values depends on other parameters. Thus, for a larger value of the rate of protein synthesis measured by $k_s$, the period varies as a function of $v_d$ from 15.9 to 62.1 h (fig. 11.9).

The fact that the period of PER oscillations increases with the rate of PER degradation is due to the longer time required to reach the threshold beyond which PER significantly represses the transcription of its gene. The lengthening of the period is also due to the fact that the amplitude of the oscillations in $M$ – and, consequently, in the various forms of PER – rises with $v_d$, since *per* mRNA has more time to accumulate before transcription shuts off.

The range of period found in the window of $v_d$ values fits reasonably well with the range found in rhythmic *per* mutants of *Drosophila*. Thus, the short- and long-period mutants, *per*[s] and *per*[l], have a period of 19 h and 28 h, respectively. The ultrashort *per*[T] mutant recently described (Konopka *et al.*, 1994), which has a period of 16 h, would also fall within the range shown in fig. 11.9. Longer periods, of up to 50 h, such as those obtained in the model at high values of $v_d$ in the situation considered in fig. 11.9, have been observed in the *per*[l] mutant at low light intensity (Konopka *et al.*, 1989). Moreover, while *frq* circadian mutants in

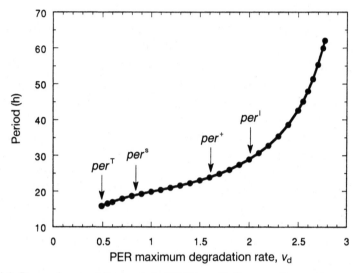

Fig. 11.9. Dependence of the period of PER oscillations on the maximum rate of PER degradation, $v_d$. Tentative vertical arrows correspond to values of parameter $v_d$ that would produce the phenotypes of the ultrashort mutant $per^T$ and of the short- and long-period *per* mutants, as well as that of wild type *per*. The data points are obtained by numerical integration of eqns (11.1a–e) for different values of parameter $v_d$ ($\mu$M/h) other parameter values are as in fig. 11.7 with $k_s = 0.78\ \mathrm{h}^{-1}$ (Goldbeter, 1995).

*Neurospora* have a period that generally ranges between 16.5 and 29 h, a quadruple mutant, affected in several genetic loci, has been isolated, with a period of 58 h (Feldman, 1982).

The results shown in fig. 11.9 support a plausible explanation for the *per* mutant phenotypes. The effect of parameter $v_d$ shows that mutations that would affect the structure of PER so as to hinder or enhance its enzymic degradation would result in a shortening or lengthening of the period, respectively; the putative positions of the $per^T$, $per^s$ and $per^l$ mutants, together with that of the wild type $per^+$, are indicated by vertical arrows in fig. 11.9. That PER could become a better or less good substrate for degradation could also be due to changes in its pattern of multiple phosphorylation, as a result of these mutations. The data of fig. 11.9 further indicate that rhythmic behaviour may disappear altogether if the effect of mutations is such that the corresponding parameter (here, the maximum rate of PER degradation) quits the window in which oscillations occur. This could well happen for some arrhythmic *per* mutants.

The above explanation in terms of variable PER degradation holds with the results of experiments in which the amount of *per* transcript was manipulated (Baylies *et al.*, 1987; see below). Baylies *et al.* concluded from that work that one possible interpretation of their results would be that "in *per^s* and *per^l*, amino-acid substitutions respectively increase or decrease the stability of the *per* protein'; a similar suggestion was made by Coté & Brody (1986). With regard to the above discussion, the stability of PER would inversely correlate with its rate of degradation. The explanation of *per* mutants in terms of the PER degradation rate is only tentative and illustrates how the model may serve to pinpoint the parameters whose changes might cause the observed period alterations.

There are, of course, other possible explanations for the observed changes in period or suppression of the *Drosophila* circadian rhythm. Thus, another important parameter that may control the oscillations is the apparent first-order rate constant $k_1$ that measures the transport of PER into the nucleus. Under the conditions of fig. 11.7, there is again a range of $k_1$ values that produce sustained oscillations; this range extends from $0.02 \text{ h}^{-1}$ to $5.5 \text{ h}^{-1}$. Here, the period of oscillations decreases as $k_1$ increases. The dependence of the period on parameter $k_1$ can be understood as for parameter $v_d$: when transport of $P_2$ into the nucleus is slower, $P_N$ accumulates at a lower rate so that it takes more time for $P_N$ to reach the threshold for repression of *per* transcription. In terms of parameter $k_1$, the *per^s* and *per^l* phenotypes might thus be respectively associated with enhanced and decreased rate of transport of PER into the nucleus.

If the transport rate into the nucleus decreases below the lower critical value, equal to $0.02 \text{ h}^{-1}$ in the case of fig. 11.7, sustained oscillations are suppressed. Such a situation could correspond to the loss of circadian rhythmicity in the recently characterized *timeless* (*tim*) mutant in *Drosophila* (Sehgal *et al.*, 1994); nuclear localization of PER appears to be blocked in that mutant (Vosshall *et al.*, 1994). A key role for transport of PER into the nucleus is consistent with the view that the negative feedback exerted by PER on transcription is at the core of the mechanism of circadian rhythmicity.

Besides $k_1$ and $v_d$, other parameters can of course affect the amplitude and period of oscillations in the PER protein and *per* mRNA. In that sense, although PER appears to be the main actor in the oscillatory mechanism, any biochemical process that alters the amount of PER (via synthesis, degradation, or transport into the nucleus) or its phosphorylation status (which may itself bear on PER degradation or transport into the nucleus) may control the properties of the oscillations or the very existence of the rhythm. The oscillatory mechanism involves not only the *per* gene product but also those gene products that impinge on the level of PER in the cytosol and in the nucleus. Clock mutants other than *per*, of which *tim* provides a recent example, should therefore exist in *Drosophila*.

### Effect of per gene dosage on circadian oscillations

Interesting results were obtained in experiments in which the dosage of the *per* gene was altered. Thus, Smith & Konopka (1982) showed that decreased *per* dosage lengthens the period whereas increased dosage has the opposite effect. The shortening effect of more than one *per* dose saturates at a level of about -1.5 h. Baylies *et al.* (1987) found, similarly, that a 3-fold drop in the amount of *per* mRNA leads to an increase in period from 24 to 27 h. Because the *per^s* ands *per^l* mutants make the same amount of *per* mRNA as the wild type, these authors suggested (see also Coté & Brody, 1986) that the two mutations respectively increase or decrease the stability of the PER protein, or its activity.

It is tempting to test the model presented above for the gene dosage effect of *per*. A question that arises, however, is how to represent in the kinetic equations a change in the amount of *per* gene. One way is to alter the rate of *per* mRNA synthesis, $v_s$, assuming that this rate is proportional to the amount of *per* gene present (such an assumption may not hold; see below). Numerical simulations show that, when the value of $v_s$ increases in the model, the period of the oscillations rises, passing from close to 24 h in the conditions of fig. 11.7, where $v_s = 0.76\ \mu M/h$, to 60.8 h for $v_s = 4\ \mu M/h$, and even to larger values when $v_s$ continues to increase. Checking the effect of a decrease in $v_s$ below the value taken in fig. 11.7 is not feasible, as the oscillations become damped when $v_s$ becomes less than 0.65 $\mu M/h$ (this means that the value of $v_s$ in fig. 11.7 is very close to the lower boundary of the oscillatory domain). The result that the period goes up with increasing parameter $v_s$ is in contrast with experimental observations, which indicate that the period decreases when the dosage of the *per* gene rises.

The reason why the period rises with parameter $v_s$ in the model governed by eqns (11.1a–e) is that the amount of *per* mRNA accumulates more and more rapidly as $v_s$ increases; this results in a more and more massive synthesis of PER protein, which eventually will shut down transcription for a prolonged period of time, bringing M down to a very low level. This occurs because in the conditions of fig. 11.7 the limiting process is the degradation of *per* mRNA, which proceeds at a quasi-constant rate when $M$ reaches values that saturate the Michaelian enzyme by which *per* mRNA is degraded. Thus, as $v_s$ progressively increases, the amplitude of $M$ and, consequently, of the levels of the various forms of PER, also increases, which makes the period rise.

Changing the Michaelian sink of *per* mRNA to a linear one – i.e. assuming that the RNA-degrading enzyme has a much larger $K_m$ – greatly improves the situation. Indeed, under these conditions, the increase in period as $v_s$ rises remains much more limited, of the order of a few hours only. Thus, an increase in $v_s$ has two antagonistic effects. On the one hand, the accumulation of $M$ is such that it will take more time to remove the newly synthesized *per* mRNA and to shut off synthesis of the PER protein – this effect enlarges the period. On the other hand, synthesizing more PER protein should shorten the time required to reach the threshold for repression by PER – this should, on the contrary, bring down the period of oscillations. Under the conditions of fig. 11.7, the former effect remains predominant.

In discussing the effect of gene dosage, we should take into account the possibility that more subtle regulatory processes may be at work, which are not encompassed by eqns (11.1a–e). Thus, Smith & Konopka (1982) suggested the existence of mechanisms that would partially compensate for a variation in *per* gene dosage; such mechanisms would regulate the amount of *per* gene product, depending on the quantity of *per* gene present.

### *Suppression of circadian oscillations by overexpression of PER*

In a recent study, Zeng *et al.* (1994) showed that the overexpression of PER in photoreceptor cells in *Drosophila* eyes suppresses circadian rhythms in these cells, but not in other *per*-expressing cells located in the brain. The experiment was performed by means of a construct containing the promoter region of the *Drosophila* opsin gene followed by a fragment of *per*.

To account for this observation, we may treat the overproduction of PER as resulting from an increase in the level of *per* mRNA, for example by increasing parameter $v_s$ (see above). However, it is only when $v_s$ increases by several orders of magnitude that oscillations are suppressed under the conditions of fig. 11.7. Another way to address this experiment is to consider that the protein made from the construct represses the replication of endogenous *per*. Including the presence of high levels of this 'additional' repressor, besides endogenous PER, is tantamount to considering that the effective value of the repression constant $K_I$ for endogenous PER has been lowered. The threshold constant $K_I$ measures the level of $P_N$ at which repression reaches its half-maximum value.

Under the conditions of fig. 11.7, in which $K_I = 1$ $\mu$M and the period is close to 24 h, sustained oscillations disappear when $K_I$ decreases below 0.2 $\mu$M; $P_N$ then settles at a constant value, just above the value of $K_I$. For intermediate values between 1 and 0.2 $\mu$M, the period of oscillations decreases as $K_I$ diminishes; thus, for $K_I = 0.5$ $\mu$M, the period is close to 17 h. If the experiment is feasible, it would be interesting to determine how the period of the oscillations varies when the expression of the construct made by Zeng *et al.* (1994) is progressively increased, before oscillations are suppressed.

### *Temperature compensation of the circadian clock*

That the period of circadian oscillations remains largely independent of temperature represents one of the most characteristic properties of these rhythms. Such temperature compensation is also observed for the circadian clock in *Drosophila* (Pittendrigh, 1954). The origin of temperature compensation remains unclear. It is intriguing that, in contrast to the wild type, *per* mutants have lost this property. Thus (see fig. 11.4), between 15 °C and 26 °C, the period of *per*[l] flies increases with temperature whereas the period of *per*[s] flies slightly decreases (Konopka *et al.*, 1989). A molecular explanation for temperature compensation of circadian oscillations in *Drosophila* has recently been proposed (Huang *et al.*, 1995). Investigating such a mechanism in a theoretical manner will require extensions of the models presented for circadian PER oscillations.

Attempts to explain the property of temperature compensation by means of theoretical models have been made (see, for example, Pavlidis & Kauzmann, 1969). A recent theoretical study of an abstract, time-delay oscillator model (Lakin-Thomas *et al.*, 1991) suggests that temperature compensation might involve changes in the amplitude of the circadian rhythm as a function of temperature. Thus, while a rise in temperature may accelerate biochemical reactions, thereby tending to shorten the period, an antagonistic effect would result from an increase in the amplitude of the oscillations. In the phase plane, the oscillating system would take roughly the same time when travelling at an increased speed along a large limit cycle than at a lower velocity along a closed trajectory of smaller amplitude.

A similar effect involving modulation of the amplitude of oscillations might be obtained in the model proposed above. That this may be the case is suggested, for example, by the significant variation in amplitude produced by a change in parameters $v_s$ or $v_d$. To obtain such a result, however, temperature should affect certain parameters more strongly than others. A general, indiscriminate increase in all kinetic parameters as a function of temperature would indeed bring down the period without altering the amplitude of oscillations. Differences in the effect of temperature on the various parameters may be expected, since some parameters relate to enzyme reactions while others pertain to transport across the nuclear membrane.

### Relevance of the model to other circadian rhythms and to the courtship song rhythm in **Drosophila**

Although it is based on the evidence that has accumulated in recent years on the role of the *per* gene in the circadian clock of *Drosophila*, the model presented above provides a prototype for circadian oscillations in other organisms. Of all the features of circadian rhythms brought to light by studies in different systems, one of the most conspicuous is the role played by protein synthesis in the oscillatory mechanism. Such a role is attested by the observation repeatedly made that inhibitors of protein synthesis alter the period of circadian oscillations or produce phase shifts in these rhythms (Karakashian & Hastings, 1963; Feldman, 1967; Jacklet, 1977; Taylor *et al.*, 1982a,b; Eskin *et al.*, 1984; Dunlap & Feldman, 1988; Morse *et al.*, 1990; Khalsa *et al.*, 1992; Takahashi *et al.*, 1993). Similar results have been obtained by means of an inhibitor of transcription (Raju *et al.*, 1991).

Protein and mRNA synthesis play a key role in the mechanism proposed for circadian oscillations in *Drosophila* (Hardin *et al.*, 1990, 1992) on which the above model is based. Regulatory feedback is needed to produce sustained oscillations. A natural candidate for such regulation is the negative control exerted by the PER protein on the synthesis of its mRNA (Hardin *et al.*, 1990, 1992). Such a negative feedback on transcription has also been observed for the *frq* gene in *Neurospora* (Aronson *et al.*, 1994) and could well prove to be a general feature of the mechanism of circadian rhythmicity in unicellular as well as multicellular organisms (Takahashi, 1993; Sassone-Corsi, 1994). The existence of a unified mechanism for the circadian clock – encompassing mRNA and protein synthesis, as well as their control by a key protein behaving as a regulator of transcription – would parallel the situation encountered for the cell cycle (Takahashi, 1992), which turns out to possess universal features among all eukaryotic organisms. Moreover, that protein phosphorylation plays a role in circadian rhythms in other organisms besides *Drosophila* is shown by experiments which indicate that an inhibitor of protein kinase activity blocks circadian rhythms in *Gonyaulax* (Comolli *et al.*, 1994)

It remains unclear how a gene involved in circadian rhythmicity in *Drosophila* can also be involved in the control of the rhythm of about 1 min period that characterizes the courtship song (Kyriacou & Hall, 1980; Kyriacou *et al.*, 1993; Hall, 1994). The *per*$^l$ and *per*$^s$ mutations affect the song rhythm and the circadian rhythm in the same direction. However, gene dosage affects more the song rhythm than the circadian clock (Smith & Konopka, 1982; Baylies *et al.*, 1987, 1993). This, together with the results of transformation experiments with modified *per* genes (Yu *et al.*, 1987a), suggests that the *per* gene product controls in a different manner the circadian and ultradian rhythms. While we can conceive how a master gene product varying in a circadian manner (e.g. a transcription factor or a transcriptional regulator, as in the case of PER in *Drosophila*) might govern the circadian variation of a large number of enzymes or ionic channels through controlling the expression of the corresponding genes, it is still unclear how such a protein could participate in the control of a rhythm of a 1 min period. Perhaps the level of a key protein other than PER, involved in song rhythm production, is altered in *per* mutants. The theoretical model discussed in this chapter deals only with circadian rhythmicity and does not address the question of the origin of the song rhythm, which awaits further experimental progress.

### *Comparison with other models for biochemical oscillations based on negative feedback*

The model for circadian oscillations in PER protein and *per* mRNA is closely related to the model proposed some 30 years ago by Goodwin (1963, 1965), who discussed the conditions in which a protein repressing the transcription of its gene can produce sustained oscillations in the levels of that protein and its mRNA. A model of the Goodwin type was used explicitly for circadian rhythms, to determine phase response curves with respect to transient perturbations (Drescher *et al.*, 1982).

Because they were not sufficiently nonlinear, the equations originally proposed by Goodwin did not give rise to limit cycle oscillations. Equations of that type were investigated for limit cycle behaviour in further detail by Griffith (1968) and, more recently, by Sinha & Ramaswamy (1988). Similar equations were also studied for the related problem of how sustained oscillations occur in a metabolic chain of enzyme reactions regulated by end-product inhibition (Morales & McKay, 1967; Walter, 1970; Hunding, 1974; Rapp, 1975, 1980; Mees & Rapp, 1978; Tyson & Othmer, 1978; Palsson & Groshans, 1988). An experimental system studied by means of related models is the oscillatory synthesis of tryptophan, in which negative control by the end product is exerted at the levels of both gene and enzyme (Bliss, Painter & Marr, 1982; Tyson, 1983; Painter & Tyson, 1984; Sen & Liu, 1990).

The main result coming from these studies was that the occurrence of sustained oscillations is favoured both by enlarging the length of the reaction chain that leads from the regulated step to the end product, and by increasing the degree of cooperativity of the negative feedback process. These results bear on the model for PER oscillations. Here, the sequence of successive phosphorylations of the PER protein can be viewed as playing a role similar to that of increasing the number of intermediate steps in the chain of enzyme reactions regulated by end-product inhibition. A similar conclusion was reached in the model analysed in chapter 10 for the mitotic oscillator. Incorporating more than two phosphorylation steps in the model enlarges accordingly the domain of sustained oscillations.

With regard to the role of nonlinear feedback control, periodic behaviour was illustrated in fig. 11.7 for a repression function characterized by a Hill coefficient of 4; however, a value of 2 or 1 for that cooperativity coefficient can also give rise to sustained oscillations. The cooperativity of repression provides a major source of nonlinearity required for sustained oscillatory behaviour. This is the reason why steep thresholds due to zero-order ultrasensitivity are not required here to generate limit cycle oscillations (see the relatively large values of the reduced Michaelis constants $K_i$ used in fig. 11.7), in contrast to the situation encountered for the phosphorylation–dephosphorylation cascade model analysed for the mitotic oscillator.

The domain of oscillations in parameter space decreases in the model when the cooperativity index $n$ goes from 4 to 2, and then to 1. This shows that if the cooperativity of repression favours periodic behaviour, multiple phosphorylation of PER, by introducing a series of delays, reinforces, and could even substitute for the effects of such cooperativity in allowing the occurence of sustained oscillations. This conclusion supports the view (Abbot *et al.*, 1995; Curtin *et al.*, 1995) that the gating of PER entry into the nucleus, possibly due to PER phosphorylation, delays the negative feedback exerted by PER on *per* transcription, and thereby, at the same time, strengthens the capability of such feedback to produce robust oscillations and contributes to raising their period up to circadian values.

Many models for biochemical oscillations rely on positive feedback, as illustrated in this book by the case of glycolytic (chapter 2), cAMP (chapter 5) or $Ca^{2+}$ (chapter 9) oscillations. Instabilities can, nevertheless, also arise from negative regulation. Together with the models discussed in chapter 10 for the mitotic oscillator in embryonic cells, for which a combination of positive and negative feedback may contribute to periodic behaviour, the model for circadian oscillations in the *period* protein in *Drosophila* supports the view that negative feedback is at the core of some of the most important biological rhythms.

# 12

# Conclusions and perspectives

## From the molecular properties of regulation to the temporal self-organization of biological systems

Glycolytic oscillations in yeast and muscle represent a classic example of periodic behaviour in biochemistry. Likewise, oscillations of cAMP in *Dictyostelium* amoebae are the prototype of intercellular communication by periodic signals of a nonelectrical nature. Besides these examples known for more than 20 years (Hess & Boiteux, 1971; Goldbeter & Caplan, 1976; Hess & Chance, 1978; Berridge & Rapp, 1979; Rapp, 1979), additional periodic phenomena in biochemistry have been characterized during the last decade. Conspicuous by their widespread nature are intracellular oscillations of $Ca^{2+}$, which occur in a large variety of cell types, either spontaneously or as a result of external stimulation. The experimental demonstration of oscillations in cytosolic $Ca^{2+}$, together with the unravelling of the molecular basis of the mitotic oscillator driving the cell division cycle in eukaryotes, represents the most significant advances in the field of biochemical oscillations in recent years. Also important is the progress currently made in elucidating the regulatory mechanisms underlying circadian rhythms. The study of theoretical models presented in this book allowed us to clarify the molecular mechanism of these oscillations that provide some of the best-known examples of biochemical and cellular rhythms.

The use of models based on experimental observations has shown how the regulatory properties at the level of enzymes and receptors can give rise to nonequilibrium, temporal self-organization in the form of sustained oscillations of the limit cycle type. In this, sustained oscillations represent examples of the dissipative structures described by Prigogine (1969). Like spatial or spatiotemporal dissipative structures, limit cycle oscillations occur beyond a critical point of instability and

491

are maintained solely by energy dissipation, in a system that exchanges matter with its environment (Glansdorff & Prigogine, 1971; Nicolis & Prigogine, 1977).

Although it relies on molecular processes only, this nonequilibrium, temporal self-organization is manifested on the macroscopic scale of cellular behaviour (Hess, 1973; Hess, Goldbeter & Lefever, 1978). The kinetic constants characterizing the catalytic activity of phosphofructokinase in yeast, the phosphorylation of the cAMP receptor in *D. discoideum*, the activity of the various kinases and phosphatases involved in the cascade regulating the activation of cdc2 kinase during the cell cycle, or the pumping and release of $Ca^{2+}$ into and out of intracellular stores give rise to periodicities of the order of seconds to tens of minutes, while the molecular events occur on much shorter time scales. The emergence of oscillations represents a phenomenon of supramolecular organization: the period of the oscillations and their very existence depend on each of the parameters of the whole system. Thus, phosphofructokinase isolated in solution would not be capable of rhythmic behaviour: to oscillate, the enzyme must be supplemented with a source of substrate and with a reaction that ensures the transformation of the reaction product. Another clear example of the role of the supramolecular kinetic organization in the onset of oscillations is given by the phosphorylation cascade controlling the periodic activation of cdc2 kinase. Elucidating the molecular bases of periodic (and chaotic) behaviour requires the identification of the molecules involved and of their regulatory interactions. The picture of the oscillatory mechanism becomes complete only when both aspects have been successfully addressed, and when a theoretical model based on these results is capable of reproducing the observed oscillations.

Most experimental observations on these biochemical periodic phenomena can be accounted for in terms of the dynamic behaviour of models containing two or three variables. In the case where two variables suffice to describe the oscillatory dynamics, phase plane analysis throws light on the close link that exists between excitable and oscillatory behaviour (Hahn *et al.*, 1974; Goldbeter & Erneux, 1978; Goldbeter, Erneux & Segel, 1978; Martiel & Goldbeter, 1987a). Thus, the mechanisms considered for glycolytic and cAMP oscillations give rise to similar phase portraits, despite the differences in reaction kinetics. There exists, moreover, a significant degree of unity between the results obtained in the analysis of the various biochemical models and those obtained in the theoretical study of reduced, two-variable models for the nerve membrane (Fitzhugh, 1961; Kokoz & Krinskii, 1973; Krinskii & Kokoz, 1973; Rinzel, 1985; Av-Ron *et al.*, 1991; Goldbeter, 1992). While the molecular basis of excitable and oscillatory behaviour is biochemical and relies on the regulation of enzymes, receptors or transport proteins, it is electrical in the case of excitablity and oscillations observed in neurons.

Clearly, because of their very simplicity, the models studied in this book cannot account for all experimental details concerning glycolytic oscillations, the oscillatory synthesis of cAMP in *Dictyostelium*, $Ca^{2+}$ oscillations, the periodic activation of cdc2 kinase, or the circadian rhythm in the *period* protein in *Drosophila*. The virtue of these models, however, is to help to clarify the molecular mechanism of these oscillatory phenomena while providing a qualitative – and, in a certain measure, quantitative – agreement with a large number of experimental observations. Extensions of these models to a larger number of variables permit us to refine the agreement with the experiments. The greater complexity of such extended models, however, makes them more unwieldy and prevents us from resorting to phase plane analysis or to other useful, simple approaches permitted by a smaller number of variables.

A model is like a map from which we expect help in finding our way in the surrounding world. To be useful and to perform its function, this map must be drawn to the appropriate scale. A passage in a story by Borges refers to the decline of a geographical school whose members, obsessed by precision and research of details, eventually drew maps on a 1:1 scale, thereby annihilating the very function of their craft. Similarly, models should be sufficiently detailed without becoming too complex: too far removed from experimental data, they would be unable to lead to testable predictions; too complicated, because of taking into account too many variables or parameters, their analysis would become so cumbersome that such models might be of little use. The study of realistic models containing few variables, as attempted here, has the additional merit of highlighting the core mechanism responsible for oscillations. A model with 15 variables makes it difficult to establish a hierarchy between the different processes involved in the onset of a rhythm.

It remains true, nevertheless, that variations on the precise nature of the instability-generating mechanism generally lead to similar results. Thus, for cAMP synthesis in *Dictyostelium*, a model differing from the one analysed in chapter 5 was proposed by Othmer *et al.* (Othmer *et al.*,

1985; Rapp *et al.*, 1985; Monk & Othmer, 1989). That model is also based on the activation of cAMP synthesis by extracellular cAMP. While this self-amplification phenomenon is counteracted by receptor desensitization in the model of Martiel & Goldbeter (1987a) (see chapter 5), Othmer *et al.* ascribe the role of counterbalance to the putative inhibition of adenylate cyclase by $Ca^{2+}$, whose influx into the cell is triggered by the cAMP signal. Relay and oscillations of cAMP also occur in that model, in agreement with experimental observations. The direct inhibition of adenylate cyclase by $Ca^{2+}$, however, has not been demonstrated in *Dictyostelium*. Moreover, the model by Othmer *et al.* does not take into account the causal link observed between covalent modification of the receptor and the process of adaptation. Receptor desensitization is included in the model proposed more recently by Tang & Othmer (1994a), which further incorporates the role of G-proteins, as also considered in section 5.8.

The analysis of models for glycolytic oscillations, for the periodic synthesis of cAMP in *Dictyostelium*, and for $Ca^{2+}$ oscillations indicates that kinetic nonlinearities play an essential role in the occurrence of the instability leading to oscillations. In these systems, such nonlinearities arise from regulatory processes (see below) and from cooperative phenomena. Other sources of nonlinearity in enzyme systems may arise from electric repulsion effects in charged membranes (Mulliert, Kellershohn & Ricard, 1990). Cooperativity and regulation, which are the most common sources of nonlinearity in biochemistry, are often closely associated, as in allosteric enzymes. Such a link further enhances the nonlinear nature of the system and thereby favours the onset of periodic behaviour. Thus glycolytic oscillations result from the regulatory and allosteric properties of phosphofructokinase. The structural basis of these properties has been elucidated by means of a crystallographic study of the enzyme (Schirmer & Evans, 1990).

The detailed analysis of the model for the product-activated allosteric enzyme allowed a precise quantification of the role played by enzyme cooperativity in glycolytic oscillations (see chapter 2). However, the analysis of a slightly modified model in which the sink of the product becomes Michaelian – i.e. saturable – instead of linear, showed (see section 2.7) that oscillations can occur even if the allosteric enzyme contains a single subunit existing in two conformational states. Enzyme cooperativity is therefore not a condition *sine qua non* for oscillations to occur: weaker nonlinearities, of the Michaelian type, distributed over several reactions of the system, can thus cooperate to raise its global degree of nonlinearity. This result highlights the role of **diffuse non-linearities** in the origin of sustained oscillations (Goldbeter & Dupont, 1990). A similar result is obtained in the two-variable model for signal-induced $Ca^{2+}$ oscillations (Dupont & Goldbeter, 1989; Goldbeter *et al.*, 1990): there also, cooperativity favours oscillations, but the latter may nevertheless occur in its absence, provided that the appropriate regulation is present and that several reaction steps possess Michaelian kinetics (Goldbeter & Dupont, 1990). The same conclusion was again reached in the model for circadian PER oscillations considered in chapter 11.

A related observation was made by Thron (1991), who investigated the conditions for oscillations in metabolic chains regulated by end-product inhibition. In such systems a certain degree of cooperativity characterizing the inhibition of the first enzyme of the chain is required for oscillatory behaviour (Morales & McKay, 1967; Walter, 1970; Hunding, 1974; Rapp, 1975; Tyson & Othmer, 1978). The degree of cooperativity required decreases, however, when the sink of the end product instead of being linear becomes Michaelian (Thron, 1991).

The analysis of the model for cAMP signalling in *Dictyostelium* also indicates that nonlinearity is essential for obtaining periodic behaviour. It turns out, however, that the precise form of this nonlinearity does not matter too much. In the case of the slime mould, oscillations and relay of cAMP occur, with very similar time courses, when the cooperativity pertains to the binding of cAMP to the receptor or to the activation of adenylate cyclase by the cAMP–receptor complex. Similar results as to relay and oscillations are again recovered (Tang & Othmer, 1994a; Halloy, 1995; Halloy & Goldbeter, 1995; see section 5.8) when the model is extended to take into account the role of G-proteins in the transduction of the signal from the receptor to adenylate cyclase.

### Biological rhythms: limit cycles or discontinuous iterative behaviour?

The examples of rhythmic behaviour analysed in this book all belong to dynamics of the limit cycle type. In the case of phosphofructokinase, like in that of cAMP synthesis in *Dictyostelium* or signal-induced $Ca^{2+}$ oscillations, the analysis of models based on experimental data indeed shows that these systems admit a nonequilibrium steady state that becomes unstable beyond a critical value of some control parameter. It is in these conditions that sustained oscillations occur, in the form of a limit cycle in the phase space.

The question arises as to whether endogenous rhythms in biology are all of the limit cycle type. Are there biological periodicities which do not correspond to the continuous dynamics associated with a limit cycle? An important example in this regard is the cell division cycle. Until recent observations, reviewed in chapter 10, allowed the clarification of the molecular bases of the phenomenon, two contrasting views of the mitotic cycle were put forward (see also Winfree, 1984). According to the first view (Kaufmann, 1974; Kaufmann & Wille, 1975), the periodic nature of mitosis is due to the existence of a continuous biochemical oscillator. Cell division would occur whenever one of the variables of the oscillating system, behaving as a mitogenic factor, exceeded some critical level. Small-amplitude oscillations could thus occur in the absence of mitosis.

The second view did not rely on the existence of an oscillator independent from mitosis and stated that the latter is an integral part of the mechanism underlying the cell division rhythm (Tyson & Sachsenmaier, 1978, 1984). The rhythm would thus have a simpler origin, due to the discontinuity of cell division. Tyson & Sachsenmaier (1978) compared such a rhythm to the effect of a thin stream of sand constantly falling on a pan on one side of a balance: when a critical weight is reached, the pan tips over and, once emptied, recovers its equilibrium position. The sand then accumulates again until the next flip. In this phenomenon, the phase of abrupt decrease that precedes the replenishment phase consists of the discontinuous reversal of the position of the pan. This image could apply to the case of the mitotic cycle: the latter could result from the accumulation of a mitotic factor up to a threshold beyond which the discontinuity of cell division would occur. No oscillation should take place in the absence of mitosis.

Experiments carried out on mitosis in *Physarum polycephalum* (Kaufmann & Wille, 1975; Tyson & Sachsenmaier, 1978) did not allow a definitive conclusion in favour of one of the two views. The difficulty came from the fact that, short of the identification of the true biochemical variables driving the cell cycle, it was difficult to demonstrate the existence of a limit cycle in *Physarum*. Experiments based on the fusion of two plasmodia taken at different phases of the mitotic cycle or on the effect of heat shocks aimed at demonstrating the existence of a singular point from which the limit cycle would be reached with an indefinite phase (Winfree, 1974, 1980, 1987). These experiments did not allow the distinction to be made between a limit cycle characterized by relaxation oscillations and a discontinuous mechanism of the type discussed above.

From a theoretical point of view (Andronov, Vitt & Khaikin, 1966), the two situations can be approached with comparable methods.

The recent experimental advances on the biochemical mechanism driving the eukaryotic cell division cycle have shed much light on the debate regarding the continuous vs discontinuous nature of the mitotic cycle. These experiments demonstrated (see chapter 10) that mitosis is brought about by the activation of cdc2 kinase. The continuous biochemical oscillator that produces the periodic activation of cdc2 kinase is driven by cyclin synthesis (Murray & Kirschner, 1989b). In fact, the mitotic oscillatory mechanism reconciles the two contrasting views of mitosis based, on one hand, on a succession of causally linked, quasi-discontinuous events leading to cell division with the subsequent return to the initial state (such a view arose particularly from genetic studies of the yeast cell cycle; see Hartwell & Weinert, 1989) and, on the other hand, on the existence of a continuous oscillator. That the two situations converge in the present case was well stressed by Murray & Kirschner (1989a) in the title of one of their articles, 'Dominoes and clocks: the union of two views of the cell cycle'. The analysis of the cascade model for the mitotic oscillator, developed in chapter 10, as well as that of other models (Novak & Tyson, 1993a,b), supports the latter conclusion: these models show that the sequence of well-separated biochemical steps leading to the activation of cdc2 kinase can repeat itself periodically in a continuous, limit cycle manner, however discontinuous the activation of each of the steps of the sequence may be.

An example of a rhythm for which the existence of a limit cycle can still be questioned is the process of ovulation. A theoretical model shows (Lacker, 1981; Lacker, Feuer & Akin, 1989) how the follicular phase of the menstrual cycle can be viewed in dynamic terms as the selection of an oocyte leading, in about two weeks, to ovulation. The latter, discontinuous process is followed by the degeneration of the luteal body, which corresponds to the luteal phase of the cycle. After the end of that phase, a new follicular phase begins. The periodicity of 28 days of the menstrual cycle in women thus follows from the succession of the follicular and luteal phases, each of which lasts 14 days. Separating these phases is the brief discontinuity of ovulation. The absence of ovulation is necessarily linked to the arrest of the hormonal rhythms observed in the course of a normal cycle.

Thus we see how certain biological rhythms can be viewed as a cyclic succession of discontinuously linked events rather than as continuous oscillations of the limit cycle type. As demonstrated by the biochemical and cellular examples discussed here, and by cardiac and neuronal rhythms, it nevertheless remains clear that most biological rhythms correspond to limit cycle oscillations.

### Positive versus negative feedback as a source of instability in biochemical and cellular systems

Multiple types of nonlinearity can give rise to nonequilibrium instabilities and the associated phenomena of temporal self-organization. In biochemical systems, these nonlinearities arise from the kinetic properties of enzymes, related to their cooperativity and regulation. In principle, instabilities may originate from either positive or negative feedback (Sel'kov, 1972a). Negative feedback is by far more prevalent in cellular regulation. Thus the inhibition of an enzyme by the end product of a biosynthesis pathway is the prototype for metabolic regulation (Cohen, 1983). The theoretical analysis of models based on enzyme regulation shows, however, that negative feedback can induce oscillations only under rather strict conditions (Goodwin, 1965; Morales & McKay, 1967; Landahl, 1969; Walter, 1970; Hunding, 1974; Rapp, 1975; Tyson & Othmer, 1978; Thron, 1991), in contrast to positive feedback whose potential for destabilization is much greater (Goldbeter & Caplan, 1976).

Some of the best-known examples of periodic behaviour in biochemistry are thus based on positive feedback. Glycolytic oscillations (chapter 2) result from the activation of phosphofructokinase by a reaction product (Ghosh & Chance, 1964; Higgins, 1964, 1967; Sel'kov, 1968a,b, 1972a; Goldbeter & Lefever, 1972). Similarly, cAMP oscillations in *D. discoideum* amoebae (chapter 5) originate from the activation of adenylate cyclase by its reaction product, cAMP, via binding of the latter to a membrane receptor (Goldbeter & Segel, 1977; Martiel & Goldbeter, 1987a). In each of these cases, positive feedback corresponds to a process of self-amplification. The role of such autocatalytic reactions in the occurrence of oscillatory behaviour in chemical systems is also well known (Field *et al.*, 1972; Nicolis & Portnow, 1973; Pacault *et al.*, 1976; Nicolis & Prigogine, 1977; Epstein, 1984).

It is again a process of positive feedback that is invoked for explaining the origin of oscillations in intracellular $Ca^{2+}$ that are observed in a variety of cells either spontaneously or in response to external stimulation by a hormone or neurotransmitter (Berridge & Galione, 1988; Berridge *et al.*, 1988; Berridge, 1989; Cuthbertson, 1989; Jacob, 1990a;

Petersen & Wakui, 1990; Tsien & Tsien, 1990; Meyer & Stryer, 1991; Tsunoda, 1991). These oscillations are also observed in several egg species after fertilization (Cuthbertson & Cobbold, 1985; Miyazaki *et al.*, 1986; Miyazaki, 1988). It appears that the external stimulus elicits the synthesis of inositol 1,4,5-trisphosphate ($IP_3$), the intracellular messenger that triggers the release of $Ca^{2+}$ from an intracellular pool sensitive to $IP_3$. This increase in cytosolic $Ca^{2+}$ in turn elicits the release of $Ca^{2+}$ from a second pool, insensitive to $IP_3$ (Berridge & Irvine, 1989; Berridge, 1993).

As shown by the study of a minimal model involving two variables (see chapter 9) the latter process known as '$Ca^{2+}$-induced $Ca^{2+}$ release' (CICR) (Endo *et al.*, 1970; Fabiato, 1983) can give rise to an instability accompanied by sustained oscillations in cytosolic $Ca^{2+}$. This model, based on positive feedback exerted by cytosolic $Ca^{2+}$, is close to that proposed by Kuba & Takeshita (1981) for $Ca^{2+}$ oscillations triggered by caffeine in certain neurons. Other models (see chapter 9) attribute the origin of $Ca^{2+}$ oscillations to a negative feedback exerted by this ion on the synthesis of $IP_3$ (Woods *et al.*, 1988; Cuthbertson, 1989), or to the existence of crossed positive feedback loops between $Ca^{2+}$ and $IP_3$ (Meyer & Stryer, 1988, 1991). In these models, $Ca^{2+}$ oscillations are accompanied by a periodic variation in $IP_3$, which is not required in the model based on CICR (Dupont & Goldbeter, 1989, 1992b; Goldbeter *et al.*, 1990; Dupont *et al.*, 1991). The mechanism of $Ca^{2+}$ oscillations due to CICR may also involve a role for negative feedback, since $Ca^{2+}$ release through the $IP_3$ receptor channel exhibits a bell-shaped dependence on cytosolic $Ca^{2+}$; this dual regulation by cytosolic $Ca^{2+}$ is incorporated into several models for $Ca^{2+}$ oscillations (De Young & Keizer, 1992; Atri *et al.*, 1993; Keizer & De Young, 1994; Li *et al.*, 1994).

Positive feedback is again encountered in neurobiology. Excitability and oscillations in neurons indeed result from the self-amplification properties of ion transport across the cell membrane. For example, the inflow of $Na^+$ into the giant squid axon leads to depolarization of the membrane and to an increase in $Na^+$ conductance. The inflow of $Na^+$ thus accelerates in an autocatalytic manner and stops only when the intracellular $Na^+$ concentration approaches the $Na^+$ concentration in the extracellular medium; a $K^+$ efflux then underlies the repolarization phase. After hyperpolarization, the membrane potential returns to its stable, resting state (Hodgkin & Huxley, 1952). For destabilizing the resting state, it is enough that a depolarizing current of small amplitude be superimposed on these processes. Then, after hyperpolarization, the cell 'spontaneously' reaches the threshold for depolarization beyond which a new action potential is produced. The cell becomes capable of generating action potentials in a repetitive, periodic manner. Such is the origin of a simple neuronal rhythm (Huxley, 1959). The analysis of models based on modified Hodgkin–Huxley equations accordingly predicts the occurrence of sustained oscillations of the limit cycle type (Connor *et al.*, 1977; Hassard, 1978; Troy, 1978; Guttman *et al.*, 1980; Aihara & Matsumoto, 1982; Rinzel & Ermentrout, 1989). Similar ionic mechanisms underlie oscillations of the membrane potential in muscle fibres (Morris & Lecar, 1981) and the spontaneous oscillatory properties of nodal tissues of the heart (Noble, 1979, 1984; DiFrancesco & Noble, 1989; Noble *et al.*, 1989; DiFrancesco, 1993). Models based on kinetic equations of the Hodgkin–Huxley type, incorporating a large number of ionic currents, account for the periodic generation of the electrical signal that triggers the rhythmic contraction of the cardiac muscle. These studies point to the primary role played in the oscillatory mechanism by depolarizing currents activated in the hyperpolarization range (Noble & Noble, 1984; DiFrancesco & Noble, 1985; Wilders, Jongsma & van Ginneken, 1991; Varghese & Winslow, 1994).

The nonlinearity of kinetic equations of the Hodgkin–Huxley type might also result, at least in part, from the cooperative allosteric properties of ion channels. The crystallographic study of these channels indeed reveals that they are often formed by multiple interacting subunits (Noda *et al.*, 1984).

We may conclude that many important biological rhythms originate from positive feedback mechanisms whose nonlinearity is further strengthened by the cooperative nature of the regulatory process. Although the detailed molecular implementation of the feedback process differs in each case, it is the self-amplification with which it is associated that gives rise to instabilities followed by sustained oscillations in biochemical systems as well as in cardiac or neural cells (Goldbeter, 1992).

That negative feedback can also be involved in the origin of an experimentally observed biological rhythm is shown by the case of the mitotic oscillator that drives the eukaryotic cell division cycle. The minimal cascade model analysed in chapter 10 corroborates the suggestion (Hunt, 1989; Murray & Kirschner, 1989a,b; Félix *et al.*, 1990) that this oscillator relies on a negative feedback loop involving cyclin and cdc2 kinase. The control of cdc2 kinase activation and cyclin degradation occurs via phosphorylation–dephosphorylation cycles. The model shows how time delays and threshold phenomena arise naturally from these reactions, and indicates that such properties, together with the negative feedback exerted by cdc2 kinase on cyclin, suffice to destabilize the system and to produce sustained oscillations in the absence of any positive feedback. The possibility exists, nevertheless, that even in the case of the mitotic oscillator positive feedback could still have a significant role in the mechanism of periodic behaviour. Thus, several models for the mitotic oscillator are based on a positive feedback exerted by cdc2 kinase on its own activation. As discussed in chapter 10, positive feedback as well as negative feedback could both contribute to the onset of sustained oscillations in the biochemical mechanism controlling mitosis in eukaryotic cells.

The periodic activation of cdc2 kinase may only represent a 'building clock' in the control of the somatic and yeast cell cycles. It appears, indeed, that in these more complex cell cycles, at least two such oscillators, of similar mechanism, may be at work; one would control the onset of mitosis at the $G_2/M$ transition, while the other would control entry into the S phase at the $G_1/S$ transition. The two oscillators would respectively involve $G_2$ and $G_1$ cyclins, and, in somatic cells, the kinases cdc2 and cdk2 as well as specific cyclin proteases. As discussed at the end of chapter 10, some form of mutual inhibition between the two control systems would ensure the proper alternation between peaks of cdc2 and cdk2, as shown by simulations of the double oscillator model (fig. 10.18). This alternation, secured by checkpoint controls, would underlie the ordering of the M and S phases during the cell cycle (Nurse, 1994).

Finally, negative feedback appears to play a prominent role in the molecular mechanism of circadian rhythms. As discussed further below, recent experiments in *Drosophila* and in *Neurospora* point to the existence of a negative feedback loop in which a key protein would repress the transcription of its gene. The best evidence in favour of such autoregulatory mechanism has been obtained for the *per* gene in *Drosophila*. Based on these observations, a theoretical model for circadian oscillations in *per* mRNA and PER protein was presented in chapter 11.

## The appearance of complexity in the temporal domain

Whereas biological complexity is usually associated with the formation of more and more intricate spatial structures in the course of morphogenesis,

it is important to realize that it can also characterize the purely temporal forms of the organization of living systems. The rich repertoire of
bifurcations associated with spatial or spatiotemporal structures finds a
no less luxurious reflection in the variety of dynamic behaviour that can
be observed in the course of time, under conditions of spatial homogeneity. Multiple attractors, bursting oscillations and chaos give a measure of the richness of dynamic behaviour in the temporal domain.

Besides their use in accounting for experimental observations on biochemical and cellular rhythms, models were also studied here in a more
abstract manner, departing somewhat from experimental constraints, in
order to determine the ways by which biological systems can acquire
more and more complex modes of temporal self-organization, starting
from simple periodic behaviour.

Thus, the model considered in chapter 3 (fig. 12.1b), which is an
extension of the two-variable model analysed in chapter 2 for glycolytic
oscillations (fig. 12.1a), allows us to propose and verify a conjecture on

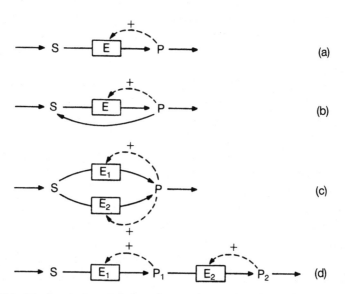

Fig. 12.1. Biochemical models based on various modes of positive feedback in
enzyme reactions. All these models admit simple periodic behaviour of the limit
cycle type. The coexistence between two stable limit cycles (birhythmicity) or
between a stable limit cycle and a stable steady state (hard excitation) is
observed in models (b) to (d). Model (d) also admits complex periodic oscillations of the bursting type, chaos, as well as the coexistence between three simultaneously stable limit cycles (trirhythmicity) (Goldbeter *et al.*, 1988).

the origin of birhythmicity in nonlinear systems. The extension of the model consists in adding to the product-activated enzyme system responsible for glycolytic oscillations a reaction for the recycling of product into substrate. This extension does not change the number of variables but allows the occurrence of new modes of dynamic behaviour.

The phenomenon of birhythmicity is the rhythmic counterpart of bistability for which many examples exist in biochemistry (Degn, 1968; Naparstek *et al.*, 1974; Eschrisch *et al.*, 1980, 1990; Hervagault *et al.*, 1983). Birhythmicity consists in the coexistence of two simultaneously stable limit cycles for the same set of parameter values. There is as yet no example of birhythmicity in biology, but the phenomenon has been observed in chemical systems (Alamgir & Epstein, 1983; Citri & Epstein, 1988). Before observing it, in fact, it is necessary to know of its possible existence, since any experimental study of birhythmicity relies, necessarily, on the demonstration of two simultaneously stable, distinct rhythms, when only one rhythm can be recorded at a given moment. The purpose of the theoretical study was to determine the conditions for the occurrence of birhythmicity and to show how the transition between the two stable rhythms can be achieved, once their existence has been demonstrated.

The phase plane analysis of the two-variable model for birhythmicity showed how the progressive deformation of one of the nullclines of the system gives rise to the appearance of a stable, small-amplitude limit cycle in the domain of existence of a stable limit cycle corresponding to large-amplitude oscillations. Numerical simulations confirm that the passage from one limit cycle to the other occurs after a chemical perturbation of adequate magnitude delivered at the appropriate phase of each of the two cycles. Alternatively, the passage from one cycle to the other can be achieved upon variation of some control parameter. Such transitions are reversible in the models considered here. If, however, one of the two limit points of the hysteresis curve is not accessible, owing to some constraint on the parameters, then the transition from one limit cycle to the other might become irreversible with respect to the variation of the control parameter, as is the case in some examples of bistability (Hervagault & Canu, 1987). The transition between the two cycles should nevertheless remain possible by means of suitable chemical perturbation, e.g. substrate or product addition at the appropriate phase.

Very interestingly, multiple periodic regimes have recently been found in a neuronal model (Canavier *et al.*, 1993). Such a multiplicity suggests a new mode of action for neuromodulators, which may change not simply the frequency of the oscillatory response but also the very type of rhythm produced by target neurons (Canavier *et al.*, 1994). An experimental demonstration of this theoretically predicted behaviour would be a fascinating addition to the already rich repertoire of nonlinear neuronal dynamics.

The origin of birhythmicity in the two-variable biochemical model is closely related to the breaking up of an instability domain into two distinct parts upon varying some control parameter. Under conditions close to those producing birhythmicity, the elevation or diminution of this parameter, starting from a value corresponding to a stable steady state, can give rise to either one of two types of oscillation differing totally from each other in amplitude and in period. In a somewhat unexpected manner, this behaviour of the biochemical model accounts for the dynamic properties of certain neurons such as those of the thalamus. These nerve cells are characterized by the existence of two distinct rhythms, with frequency 6 and 10 Hz, depending on whether the membrane potential is slightly depolarized or hyperpolarized starting from a value corresponding to a stable, resting nonoscillatory state (Jahnsen & Llinas, 1984b; Llinas, 1988; Steriade *et al.*, 1990). Despite the obvious differences in the nature of the mechanisms responsible for oscillations in the two systems, the existence of multiple domains of oscillations can be explained by a common structure of the nullclines in the phase plane. The same remark holds for the phenomenon of multi-threshold excitability observed in these neurons as in the biochemical model.

The generality of the results obtained for the two-variable model of fig. 12.1b is attested by the fact that they are also recovered in another two-variable model (fig. 12.1c) based on the coupling in parallel of two reactions with positive feedback, catalysed by two isozymes differing in their kinetic properties. This model can be viewed as the simplest example of coupling between two biochemical oscillators.

## The interaction between two instability-generating mechanisms as a source of complex oscillatory behaviour

A further extension, considered in chapter 4, consists in determining the consequences of coupling two instability-generating mechanisms within the same system. The simplest prototype for such a situation is that of a coupling in series of two enzyme reactions subjected to positive feedback (fig. 12.1d). This system comprises three variables and possesses all the attributes necessary for complex oscillatory phenomena. Indeed, in addition to simple periodic behaviour or to the evolution towards a single, stable steady state observed in the model described in fig. 12.1a, which contains a single mechanism of instability, the coupling between two instability-generating mechanisms permits the system to acquire the capability of 'choosing' between two simultaneously stable attractors: these attractors can both be limit cycles (birhythmicity), or a stable limit cycle and a stable steady state. An identical phenomenon is observed in a chemical system formed by two exothermic reactions coupled in series when the system temperature is allowed to vary (Cohen & Keener, 1976; Doedel & Heinemann, 1983). The latter system, much as the biochemical model, relies on the coupling in series of two reactions that can each give rise to oscillatory behaviour.

While the coexistence between two limit cycles or between a limit cycle and a stable steady state is also shared by the two-variable models of fig. 12.1b and c, new modes of complex dynamic behaviour arise because of the presence of a third variable in the multiply regulated system. The coexistence between three simultaneously stable limit cycles, i.e. trirhythmicity, is the first of these. Moreover, the interaction between two instability-generating mechanisms allows the appearance of complex periodic oscillations, of the bursting type, as well as chaos. The system also displays the property of final state sensitivity (Grebogi *et al.*, 1983a) when two stable limit cycles are separated by a regime of unstable chaos.

The route followed by the system to aperiodic oscillations is through a cascade of period-doubling bifurcations as described by Feigenbaum (1978) in a totally different context. This illustrates again the universality of nonequilibrium dynamic behaviour in nonlinear systems, regardless of the evolution laws governing these systems.

The particular interest of bursting oscillations stems from the fact that they are commonly encountered in neurobiology (Alving, 1968; Gola, 1974; Chaplain, 1976; Meech, 1979; Gorman & Hermann, 1982; Johnston & Brown, 1984; Adams & Benson, 1985, 1989; Alonso *et al.*, 1994). Numerous theoretical studies have been devoted to the origin of bursting in electrically excitable cells (Plant & Kim, 1976; Plant, 1978; Carpenter, 1979; Chay & Keizer, 1983; Hindmarsh & Rose, 1984, 1989; Ermentrout & Koppell, 1986; Rinzel & Lee, 1986; Sherman *et al.*, 1988; Rose & Hindmarsh, 1989a,b,c; Canavier *et al.*, 1991; Chay, 1993; Destexhe *et al.*, 1993; Bertram, 1994; Bertram *et al.*, 1995; Destexhe &

Sejnowski, 1995), and to the transition from bursting to chaos (see, e.g. Chay & Rinzel, 1985; Canavier, Clark & Byrne, 1990). The idea that often emerges from these experimental and theoretical studies is that bursting results from the cyclic variation of an ionic conductance (Junge & Stephens, 1973; Gola, 1974) that periodically brings the membrane potential beyond the threshold for excitability: a burst of rapid spikes thus rides on top of the wave due to the slow oscillation of the membrane potential. The high-frequency spikes occur as long as the potential exceeds the threshold for spontaneous generation of action potentials (Ermentrout & Koppell, 1986).

The biochemical three-variable prototype allows a detailed analysis of the transition from simple periodic behaviour to bursting oscillations. Two complementary approaches were followed to this end. The first, proposed by others (Rinzel, 1987; Rinzel & Lee, 1986) for the study of bursting in electrically excitable cells, consists in reducing the number of variables by considering that one of these behaves as a parameter whose slow variation governs the dynamics of a rapid, two-variable subsystem. Analysing the behaviour of that rapid subsystem as a function of the third variable taken as control parameter allows us to explain the transition between simple and complex oscillations when the slow variation of the control parameter is taken into account. Moreover, different modes of bursting can thus be demonstrated (see fig. 4.18). If this analysis explains the origin of various patterns of bursting, it cannot throw light on the origin of more complex modes of periodic behaviour such as that represented at the bottom of fig. 4.18, nor on the evolution toward chaos.

The second approach, successfully followed in the analysis of complex oscillations observed in the model of the multiply regulated biochemical system, relies on a further reduction that permits the description of the dynamics of the three-variable system in terms of a single variable only, by means of a Poincaré section of the original system. Based on the one-dimensional map thus obtained from the differential system, a piecewise linear map can be constructed for bursting. The fit between the predictions of this map and the numerical observations on the three-variable differential system is quite remarkable. This approach allows us to understand the mechanism by which a pattern of bursting with $n$ peaks per period transforms into a pattern with $(n + 1)$ peaks.

Even more interestingly the analysis reveals that these simple patterns of bursting are separated by complex patterns of the type represented at the bottom of fig. 4.18. Moreover, the alternation of simple and complex patterns of bursting as a function of a control parameter is characterized by a property of self-similarity (fig. 4.28).

The form of the piecewise linear map analysed for that model closely resembles that of the one-dimensional map obtained in a model for electrical bursting in pancreatic β-cells (Chay & Rinzel, 1985). This suggests the generality of the results obtained with the model for the multiply regulated system. Similar results with regard to bursting should be obtained in any system in which the synthesis of a peak in the course of a burst consumes a quasi-constant quantity of a precursor whose slow accumulation separates one bursting phase from the next.

The piecewise linear map does not account, however, for the appearance of chaotic behaviour. A slight modification of the unidimensional map, taking into account some previously neglected details of the Poincaré section of the differential system, shows how chaos may appear besides complex periodic oscillations of the bursting type.

Similar complex oscillatory phenomena have been observed in a closely related model containing two regulated enzyme reactions coupled in a different manner (Li, Ding & Xu, 1984). An additional indication of the generality of the results obtained in the multiply regulated biochemical system is given by the study of the model for the synthesis of cAMP in *Dictyostelium* amoebae. In addition to simple periodic oscillations and excitability (see chapter 5), this realistic model based on experimental observations also predicts the appearance of more complex oscillatory phenomena in the form of birhythmicity, bursting and chaos (chapter 6).

The study of the biochemical prototype with multiple regulation throws light on the origin of complex oscillatory behaviour in the model for the cAMP signalling system: in both cases two instability-generating mechanisms interact within the same system. For the synthesis of cAMP, the two oscillatory mechanisms are coupled in parallel (fig. 6.25) and share the same positive feedback loop but differ in the process limiting autocatalysis. In each case, complex behaviour appears when conditions are such that the two oscillatory mechanisms are active at the same time.

This conjecture suggests an empirical procedure for the search for complex oscillatory phenomena in parameter space (Goldbeter *et al.*, 1988). If two instability-generating mechanisms coexist within the same system, we may expect that such a coexistence is reflected in the formation of distinct instability domains in parameter space. A typical

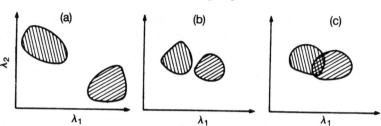

Fig. 12.2. Empirical method for determining the domain of complex oscillatory behaviour in parameter space when two distinct instability domains are observed (see text) (Goldbeter *et al.*, 1988).

situation is schematized in fig. 12.2a where two instability domains, represented by the dashed areas, can be distinguished in the plane formed by parameters $\lambda_1$, $\lambda_2$. Within each of these domains, the system oscillates primarily as a result of one or the other mechanism responsible for instability of the steady state. By changing the value of a third parameter, $\lambda_3$, we can bring the two instability domains closer to each other (fig. 12.2b) until they overlap (fig. 12.2c). It is then that complex oscillatory phenomena are likely to occur: the two oscillatory mechanisms are indeed simultaneously active, and their interaction gives rise to bursting oscillations, birhythmicity or chaos.

The empirical procedure suggested by fig. 12.2 consists in bringing closer together the instability domains in parameter space, provided that several such domains exist. This procedure was followed successfully in the search for complex oscillations in the model for cAMP synthesis.

The peroxidase reaction provides another prototype for periodic behaviour and chaos in an enzyme reaction. As noted by Steinmetz *et al.* (1993), in view of its mechanism based on free radical intermediates, this reaction represents an important bridge between chemical oscillations of the Belousov–Zhabotinsky type, and biological oscillators. In view of the above discussion, it is noteworthy that the model proposed by Olsen (1983), and further analysed by Steinmetz *et al.* (1993), also contains two parallel routes for the autocatalytic production of a key intermediate species in the reaction mechanism. As shown by experiments and accounted for by theoretical studies, the peroxidase reaction possesses a particularly rich repertoire of dynamic behaviour (Larter *et al.*, 1993) ranging from bistability (Degn, 1968; Degn *et al.*, 1979) to periodic oscillations (Yamazaki *et al.*, 1965; Nakamura *et al.*, 1969;

Olsen, 1978), chaos (Olsen & Degn, 1977; Olsen, 1979, 1983; Aguda & Larter, 1991; Geest *et al.*, 1992; Steinmetz *et al.*, 1993), and hard excitation (Aguda, Hofmann-Frisch & Olsen, 1990).

It is likely that complex oscillations sometimes observed in cytosolic $Ca^{2+}$ (Woods *et al.*, 1987; Cuthbertson, 1989) also result from the interaction between two instability-generating mechanisms (Dupont & Goldbeter, 1989; Goldbeter *et al.*, 1990). This conjecture is corroborated by preliminary results (Dupont, 1993) showing that oscillations of the bursting type can occur when the CICR mechanism producing rapid $Ca^{2+}$ oscillations (see chapter 9) is coupled to a mechanism generating slower oscillations in the level of $IP_3$. Preliminary results of a study under way (J. Borghans, G. Dupont & A. Goldbeter, unpublished data) indicate, however, that complex oscillations may also arise, from another mechanism, in a two-pool model in which the level of $IP_3$ is held constant.

That the interplay between two instability-generating mechanisms can lead to chaos, bursting and/or birhythmicity is further corroborated by theoretical studies of chemical systems, in which complex oscillatory phenomena have also been studied, both from an experimental and a theoretical point of view. Close consideration of the models proposed for these phenomena often reveals the existence of multiple instability-generating steps in the reaction mechanism (Janz *et al.*, 1980; Rinzel & Troy, 1982; Scott, 1991; Györgyi & Field, 1993; Zhang *et al.*, 1993); for an example in combustion, see Johnson *et al.* (1991). Like the coupling in series of two allosteric enzymes activated by their reaction product, the coupling of two consecutive exothermic reactions has been shown to give rise to complex oscillatory phenomena (Cohen & Keener, 1976; Doedel & Heinemann, 1983; Elnashaie, Abashar & Teymour, 1993). Similar results have been obtained recently (Fuchikami *et al.*, 1993) for a model of an electrically excitable membrane with two distinct autocatalytic ion channels.

## Periodic behaviour versus chaos

In each of the models studied here, the domains of chaos and birhythmicity are much more reduced than the domain of complex periodic oscillations, which is itself smaller than the domain where simple periodic oscillations occur. This observation, which is corroborated by results obtained on the occurrence of complex oscillations in multi-looped negative feedback systems (Glass & Malta, 1990), accounts for the regular nature of most biological rhythms observed in nature (Goldbeter, 1993b), save the rhythms subjected to significant variations in external conditions and the rhythms that reflect the interaction of a large number of oscillators, as is the case, for example, with some cerebral rhythms. For the last of these, the absence or the reduction of irregular behaviour can even be associated with physiological disorders (Babloyantz & Destexhe, 1986).

The smallness of the domain of chaos in the model for cAMP synthesis in *Dictyostelium* could also explain why the mutant *HH201*, derived from the mutant *Fr17*, behaves in a rather regular manner in cell suspensions (see fig. 6.14), whereas the mutant *Fr17* is known to aggregate on agar in an aperiodic manner – which behaviour could be interpreted in terms of autonomous chaos (chapter 6). Theoretical studies (Halloy *et al.*, 1990; Li *et al.*, 1992a,b) suggest another explanation for these observations. These studies are based on the coupling of two populations of amoebae in which the synthesis of cAMP is described by the model analysed in chapter 5. One of the populations functions in the periodic domain while the other behaves in a chaotic manner (see chapter 6). The coupling between amoebae of the two populations occurs through the cAMP secreted by the cells into the extracellular medium in the mixed suspension.

The most striking result of these studies is that the behaviour of the mixed population resulting from such a coupling markedly depends on the relative proportions of cells behaving initially in a periodic or chaotic manner. Thus, it suffices that a small percentage of cells initially oscillate in a periodic manner to suppress the chaotic behaviour of the remaining cells and to transform it into periodic oscillations. These results, whose relevance extends beyond the particular case of *Dictyostelium* and pertains to other examples of rhythmic intercellular communications, provide a plausible explanation for the regular behaviour of the mutant *HH201* in cell suspensions (fig. 6.17): if a small fraction of the mutant amoebae present in the cell suspension are capable of oscillating in a regular manner while the remaining cells on their own oscillate aperiodically, the presence of the small amount of periodic cells would suffice to suppress chaos in the rest of the mixed suspension. In contrast, cells aggregating on agar should not experience the strong coupling present in cell suspensions. Any centre behaving in an aperiodic manner would then be capable of sending its chaotic signals to surrounding cells.

The suppression of chaos by a tiny proportion of periodic

*Dictyostelium* cells turns out to be equivalent to the transformation of aperiodic into periodic oscillations by the low-amplitude, sinusoidal forcing of a strange attractor. These results can be related to the more general problem of 'controlling' chaos (Ott *et al.*, 1990; Petrov *et al.*, 1993; Shinbrot *et al.*, 1993). In chemistry and biology, the control of chaos by stabilizing one of the infinitely large number of unstable periodic orbits through external perturbation of a parameter has been achieved in the Belousov–Zhabotinsky reaction (Peng *et al.*, 1991; Petrov *et al.*, 1993) and for an *in vitro* preparation of cardiac tissue (Garfinkel *et al.*, 1993; see also Glass & Zheng, 1994). Here, in contrast, the stabilization of the periodic orbit is achieved through coupling chaos with periodic behaviour. The possibility of applying chaos-control techniques to brain disorders such as epilepsy has recently been investigated (Schiff *et al.*, 1994; Jergen & Schiff, 1995; see also Glanz, 1994).

More generally, the question arises as to the applicability of chaos in various fields ranging from physics, chemistry and biology to economics (see Ruelle, 1994). The present discussion was restricted to the occurrence and possible roles of chaos in biochemical and cellular systems. In the case of *Dictyostelium* cells, periodic signals of cAMP turn out to be more efficient than chaotic ones. It is still unclear whether chaotic behaviour is associated with pathological or healthy conditions in physiology (Pool, 1989). Besides the nonlinear dynamics of the brain (Skarda & Freeman, 1987; Rapp *et al.*, 1989), a case in point is the heart, whose rate exhibits a variability that has been analysed in terms of chaos; the source of this variability appears to lie in neurohumoral control of the heart (see Goldberger & Rigney, 1989; Kaplan & Talajic, 1991). Suppressing this control may prove more detrimental than suppressing chaotic dynamics per se, which may just be a marker of the multiplicity of controls exerted on the heart rhythm. At the cellular level, the latter rhythm, as shown by the study of models for electrical activity in cardiac pacemaker tissues (Noble & Noble, 1984; DiFrancesco & Noble, 1985), appears to be of the limit cycle type. What can be said at this point is that the function of chaos in physiology remains largely unclear, compared to the well-established roles of periodic phenomena.

### Link with the evolution to chaos through the periodic forcing of an oscillator

The appearance of complex autonomous behaviour due to the interaction between two endogenous oscillatory mechanisms can be related to the numerous studies devoted to the evolution toward bursting or chaos in oscillating systems subjected to periodic forcing (Holden, 1986; Glass & Mackey, 1988). Chaos was thus observed in oscillating yeast extracts driven by a sinusoidal source of glycolytic substrate (Markus, Kuschmitz & Hess, 1984, 1985; Hess & Markus, 1985), in experimental and theoretical studies of an excitable neuron subjected to sinusoidal stimulation (Hayashi, Nakao & Hirakawa, 1982; Matsumoto *et al.*, 1984) and of periodically stimulated cardiac cells (Guevara *et al.*, 1981; Glass *et al.*, 1983, 1984; Lewis & Guevara, 1990).

Whereas chaos and complex oscillations in the models studied here result from the interaction between two endogenous oscillators, in these experiments and in the associated models they result from the coupling between an endogenous oscillator and a periodic external source, as indicated by the analysis of normal forms for such a situation (Baesens & Nicolis, 1983). Evidence for autonomous chaos has nevertheless been obtained in molluscan neurons (Holden *et al.*, 1982; Hayashi & Ishizuka, 1992).

In forced systems, the facility of controlling the amplitude and the period of the external input allows us to obtain chaos much more easily than in the autonomous case in which it is difficult to control independently the period and the amplitude of the two endogenous oscillators. This could be a reason why in autonomous systems the domains of complex oscillatory behaviour remain relatively reduced as compared with the domains of simple periodic behaviour.

### From oscillations to spatiotemporal organization

Systems that can self-organize in time can also, for appropriate values of the parameters, present patterns of spatiotemporal organization in the form of propagating waves of chemical concentration. That aspect of nonlinear dynamics was not emphasized in this book, which is devoted primarily to the analysis of temporal organization in regulated biochemical systems. The link between excitability and oscillations on the one hand, and wavelike phenomena on the other, has nevertheless been stressed, particularly with regard to cAMP signalling in *Dictyostelium* and to $Ca^{2+}$ signalling (Meyer, 1991; Dupont & Goldbeter, 1992b, 1994). In both systems sustained oscillations occur, while concentric (target) or spiral waves are also observed, either at the cellular level, in the case of $Ca^{2+}$, or at the supracellular level, in *Dictyostelium*. The propagation of intercellular $Ca^{2+}$ waves has also been described in some cell types (see chapter 9). In addition, three-dimensional scroll waves of chemotactic movement have been observed at the tip of the slug after *D. discoideum* aggregation (Siegert & Weijer, 1992).

The fact that temporal as well as spatial patterns can arise in the same systems, depending on the experimental conditions or the choice of parameter values, results from the fact that the same kinetic, instability-producing mechanism is at work. The same regulatory mechanism, often based on positive or negative feedback, that destabilizes the system in spatially homogeneous conditions, thus leading to oscillations or to the transition between multiple steady states, is indeed capable of giving rise to spatial or spatio-temporal organization when it interacts with transport processes such as diffusion (Prigogine *et al.*, 1969; Nicolis & Prigogine, 1977; Murray, J.D., 1989). Such a link between oscillations and waves has long been observed and analysed in chemical systems such as the Belousov–Zhabotinsky reaction (Zaikin & Zhabotinsky, 1970; Winfree, 1972b, 1987, 1991b; Tyson & Fife, 1980; Keener & Tyson, 1986; Scott & Showalter, 1992). A similar link between oscillatory activity in time, excitability and wavelike patterns exists in the heart (see, e.g. Davidenko *et al.*, 1990, 1992); the formation of spiral patterns in cardiac tissue is thought to play a key role in the onset of fibrillation (Winfree, 1987).

The variety of patterns of temporal organization discussed in chapters 3, 4 and 6 finds its counterpart in the richness of spatial or spatiotemporal patterns. This analogy extends to the existence of multiply stable solutions. Reminiscent of the phenomenon of birhythmicity is the observation of multiple, simultaneously stable, stationary spatial patterns in a theoretical model of morphogenetic interest (Maini, 1990). Likewise, when diffusion is incorporated into the two-variable model for birhythmicity analysed in chapter 3, two types of wave with different amplitudes can be shown to propagate within the system (Pérez-Iratxeta *et al.*, 1995); which one of these two stable waves occurs depends on the choice of initial conditions. Experimental examples of multiple spatial structures range from alternative stable rotors in chemical media (Winfree, 1991a) to the propagation of two distinct action potentials in a neuronal preparation (Hochner & Spira, 1986).

## The ontogenesis of biological rhythms

The periodic synthesis of cAMP in *D. discoideum* amoebae shares with other biological rhythms the property of appearing at a precise moment in development. The cAMP signalling system thus provides a useful model for the ontogenesis of biological rhythms. This model is particularly appropriate given that the unicellular nature of the organism facilitates the experimental approach of the question.

Moreover, as seen in chapter 7, the cAMP signalling system undergoes a series of successive transitions in the hours that follow starvation. At first insensitive to extracellular cAMP stimuli, the system becomes excitable in that it amplifies in a pulsatory manner cAMP signals whose amplitude exceeds a threshold. A few hours later, the system becomes capable of generating in an autonomous, periodic manner pulsatile signals of cAMP. This capability, however, lasts for only several hours before cells lose their oscillatory properties and become excitable again. Thus, any explanation of the appearance of the cAMP rhythm in the course of development must account for the entire sequence of transitions absence of relay → relay → oscillations. This multiplicity of transitions puts additional constraints on the theoretical explanation but thereby strengthens it if it can account for the entirety of the sequence.

The analysis of the model based on receptor desensitization showed how the sequential developmental transitions in cAMP signalling result from the continuous increase observed (Loomis, 1979), during the hours following starvation, in the activity of certain enzymes such as adenylate cyclase and phosphodiesterase, and in the quantity of cAMP receptor. In parameter space, the system of intercellular communication follows a developmental path (Goldbeter & Segel, 1980) along which it successively crosses a domain where the steady state is stable but nonexcitable, then a domain of excitability, before the system enters a domain of autonomous, sustained oscillations around an unstable steady state. The first amoebae to enter the oscillatory domain along this developmental path begin spontaneously to secrete pulsatile cAMP signals in a periodic manner. These cells become the aggregation centres, while other cells are capable only of relaying the signals.

In *D. discoideum*, continuous changes in the activity of certain enzymes and in the level of membrane receptor would thus underlie the appearance of the cAMP rhythm. Similar changes could explain the appearance of other biological rhythms in the course of development. In nerve cells, the simple increase in an ionic conductance, due for example to progressive incorporation into the membrane of a protein serving as ion channel, could thus modify the dynamic properties of a neuron by promoting the passage through a Hopf bifurcation corresponding to the onset of periodic behaviour (Holden & Yoda, 1981).

Analogous variations could underlie the appearance of bursting oscillations in the neuron R15 in the course of *Aplysia* development (Ohmori,
1981). In a similar manner, the electrical activity of cardiac cells could
acquire its periodic character at a precise moment in embryogenesis
(DeHaan, 1980; Fukii, Hirota & Kamino, 1981); this bifurcation would
signal the onset of the heart beat.

Other biological rhythms appear within cellular networks. There are
numerous neural networks whose activity is periodic (Friesen & Stent,
1978; Rose & Benjamin, 1981; Getting, 1983, 1989; Grillner & Wallén,
1985; Selverston & Moulins, 1985; Friesen, 1989; Grillner *et al.*, 1989;
Jacklet, 1989a). The origin of the rhythm sometimes lies in the connections between neurons, as in the case of recurrent cyclic inhibition
(Kling & Szekely, 1968). In that case, the formation of inhibitory (or
activating) connections in the course of development could control the
emergence of the rhythm. The oscillatory properties observed in simple
neural networks (Rose & Benjamin, 1981; Selverston & Moulins, 1985;
Benjamin & Elliott, 1989) or in the central nervous system (Llinas,
1988), however, are often encountered at the (isolated) cellular level.
The ontogenesis of the global rhythm is then closely linked to the development of the cellular oscillatory behaviour.

### Temporal coding in intercellular communication: from cAMP signalling in *Dictyostelium* to pulsatile hormone secretion

A last point addressed in the study of the mechanism of intercellular
communication in *D. discoideum* pertains to the physiological function
of cAMP oscillations. Whereas this function is still a matter of debate
regarding glycolytic oscillations in yeast, the role of cAMP oscillations
in *Dictyostelium* appears to be much clearer. The mechanism for the
production of pulsatile signals of cAMP is not only a model for understanding the ontogenesis of biological rhythms at the molecular level
but also provides a prototype for assessing in higher organisms the role
of rhythmic, pulsatile hormone secretion in intercellular communication.

The periodic secretion of cAMP signals first controls the aggregation
of the amoebae after starvation. This essential phase of the life cycle
eventually leads to the formation of a fruiting body (chapter 5). The
periodic signals of cAMP also control the process of cell differentiation
during the hours that separate the beginning of starvation from aggregation. The experiments indicate (Darmon *et al.*, 1975; Gerisch *et al.*,

1975; Juliani & Klein, 1978) that cAMP signals of 5 min periodicity induce the synthesis of proteins required for the intercellular communication machinery (adenylate cyclase, phosphodiesterase, cAMP receptor, ...). Constant signals or cAMP pulses delivered every 2 min are without effect (Wurster, 1982), like cAMP stimuli applied in a random manner (Nanjundiah, 1988).

The model based on receptor desensitization allows us to account for the efficiency of cAMP signals of 5 min periodicity and for the failure of the other types of signal mentioned above. As indicated in chapter 8, constant cAMP signals induce the desensitization of the receptor, while the latter has enough time to resensitize between two stimuli when the interval between these is of the order of 5 min. When the interval is shorter, the receptor only partly resensitizes, which results in a significant decrease in the synthesis of intracellular cAMP in response to extracellular cAMP stimulation (fig. 8.1). A more thorough analysis of this model (Li & Goldbeter, 1990; see chapter 8) explains the efficiency of signals delivered at 5 min intervals, as well as the lack of effect of signals delivered at shorter intervals or in a constant manner.

These results are based on the hypothesis that cAMP signals control differentiation via binding to the cAMP receptor and the subsequent triggering of intracellular cAMP synthesis (Kimmel, 1987; Mann & Firtel, 1987); the latter would exert its effect on development through the control of a cAMP-dependent protein kinase (Schaller *et al.*, 1984; Simon *et al.*, 1989; Firtel & Chapman, 1990). Similar conclusions on the efficiency of pulsatile signals of adequate period are obtained when we consider that the response is mediated by the synthesis of other second messengers such as $IP_3$ or diacylglycerol, which also act on *Dictyostelium* development (Ginsburg & Kimmel, 1989) and are synthesized following binding of cAMP to its receptor (Europe-Finner & Newell, 1987).

The intercellular communication by pulsatile cAMP signals in *Dictyostelium* amoebae can be viewed as a primitive hormonal communication system. In higher organisms, many hormones are in fact secreted in a pulsatile manner into the circulation (Crowley & Hofler, 1987; Wagner & Filicori, 1987; Leng, 1988). The question arises as to the physiological function of pulsatile patterns of hormone secretion. As recalled in section 8.3, the prototype of this phenomenon is the secretion of the hormone GnRH by the hypothalamus with a frequency of one 6 min pulse per hour in humans and primates (Dierschke *et al.*, 1970; Carmel *et al.*, 1976; Crowley *et al.*, 1985). The effect of GnRH is to induce the secretion of the gonadotropic hormones LH and FSH by the pituitary. The function of the rhythm of GnRH secretion was elucidated in a series of remarkable experiments conducted by Knobil and coworkers in the rhesus monkey (Belchetz *et al.*, 1978; Knobil, 1980; Wildt *et al.*, 1981; Pohl *et al.*, 1983). These experiments indicate that the constant signal of GnRH is without effect (figs. 8.2 and 8.3) and that the response of target cells in the pituitary closely depends on the frequency of stimulation by GnRH (fig. 8.4). Knobil (1981) could thus conclude that the temporal pattern of the hormone is as important, if not more, than its level in the blood.

The observations on the role of the frequency of secretion of GnRH have since been incorporated into the treatment of several reproductive disorders in men and women (Wagner, 1985; Santoro *et al.*, 1986). Chronotherapy relying on the use of pulsatile GnRH signals is thus used successfully for solving certain cases of sterility (Leyendecker *et al.*, 1980; Reid *et al.*, 1981). Moreover, the rhythm of GnRH secretion also plays a primary role in controlling the maturation of the reproductive system at puberty (Wildt *et al.*, 1980; Karsch, 1987).

The number of hormones for which it has been shown that the physiological effect is maximum when secreted in a periodic, pulsatile manner continues to increase. To the hormone GnRH can be added growth hormone (GH: Clark *et al.*, 1985; Bassett & Gluckman, 1986; Hindmarsh *et al.*, 1990; Waxman *et al.*, 1991, 1995) and the factor inducing its secretion, GHRH (Borges *et al.*, 1984), ACTH and cortisol (Van Cauter, 1987), insulin and glucagon (Matthews *et al.*, 1983; Weigle *et al.*, 1984; Komjati *et al.*, 1986; Weigle, 1987; Lefèbvre *et al.*, 1987; Paolisso *et al.*, 1991), to cite but a few examples. Besides these ultradian rhythms, many hormones also vary in a circadian manner (Turek, Swann & Earnest, 1984; Van Cauter & Aschoff, 1989); thus there exists a temporal hierarchy of hormonal rhythms.

Some observations suggest that the conclusions might extend to growth factors. Brewitt & Clark (1988) studied the effect of stimulation of the lens growth by platelet-derived growth factor (PDGF). These authors showed that the growth of this epithelial tissue *in vitro* is stimulated only if PDGF is applied in a pulsatile manner, with a periodicity of several hours. As in *Dictyostelium* or in the case of GnRH, constant stimulation is ineffective. A systematic study of the effect of the frequency of stimulation by PDGF was not carried out, but these preliminary results suggest that the efficiency of pulsatile signals goes beyond the case of hormones and also pertains to certain growth factors whose role in cell differentiation or proliferation is crucial. Such a possibility is also raised by simulations carried out with the model presented in chapter 10, with regard to the effect exerted on cell division by desensitization or down regulation of a receptor for a growth factor that would control the mitotic oscillator. Further experimental studies are needed to clarify the possible role of pulsatile patterns of growth factors in the control of cell proliferation.

The efficiency of pulsatile hormone secretion stems from the fact that target cells often desensitize under constant stimulation. For GnRH, this explanation (Knobil, 1980, 1981) is corroborated by the observation of desensitization of pituitary cells subjected to prolonged stimulation by the hormone. This desensitization occurs, at least partly, at the level of the GnRH receptor located on the membrane of target cells (Smith & Vale, 1981; Zilberstein *et al.*, 1983; Adams *et al.*, 1986; Conn *et al.*, 1986). It appears, however, that desensitization also involves a process located beyond the receptor, which appears to be related to the progressive inactivation of $Ca^{2+}$ channels (Stojilkovic *et al.*, 1989).

A close link therefore exists between the function of the cAMP rhythm in *D. discoideum* and that of the hormonal rhythm of GnRH secretion. In both cases, pulsatile signalling represents an optimal mode of intercellular communication, allowing the avoidance of desensitization in target cells.

The results in chapter 8 on the effect of periodic stimulation were obtained first in the model for cAMP signalling and then in a general model of a receptor subjected to reversible desensitization. These results establish the existence of an optimal frequency of pulsatile stimulation maximizing cellular responsiveness (fig. 8.13). That the response of target cells closely depends on the frequency of the pulsatile stimulus also indicates that receptor desensitization provides a mechanism for the frequency encoding of the extracellular signal. The optimum values of the stimulus duration and of the interval between successive stimuli are dictated by the kinetic parameters governing desensitization and resensitization of the receptor. Each hormone could thus be characterized by an optimal frequency of pulsatile secretion, which would be determined by the rate constants of receptor desensitization and resensitization in target cells. A similar conclusion holds regardless of the precise molecular mechanism of desensitization, be it via covalent modification of the receptor, ligand-induced conformational transition to a less active state, or some other process such as inactivation of an ionic channel.

In the case of *Dictyostelium* amoebae, the adequacy of the frequency of cAMP signalling with the responsiveness of target cells is ensured by the fact that the same mechanism governs the periodic secretion and the relay of cAMP signals. In the case of GnRH, the mechanism for the rhythmic production of the hormonal signal is of neuronal nature (Wilson *et al.*, 1984; Lincoln *et al.*, 1985) and was probably selected in the course of evolution, together with the kinetic properties of the GnRH receptor on target cells, in such a manner that the frequency of pulsatile signalling ensures maximum responsiveness in pituitary cells with regard to LH and FSH secretion.

If biological rhythms appear to be an optimal mode of intercellular communication compared to constant signals, it is likely that the same conclusion extends to the superiority of periodic signals with respect to those delivered in a random or chaotic manner. Experiments carried out in *D. discoideum* amoebae (Nanjundiah, 1988) and theoretical results reported in chapter 8 corroborate this viewpoint. With a stochastic or aperiodic signal, certain intervals between two successive stimuli will be shorter, and others longer, than the optimal interval determined for the periodic stimulus. Such a situation is less favourable than that corresponding to the optimal, periodic signal, which each time will elicit a quasi-maximal effect. The mean cellular responsiveness will thus be greater for the optimal periodic signal than for random or chaotic stimuli (Li & Goldbeter, 1992).

Using periodic stimuli to optimize intercellular communication or other cellular processes has counterparts in far distant fields. An example is the possibility of improving reactor perfomance in chemical engineering (Gaitonde & Douglas, 1969), by using positive feedback control systems to make stable chemical reactors generate periodic outputs of higher average yield. Similar ideas have been raised by Richter & Ross (1980) in relation to the possible role of glycolytic oscillations.

## Towards a generalized chronopharmacology

The above results possess therapeutic implications, as indicated by treatments of reproductive disorders by the periodic injection of the hormone GnRH. Whereas the pulsatile chronotherapy is based in that case on the physiological rhythm of GnRH secretion, it is tempting to extrapolate these results to the case of artificial ligands. For numerous such substances, the aim is currently to reach a constant level of the drug in the circulation, by means of devices such as cutaneous patches whose function is to ensure constant drug delivery. However, as many drugs exert their effects via binding to receptors on target cells, and as many of these receptors undergo desensitization in the continuous presence of the ligand, it might be more advantageous to deliver the drug in a pulsatile manner, at the optimal frequency. This might permit bypassing of desensitization in target cells, increase in their responsiveness, and also avoidance of side-effects possibly by use of smaller amounts of drug in the circulation, by means of devices such as cutaneous patches whose function is to ensure constant drug delivery. However, as many drugs exert their effects via binding to receptors on target cells, and as many of these receptors undergo desensitization in the continuous presence of the ligand, it might be more advantageous to deliver the drug in a pulsatile manner, at the optimal frequency. This might permit bypassing of desensitization in target cells, increase in their responsiveness, and also avoidance of side-effects possibly by use of smaller amounts of drug. An opposite goal may nevertheless be desirable in some instances. Thus, constant GnRH stimulation leading to desensitization of pituitary cells is used in certain therapeutic treatments, which aim at suppressing LH and FSH secretion in certain types of cancer.

The efficiency of pulsatile hormonal stimulation is well illustrated by the results of treatment with GH (see chapter 8): when the hormone is delivered at the appropriate frequency, the same physiological effect is obtained at smaller doses of GH than that reached with much higher doses at nonoptimal frequencies (Clark *et al.*, 1985; Hindmarsh *et al.*, 1990). The optimal frequency chosen for drug delivery would depend on the kinetics of desensitization and resensitization of target cells. Such a view of a **generalized chronopharmacology** (Goldbeter, 1991b) would enlarge the current field of this discipline (Lemmer, 1989), which takes into account mainly the most important occurrence of circadian rhythms in the responsiveness of the organism toward a large variety of drugs. Technical aspects do not any more represent significant limiting factors since recent advances in pulsed and programmed drug delivery (Theeuwes, 1989; Theeuwes *et al.*, 1991) now allow the implementation of a large variety of chronotherapeutic approaches.

At present, such techniques are being used in the circadian chronomodulation of anticancer drugs (Levi *et al.*, 1994). These therapeutic approaches are based on the observation that the antitumoral efficacy of the drugs, as well as their toxicity for the organism, vary with 24 h periodicity. In the case of 5-FU (see chapter 11), such observations can be related to the fact that the activity of enzymes involved in the utilization and degradation of the drug vary, with opposite phases, in a circadian manner.

Also belonging to cancer chronotherapy, but relying on differences in cycling time of normal and tumour cells, is the strategy that aims at finding optimum, noncircadian periodic regimens of anticancer drug administration. The goal of these approaches (Dibrov *et al.*, 1985; Agur, 1986; Agur, Arnon & Schechter, 1988; Webb, 1990), which are mostly theoretical so far, is again to maximize antitumoural activity while reducing cytotoxicity to normal tissues.

## Generation of long-period oscillations: from circhoral hormone secretion to circadian rhythms

Whereas the pulsatile secretion of insulin, with a periodicity close to 13 min, may be linked to glycolytic oscillations (Corkey *et al.*, 1988; Chou *et al.*, 1990, 1992), the origin of pusatile secretion for other hormones remains unclear. How GnRH-secreting neurons in the arcuate nuclei of the hypothalamus can generate rhythms of 1 h period, while neuronal oscillations are generally much faster, remains one of the prominent questions in the field of cellular rhythms. A plausible, although hypothetical mechanism for such long-period neuronal bursting was outlined in chapter 8 in terms of a coupling between the regular ionic mechanisms of the Hodgkin–Huxley type and a slow, reversible covalent modification of an ionic channel. An alternative mechanism was mentioned that bears some resemblance to the mechanism generating cAMP oscillations in *Dictyostelium*: according to that view, GnRH pulses would originate from a positive autocrine loop due to the presence of GnRH receptors in the membrane of GnRH-secreting cells (Krsmanovic *et al.*, 1993). A negative autocrine loop might also be involved in the oscillatory mechanism (Bourguignon *et al.*, 1994).

The question of how to generate hourly hormonal rhythms pertains of course to the origin of circadian rhythms that are observed in unicellular (Vanden Driessche, 1980; Schweiger *et al.*, 1986; Edmunds, 1988) up to the most evolved organisms (Moore-Ede *et al.*, 1982). In mammals, the suprachiasmatic nuclei contain the neurons that generate these periodicities, so that the key to both circadian rhythms and the GnRH rhythm controlling reproduction lies in neurons of the hypothalamus.

### Towards a model for circadian oscillations of PER in *Drosophila*

Studies based on genetics and molecular biology have recently led to significant advances in the understanding of the function of genes involved in the origin of circadian rhythmicity, particularly the *per* gene in *Drosophila* (Hall & Rosbach, 1988; Takahashi, 1992, 1993; Dunlap, 1993; Stehle *et al.*, 1993; Takahashi *et al.*, 1993; Young, 1993). A key progress came from the demonstration of a circadian rhythm in *per* mRNA, followed after several hours by a similar rhythm in PER. The latter protein has recently been shown to interact with transcription factors. This, together with the nuclear localization of PER, has suggested a mechanism for circadian rhythmicity in which PER represses the transcription of its gene (Hardin *et al.*, 1990, 1992). Recent findings (Zeng *et al.*, 1994) support the view that each *per*-expressing cell in *Drosophila* contains a circadian oscillator involving PER.

A theoretical model for circadian oscillations in PER and *per* mRNA was presented in chapter 11. The model is based on negative feedback by PER at the transcription level and on the observation (Edery *et al.*, 1994) that PER is multiply phosphorylated. The role of this covalent modification is still unclear; a plausible assumption retained for the model was that PER phosphorylation in the cytosol targets the protein for both translocation into the nucleus and degradation. The former assumption holds with the results of recent experiments (Curtin *et al.*, 1995), which indicate that the entry of PER into the nucleus is delayed, possibly due to phosphorylation.

An alternative model for circadian oscillations in *Drosophila* has been developed independently by Abbot *et al.* (1995); that model, also based on the negative feedback exerted by PER, takes into account a larger number of phosphorylated residues and stresses the dual role of the delays introduced by multiple phosphorylation, which is to strengthen the destabilizing effect of negative feedback, and to contribute to lengthen the period of the rhythm up to circadian values.

The more minimal model presented in chapter 11 clarifies the conditions in which negative feedback at the transcription level can give rise to sustained oscillations in PER and *per* mRNA. The phase difference between the two curves results from the successive phosphorylations of PER. Changing the value of parameters such as the maximum rate of PER degradation markedly affects the period of PER oscillations, suggesting a possible molecular explanation for the long- and short-period *Drosophila* clock mutants isolated by Konopka & Benzer (1971). The model provides a molecular basis for oscillations of the limit cycle type, which have long been suggested for circadian rhythms.

Because DNA transcription and protein synthesis are at the core of this model and of that proposed by Abbot *et al.* (1995), the results bear on the origin of circadian rhythms in other organisms. These processes are indeed repeatedly implicated in the generation of circadian rhythmicity (Takahashi *et al.*, 1993; see chapter 11), so that – as for cdc2 kinase in the case of the mitotic oscillator controlling the eukaryotic cell cycle – a universal mechanism for the circadian clock might well emerge, based on autoregulated, negative feedback control of gene expression (Takahashi, 1992; Sassone-Corsi, 1994). In support of this view, recent results (Aronson *et al.*, 1994) indicate that the *frq* gene product in *Neurospora* plays a role similar to that of PER in *Drosophila*, so that a similar mechanism governs circadian rhythmicity in the two organisms.

This book began with an overview of early developments in the field of biochemical oscillations. In a somewhat cyclical fashion the discussion of the model for PER circadian oscillations closes a loop, as it brings us back to the early studies of Goodwin (1963, 1965), who first suggested that repression of mRNA synthesis by a protein product could produce oscillations (see chapter 11). This theoretical prediction was made at a time when no experimental data were available to support such a view.

## A variety of functions for biochemical and cellular rhythms

The roles of biological rhythms are manifold. Well-known is the function of circadian rhythms in adaptation to the environment. If these endogenous rhythms and the periodic alternation of day and night with which they coincide did not exist, how would we see the passage of time? Also clear is the role of other periodic phenomena in the generation of rhythmic behaviour in repetitive machines such as the heart or the respiratory system. The functions of other ultradian rhythms have recently been reviewed (Lloyd & Stupfel, 1991). The preceding discussion suggests that the provision of an optimal mode of intercellular communication could well be a most important function of biological rhythms (Goldbeter, 1988a,b). Frequency-encoded, rhythmic processes would be more accurate and versatile than those encoded by the sole amplitude of the intercellular signal (Rapp, Mees & Sparrow, 1981; Rapp, 1987).

Communication by periodic, pulsatile signals is not limited to pulsatile hormone secretion or cAMP signalling in *Dictyostelium* cells. Thus frequency coding is also encountered in neural rhythms, for example in the control of neurosecretion (Cazalis, Dayanithi & Nordmann, 1985; Whim & Lloyd, 1989). Although the rhythmic properties of neurons, reflected by the rhythmic nature of the EEG itself, have been known for long (see, e.g. Fessard, 1936), there has beeen a recent surge of interest in the coherent oscillations observed in the brain. Thus, the role and the origin of synchronous oscillatory activity in the cortex are being investigated (see Basar & Bullock, 1992, and the recent review by Singer, 1993). One area of research is the encoding of sensory, e.g. olfactory or visual, information in terms of synchronous oscillations of neuronal assemblies (Eckhorn *et al.*, 1988; Gray *et al.*, 1989; Delaney *et al.*, 1994; Laurent & Davidowitz, 1994). The importance of temporal coding in the brain (Buzsaki *et al.*, 1994; Hopfield, 1995) becomes progressively clearer. Of key importance for this function are the electrophysiological properties of neurons (Llinas, 1988; Basar & Bullock, 1992), which make oscillatory behaviour a hallmark of both these cells and neural networks.

Frequency coding is also encountered in the transduction of certain extracellular signals that induce oscillations in cytosolic $Ca^{2+}$. As discussed in chapter 9, the frequency of the latter rhythm increases with the degree of stimulation (Woods *et al.*, 1987; Berridge *et al.*, 1988; Berridge, 1989; Cuthbertson, 1989). Different hormones could exert their effect via the triggering of intracellular $Ca^{2+}$ oscillations, but the specificity of the response could be achieved by differences in the frequency or waveform of periodic $Ca^{2+}$ evolution. The specificity of the 'signature' of each agonist would be enhanced by a simultaneous action of the ligand on more than one parameter of the system: thus, the signal could induce the synthesis of $IP_3$ and modulate at the same time the rate of another process such as cytosolic $Ca^{2+}$ extrusion. Such a dual effect would produce oscillations whose frequency and form would be specific to the agonist (Goldbeter *et al.*, 1990).

The experimental study of the effect of $Ca^{2+}$ oscillations on cell metabolism is only beginning (Pralong *et al.*, 1994). A recent study showed that certain aspects of neuronal differentiation may be governed by the frequency of $Ca^{2+}$ transients (Gu & Spitzer, 1995). A plausible mechanism for the frequency encoding of periodic variations in cytosolic $Ca^{2+}$ is based on the phosphorylation of one or more cellular proteins by a $Ca^{2+}$-activated kinase (Berridge *et al.*, 1988). The kinetic analysis of such a process shows that the mean level of phosphorylated protein increases with the frequency of $Ca^{2+}$ oscillations, which itself rises with the intensity of stimulation (Goldbeter *et al.*, 1990; Dupont & Goldbeter, 1992a). Such a frequency encoding based on protein phosphorylation provides an alternative mechanism to that based on reversible receptor desensitization. These two molecular processes show that there exist numerous means to encode an external signal in terms of the frequency of the first (hormone, cAMP in *Dictyostelium*) or second messenger ($Ca^{2+}$). Also to be mentioned is the possibility that the $Ca^{2+}$-calmodulin-dependent protein kinase may itself be activated to a degree that depends on the frequency of $Ca^{2+}$ pulses (Meyer *et al.*, 1992; Michelson & Schulman, 1994).

Periodic signals are also used in insect communication, by means of pheromones (Bossert, 1969; Conner, 1985; Conner, Webster & Itagaki, 1985) or bioluminescence (Lloyd, 1983). In addition, auditory stimuli are used in the courtship song of *Drosophila* males; this song consists in a rhythmic variation of intervals separating series of rapid wing movements. As recalled in chapter 11, the period of this rhythm, of the order of 40 s, is species specific and is controlled by the same *per* gene that governs the period of circadian rhythms in the fly (Kyriacou & Hall, 1980, 1986).

To conclude, let us stress the roles of rhythmic processes in cell biology and physiology. These roles are illustrated, of course, by cardiac and neuronal oscillations, by the circadian sleep–wake cycle, and by the biochemical clock controlling the onset of mitosis. Other rhythms govern, in a precise window of periods, communications between individuals in some microorganism or insect species, as well as communications between endocrine organs in higher organisms. Thus, in the human and other mammalian species, cellular rhythms are involved in many of the most important physiological functions. Such rhythms underlie the periodic operation of the heart, respiration, the functioning of the brain, adaptation to the environment, the control of cell division, and communications between the hypothalamus and the pituitary, which allow the secretion of the hormones that govern reproduction. Elucidating the mechanism of these biochemical and cellular rhythms is therefore essential for understanding the nonlinear, dynamic aspects of life.

# References

Abbott, L.F., H. Zeng & M. Rosbash. 1995. A model of circadian rhythm generation in *Drosophila*. Submitted for publication.

Abramson, J.J., S. Milne, E. Buck & I.N. Pessah. 1993. Porphyrin induced calcium release from skeletal muscle sarcoplasmic reticulum. *Arch. Biochem. Biophys.* **301**:396–403.

Adair, G.S. 1925. The hemoglobin system. VI. The oxygen dissociation curve of hemoglobin. *J. Biol. Chem.* **63**:529–45.

Adams, T.E., S. Cumming & B.M. Adams. 1986. Gonadotropin-releasing hormone (GnRH) receptor dynamics and gonadotrope responsiveness during and after continuous GnRH stimulation. *Biol. Reprod.* **35**:881–9.

Adams, W.B. & J.A. Benson. 1985. The generation and modulation of endogenous rhythmicity in the *Aplysia* bursting pacemaker neurone *R15*. *Progr. Biophys. Mol. Biol.* **46**:1–49.

Adams, W.B. & J.A. Benson. 1989. Rhythmic neuronal burst generation: Experiment and theory. In: *Cell to Cell Signalling: From Experiments to Theoretical Models.* A. Goldbeter, ed. Academic Press, London, pp. 29–45.

Aguda, B.D., L.L. Hofmann-Frisch & L.F. Olsen. 1990. Experimental evidence for the coexistence of oscillatory and steady states in the peroxidase–oxidase reaction. *J. Am. Chem. Soc.* **112**:6652–6.

Aguda, B.D. & R. Larter. 1991. Periodic–chaotic sequences in a detailed mechanism of the peroxidase–oxidase reaction. *J. Am. Chem. Soc.* **113**:7913–16.

Agur, Z. 1986. The effect of drug schedule on responsiveness to chemotherapy. *Ann. N.Y. Acad. Sci.* **504**:274–7.

Agur, Z., R. Arnon & B. Schechter. 1988. Reduction of cytotoxicity to normal tissue by new regimens of cell-cycle phase-specific drugs. *Math. Biosci.* **92**:1–15.

Aihara, K. & G. Matsumoto. 1982. Temporally coherent organization and instabilities in squid giant axon. *J. Theor. Biol.* **95**:697–720.

Alamgir, M. & I.R. Epstein. 1983. Birhythmicity and compound oscillations in coupled chemical oscillators: Chlorite–Bromate–Iodide system. *J. Am. Chem. Soc.* **105**:2500–1.

Alamgir, M. & I.R. Epstein. 1984. Experimental study of complex dynamical behavior in coupled chemical oscillators. *J. Phys. Chem.* **88**:2848–51.

Alcantara, F. & M. Monk. 1974. Signal propagation during aggregation in the slime mould *Dictyostelium discoideum*. *J. Gen. Microbiol.* **85**:321–34.

Allbritton, N.L., T. Meyer & L. Stryer. 1992. Range of messenger action of calcium ion and inositol 1,4,5-trisphosphate. *Science* **258**:1812–15.

Alonso, A., M.-P. Faure & A. Beaudet. 1994. Neurotensin promotes oscillatory bursting behavior and is internalized in basal forebrain cholinergic neurons. *J. Neurosci.* **14**:5778–92.

Alving, B.O. 1968. Spontaneous activity in isolated somata of *Aplysia* pacemaker neurons. *J. Gen. Physiol.* **51**:29–45.

Amon, A., S. Irniger & K. Nasmyth. 1994. Closing the cell cycle in yeast: G2 cyclin proteolysis initiated at mitosis persists until the activation of G1 cyclins in the next cycle. *Cell* **77**:1037–50.

Amundson, J. & D. Clapham. 1993. Calcium waves. *Curr. Opin. Neurobiol.* **3**:375–82.

an der Heiden, U. 1979. Delays in physiological systems. *J. Math. Biol.* **8**:345–64.

Andrés, V., V. Schulz & K. Tornheim. 1990. Oscillatory synthesis of glucose 1,6-bisphosphate and frequency modulation of glycolytic oscillations in skeletal muscle extracts. *J. Biol. Chem.* **265**: 21441–7.

Andronov, A.A., A.A. Vitt & S.E. Khaikin. 1966. *Theory of Oscillators.* Pergamon Press, Oxford.

Aon, M.A., S. Cortassa, H.V. Westerhoff & K. Van Dam. 1992. Synchrony and mutual stimulation of yeast cells during fast glycolytic oscillations. *J. Gen. Microbiol.* **138**:2219–27.

Armstrong, D. 1989. Calcium channel regulation by calcineurin, a $Ca^{2+}$-activated phosphatase in mammalian brain. *Trends Neurosci.* **12**:117–22.

Aronson, B.D., K.A. Johnson, J.L. Loros & J.C. Dunlap. 1994. Negative feedback defining a circadian clock: Autoregulation of the clock gene *frequency. Science* **263**:1578–84.

Aronson, D.G., E.J. Doedel & H.G. Othmer. 1987. An analytical and numerical study of the bifurcations in a system of linearly coupled oscillators. *Physica* **25D**:20–104.

Aschoff, J. (ed.) 1981. *Biological Rhythms.* Handbook of Behavioral Neurobiology, Vol. 4. Plenum Press, New York.

Asplin, C.M., A.C.S. Faria, E.C. Carlsen, V.A. Vaccaro, R.E. Barr, A. Iranmanesh, M.M. Lee, J.D. Veldhuis & W.S. Evans. 1989. Alterations in the pulsatile mode of growth hormone release in men and women with insulin-dependent diabetes mellitus. *J. Clin. Endocrinol. Metab.* **69**:239–45.

Atkinson, D.E. 1968. The energy charge of the adenylate pool as a regulatory parameter. Interaction with feedback modifiers. *Biochemistry* **7**:4030–4.

Atkinson, D.E. 1977. *Cellular Energy Metabolism and its Regulation.* Academic Press, New York.

Atri, A., J. Amundson, D. Clapham & J. Sneyd. 1993. A single pool model for intracellular calcium oscillations and waves in the *Xenopus laevis* oocyte. *Biophys. J.* **65**:1727–39.

Av-Ron, E., H. Parnas & L.A. Segel. 1991. A minimal biophysical model for an excitable and oscillatory neuron. *Biol. Cybern.* **65**:487–500.

Avgerinos P.C., T.H. Schurmeyer, P.W. Gold, T.P. Tornai, D.L. Loriaux, R.J. Sherins, G.B. Cutler, Jr & G.P. Chrousos. 1986. Pulsatile administration of human corticotropin-releasing hormone in patients with secondary adrenal insufficiency: restoration of the normal cortisol secretory pattern. *J. Clin. Endocr. Metabol.* **62,** 816–21.

Babloyantz, A. & A. Destexhe. 1986. Low-dimensional chaos in an instance of epilepsy. *Proc. Natl. Acad. Sci. USA* **83**:3513–17.

528      *References*

Backx, P.H., P.P. de Tombe, J.H.K. van Deen, B.J. Mulder & H.E.D.J. ter Keurs. 1989. A model of propagating calcium-induced calcium release mediated by calcium diffusion. *J. Gen. Physiol.* **93**:963–77.

Badola, P., V.R. Kumar & B.D. Kulkarni. 1991. Effects of coupling nonlinear systems with nonlinear dynamics. *Phys. Lett. A* **155**:365–72.

Baesens, C. & G. Nicolis. 1983. Complex bifurcations in a periodically forced normal form. *Z. Phys. B* **52**:345–54.

Bar-Eli, K. 1984. Coupling of chemical oscillators. *J. Phys. Chem.* **88**:3616–22.

Barchilon, M. & L.A. Segel. 1988. Adaptation, oscillations and relay in a model for cAMP secretion in cellular slime molds. *J. Theor. Biol.* **133**:437–46.

Bargiello, T.A., F.R. Jackson & M.W. Young. 1984. Restoration of circadian behavioural rhythms by gene transfer in *Drosophila*. *Nature* **312**:752–4.

Bargiello, T.A., L. Saez, M.K. Baylies, G. Casic, M.W. Young & D.C. Spray. 1987. The *Drosophila* clock gene *per* affects intercellular junctional communication. *Nature* **328**:686–91.

Barker, C.J., T. Nilsson, C.J. Kirk, R.H. Michell & P.O. Berggren. 1994. Simultaneous oscillations of cytoplasmic free $Ca^{2+}$ concentration and $Ins(1,4,5)P_3$ concentration in mouse pancreatic β-cells. *Biochem. J.* **297**:265–8.

Basar, E. & T.H. Bullock (eds.) 1992. *Induced Rhythms in the Brain.* Birkhäuser, Boston, MA.

Bassett, N.S. & P.D. Gluckman. 1986. Pulsatile growth hormone secretion in the ovine fetus and neonatal lamb. *J. Endocr.* **109**:307–12.

Baylies, M.K., T.A. Bargiello, F.R. Jackson & M.W. Young. 1987. Changes in abundance or structure of the *per* gene product can alter periodicity of the *Drosophila* clock. *Nature* **326**:390–2.

Baylies, M.K., L. Weiner, L.B. Vosshall, L. Saez & M.W. Young. 1993. Genetic, molecular, and cellular studies of the *per* locus and its products in *Drosophila melanogaster*. In: *Molecular Genetics of Biological Rhythms*, M.W. Young, ed. Marcel Dekker, New York, pp. 123–53.

Belchetz, P.E., T.M. Plant, Y. Nakai, E.J. Keogh & E. Knobil. 1978. Hypophysial responses to continuous and intermittent delivery of hypothalamic gonadotropin-releasing hormone. *Science* **202**:631–3.

Belousov, B.P. 1959. A periodic chemical reaction and its mechanism. *Sb. Ref. Radiats. Med.* Medgiz, Moscow, pp. 145–7.

Benjamin, P.R. & C.J.H. Elliott. 1989. Snail feeding oscillator: The central pattern generator and its control by modulatory interneurons. In: *Neuronal and Cellular Oscillators*. J.W. Jacklet, ed. Marcel Dekker, New York and Basel, pp. 173–214.

Bergé, P., Y. Pomeau & C. Vidal. 1984. *L'Ordre dans le Chaos.* Hermann, Paris.

Berridge, M.J. 1988. Inositol lipids and calcium signalling. *Proc. R. Soc. London B* **234**:359–79.

Berridge, M.J. 1989. Cell signalling through cytoplasmic calcium oscillations. In: *Cell to Cell Signalling: From Experiments to Theoretical Models*. A. Goldbeter, ed. Academic Press, London, pp. 449–59.

Berridge, M.J. 1990. Calcium oscillations. *J. Biol. Chem.* **265**:9583–6.

Berridge, M.J. 1993. Inositol trisphosphate and calcium signalling. *Nature* **361**:315–25.

Berridge, M.J. (ed.) 1995. $Ca^{2+}$ *Waves, Gradients and Oscillations*. CIBA Found. Symp. Wiley, Chichester.

Berridge, M.J, P.H. Cobbold & K.S.R. Cuthbertson. 1988. Spatial and temporal aspects of cell signalling. *Phil. Trans. R. Soc. Lond. B* **320**:325–43.

Berridge, M.J. & G. Dupont. 1994. Spatial and temporal signalling by calcium. *Curr. Opin. Cell Biol.* **6**:267–74.

Berridge, M.J. & A. Galione. 1988. Cytosolic calcium oscillators. *FASEB J.* **2**:3074–82.

Berridge, M.J. & R.F. Irvine. 1989. Inositol phosphates and cell signalling. *Nature* **341**:197–205.

Berridge, M.J. & P.E. Rapp. 1979. A comparative survey of the function, mechanism and control of cellular oscillations. *J. Exp. Biol.* **81**:217–79.

Bertram, R. 1994. Reduced system analysis of the effects of serotonin on a molluscan burster neuron. *Biol. Cybern.* **70**:359–68.

Bertram, R., M.J. Butte, T. Kiemel & A. Sherman. 1995. Topological and phenomenological classification of bursting oscillations. *Bull. Math. Biol.* **57**:413–39.

Best, E. 1979. Null space in the Hodgkin–Huxley equations. A critical test. *Biophys. J.* **27**:87–104.

Betz, A. & J.U. Becker. 1975. Phase dependent phase shifts induced by pyruvate and acetaldehyde in oscillating NADH of yeast cells. *J. Interdisciplin. Cycle Res.* **6**:167–73.

Betz, A. & B. Chance. 1965. Phase relationship of glycolytic intermediates in yeast cells with oscillatory metabolic control. *Arch. Biochem. Biophys.* **109**:585–94.

Betz, A. & C. Moore. 1967. Fluctuating metabolite levels in yeast cells and extracts, and the control of phosphofructokinase activity *in vitro*. *Arch. Biochem. Biophys.* **120**:268–73.

Betz, A. & E. Sel'kov. 1969. Control of phosphofructokinase activity in conditions simulating those of glycolyzing yeast extracts. *FEBS Lett.* **3**:5–9.

Bezprozvanny, I. & B.E. Ehrlich. 1994. Inositol (1,4,5)-trisphosphate (Ins$P_3$)-gated Ca channels from cerebellum: conduction properties for divalent cations and regulation by intraluminal calcium. *J. Gen. Physiol.* **104**:821–56.

Bezprozvanny I., J. Watras & B. Ehrlich. 1991. Bell-shaped calcium-response curves of Ins$(1,4,5)P_3$- and calcium-gated channels from endoplasmic reticulum of cerebellum. *Nature* **351**:751–4.

Bick, T., M.B.H. Youdim & Z. Hochberg. 1989a. Adaptation of liver membrane somatogenic and lactogenic growth hormone (GH) binding to the spontaneous pulsation of GH secretiion in the male rat. *Endocrinology* **125**:1711–17.

Bick, T., M.B.H. Youdim & Z. Hochberg. 1989b. The dynamics of somatogenic and lactogenic growth hormone binding: Internalization to Golgi fractions in the male rat. *Endocrinology* **125**:1718–22.

Binkley, S. 1983. Rhythms in ocular and pineal $N$-acetyltransferase: A portrait of an enzyme clock. *Comp. Biochem. Physiol.* **75A**:123–9.

Binkley, S., J.B. Riebman & K.B. Reilly. 1978. The pineal gland: a biological clock *in vitro*. *Science* **202**:1198–201.

Birnbaumer, L., J. Abramowitz & A.M. Brown. 1990. Receptor–effector coupling by G proteins. *Biochim. Biophys. Acta* **1031**:163–224.

Blangy, D., H. Buc & J. Monod. 1968. Kinetics of the allosteric interactions of phosphofructokinase from *Escherichia coli*. *J. Mol. Biol.* **31**:13–35.

Bliss, R.D., P.R. Painter & A.G. Marr. 1982. Role of feedback inhibition in stabilizing the classical operon. *J. Theor. Biol.* **97**:177–93.

Block, G.D. & S.F. Wallace. 1982. Localization of a circadian pacemaker in the eye of a mollusc, *Bulla*. *Science* **217**:155–7.

Blum, J.J. 1985. The role of microaggregation in hormone-receptor-effector interactions. In: *Receptors*. Vol. II. P.M. Conn, ed. Academic Press, San Diego, pp. 57–88.

Blum, J.J. & P.M. Conn. 1982. Gonadotropin-releasing hormone stimulation of luteinizing hormone release: a ligand–receptor–effector model. *Proc. Natl. Acad. Sci. USA* **79**:7307–11.

Boeynaems, J.M. & J.E. Dumont. 1980. *Receptor Theory*. Elsevier/North-Holland, Amsterdam.

Boitano, S., E.R. Dirksen & M.J. Sanderson. 1992. Intercellular propagation of calcium waves mediated by inositol trisphosphate. *Science* **258**:292–5.

Boiteux, A., A. Goldbeter & B. Hess. 1975. Control of oscillating glycolysis of yeast by stochastic, periodic, and steady source of substrate: a model and experimental study. *Proc. Natl. Acad. Sci. USA* **72**:3829–33.

Boiteux, A., B. Hess & E.E. Sel'kov. 1980. Creative functions of instability and oscillations in metabolic systems. *Curr. Topics Cell. Reg.* **17**:171–203.

Boiteux, A. & S.C. Müller. 1983. The effect of fructose 2,6-bisphosphate on the regulation of glycolytic flux in yeast extract. Abstract S-04/TH-045. 15th FEBS Meeting, Brussels, Belgium.

Bonner, J.T. 1947. Evidence for the formation of cell aggregates by chemotaxis in the development of the slime mold *Dictyostelium discoideum*. *J. Exp. Zool.* **106**:1–26.

Bonner, J.T. 1967. *The Cellular Slime Molds*. Princeton Univ. Press, Princeton, NJ.

Borges, J.L.C., R.M. Blizzard, W.S. Evans, R. Furlanetto, A.D. Rogol, D.L. Kaiser, J. Rivier, W. Vale & M.O. Thorner. 1984. Stimulation of growth hormone (GH) and somatomedin C in idiopathic GH-deficient subjects by intermittent pulsatile administration of synthetic human pancreatic tumor GH-releasing factor. *J. Clin. Endocr. Metabol.* **59**:1–6.

Bossert, W.H. 1969. Temporal patterning in olfactory communication. *J. Theor. Biol.* **18**:157–70.

Both, R., W. Finger & R.A. Chaplain. 1976. Model predictions of the ionic mechanisms underlying the beating and bursting pacemaker characteristics of molluscan neurons. *Biol. Cybern* **23**:1–11.

Boukalouch, M., J. Elezgaray, A. Arneodo, J. Boissonade & P. De Kepper. 1987. Oscillatory instability induced by mass interchange between two coupled steady-state reactors. *J. Phys. Chem.* **91**:5843–5.

Bourguignon, J.P., M.L. Alvarez Gonzalez, A. Gerard & P. Franchimont. 1994. Gonadotropin releasing hormone inhibitory autofeedback by subproducts antagonist at *N*-methyl-D-aspartate receptors: A model of autocrine regulation of peptide secretion. *Endocrinology* **134**:1589–92.

Bourne, H.R., D.A. Sanders & F.M. McCormick. 1990. The GTPase superfamily: a conserved switch for diverse cell functions. *Nature* **348**:125–32.

Bowers, C.Y., D.K. Alster & J.M. Frentz. 1992. The growth hormone-releasing activity of a synthetic hexapeptide in normal men and short statured children after oral administration. *J. Clin. Endocrinol. Metab.* **74**:292–8.

Boyce, W.E. & R.G. Di Prima. 1969. *Elementary Differential Equations and Boundary Value Problems*. Wiley, New York.

Braiman, Y. & I. Goldhirsch. 1991. Taming chaotic dynamics with weak periodic perturbations. *Phys. Rev. Lett.* **66**:2545–8.

Bray, W.C. 1921. A periodic chemical reaction and its mechanism. *J. Am. Chem. Soc.* **43**:1262.

Brenner, M. & S. Thoms. 1984. Caffeine blocks activation of cAMP synthesis in *Dictyostelium discoideum. Dev. Biol.* **101**:136–46.

Bretschneider, T., F. Siegert & C.J. Weijer. 1995. Three-dimensional scroll waves of cAMP could direct cell movement and gene expression in *Dictyostelium* slugs. *Proc. Natl. Acad. Sci. USA* **92**, in press.

Brewitt, B. & J.I. Clark. 1988. Growth and transparency in the lens, an epithelial tissue, stimulated by pulses of PDGF. *Science* **242**:777–9.

Brizuela, L., G. Draetta & D. Beach. 1989. Activation of human CDC2 protein as a histone H1 kinase is associated with complex formation with the p62 subunit. *Proc. Natl. Acad. Sci. USA* **86**:4362–6.

Brooker, G., T. Seki, D. Croll & C. Wahlestedt. 1990. Calcium wave evoked by activation of endogenous or exogenously expressed receptors in *Xenopus* oocytes. *Proc. Natl. Acad. Sci. USA* **87**:2813–17.

Bumann, J., D. Malchow & B. Wurster. 1986. Oscillations of $Ca^{++}$ concentration during the cell differentiation of *Dictyostelium discoideum.* Their relation to oscillations in cyclic AMP and other components. *Differentiation* **31**:85–91.

Bünning, E. 1973. *The Physiological Clock. Circadian Rhythms and Biological Chronometry.* Third edn. Springer, Berlin.

Busa, W.B., J.E. Ferguson, S.K. Joseph, J.R. Williamson & R. Nuccitelli. 1985. Activation of frog (*Xenopus laevis*) eggs by inositol trisphosphate. I. Characterization of $Ca^{2+}$ release from intracellular stores. *J. Cell Biol.* **101**:677–82.

Busa, W.B. & R. Nuccitelli.1985. An elevated free cytosolic $Ca^{2+}$ wave follows fertilization in eggs of the frog *Xenopus laevis. J. Cell Biol.* **100**:1325–9.

Buzsaki, G., R. Llinas, W. Singer, A. Berthoz & Y. Christen (eds.) 1994. *Temporal Coding in the Brain.* Springer, Berlin.

Camacho, P. & J.D. Lechleiter. 1993. Increased frequency of calcium waves in *Xenopous laevis* oocytes that express a calcium-ATPase. *Science* **260**:226–9.

Canavier, C.C., D.A. Baxter, J.W. Clark & J.H. Byrne. 1991. Simulation of the bursting activity of neuron R15 in *Aplysia*: Role of ionic currents, calcium balance, and modulatory transmitters. *J. Neurophysiol.* **66**:2107–24.

Canavier, C.C., D.A. Baxter, J.W. Clark & J.H. Byrne. 1993. Nonlinear dynamics in a model neuron provide a novel mechanism for transient synaptic inputs to produce long-term alterations of postsynaptic activity. *J. Neurophysiol.* **69**:2252–7.

Canavier, C.C., D.A. Baxter, J.W. Clark & J.H. Byrne. 1994. Multiple modes of activity in a model neuron suggest a novel mechanism for the effects of neuromodulators. *J. Neurophysiol.* **72**:872–82.

Canavier, J.W. Clark & J.H. Byrne. 1990. Routes to chaos in a model of a bursting neuron. *Biophys. J.* **57**:1245–51.

Capogrossi, M., S. Houser, A. Bahinski & E. Lakatta. 1987. Synchronous occurrence of spontaneous localized calcium release from the sarcoplasmic reticulum generates action potentials in rat cardiac ventricular myocytes at normal resting membrane potential. *Circ. Res.* **61**:498–503.

Carafoli, E. & M. Crompton. 1978. The regulation of intracellular calcium. *Curr. Top. Membr. Transp.* **10**:151–216.

Carlier, M.-F., R. Melki, D. Pantaloni, T.L. Hill & Y. Chen. 1987. Synchronous oscillations in microtubule polymerization. *Proc. Natl. Acad. Sci. USA* **84**:5257–61.

Carmel, P.W., S. Araki & M. Ferin. 1976. Pituitary stalk portal blood collection in rhesus monkeys: evidence of pulsatile release of gonadotropin-releasing hormone (GnRH). *Endocrinology* **99**:243–8.

Carpenter, D.O. (ed.) 1982. *Cellular Pacemakers*. Wiley, New York.

Carpenter, G.A. 1979. Bursting phenomena in excitable membranes. *SIAM J. Appl. Math.* **36**:334–72.

Carpenter, G.A. & S. Grossberg. 1985. A neural theory of circadian rhythms: split rhythms, after-effects and motivational interactions. *J. Theor. Biol.* **113**:163–223.

Carter, P.J., H.G. Nimmo, C.A. Fewson & M.B. Wilkins. 1991. Circadian rhythms in the activity of a plant protein kinase. *EMBO J.* **10**:2063–8.

Cazalis, M., G. Dayanithi & J.J. Nordmann. 1985. The role of patterned burst and interburst interval on the excitation-coupling mechanism in the isolated rat neural lobe. *J. Physiol.* **369**:45–60.

Chance, B., B. Hess & A. Betz. 1964. DPNH oscillations in a cell-free extract of *S. carlsbergensis*. *Biochem. Biophys. Res. Commun.* **16**:182–7.

Chance, B., E.K. Pye, A.K. Ghosh & B. Hess (eds.) 1973. *Biological and Biochemical Oscillators*. Academic Press, New York.

Chance, B., K. Pye & J. Higgins. 1967. Waveform generation by enzymatic oscillators. *IEEE Spectrum* **4**:79–86.

Chance, B., B. Schoener & S. Elsaesser. 1964. Control of the waveform of oscillations of the reduced pyridine nucleotide level in a cell-free extract. *Proc. Natl. Acad. Sci. USA* **52**:337–41.

Chance, B. & T. Yoshioka. 1966. Sustained oscillations of ionic constituents of mitochondria. *Arch. Biochem. Biophys.* **117**:451–65.

Changeux, J.P. 1981. The acetylcholine receptor: an 'allosteric' membrane protein. In: *The Harvey Lectures 1979–1980*. Academic Press, New York, pp. 85–254.

Chaplain, R.A. 1976. Metabolic regulation of the rhythmic activity in pacemaker neurons. II. Metabolically induced conversion of beating in bursting pacemaker activity in isolated *Aplysia* neurons. *Brain Res.* **106**:307–19.

Charles, A.C., J.E. Merrill, E.R. Dirksen & M.J. Sanderson. 1991. Intercellular signaling in glial cells: calcium waves and oscillations in response to mechanical stimulation and glutamate. *Neuron* **6**:983–92.

Charles, A.C., C.C.G. Naus, D. Zhu, G.M. Kidder, E.R. Dirksen & M.J. Sanderson. 1992. Intercellular signaling via gap junctions in glioma cells. *J. Cell Biol.* **118**:195–201.

Chay, T.R. 1993. The mechanism of intracellular $Ca^{2+}$ oscillation and electrical bursting in pancreatic β–cells. *Adv. Biophys.* **29**:75–103.

Chay, T.R. & J. Keizer. 1983. Minimal model for membrane oscillations in the pancreatic β-cell. *Biophys. J.* **42**:181–90.

Chay, T.R. & J. Rinzel. 1985. Bursting, beating, and chaos in an excitable membrane model. *Biophys. J.* **47**:357–66.

Cheer, A., J.P. Vincent, R. Nuccitelli & G. Oster. 1987. Cortical activity in vertebrate eggs. I: The activation wave. *J. Theor. Biol.* **124**:377–404.

Chisholm, R.L., S. Hopkinson & H.F. Lodish. 1987. Superinduction of the *Dictyostelium discoideum* cell surface cAMP receptor by pulses of cAMP. *Proc. Natl. Acad. Sci. USA* **84**:1030–4.

Chou, H.-F., N. Berman & E. Ipp. 1990. Evidence for oscillatory glycolysis in islets: regular cycling of secreted lactate. *Diabetes* **39** (Suppl. 1):133A.

Chou, H.-F., N. Berman & E. Ipp. 1992. Oscillations of lactate released from islets of Langerhans: evidence for oscillatory glycolysis in β-cells. *Am. J. Physiol.* **262**:E800–E805.

Chovnik, A. (ed.) 1960. Biological clocks. *Cold Spring Harbor Symp. Quant. Biol.* **25**. Cold Spring Harbor Laboratory Press, New York.

Citri, O. & I.R. Epstein. 1988. Mechanistic study of a coupled chemical oscillator: The bromate–chlorite–iodide reaction. *J. Phys. Chem.* **92**:1865–71.

Clark, R.G., J.O. Jansson, O. Isaksson & I.C.A.F. Robinson. 1985. Intravenous growth hormone: Growth responses to patterned infusions in hypophysectomised rats. *J. Endocr.* **104**:53–61.

Cobbold, P.H. & K.S.R. Cuthbertson. 1990. Calcium oscillations: Phenomena, mechanisms and significance. *Semin. Cell Biol.* **1**:311–21.

Cobbold P.H., A. Sanchez-Bueno & C.J. Dixon. 1991. The hepatocyte calcium oscillator. *Cell Calcium* **12**:87–95.

Cohen, D.S. & J.P. Keener. 1976. Multiplicity and stability of oscillatory states in a continuous stirred tank reactor with exothermic consecutive reactions $A \rightarrow B \rightarrow C$. *Chem. Eng. Sci.* **31**:115–22.

Cohen, M.H. & A. Robertson. 1971. Wave propagation in the early stages of aggregation of cellular slime molds. *J. Theor. Biol.* **31**:101–18.

Cohen, M.S. 1977. The cyclic AMP control system in the development of *Dictyostelium discoideum*. I. Cellular dynamics. *J. Theor. Biol.* **69**:57–85.

Cohen, P. 1983. *Control of Enzyme Activity.* Chapman & Hall, London.

Collas, P., T. Chang, C. Long & J.M. Robl. 1995. Inactivation of histone H1 kinase by $Ca^{2+}$ in rabbit oocytes. *Mol. Reprod. Devel.* **40**:253–8.

Collatz, K.-G. & M. Horning. 1990. Age dependent changes of a biochemical rhythm – The glycolytic oscillator of the blowfly *Phormia terraenovae*. *Comp. Biochem. Physiol.* **96B**:771–4.

Collet, P. & J.P. Eckmann. 1980. *Iterated Maps on the Interval as Dynamical Systems.* Birkhäuser, Basel and Boston.

Combettes, L. & P. Champeil. 1994. Calcium and inositol 1,4,5-trisphosphate–induced $Ca^{2+}$ release. *Science* **265**:813.

Comolli, J., W. Taylor & J.W. Hastings. 1994. An inhibitor of protein phosphorylation stops the circadian oscillator and blocks light-induced phase shifting in *Gonyaulax polyedra*. *J. Biol. Rhythms* **9**:13–26.

Conn, P. M., C.A. McArdle, W.V. Andrews & W.R. Huckle. 1987. The molecular basis of gonadotropin-releasing hormone (GnRH) action in the pituitary gonadotrope. *Biol. Reprod.* **36**:17–35.

Conn, P.M., D.C. Rogers & R. McNeil. 1982. Potency enhancement of a GnRH agonist: GnRH-receptor microaggregation stimulates gonadotropin release. *Endocrinology* **111**:335–7.

Conn, P.M., D. Staley, C. Harris, W.V. Andrews, W.C. Gorospe, C.A. McArdle, W.R. Huckle & J. Hanson. 1986. Mechanism of action of gonadotropin releasing hormone. *Annu. Rev. Physiol.* **48**:495–13.

Conner, W.E. 1985. Temporally patterned chemical communication: is it feasible? In: *Perspectives in Ethology.* Vol. 6: *Mechanisms.* P.P.G. Bateson & P.H. Klopfer, eds. Plenum Press, New York and London, pp. 287–301.

Conner, W.E., R.P. Webster & H. Itagaki. 1985. Calling behavior in arctiid moths: the effects of temperature and wind speed on the rhythmic exposure of the sex attractant gland. *J. Insect Physiol.* **31**:815–20.

Connor, J.A., D. Walter & R. McKown. 1977. Neural repetitive firing:

534 *References*

Modifications of the Hodgkin–Huxley axon suggested by experimental results from crustacean axons. *Biophys. J.* **18**:81–102.

Corkey, B.E., K. Tornheim, J.T. Deeney, M.C. Glennon, J.C. Parker, F.M. Matschinsky, N.B. Ruderman & M. Prentki. 1988. Linked oscillations of free $Ca^{2+}$ and the ATP/ADP ratio in permeabilized RINm5F insulinoma cells supplemented with a glycolyzing cell-free muscle extract. *J. Biol. Chem.* **263**:4254–8.

Cornell-Bell, A.H., S.M. Finkbeiner, M.S. Cooper & S.J. Smith. 1990. Glutamate induces calcium waves in cultured astrocytes: Long-range glial signaling. *Science* **247**:470–3.

Cornish-Bowden, A. & D.E. Koshland, Jr. 1975. Diagnostic uses of the Hill (Logit and Nernst) plots. *J. Mol. Biol.* **95**:201–12.

Coté, G.G. & S. Brody. 1986. Circadian rhythms in *Drosophila melanogaster*: analysis of period as a function of gene dosage at the *per* (period) locus. *J. Theor. Biol.* **121**:487–503.

Coukell, M.B. 1975. Parasexual genetic analysis of aggregation-deficient mutants of *Dictyostelium discoideum. Mol. Gen. Genet.* **142**:119–35.

Coukell, M.B. 1981. Apparent positive cooperativity at a surface cAMP receptor in *Dictyostelium. Differentiation* **20**:29–35.

Coukell, M.B. & F.K. Chan. 1980. The precocious appearance and activation of an adenylate cyclase in a rapid developing mutant of *Dictyostelium discoideum. FEBS Lett.* **110**:39–42.

Cross, F., J. Roberts & H. Weintraub. 1989. Simple and complex cell cycles. *Annu. Rev. Cell Biol.* **5**:341–95.

Cross, F.R. & A.H. Tinkelenberg. 1991. A potential positive feedback loop controlling *CLN1* and *CLN2* gene expression at the start of the yeast cell cycle. *Cell* **65**:875–83.

Crowley, M.F. & I.R. Epstein. 1989. Experimental and theoretical studies of a coupled chemical oscillator: phase death, multistability, and in-phase and out-of-phase entrainment. *J. Phys. Chem.* **93**:2496–502.

Crowley, M.F. & R.J. Field. 1986. Electrically-coupled Belousov–Zhabotinskii oscillators. 1. Experiments and simulations. *J. Phys. Chem.* **90**:1907–15.

Crowley, W.F. Jr, M. Filicori, D.I. Spratt & N.F. Santoro. 1985. The physiology of gonadotropin-releasing hormone (GnRH) secretion in men and women. *Rec. Progr. Horm. Res.* **41**:473–531.

Crowley, W.F. Jr & J.G. Hofler (eds.) 1987. *The Episodic Secretion of Hormones.* Wiley, New York.

Curtin, K.D., Z.J. Huang & M. Rosbash. 1995. Temporally regulated nuclear entry of the Drosophila *period* protein contributes to the circadian clock. *Neuron* **14**:365–72.

Cuthbertson, K.S.R. 1989. Intracellular calcium oscillators. In: *Cell to Cell Signalling: From Experiments to Theoretical Models.* A. Goldbeter, ed. Academic Press, London, pp. 435–47.

Cuthbertson, K.S.R. & T.R. Chay. 1991. Modelling receptor-controlled intracellular calcium oscillations. *Cell Calcium* **12**:97–109.

Cuthbertson, K.S.R. & P.H. Cobbold. 1985. Phorbol ester and sperm activate mouse oocytes by inducing sustained oscillations in cell $Ca^{2+}$. *Nature* **316**:541–2.

Cuthbertson, K.S.R. & P.H. Cobbold (eds.) 1991. Oscillations in cell calcium. *Cell Calcium* **12**, nos. 2–3.

Cyert, M.S. & M.W. Kirschner. 1988. Regulation of MPF activity in vitro. *Cell* **53**:185–95.

Czeisler, C.A., M.P. Johnson, J.F. Duffy, E.N. Brown, J.M. Ronda & R.E. Kronauer. 1990. Exposure to bright light and darkness to treat physiologic maladaptation to night work. *New Engl. J. Med.* **322**:1253–9.

Czeisler, C.A., R.E. Kronauer, J.S. Allan, J.F. Duffy, M.E. Jewett, E.N. Brown & J.M. Ronda. 1989. Bright light induction of strong (type 0) resetting of the human circadian pacemaker. *Nature* **244**:1328–33.

Daan, S., Beersma, D.G.M. & A.A. Borbely. 1984. Timing of human sleep: Recovery process gated by a circadian pacemaker. *Am. J. Physiol.* **246**:R161–R178.

Daan, S. & C. Berde. 1978. Two coupled oscillators: Simulations of the circadian pacemaker in mammalian activity rhythms. *J. Theor. Biol.* **70**:297–313.

Daher, G.C., B.E. Harris, E.M. Willard & R.B. Diasio. 1991. Biochemical basis for circadian-dependent metabolism of fluoropyrimidines. *Ann. N.Y. Acad. Sci.* **618**:350–61.

Dalton, S. 1992. Cell cycle regulation of the human *cdc2* gene. *EMBO J.* **11**:1797–804.

Darmon, M., J. Barra & P. Brachet. 1978. The role of phosphodiesterase in aggregation of *Dictyostelium discoideum. J. Cell Sci.* **31**:233–43.

Darmon, M. & P. Brachet. 1978. Chemotaxis and differentiation during the aggregation of *Dictyostelium discoideum* amoebae. In: *Taxis and Behavior. Receptors and Recognition. Ser. B,* Vol. 5. G.L. Hazelbauer, ed. Chapman & Hall, London, pp. 103–39.

Darmon, M., P. Brachet & L.H. Pereira da Silva. 1975. Chemotactic signals induce cell differentiation in *Dictyostelium discoideum. Proc. Natl. Acad. Sci. USA* **72**:3163–6.

Das, J. & H.G. Busse. 1985. Long term oscillation in glycolysis. *J. Biochem.* **97**:719–27.

Davidenko, J.M., P. Kent, D.R. Chialvo, D.C. Michaels & J. Jalife. 1990. Sustained vortex-like waves in normal isolated ventricular muscle. *Proc. Natl. Acad. Sci. USA* **87**:8785–9.

Davidenko, J.M., A.V. Pertsov, R. Salomonsz, W. Baxter & J. Jalife. 1992. Stationary and drifting spiral waves of excitation in isolated cardiac muscle. *Nature* **355**:349–51.

De Bondt, H.L., J. Rosenblatt, J. Jancarik, H.D. Jones, D.O. Morgan & S.H. Kim. 1993. Crystal structure of cyclin-dependent kinase 2. *Nature* **363**:595–602.

De Kepper, P. 1976. Etude d'une réaction chimique périodique. Transitions et excitabilité. *C.R. Acad. Sci. (Paris) Sér. C* **283**:25–8.

De Kepper, P., V. Castets, E. Dulos & J. Boissonnade. 1991. Turing-type chemical patterns in the chlorite–iodide–malonic acid reaction. *Physica* **49D**:161–69.

De Young, G.W. & J. Keizer. 1992. A single-pool inositol 1,4,5-trisphosphate-receptor-based model for agonist-stimulated oscillations in $Ca^{2+}$ concentration. *Proc. Natl. Acad. Sci. USA* **89**:9895–9.

Decroly, O. 1987a. Du comportement périodique simple aux oscillations complexes dans les systèmes biochimiques: Bursting, chaos et attracteurs multiples. Thèse de Doctorat en Sciences, Université Libre de Bruxelles.

Decroly, O. 1987b. Interplay between two periodic enzyme reactions as a source for complex oscillatory behaviour. In: *Chaos in Biological Systems*. H. Degn, A.V. Holden & L.F. Olsen, eds. Plenum Press, New York and London, pp. 49–58.

Decroly, O. & A. Goldbeter. 1982. Birhythmicity, chaos, and other patterns of temporal self-organization in a multiply regulated biochemical system. *Proc. Natl. Acad. Sci. USA* **79**:6917–21.

Decroly, O. & A. Goldbeter. 1984a. Coexistence entre trois régimes périodiques stables dans un système biochimique à régulation multiple. *C.R. Acad. Sci. (Paris) Sér. II* **298**:779–82.

Decroly, O. & A. Goldbeter. 1984b. Multiple periodic regimes and final state sensitivity in a biochemical system. *Phys. Lett.* **105A**:259–62.

Decroly, O. & A. Goldbeter. 1985. Selection between multiple periodic regimes in a biochemical system: Complex dynamic behaviour resolved by use of one-dimensional maps. *J. Theor. Biol.* **113**:649–71.

Decroly, O. & A. Goldbeter. 1987. From simple to complex oscillatory behaviour: Analysis of bursting in a multiply regulated biochemical system. *J. Theor. Biol.* **124**:219–50.

Degn, H. 1968. Bistability caused by substrate inhibition of peroxidase in an open reaction system. *Nature* **217**:1047–50.

Degn, H., H. Olsen & J.W. Perram. 1979. Bistability, oscillations and chaos in an enzyme reaction. *Ann. N.Y. Acad. Sci.* **316**:623–37.

DeHaan, R.L. 1980. Differentiation of excitable membranes. *Curr. Top. Dev. Biol.* **16**:117–64.

Delaney, K.R., A. Gelperin, M.S. Fee, J.A. Flores, R. Gervais, D.W. Tank & D. Kleinfeld. 1994. Waves and stimulus-modulated dynamics in an oscillating olfactory network. *Proc. Natl. Acad. Sci. USA* **91**:669–73.

DeLisle, S. & M. Welsh.1992. Inositol trisphosphate is required for the propagation of calcium waves in *Xenopus* oocytes. *J. Biol. Chem.* **267**:7963–6.

Demongeot, J. & F. J. Seydoux. 1979. Oscillations glycolytiques. Modélisation d'un système minimum à partir des données physiologiques et moléculaires. In: *Elaboration et Justification des Modèles. Applications en Biologie*. P. Delattre & M. Thellier, eds. Maloine, Paris. Vol. II, pp. 519–36.

Destexhe, A. & A. Babloyantz. 1991. Pacemaker-induced coherence in cortical networks. *Neural Computation* **3**:145–54.

Destexhe, A., A. Babloyantz & T.J. Sejnowski. 1993. Ionic mechanisms for intrinsic slow oscillations in thalamic relay neurons. *Biophys. J.* **65**:1538–52.

Destexhe, A. & T.J. Sejnowski. 1995. Modeling the mechanisms of thalamic rhythmicity. In: *Thalamus*, M. Steriade, E.G. Jones & D.A. McCormick, eds. Elsevier, Amsterdam (in press).

Devreotes, P.N. 1982. Chemotaxis. In: *The Development of* Dictyostelium discoideum. W.F. Loomis, ed. Academic Press, New York, pp. 117–68.

Devreotes, P.N. 1989. *Dictyostelium discoideum*: A model system for cell–cell interactions in development. *Science* **245**:1054–8.

Devreotes, P.N., M. Potel & S. MacKay. 1983. Quantitative analysis of cyclic AMP waves mediating aggregation in *Dictyostelium discoideum. Dev. Biol.* **96**:405–15.

Devreotes, P.N. & J.A. Sherring. 1985. Kinetics and concentration dependence of reversible cAMP-induced modification of the surface cAMP receptor in *Dictyostelium. J. Biol. Chem.* **260**:6378–84.

Devreotes, P.N. & T.L. Steck. 1979. Cyclic 3′,5′ AMP relay in *Dictyostelium discoideum.* II. Requirements for the initiation and termination of the response. *J. Cell Biol.* **80**:300–9.

Dibrov, B.F., A.M. Zhabotinsky, Yu. A. Neyfakh, M.P. Orlova & L.I. Churikova. 1985. Mathematical model of cancer chemotherapy. Periodic schedules of phase-specific cytotoxic agent administration increasing the selectivity of therapy. *Math. Biosci.* **73**:1–31.

Dierschke, D.J., A.N. Bhattacharya, L.E. Atkinson & E. Knobil. 1970. Circhoral oscillations of plasma LH levels in the ovariectomized rhesus monkey. *Endocrinology* **87**:850–3.

DiFrancesco, D. 1993. Pacemaker mechanisms in cardiac tissue. *Annu. Rev. Physiol.* **55**:455–72.

DiFrancesco, D. & D. Noble. 1985. A model of cardiac electrical activity incorporating ionic pumps and concentration changes. *Phil. Trans. R. Soc. Lond. B* **307**:353–98.

DiFrancesco, D. & D. Noble. 1989. Current $i_f$ and its contribution to cardiac pacemaking. In: *Neuronal and Cellular Oscillators.* J.W. Jacklet, ed. Marcel Dekker, New York and Basel, pp. 31–57.

Dinauer, M., S. MacKay & P.N. Devreotes. 1980a. Cyclic 3',5' AMP relay in *Dictyostelium discoideum.* III. The relationship of cAMP synthesis and secretion during the cAMP signaling response. *J. Cell Biol.* **86**:537–44.

Dinauer, M., T.L. Steck & P.N. Devreotes. 1980b. Cyclic 3',5' AMP relay in *Dictyostelium discoideum.* IV. Recovery of the cAMP signaling response after adaptation to cAMP. *J. Cell Biol.* **86**:545–53.

Dinauer, M., T.L. Steck & P.N. Devreotes. 1980c. Cyclic 3',5' AMP relay in *Dictyostelium discoideum.* V. Adaptation of the cAMP signaling response during cAMP stimulation. *J. Cell Biol.* **86**:554–61.

Ding, J.M., D. Chen, E.T. Weber, L.E. Faiman, M.A. Rea & M.U. Gillette. 1994. Resetting the biological clock: Mediation of nocturnal circadian shifts by glutamate and NO. *Science* **266**:1713–17.

Dirick, L. & K. Nasmyth. 1991. Positive feedback in the activation of G1 cyclins in yeast. *Nature* **351**:754–7.

Dissing, S. (ed.) 1993. $Ca^{2+}$ signalling: waves and gradients. *Cell Calcium* **14**, no. 10.

Doedel, E.J. 1981. AUTO: A program for the automatic bifurcation analysis of autonomous systems. *Congr. Num.* **30**:265–84.

Doedel, E.J. & R.F. Heinemann. 1983. Numerical computation of periodic solution branches and oscillatory dynamics of the stirred tank reactor with A → B → C reactions. *Chem. Eng. Sci.* **38**:1493–9.

Dolmetsch, R.E. & R.S. Lewis. 1994. Signaling between intracellular $Ca^{2+}$ stores and depletion-activated $Ca^{2+}$ channels generates $[Ca^{2+}]_i$ oscillations in T lymphocytes. *J. Gen. Physiol.* **103**:365–88.

Dolor, R., L. Hurwitz, Z. Mirza, H. Strauss & R. Whorton. 1992. Regulation of extracellular $Ca^{2+}$ entry in endothelial cells: role of intracellular calcium pool. *J. Am. Physiol.* **262**:C171–C181.

Dorée, M., T. Lorca & A. Picard. 1991. Involvement of protein phosphatases in the control of cdc2 kinase activity during entry into and exit from M-phase of the cell cycle. *Adv. Protein Phosphat.* **6**:19–34.

Dowse, H.B., J.C. Hall & J.M. Ringo. 1987. Circadian and ultradian rhythms in period mutants of *Drosophila melanogaster.* *Behav. Genet.* **17**:19–35.

Dowse, H.B. & J.M. Ringo. 1993. Is the circadian clock a 'meta-oscillator'? Evidence from studies of ultradian rhythms in *Drosophila.* In: *Molecular Genetics of Biological Rhythms*, M.W. Young, ed. Marcel Dekker, New York, pp. 195–220.

Draetta, G. 1990. Cell cycle control in eukaryotes: Molecular mechanism of cdc2 activation. *Trends Biochem. Sci.* **15**:378–83.

Draetta, G. & D. Beach. 1989. The mammalian cdc2 protein kinase: Mechanisms of regulation during the cell cycle. *J. Cell Sci. Suppl.* **12**:21–7.

Drescher, K., G. Cornelius & L. Rensing. 1982. Phase response curves obtained by perturbing different variables of a 24 hr model oscillator based on translational control. *J. Theor. Biol.* **94**:345–53.

Ducommun, B., P. Brambilla, M.A. Félix, B.R. Franza Jr, E. Karsenti & G. Draetta. 1991. cdc2 phosphorylation is required for its interaction with cyclin. *EMBO J.* **10**:3311–19.

Dulic, V., E. Lees & S.I. Reed. 1992. Association of human cyclin E with a periodic $G_1$-S phase protein kinase. *Science* **257**:1958–61.

Dunlap, J.C. 1993. Genetic analysis of circadian clocks. *Annu. Rev. Physiol.* **55**:683–728.

Dunlap, J.C. & J.F. Feldman. 1988. On the role of protein synthesis in the circadian clock of *Neurospora crassa*. *Proc. Natl. Acad. Sci. USA* **85**:1096–100.

Dunlap, J.C., Q. Liu, K.A. Johnson & J.J. Loros. 1993. Genetic and molecular dissection of the *Neurospora* clock. In: *Molecular Genetics of Biological Rhythms*, M.W. Young, ed. Marcel Dekker, New York, pp. 37–54.

Dupont, G. 1993. Modélisation des oscillations et des ondes de calcium intracellulaire. Thèse de Doctorat en Sciences, Université Libre de Bruxelles.

Dupont, G., M.J. Berridge & A. Goldbeter. 1990. Latency correlates with period in a model for signal-induced $Ca^{2+}$ oscillations based on $Ca^{2+}$-induced $Ca^{2+}$ release. *Cell Regul.* **1**:853–61.

Dupont, G., M.J. Berridge & A. Goldbeter. 1991. Signal-induced $Ca^{2+}$ oscillations: Properties of a model based on $Ca^{2+}$-induced $Ca^{2+}$ release. *Cell Calcium* **12**:73–85.

Dupont, G. & A. Goldbeter. 1989. Theoretical insights into the origin of signal-induced calcium oscillations. In: *Cell to Cell Signalling: From Experiments to Theoretical Models*. A. Goldbeter, ed. Academic Press, London, pp. 461–74.

Dupont, G. & A. Goldbeter. 1992a. Protein phosphorylation driven by intracellular calcium oscillations: A kinetic analysis. *Biophys. Chem.* **42**:257–70 and (*Erratum*) **54**:291 (1995).

Dupont, G. & A. Goldbeter. 1992b. Oscillations and waves of cytosolic calcium: Insights from theoretical models. *BioEssays* **14**:485–93.

Dupont, G. & A. Goldbeter. 1993. One-pool model for $Ca^{2+}$ oscillations involving $Ca^{2+}$ and inositol 1,4,5-trisphosphate as co-agonists for $Ca^{2+}$ release. *Cell Calcium* **14**:311–22.

Dupont, G. & A. Goldbeter. 1994. Properties of intracellular $Ca^{2+}$ waves generated by a model based on $Ca^{2+}$-induced $Ca^{2+}$ release. *Biophys. J.* **67**:2191–204.

Durston, A.J. 1973. *Dictyostelium discoideum* aggregation fields as excitable media. *J. Theor. Biol.* **42**:483–504.

Durston, A.J. 1974a. Pacemaker activity during aggregation in *Dictyostelium discoideum*. *Dev. Biol.* **37**:225–35.

Durston, A.J. 1974b. Pacemaker mutants of *Dictyostelium discoideum*. *Dev. Biol.* **38**:308–19.

Duysens, L.N.M. & J. Amesz. 1957. Fluorescence spectrophotometry of reduced phosphopyridine nucleotide in intact cells in the near-ultraviolet and visible region. *Biochim. Biophys. Acta* **24**:19–26.

Dynnik, V.V. & E.E. Sel'kov. 1973. On the possibility of self-oscillations in the lower part of the glycolytic system. *FEBS Lett.* **37**:342–6.

Dynnik, V.V., E.E. Sel'kov & I.A. Ovtchinnikov. 1977. Self-oscillations in the lower part of the glycolytic system. Two alternative mechanisms. *Studia Biophys.* **63**:9–23.

Eberhard, D.A. & R.W. Holz. 1988. Intracellular $Ca^{2+}$ activates phospholipase C. *Trends Neurosci.* **12**:517–20.

Eckhorn, R., R. Bauer, W. Jordan, M. Brosch, W. Kruse, M. Munk & H.J. Reitboeck. 1988. Coherent oscillations: A mechanism for feature linking in the visual cortex? Multiple electrode and correlation analyses in the cat. *Biol. Cybern.* **60**:121–30.

Eden, S. 1979. Age- and sex-related differences in episodic growth hormone secretion in the rat. *Endocrinology* **105**:555–60.

Edery, I., L.J. Zwiebel, M.E. Dembinska & M. Rosbash. 1994. Temporal phosphorylation of the *Drosophila* period protein. *Proc. Natl. Acad. Sci. USA* **91**:2260–4.

Edgar, B.A. & P.H. O'Farrell. 1989. Genetic control of cell division patterns in the *Drosophila* embryo. *Cell* **57**:177–87.

Edgar, B.A., F. Sprenger, R.J. Duronio, P. Leopold & P.H. O'Farrell. 1994. Distinct molecular mechanisms regulate cell cycle timing at successive stages of *Drosophila* embryogenesis. *Genes Dev.* **8**:440–52.

Edmunds, L.N. Jr. 1988. *Cellular and Molecular Bases of Biological Clocks. Models and Mechanisms for Circadian Timekeeping.* Springer, New York.

Edmunds, L.N. Jr & D.L. Laval-Martin. 1984. Cell division cycles and circadian oscillators. In: *Cell Cycle Clocks.* L.N. Edmunds Jr, ed. Marcel Dekker, New York, pp. 295–324.

Elledge, S.J., R. Richman, F.L. Hall, R.T. Williams, N. Lodgson & J. W. Harper. 1992. *CDK2* encodes a 33-kDa cyclin A-associated protein kinase and is expressed before *CDC2* in the cell cycle. *Proc. Natl. Acad. Sci. USA* **89**:2907–11.

Elnashaie, S.S., M.E. Abashar & F.A. Teymour. 1993. Bifurcation, instability and chaos in fluidized bed catalytic reactors with consecutive exothermic chemical reactions. *Chaos, Solitons and Fractals* **3**:1–33.

Endo, M., M. Tanaka & Y. Ogawa. 1970. Calcium induced release of calcium from the sarcoplasmic reticulum of skinned skeletal muscle cells. *Nature* **228**:34–6.

Epstein, I.R. 1983. Oscillations and chaos in chemical systems. *Physica* **7D**:47–56.

Epstein, I.R. 1984. The search for new chemical oscillators. In: *Chemical Instabilities.* G. Nicolis & F. Baras, eds. D. Reidel, Dordrecht, pp. 3–18.

Erle, D. 1981. Nonuniqueness of stable limit cycles in a class of enzyme catalyzed reactions. *J. Math. Anal. Applic.* **82**:386–91.

Erle, D., K.H. Mayer & T. Plesser. 1979. The existence of stable limit cycles for enzyme catalyzed reactions with positive feedback. *Math. Biosci.* **44**:191–208.

Ermentrout, G.B. & N. Kopell. 1986. Parabolic bursting in an excitable system coupled with a slow oscillation. *SIAM J. Appl. Math.* **46**:233–53.

Eschrich, K., W. Schellenberger & E. Hofmann. 1980. *In vitro* demonstration of alternative stationary states in an open enzyme system containing phosphofructokinase. *Arch. Biochem. Biophys.* **205**:114–21.

Eschrich, K., W. Schellenberger & E. Hofmann. 1983. Sustained oscillations in a reconstituted system containing phosphofructokinase and fructose-1,6-bisphosphate. *Arch. Biochem. Biophys.* **222**:657–60.

Eschrich, K., W. Schellenberger & E. Hofmann. 1990. A hysteretic cycle in glucose 6-phosphate metabolism observed in a cell-free yeast extract. *Eur. J. Biochem.* **188**:697–703.

Esguerra, M., J. Wang, C.D. Foster, J.P. Adelman, R.A. North & I.B. Levitan. 1994. Cloned $Ca^{2+}$-dependent $K^+$ channel modulated by a functionally associated protein kinase. *Nature* **369**:563–5.

Eskin, A., S.J. Yeung & M.R. Klass. 1984. Requirement for protein synthesis in the regulation of a circadian rhythm by serotonin. *Proc. Natl. Acad. Sci. USA* **81**:7637–41.

Europe-Finner, G.N. & P.C. Newell. 1987. Cyclic AMP stimulates accumulation of inositol trisphosphate in *Dictyostelium. J. Cell Sci.* **87**:221–9.

Evans, T., E.T. Rosenthal, J. Youngblom, D. Distel & T. Hunt. 1983. Cyclin: A protein specified by maternal mRNA in sea urchin eggs that is destroyed at each cleavage division. *Cell* **33**:389–96.

Ewer, J., B. Frisch, M.J. Hamblen-Coyle, M. Rosbash & J.C. Hall. 1992. Expression of the *period* clock gene within different cell types in the brain of *Drosophila* adults and mosaic analysis of these cells' influence on circadian behavioral rhythms. *J. Neurosci.* **12**:3321–49.

Fabiato, A. 1983. Calcium-induced release of calcium from the cardiac sarcoplasmic reticulum. *Am. J. Physiol.* **245**:C1–C14.

Fabiato, A. 1985. Simulated calcium current can both cause calcium loading in and trigger release of calcium from the sarcoplasmic reticulum of single skinned cardiac cells. *J. Gen. Physiol.* **85**:291–320.

Fabiato, A. & F. Fabiato. 1975. Contractions induced by a calcium-triggered release of calcium from the sarcoplasmic reticulum of single skinned cardiac cells. *J. Physiol. (Lond.)* **249**:469–95.

Fantes, P.A., W.D. Grant, R.H. Pritchard, P.E. Sudbery & A.E. Wheals. 1975. The regulation of cell size and the control of mitosis. *J. Theor. Biol.* **50**:213–44.

Farnham, C.J.M. 1975. Cytochemical localization of adenylate cyclase and 3',5'-nucleotide phosphodiesterase in *Dictyostelium. Exp. Cell Res.* **91**:36–46.

Faure, M., G.J. Podgorski, J. Franke & R.H. Kessin. 1989. Rescue of a *Dictyostelium discoideum* mutant defective in cyclic nucleotide phosphodiesterase. *Dev. Biol.* **131**:366–72.

Feigenbaum, M.J. 1978. Quantitative universality for a class of nonlinear transformations. *J. Stat. Phys.* **19**:25–52.

Feldman, J.F. 1967. Lengthening of the period of a biological clock in *Euglena* by cycloheximide, an inhibitor of protein synthesis. *Proc. Natl. Acad. Sci. USA* **57**:1080–7.

Feldman, J.F. 1982. Genetic approaches to circadian clocks. *Annu. Rev. Plant Physiol.* **33**:583–608.

Feldman, J.F. & J.C. Dunlap. 1983. *Neurospora crassa*: A unique system for studying circadian rhythms. *Photochem. Photobiol. Rev.* **7**:319–68.

Félix, M.A., J.C. Labbé, M. Dorée, T. Hunt & E. Karsenti. 1990. Triggering of cyclin degradation in interphase extracts of amphibian eggs by cdc2 kinase. *Nature* **346**:379–82.

Félix, M.A., J. Pines, T. Hunt & E. Karsenti. 1989. A post-ribosomal supernatant from activated *Xenopus* egg that displays post-transcriptionally regulated oscillation of its $cdc2^+$ mitotic kinase activity. *EMBO J.* **8**:3059–69.

Fessard, A. 1936. *Propriétés Rythmiques de la Matière Vivante.* Hermann, Paris.

Fewtrell, C. 1993. $Ca^{2+}$ oscillations in non-excitable cells. *Annu. Rev. Physiol.* **55**:427–54.

Field, R.J. & M. Burger (eds.) 1985. *Oscillations and Traveling Waves in Chemical Systems.* Wiley, New York.

Field, R.J. & L. Györgyi (eds.) 1993. *Chaos in Chemistry and Biochemistry.* World Scientific, Singapore.

Field, R.J., E. Köros & R.M. Noyes. 1972. Oscillations in chemical systems. II. Thorough analysis of temporal oscillation in the bromate–cerium–malonic acid system. *J. Am. Chem. Soc.* **94**:8649–64.

Filicori, M. 1989. The critical role of signal quality: Lessons from pulsatile GnRH pathophysiology and clinical applications. In:*Cell to Cell Signalling: From Experiments to Theoretical Models*, A. Goldbeter, ed. Academic Press, London, pp. 395–405.

Finch, E.A. & S.M. Goldin. 1994. Calcium and inositol 1,4,5-trisphosphate–induced $Ca^{2+}$ release. *Science* **265**:813–15 and (*Erratum*) **266**:353 (1994).

Finch, E.A., T. Turner & S.M. Goldin. 1991. Calcium as a coagonist of inositol 1,4,5-trisphosphate-induced calcium release. *Science* **252**:443–6.

Firtel, R.A. & A.L. Chapman. 1990. A role for cAMP-dependent protein kinase A in early *Dictyostelium* development. *Genes Dev.* **4**:18–28.

Fisher, R.P. & D.O. Morgan. 1994. A novel cyclin associates with MO15/CDK7 to form the CDK-activating kinase. *Cell* **78**:713–24.

Fitzhugh, R. 1961. Impulses and physiological states in theoretical models of nerve membranes. *Biophys. J.* **1**:445–66.

Foe, V.E. 1989. Mitotic domains reveal early commitment of cells in *Drosophila. Development* **107**:1–22.

Foe, V., G.M. Odell & B.A. Edgar. 1993. Mitosis and morphogenesis in the *Drosophila* embryo. Point and counterpoint. In: *The Development of* Drosophila melanogaster, Vol. 1, M. Bate & A. Martinez Arias, eds. Cold Spring Harb. Laboratory Press, Cold Spring Harbor, New York, pp. 149–300.

Foskett J. & D. Wong. 1991. Free cytoplasmic $Ca^{2+}$ concentration oscillations in thapsigargin-treated parotid acinar cells are caffeine- and ryanodine-sensitive. *J. Biol. Chem.* **266**:14535–8.

Frenkel, R. 1968. Control of reduced diphosphopyridine nucleotide oscillations in beef heart extracts. I. Effect of modifiers of phosphofructokinase activity. *Arch. Biochem. Biophys.* **125**:151–6.

Friel, D.D. & R.W. Tsien. 1992. Phase-dependent contributions from $Ca^{2+}$ entry and $Ca^{2+}$ release to caffeine-induced $[Ca^{2+}]_i$ oscillations in bullfrog sympathetic neurons. *Neuron* **8**:1109–25.

Friesen, W.O. 1989. Neuronal control of leech swimming movements. In: *Neuronal and Cellular Oscillators.* J.W. Jacklet, ed. Marcel Dekker, New York and Basel, pp. 269–316.

Friesen, W.O., G.D. Block & C.G. Hocker. 1993. Formal approaches to understanding biological oscillators. *Annu. Rev. Physiol.* **55**:661–81.

Friesen, W.O. & G.S. Stent. 1978. Neural circuits for generating rhythmic movements. *Annu. Rev. Bioeng.* **7**:37–61.

Fuchikami, N., N. Sawashima, M. Naito & T. Kambara. 1993. Model of chemically excitable membranes generating autonomous chaotic oscillations. *Biophys. Chem.* **46**:249–59.

Fukii, S., A. Hirota & K. Kamino. 1981. Optical recording of development of electrical activity in embryonic chick heart during early phases of cardiogenesis. *J. Physiol. (Lond.)* **311**:147–60.

Gabrielli, B.G., L.M. Roy, J. Gautier, M. Philippe & J.L. Maller. 1992. A *cdc2*-related kinase oscillates in the cell cycle independently of cyclins G2/M and *cdc2*. *J. Biol. Chem.* **267**:1969–75.

Gaitonde, N.Y. & J.M. Douglas. 1969. The use of positive feedback control systems to improve reactor performance. *AIChE J.* **15**:902–10.

Galaktionov, K. & D. Beach. 1991. Specific activation of cdc25 tyrosine phosphatases by B-type cyclins: evidence for multiple roles of mitotic cyclins. *Cell* **67**:1181–94.

Galione, A. 1992. $Ca^{2+}$-induced $Ca^{2+}$ release and its modulation by cyclic ADP-ribose. *Trends Pharmacol. Sci.* **13**:304–6.

Galione, A. 1993. Cyclic ADP-ribose: A new way to control calcium. *Science* **259**:325–6.

Galione, A., A. McDougall, W.B. Busa, N. Willmott, I. Gillot & M. Whitaker. 1993. Redundant mechanisms of calcium-induced calcium release underlying calcium waves during fertilization of sea urchin eggs. *Science* **261**:348–52.

Gander, P.H., R.E. Kronauer, C.A. Czeisler & M.C. Moore-Ede. 1984a. Simulating the action of zeitgebers on a coupled two-oscillator model of the human circadian system. *Am. J. Physiol.* **247**:R418–R426.

Gander, P.H., R.E. Kronauer, C.A. Czeisler & M.C. Moore-Ede. 1984b. Modeling the action of zeitgebers on the human circadian system: comparisons of simulations and data. *Am. J. Physiol.* **247**:R427–R444.

Garfinkel, A., M.L. Spano, W.L. Ditto & J.N. Weiss. 1993. Controlling cardiac chaos. *Science* **257**:1230–5.

Gautier, J., T. Matsukawa, P. Nurse & J. Maller. 1989. Dephosphorylation and activation of *Xenopus* p34$^{cdc2}$ protein kinase during the cell cycle. *Nature* **339**:626–9.

Gautier, J., J. Minshull, M. Lohka, M. Glotzer, T. Hunt & J.L. Maller. 1990. Cyclin is a component of maturation-promoting factor from *Xenopus*. *Cell* **60**:487–94.

Geest, T., C.G. Steinmetz, R. Larter & L.F. Olsen. 1992. Period-doubling bifurcations and chaos in an enzyme reaction. *J. Phys. Chem.* **96**:5678–80.

Gelato, M.C. & G.R. Merriam. 1986. Growth hormone releasing hormone. *Annu. Rev. Physiol.* **48**:569–91.

Geller, J. & M. Brenner. 1978. The effect of 2,4-dinitrophenol on *Dictyostelium discoideum* oscillations. *Biochem. Biophys. Res. Commun.* **81**:814–21.

Gerhard, M., H. Schuster & J.J. Tyson. 1990. A cellular automaton model of excitable media including curvature and dispersion. *Science* **247**:1563–6.

Gerisch, G. 1968. Cell aggregation and differentiation in *Dictyostelium*. *Curr. Top. Dev. Biol.* **3**:157–97.

Gerisch, G. 1971. Periodische Signale steuern die Musterbildung in Zellverbänden. *Naturwissenschaften* **58**:430–8.

Gerisch, G. 1982. Chemotaxis in *Dictyostelium*. *Annu. Rev. Physiol.* **44**:535–52.

Gerisch, G. 1987. Cyclic AMP and other signals controlling cell development and differentiation in *Dictyostelium*. *Annu. Rev. Biochem.* **56**:853–79.

Gerisch, G., H. Fromm, A. Huesgen & U. Wick. 1975. Control of cell contact sites by cAMP pulses in differentiating *Dictyostelium* cells. *Nature* **255**:547–9.

Gerisch, G. & B. Hess. 1974. Cyclic-AMP controlled oscillations in suspended *Dictyostelium* cells: their relation to morphogenetic cell interactions. *Proc. Natl. Acad. Sci. USA* **71**:2118–22.

Gerisch, G., H. Kuczka & H.H. Heunert. 1963. Entwicklung von *Dictyostelium*. Film *C876T/1963*. Institut für den Wissenschaftlichen Film, Göttingen.

Gerisch, G. & D. Malchow. 1976. Cyclic AMP receptors and the control of cell aggregation in *Dictyostelium*. In: *Advances in Cyclic Nucleotide Research*. Vol. 7. P. Greengard & G.A. Robison, eds. Raven Press, New York, pp. 49–68.

Gerisch, G., D. Malchow, W. Roos & U. Wick. 1979. Oscillations of cyclic nucleotide concentrations in relation to the excitability of *Dictyostelium* cells. *J. Exp. Biol.* **81**:33–47.

Gerisch, G. & U. Wick. 1975. Intracellular oscillations and release of cyclic AMP from *Dictyostelium* cells. *Biochem. Biophys. Res. Commun.* **65**:364–70.

Getting, P.A. 1983. Mechanisms of pattern generation underlying swimming in *Tritonia*. II. Network reconstruction. *J. Neurophysiol.* **46**:64–79.

Getting, P.A. 1989. A network oscillator underlying swimming in *Tritonia*. In: *Neuronal and Cellular Oscillators*. J.W. Jacklet, ed. Marcel Dekker, New York and Basel, pp. 215–36.

Ghosh, A.K. & B. Chance. 1964. Oscillations of glycolytic intermediates in yeast cells. *Biochem. Biophys. Res. Commun.* **16**:174–81.

Ghosh, A.K., B. Chance & E.K. Pye. 1971. Metabolic coupling and synchronization of NADH oscillations in yeast cell populations. *Arch. Biochem. Biophys.* **145**:319–31.

Giannini G., E. Clementi, R. Ceci, G. Marziali & V. Sorrentino. 1992. Expression of a ryanodine receptor-$Ca^{2+}$ channel that is regulated by TGF-β. *Science* **257**:91–4.

Gilbert, D.A. 1974. The nature of the cell cycle and the control of cell proliferation. *Biosystems* **5**:197–206.

Gilbert, D.A. 1978. The relationship between the transition probability and oscillator concepts of the cell cycle and the nature of the commitment to replication. *Biosystems* **10**:235–40.

Gilkey, J.C., L.F. Jaffe, E.B. Ridgway & G.T. Reynolds.1978. A free calcium wave traverses the activating egg of the medaka, *Oryzias latipes. J. Cell Biol.* **76**:448–66.

Gilman, A.G. 1984. G proteins and dual control of adenylate cyclase. *Cell* **36**:577–9.

Gilman, A.G. 1987. G proteins: Transducers of receptor-generated signals. *Annu. Rev. Biochem.* **56**:615–49.

Gilmour, R.F. Jr, J.J. Heger, E.N. Prystowsky & D.P. Zypes. 1983. Cellular electrophysiologic abnormalities of diseased human ventricular myocardium. *Am. J. Cardiol.* **51**:137–44.

Gingle, A.R. & A. Robertson. 1976. The development of the relaying competence in *Dictyostelium discoideum. J. Cell Sci.* **20**:21–7.

Ginsburg, G. & A.R. Kimmel. 1989. Inositol trisphosphate and diacylglycerol can differentially modulate gene expression in *Dictyostelium. Proc. Natl. Acad. Sci. USA* **86**:9332–6.

Girard, S. & D. Clapham. 1993. Acceleration of intracellular calcium waves in *Xenopus* oocytes by calcium influx. *Science* **260**:229–32.

Girard S., A. Lückhoff, J. Lechleiter, J. Sneyd & D. Clapham.1992. A two-dimensional model of calcium waves reproduces the patterns observed in *Xenopus* oocytes. *Biophys. J.* **61**:509–17.

Giuliano, G., N.E. Hoffman, K. Ko, P.A. Scolnik & A.R. Cashmore. 1988. A light-entrained circadian clock controls transcription of several plant genes. *EMBO J.* **7**:3635–42.

Glansdorff, P. & I. Prigogine. 1970. *Structure, Stabilité et Fluctuations*. Masson, Paris.

Glansdorff, P. & I. Prigogine. 1971. *Thermodynamic Theory of Structure, Stability and Fluctuations*. Wiley, New York.

Glanz, J. 1994. Do chaos-control techniques offer hope for epilepsy? *Science* **265**:1174.

Glass, L., M.R. Guevara, J. Belair & A. Shrier. 1984. Global bifurcations of a periodically forced biological oscillator. *Phys. Rev.* **29A**:1348–57.

Glass, L., M.R. Guevara, A. Shrier & R. Perez. 1983. Bifurcation and chaos in a periodically stimulated cardiac oscillator. *Physica* **7D**:89–101.

Glass, L. & S.A. Kauffman. 1973. The logical analysis of continuous non-linear biochemical control networks. *J. Theor. Biol.* **39**:103–29.

Glass, L. & M.C. Mackey. 1979. Pathological conditions resulting from instabilities in physiological control systems. *Ann. N.Y. Acad. Sci.* **316**:214–35.

Glass, L. & M.C. Mackey. 1988. *From Clocks to Chaos: The Rhythms of Life*. Princeton Univ. Press, Princeton, NJ.

Glass, L. & C.P. Malta. 1990. Chaos in multi-looped negative feedback systems. *J. Theor. Biol.* **145**:217–23.

Glass, L. & W. Zheng. 1994. Bifurcations in flat-topped maps and the control of cardiac chaos. *Internat. J. Bifurc. Chaos* **4**:1061–7.

Glazer, P.M. & P.C. Newell. 1981. Initiation of aggregation by *Dictyostelium discoideum* in mutant populations lacking pulsatile signalling. *J. Gen. Microbiol.* **125**:221–32.

Glotzer, M., A.W. Murray & M.W. Kirschner. 1991. Cyclin is degraded by the ubiquitin pathway. *Nature* **349**:132–8.

Gola, M. 1974. Neurones à ondes-salves des mollusques. Variations cycliques lentes de conductances ioniques. *Pflügers Arch.* **352**:17–36.

Goldberger, A.L. & D.R. Rigney. 1989. On the non-linear motions of the heart: Fractals, chaos and cardiac dynamics. In: *Cell to Cell Signalling: From Experiments to Theoretical Models*, A. Goldbeter, ed. Academic Press, London, pp. 541–50.

Goldbeter, A. 1973. Patterns of spatiotemporal organization in an allosteric enzyme model. *Proc. Natl. Acad. Sci. USA* **70**:3255–9.

Goldbeter, A. 1974. Modulation of the adenylate energy charge by sustained metabolic oscillations. *FEBS Lett.* **43**:327–30.

Goldbeter, A. 1975. Mechanism for oscillatory synthesis of cyclic AMP in *Dictyostelium discoideum*. *Nature* **253**:540–2.

Goldbeter, A. 1976. Kinetic cooperativity in the concerted model for allosteric enzymes. *Biophys. Chem.* **4**:159–69.

Goldbeter, A. 1977. On the role of enzyme cooperativity in metabolic oscillations. Analysis of the Hill coefficient in a model for glycolytic periodicities. *Biophys. Chem.* **6**:95–9.

Goldbeter, A. 1980. Models for oscillations and excitability in biochemical systems. In: *Mathematical Models in Molecular and Cellular Biology*. L.A. Segel, ed. Cambridge Univ. Press, Cambridge, pp. 248–91.

Goldbeter, A. 1981. Bifurcations and the control of developmental transitions: Evolution of the cyclic AMP signalling system in *Dictyostelium discoideum*. In: *Mathematical Biology. Towards a Molecular Science*. T. Burton, ed. Pergamon Press, New York, pp. 79–95.

Goldbeter, A. 1987a. Adaptation, periodic signaling and receptor modification. In: *Molecular Mechanisms of Desensitization*. T.M. Konijn, P.J.M. Van Haastert, H. Van der Wel & M.D. Houslay, eds. Springer, Berlin, pp. 43–62.

Goldbeter, A. 1987b. Periodic signaling and receptor desensitization: from cAMP oscillations in *Dictyostelium* cells to pulsatile patterns of hormone secretion. In: *Temporal Disorder in Human Oscillatory Systems*. L. Rensing, U. an der Heiden & M.C. Mackey, eds. Springer, Berlin, pp. 15–23.

Goldbeter, A. 1988a. Periodic signaling as an optimal mode of intercellular communication. *News Physiol. Sci.* **3**:103–5.

Goldbeter, A. 1988b. Temps et rythmes biologiques. In: *Redécouvrir le Temps*. J.P. Boon & A. Nysenholcz, eds. *Revue de l'Université de Bruxelles*, pp. 93–102.

Goldbeter, A. 1991a. A minimal cascade model for the mitotic oscillator involving cyclin and cdc2 kinase. *Proc. Natl. Acad. Sci. USA* **88**:9107–11.

Goldbeter, A. 1991b. Du codage par fréquence des communications intercellulaires à l'ébauche d'une chronopharmacologie généralisée. *Bull. Acad. Méd. Belg.* **146**:113–27.

Goldbeter, A. 1992. Comparison of electrical oscillations in neurons with induced or spontaneous cellular rhythms due to biochemical regulation. In: *Induced Rhythms of the Brain*. E. Basar & T.H. Bullock, eds. Birkhäuser, Boston, MA, pp. 309–24.

Goldbeter, A. 1993a. Modeling the mitotic oscillator driving the cell division cycle. *Comments Theor. Biol.* **3**:75–107.

Goldbeter, A. 1993b. From periodic behavior to chaos in biochemical systems. In: *Chaos in Chemistry and Biochemistry*. R.J. Field & L. Györgyi, eds. World Scientific, Singapore, pp. 249–83.

Goldbeter, A. 1995. A model for circadian oscillations in the *Drosophila period* protein (PER). *Proc. R. Soc. Lond.* B**261**, 319–24.

Goldbeter, A. & S.R. Caplan. 1976. Oscillatory enzymes. *Annu. Rev. Biophys. Bioeng.* **5**:449–76.

Goldbeter, A. & O. Decroly. 1983. Temporal self-organization in biochemical systems: periodic behavior *versus* chaos. *Am. J. Physiol.* **245**:R478–R483.

Goldbeter, A., O. Decroly, Y.X. Li, J.L. Martiel & F. Moran. 1988. Finding complex oscillatory phenomena in biochemical systems. An empirical approach. *Biophys. Chem.* **29**:211–17.

Goldbeter, A., O. Decroly & J.L. Martiel. 1984. From excitability and oscillations to birhythmicity and chaos in biochemical systems. In: *Dynamics of Biochemical Systems*. J. Ricard & A. Cornish-Bowden, eds. Plenum Press, New York, pp. 173–212.

Goldbeter, A. & G. Dupont. 1990. Allosteric regulation, cooperativity and biochemical oscillations. *Biophys. Chem.* **37**:341–53.

Goldbeter, A. & G. Dupont. 1991. Phosphorylation and the frequency encoding of signal-induced calcium oscillations. In: *Cellular Regulation by Protein Phosphorylation*. L. Heilmeyer, ed. Springer, Berlin, pp. 35–9.

Goldbeter, A. & G. Dupont. 1992. Wavelike propagation of cAMP and $Ca^{2+}$

546 *References*

signals: Link with excitability and oscillations. In: *Oscillations and Morphogenesis*. L. Rensing, ed. Marcel Dekker, New York, pp. 195–209.

Goldbeter, A., G. Dupont & M.J. Berridge. 1990. Minimal model for signal-induced $Ca^{2+}$ oscillations and for their frequency encoding through protein phosphorylation. *Proc. Natl. Acad. Sci. USA* **87**:1461–5.

Goldbeter, A. & T. Erneux. 1978. Oscillations entretenues et excitabilité dans la réaction de la phosphofructokinase. *C.R. Acad. Sci. (Paris) Sér. C* **286**:63–6.

Goldbeter, A., T. Erneux & L.A. Segel. 1978. Excitability in the adenylate cyclase reaction in *Dictyostelium discoideum. FEBS Lett.* **89**:237–41.

Goldbeter, A. & J.M. Guilmot. 1995. Arresting the mitotic oscillator and the control of cell proliferation: Insights from a cascade model for cdc2 kinase activation. *Experientia*, in press.

Goldbeter, A. & D.E. Koshland Jr. 1981. An amplified sensitivity arising from covalent modification in biological systems. *Proc. Natl. Acad. Sci. USA* **78**:6840–4.

Goldbeter, A. & D.E. Koshland Jr. 1982a. Sensitivity amplification in biochemical systems. *Quart. Rev. Biophys.* **15**:555–91.

Goldbeter, A. & D.E. Koshland Jr. 1982b. Simple molecular model for sensing and adaptation based on receptor modification, with application to bacterial chemotaxis. *J. Mol. Biol.* **161**:395–416.

Goldbeter, A. & D.E. Koshland Jr. 1984. Ultrasensitivity in biochemical systems controlled by covalent modification. Interplay between zero-order and multistep effects. *J. Biol. Chem.* **259**:14441–7.

Goldbeter, A. & D.E. Koshland Jr. 1987. Energy expenditure in the control of biochemical systems by covalent modification. *J. Biol. Chem.* **262**:4460–71.

Goldbeter, A. & R. Lefever. 1972. Dissipative structures for an allosteric model. Application to glycolytic oscillations. *Biophys. J.* **12**:1302–15.

Goldbeter, A. & X.Y. Li. 1989. Frequency coding in intercellular communication. In *Cell to Cell Signalling: From Experiments to Theoretical Models*. A. Goldbeter, ed. Academic Press, London, pp. 415–32.

Goldbeter, A. & J.L. Martiel. 1980. Role of receptor desensitization in the mechanism of cAMP oscillations in *Dictyostelium. Fed. Proc.* **39**:1804. (Abstr.)

Goldbeter, A. & J.L. Martiel. 1983. A critical discussion of plausible models for relay and oscillations of cyclic AMP in *Dictyostelium discoideum. Lect. Notes Biomath.* **49**:173–88.

Goldbeter, A. & J.L. Martiel. 1985. Birhythmicity in a model for the cyclic AMP signaling system of the slime mold *Dictyostelium discoideum. FEBS Lett.* **191**:149–53.

Goldbeter, A. & J.L. Martiel. 1987. Periodic behaviour and chaos in the mechanism of intercellular communication governing aggregation of *Dictyostelium* amoebae. In: *Chaos in Biological Systems*. H. Degn, A.V. Holden & L.F. Olsen, eds. Plenum Press, New York, pp. 79–89.

Goldbeter, A. & J.L. Martiel. 1988. Developmental control of a biological rhythm: the onset of cAMP oscillations in *Dictyostelium* cells. In: *From Chemical to Biological Organization*. M. Markus, S. Müller & G. Nicolis, eds. Springer, Berlin, pp. 248–54.

Goldbeter, A. & F. Moran. 1987. Complex patterns of excitability and oscillations in a biochemical system. In: *The Organization of Cell Metabolism*. R.Welch & J. Clegg, eds. Plenum Press, New York, pp. 291–306.

Goldbeter, A. & F. Moran. 1988. Dynamics of a biochemical system with multiple oscillatory domains as a clue for multiple modes of neuronal oscillations. *Eur. Biophys. J.* **15**:277–87.

Goldbeter, A. & G. Nicolis. 1976. An allosteric enzyme model with positive feedback applied to glycolytic oscillations. In: *Progress in Theoretical Biology.* Vol. 4. F. Snell & R. Rosen, eds. Academic Press, New York. pp. 65–160.

Goldbeter, A. & L.A. Segel. 1977. Unified mechanism for relay and oscillations of cyclic AMP in *Dictyostelium discoideum. Proc. Natl. Acad. Sci. USA* **74**:1543–7.

Goldbeter, A. & L.A. Segel. 1980. Control of developmental transitions in the cyclic AMP signaling system of *Dictyostelium discoideum. Differentiation* **17**:127–35.

Goldbeter, A. & D. Venieratos. 1980. Analysis of the role of enzyme cooperativity in the mechanism of metabolic oscillations. *J. Mol. Biol.* **138**:137–44.

Goldbeter, A. & B. Wurster. 1989. Regular oscillations in suspensions of a putatively chaotic mutant of *Dictyostelium discoideum. Experientia* **45**:363–5.

Goodner, C.J., I.R. Sweet & H.C. Harrison Jr. 1988. Rapid reduction and return of surface insulin receptors after exposure to brief pulses of insulin in perifused rat hepatocytes. *Diabetes* **37**:1316–23.

Goodner, C.J., B.C. Walike, D.J. Koerker, J.W. Ensinck, A.C. Brown, E.W. Chideckel, J. Palmer & L. Kalnasy. 1977. Insulin, glucagon, and glucose exhibit synchronous, sustained oscillations in fasting monkeys. *Science* **195**:177–9.

Goodwin, B.C. 1963. *Temporal Organization in Cells: A Dynamic Theory of Cellular Control Processes.* Academic Press, New York.

Goodwin, B.C. 1965. Oscillatory behavior in enzymatic control processes. *Adv. Enzyme Regul.* **3**:425–38.

Gorman, A.L.F. & A. Hermann. 1982. Quantitative differences in the currents of bursting and beating molluscan pacemaker neurones. *J. Physiol. (Lond.)* **333**:681–99.

Gottmann, K. & C.J. Weijer. 1986. In situ measurements of external pH and optical density oscillations in *Dictyostelium discoideum* aggregates. *J. Cell Biol.* **102**:1623–9.

Gould, K.L., S. Moreno, D.J. Owen, S. Sazer & P. Nurse. 1991. Phosphorylation at Thr167 is required for *Schizosaccharomyces pombe* p34[cdc2] function. *EMBO J.* **10**:3297–309.

Gould, K. & P. Nurse. 1989. Tyrosine phosphorylation of the fission yeast *cdc2*[+] protein kinase regulates entry into mitosis. *Nature* **342**:39–45.

Gray, C.M., P. König, A.K. Engel & W. Singer. 1989. Oscillatory responses in cat visual cortex exhibit inter-columnar synchronization which reflects global stimulus properties. *Nature* **338**:334–7.

Grebogi, C., S.W. McDonald, E. Ott & J.A. Yorke. 1983a. Final state sensitivity: an obstruction to predictability. *Phys. Lett.* **99A**:415–18.

Grebogi, C., E. Ott & J.A. Yorke. 1983b. Crises, sudden changes in chaotic attrractors, and transient chaos. *Physica* **7D**:181–200.

Greengard, P. 1978. Phosphorylated proteins as physiological effectors. *Science* **199**:146–52.

Grieco, D., E.V. Avvedimento & M.E. Gottesman. 1994. A role for cAMP-dependent protein kinase in early embryonic divisions. *Proc. Natl. Acad. Sci. USA* **91**:9896–900.

Griffith, J.S. 1968. Mathematics of cellular control processes. I. Negative feedback to one gene. *J. Theor. Biol.* **20**:202–8.

Grillner, S., J. Christenson, L. Brodin, P. Wallén, R.H. Hill, A. Lansner & O. Ekeberg. 1989. Locomotor system in lamprey: Neuronal mechanisms controlling spinal rhythm generation. In: *Neuronal and Cellular Oscillators*. J.W. Jacklet, ed. Marcel Dekker, New York and Basel, pp. 215–36.

Grillner, S. & P. Wallén. 1985. Central pattern generators for locomotion, with special reference to vertebrates. *Annu. Rev. Neurosci.* **8**:233–61.

Gross, J.D. 1994. Developmental decisions in *Dictyostelium discoideum*. *Microbiol. Rev.* **58**:330–51.

Gross, J.D., M.J. Peacey & D.J. Trevan. 1976. Signal emission and signal propagation during early aggregation in *Dictyostelium discoideum*. *J. Cell Sci.* **22**:645–56.

Grutsch, J.F. & A. Robertson. 1978. The cAMP signal from *Dictyostelium discoideum* amoebae. *Dev. Biol.* **66**:285–93.

Gu, X. & N.C. Spitzer. 1995. Distinct aspects of neuronal differentiation encoded by frequency of spontaneous $Ca^{2+}$ transients. *Nature* **375**:784–7.

Gu, Y., C.W. Turck & D.O. Morgan. 1993. Inhibition of CDK2 activity *in vivo* by an associated 20K regulatory subunit. *Nature* **366**:707–10.

Guckenheimer, J. & P. Holmes. 1983. *Nonlinear Oscillations, Dynamical Systems, and Bifurcations of Vector Fields*. Springer, New York.

Guevara, M.R., L. Glass & A. Shrier. 1981. Phase locking, period-doubling bifurcations, and irregular dynamics in periodically stimulated cardiac cells. *Science* **214**:1350–3.

Guidi, G.M., J. Halloy & A. Goldbeter. 1995. Chaos suppression by periodic forcing: Insights from *Dictyostelium* cells, from a multiply regulated biochemical system, and from the Lorenz model. In: *Chaos and Complexity*. J. Trân Thanh Vân et al., eds. Editions Frontières, Gif-sur-Yvette, France, pp. 135–46.

Guillevic, E. 1990. *Le Chant*. Gallimard, Paris.

Guilmot, J.M. & A. Goldbeter. 1995. Role of phosphorylation-dephosphorylation thresholds in a minimal cascade model for the mitotic oscillator. Submitted for publication.

Gumowski, I. & C. Mira. 1980. *Dynamique Chaotique*. Ed. CEPADUES, Toulouse.

Gundersen, R.E., R. Johnson, P. Lilly, G. Pitt, M. Pupillo, T. Sun, R. Vaughan & P.N. Devreotes. 1989. Reversible phosphorylation of G-protein-coupled receptors controls cAMP oscillations in *Dictyostelium*. In: *Cell to Cell Signalling: From Experiments to Theoretical Models*. A. Goldbeter, ed. Academic Press, London, pp. 477–88.

Guttman, R., S. Lewis & J. Rinzel. 1980. Control of repetitive firing in squid axon membrane as a model for a neurone oscillator. *J. Physiol. (Lond.)* **305**:377–95.

Györgyi, L. & R.J. Field. 1993. Modeling and interpretation of chaos in the Belousv–Zhabotinsky reaction. In: *Chaos in Chemistry and Biochemistry*. R.J. Field & L. Györgyi, eds. World Scientific, Singapore, pp. 47–85.

Hahn, H.S., A. Nitzan, P. Ortoleva & J. Ross. 1974. Threshold excitations, relaxation oscillations, and effect of noise in an enzyme reaction. *Proc. Natl. Acad. Sci. USA* **71**:4067–71.

Hajnoczky, G. & A.P. Thomas. 1994. The inositol trisphosphate calcium channel is inactivated by inositol trisphosphate. *Nature* **370**:474–7.

Halberg, F. 1960. Temporal coordination of physiologic functions. *Cold Spring Harbor Symp. Quant. Biol.* **25**:289–310.

Halevy, O., B.G. Novitch, D.B. Spicer, S.X. Skapek, J. Rhee, G.J. Hannon, D. Beach & A.B. Lassar. 1995. Correlation of terminal cell cycle arrest of skeletal muscle with induction of p21 by MyoD. *Science* **267**:1018–21.

Hall, J.C. 1986. Learning and rhythms in courting, mutant *Drosophila*. *Trends Neurosci.* **9**:414–18.

Hall, J.C. 1994. The mating of a fly. *Science* **264**:1702–14.

Hall, J.C. & M. Rosbash. 1988. Mutations and molecules influencing biological rhythms. *Annu. Rev. Neurosci.* **11**:373–93.

Halloy, J. 1995. Analyse des compartements dynamiques d'un oscillateur cellulaire: Le système de communication par signaux d'AMP cyclique chez *Dictyostelium discoideum*. Thèse de doctorat en Sciences, Université Libre de Bruxelles.

Halloy, J. & A. Goldbeter. 1995. Incorporation of G-proteins in a model for cAMP signaling in *Dictyostelium* based on receptor desensitization. Submitted for publication.

Halloy, J., Y.X. Li, J.L. Martiel, B. Wurster & A. Goldbeter. 1990. Coupling chaotic and periodic cells results in a period-doubling route to chaos in a model for cAMP oscillations in *Dictyostelium* suspensions. *Phys. Lett. A.* **151**:33–6 and (*Erratum*) **159**:442 (1991).

Hara, K., P. Tydeman & M. Kirschner. 1980. A cytoplasmic clock with the same period as the division cycle in *Xenopus* eggs. *Proc. Natl. acad. Sci. USA* **77**:462–6.

Hardin, P.E., J.C. Hall & M. Rosbash. 1990. Feedback of the *Drosophila period* gene product on circadian cycling of its messenger RNA levels. *Nature* **343**:536–40.

Hardin, P.E., J.C. Hall & M. Rosbash. 1992. Circadian oscillations in period gene mRNA levels are transcriptionally regulated. *Proc. Natl. Acad. Sci. USA* **89**:11711–15.

Harootunian, A.T., J.P.Y. Kao, S. Paranjape & R.Y. Tsien. 1991. Generation of calcium oscillations in fibroblasts by positive feedback between calcium and $IP_3$. *Science* **251**:75–8.

Harootunian, A.T., J.P.Y. Kao & R.Y. Tsien. 1988. Agonist-induced calcium oscillations in depolarized fibroblasts and their manipulation by photoreleased Ins(1,4,5)$P_3$, $Ca^{2+}$, and $Ca^{2+}$ buffer. *Cold Spring Harbor Symp. Quant. Biol.* **53**:935–43.

Harris, B.E., R. Song, S.-J. Soong & R.B. Diasio. 1990. Relationship between dihydropyrimidine dehydrogenease activity and plasma 5-fluorouracil levels with evidence for circadian variation of enzyme activity and plasma drug levels in cancer patients receiving 5-fluorouracil by protracted continuous infusion. *Cancer Res.* **50**:197–201.

Hartwell, L.H. & T.A. Weinert. 1989. Checkpoints: Controls that ensure the order of cell cycle events. *Science* **246**:629–34.

Hartwig, R., R. Schweiger, M. Schweiger & H.G. Schweiger. 1985. Identification of a high molecular weight polypeptide that may be part of the circadian clockwork in *Acetabularia*. *Proc. Natl. Acad. Sci. USA* **82**:6899–902.

Harwood, A.J., N.A. Hopper, M.N. Simon, S. Bouzid, M. Veron & J.G.

Williams. 1992. Multiple roles for cAMP-dependent protein kinase during *Dictyostelium* development. *Dev. Biol.* **149**:90–9.

Hassard, B. 1978. Bifurcation of periodic solutions of the Hodgkin–Huxley model for the squid giant axon. *J. Theor. Biol.* **71**:401–20.

Hasslacher, B., R. Kapral & A. Lawniczak. 1993. Molecular Turing structures in the biochemistry of the cell. *Chaos* **3**:7–13.

Hastings, J.W. & B.M. Sweeney. 1958. A persistent diurnal rhythm of luminescence in *Gonyaulax polyedra. Biol. Bull.* **115**:440–58.

Hayashi, C. 1964. *Nonlinear Oscillations in Physical Systems.* McGraw-Hill, New York.

Hayashi, H. & S. Ishizuka. 1992. Chaotic nature of bursting discharges in the *Onchidium* pacemaker neuron. *J. Theor. Biol.* **156**:269–91.

Hayashi, H., M. Nakao & K. Hirakawa. 1982. Chaos in the self-sustained oscillation of an excitable membrane under sinusoidal stimulation. *Phys. Lett.* **88A**:265–6.

Hayles, J., D. Fisher, A. Woollard & P. Nurse. 1994. Temporal order of S phase and mitosis in fission yeast is determined by the state of the $p34^{cdc2}$–mitotic B cyclin complex. *Cell* **78**:813–22.

Heichman, K.A. & J.M. Roberts. 1994. Rules to replicate by. *Cell* **79**:557–62.

Heineken, F., H. Tsuchiya & R. Aris. 1967. On the mathematical status of the pseudo-steady state hypothesis of biochemical kinetics. *Math. Biosci.* **1**:95–113.

Henderson, E.J. 1975. The cyclic adenosine $3':5'$-monophosphate receptor of *Dictyostelium discoideum.* Binding characteristics of aggregation-competent cells and variation of binding levels during the life cycle. *J. Biol. Chem.* **250**:4730–6.

Hengst, L., V. Dulic, J.M. Slingerland, E. Lees & S.I. Reed. 1994. A cell cycle-regulated inhibitor of cyclin-dependent kinases. *Proc. Natl. Acad. Sci. USA* **91**:5291–5.

Henquin, J.-C. 1988. ATP-sensitive $K^+$ channels may control glucose-induced electrical activity in pancreatic B–cells. *Biochem. Biophys. Res. Commun.* **156**:769–75.

Hers, H.G. 1984. The discovery and the biological role of fructose 2,6-bisphosphate. *Biochem. Soc. Trans.* **12**:729–35.

Hershko, A., D. Ganoth, V. Sudakin, A. Dahan, L.H. Cohen, F.C. Luca, J.V. Ruderman & E. Eytan. 1994. Components of a system that ligates cyclin to ubiquitin and their regulation by the protein kinase cdc2. *J. Biol. Chem.* **269**:4940–6.

Hervagault, J.F. & S. Canu. 1987. Bistability and irreversible transitions in a simple substrate cycle. *J. Theor. Biol.* **127**:439–49.

Hervagault, J.F., M.C. Duban, J.P. Kernevez & D. Thomas. 1983. Multiple steady states and oscillatory behavior of a compartmentalized phosphofructokinase system. *Proc. Natl. Acad. Sci. USA* **52**:5455–9.

Hess, B. 1968. Biochemical regulations. In: *Systems Theory and Biology.* M.D. Mesarovic, ed. Springer, New York, pp. 88–114.

Hess, B. 1973. Organization of glycolysis: oscillatory and stationary control. *Symp. Soc. Exp. Biol.* **27**:105–31.

Hess, B. & A. Boiteux. 1968a. Mechanism of glycolytic oscillation in yeast. I. Aerobic and anaerobic growth conditions for obtaining glycolytic oscillations. *Hoppe Seyler's Z. Physiol. Chem.* **349**:1567–74.

Hess, B. & A. Boiteux. 1968b. Control of glycolysis. In: *Regulatory Functions of Biological Membranes.* J. Järnefelt, ed. Elsevier, Amsterdam, pp. 148–62.

Hess, B. & A. Boiteux. 1971. Oscillatory phenomena in biochemistry. *Annu. Rev. Biochem.* **40**:237–58.

Hess, B. & A. Boiteux. 1973. Substrate control of glycolytic oscillations. In: *Biological and Biochemical Oscillators*. B. Chance, E.K. Pye, A.K. Ghosh & B. Hess, eds. Academic Press, New York, pp. 229–41.

Hess, B., A. Boiteux & J. Krüger. 1969. Cooperation of glycolytic enzymes. *Adv. Enzyme Regul.* **7**:149–67.

Hess, B., K. Brand & K. Pye. 1966. Continuous oscillations in a cell-free extract of *S. carlsbergensis. Biochem. Biophys. Res. Commun.* **23**:102–8.

Hess, B. & B. Chance. 1978. Oscillating enzyme reactions. In: *Theoretical Chemistry. Periodicities in Chemistry and Biology.* Vol. 4. H. Eyring & D. Henderson, eds. Academic Press, New York, pp. 159–79.

Hess, B., A. Goldbeter & R. Lefever. 1978. Temporal, spatial and functional order in regulated biochemical and cellular systems. *Adv. Chem. Phys.* **38**:363–413.

Hess, B. & M. Markus. 1985. The diversity of biochemical time patterns. *Ber. Bunsenges. Phys. Chem.* **89**:642–51.

Higgins, J. 1964. A chemical mechanism for oscillation of glycolytic intermediates in yeast cells. *Proc. Natl. Acad. Sci. USA* **51**:989–94.

Higgins, J. 1967. Theory of oscillating reactions. *Ind. Eng. Chem.* **59**:19–62.

Hill, A.V. 1910. The possible effects of the aggregation of the molecules of haemoglobin on its dissociation curves. *J. Physiol. (Lond.)* **40**, iv–vii.

Hindmarsh, J.L. & R.M. Rose. 1982. A model of the nerve impulse using two first order differential equations. *Nature* **296**:162–4.

Hindmarsh, J.L. & R.M. Rose. 1984. A model of neuronal bursting using three coupled first order differential equations. *Proc. R. Soc. Lond. B* **221**:87–102.

Hindmarsh, J.L. & R.M. Rose. 1989. A three-dimensional model of a thalamic neurone. In: *Cell to Cell Signalling: From Experiments to Theoretical Models.* A. Goldbeter, ed. Academic Press, London, pp. 17–28.

Hindmarsh, P.C., R. Stanhope, M.A. Preece & C.G.D. Brook. 1990. Frequency of administration of growth hormone – An important factor in determining growth response to exogenous growth hormone. *Horm. Res.* **33** (Suppl. 4):83–9.

Hinds, P.W., S.F. Dowdy, E.N. Eaton, A. Arnold & R.A. Weinberg. 1994. Function of a human cyclin gene as an oncogene. *Proc. Natl. Acad. Sci. USA* **91**:709–13.

Hochner, B. & M.E. Spira. 1986. Two distinct propagating regenerative potentials in a single ethanol-treated axon. *Brain Res.* **398**:164–8.

Hocker, C.G., I.R. Epstein, K. Kustin & K. Tornheim. 1994. Glycolytic pH oscillations in a flow reactor. *Biophys. Chem.* **51**:21–35.

Hodgkin, A.L. & A.F. Huxley. 1952. A quantitative description of membrane currents and its application to conduction and excitation in nerve. *J. Physiol. (Lond.)* **117**:500–44.

Höfer, T., J.A. Sherratt & P.K. Maini. 1995. *Dictyostelium discoideum*: Cellular self-organization in an excitable biological medium. *Proc. R. Soc. Lond. B* **259**:249–57.

Hoffmann, I., P.R. Clarke, M.J. Marcote, E. Karsenti & G. Draetta. 1993. Phosphorylation and activation of human cdc25-C by cdc2-cyclin B and its involvement in the self-amplification of MPF at mitosis. *EMBO J.* **12**:53–63.

Hoffmann, I., G. Draetta & E. Karsenti. 1994. Activation of the phosphatase activity of human cdc25A by a cdk2–cyclin E dependent phosphorylation at the $G_1$/S transition. *EMBO J.* **13**:4302–10.

Hofmann, E. 1978. Phosphofructokinase – a favourite of enzymologists and of students of metabolic regulation. *Trends Biochem. Sci.* **3**:145–7.

Hofmann, E., K. Eschrich & W. Schellenberger. 1985. Temporal organization of the phosphofructokinase/fructose 1,6-bisphosphatase cycle. *Adv. Enzyme Regul.* **23**:331–62.

Holden, A.V. (ed.) 1986. *Chaos.* Manchester Univ. Press, Manchester.

Holden, A.V., W. Winlow & P.G. Haydon. 1982. The induction of periodic and chaotic activity in a molluscan neurone. *Biol. Cybern.* **43**:169–73.

Holden, A.V. & M. Yoda. 1981. Ionic channels density of an excitable membrane can act as bifurcation parameter. *Biol. Cybern.* **42**:29–38.

Holden, L. & T. Erneux. 1993. Understanding bursting oscillations as periodic slow passages through bifurcation and limit points. *J. Math. Biol.* **31**:351–65.

Holl, R.W., M.O. Thorner, G.L. Mandell, J.A. Sullivan, Y.N. Sinha & D.A. Leong. 1988. Spontaneous oscillations of intracellular calcium and growth hormone secretion. *J. Biol. Chem.* **263**:9682–5.

Hommes, F.A. 1964. Oscillatory reduction of pyridine nucleotides during anaerobic glycolysis in brewer's yeast. *Arch. Biochem. Biophys.* **108**:36–46.

Hopfield, J.J. 1995. Pattern recognition computation using action potential timing for stimulus representation. *Nature* **376**:33–6.

Horning, M. & K.-G. Collatz. 1990. First description of the glycolytic oscillator of an insect, the blowfly *Phormia terraenovae. Comp. Biochem. Physiol.* **95**:613–18.

Horsthemke, W. & R. Lefever. 1984. *Noise-Induced Transitions. Theory and Applications in Physics, Chemistry and Biology.* Springer, Berlin.

Hoth, M. & R. Penner.1992. Depletion of intracellular calcium stores activates a calcium current in mast cells. *Nature* **355**:353–5.

Hounsgaard, J., H. Hultborn, B. Jespersen & O. Kiehn. 1988. Bistability of α-motoneurones in the decerebrate cat and in the acute spinal cat after intravenous 5-hydroxytryptophan. *J. Physiol.* **405**:345–67.

Hrushesky, W.J.M. (ed.) 1994. *Circadian Cancer Therapy.* CRC Press, Boca Raton, FL.

Huang, Z.J., K.D. Curtin & M. Rosbash. 1995. PER protein interactions and temperature compensation of a circadian clock in *Drosophila. Science* **267**:1169–72.

Huang, Z.J., I. Edery & M. Rosbash. 1993. PAS is a dimerization domain common to *Drosophila* Period and several transcription factors. *Nature* **364**:259–62.

Hudson, J.L. & M. Mankin. 1981. Chaos in the Belousov–Zhabotinskii reaction. *J. Chem. Phys.* **74**:6171–7.

Hunding, A. 1974. Limit cycles in enzyme systems with nonlinear negative feedback. *Biophys. Struct. Mechan.* **1**:47–54.

Hunt, T. 1989. Under arrest in the cell cycle. *Nature* **342**:483–4.

Hunt, T. 1992. Cell cycle arrest and c-*mos. Nature* **355**:587–8.

Hunter, T. (ed.) 1992. *Regulation of the Eukaryotic Cell Cycle.* CIBA Found. Symp. Wiley, New York.

Hunter, T. & J. Pines. 1991. Cyclins and cancer. *Cell* **66**:1071–4.

Hunter, T. & J. Pines. 1994. Cyclins and cancer. II: Cyclin D and CDK inhibitors come of age. *Cell* **79**:573–82.

Huxley, A.H. 1959. Ion movements during nerve activity. *Ann. N.Y. Acad. Sci.* **81**:221–46.

Hyver, C. & H. Le Guyader. 1990. MPF and cyclin: Modelling of the cell cycle minimum oscillator. *Biosystems* **24**:85–90.

Ibsen, K.H. & K.W. Schiller. 1967. Oscillations of nucleotides and glycolytic intermediates in aerobic suspensions of Ehrlich ascites tumor cells. *Biochem. Biophys. Acta* **131**:405–7.

Iino, M. & M. Endo. 1992. Calcium-dependent immediate feedback control of inositol 1,4,5-trisphosphate-induced $Ca^{2+}$ release. *Nature* **360**:76–8.

Inouye, S.-I.T. & H. Kawamura. 1979. Persistence of circadian rhythmicity in a mammalian hypothalamic 'island' containing the suprachiasmatic nucleus. *Proc. Natl. Acad. Sci. USA* **76**:5962–6.

Isgaard, J., L. Carlsson, O.G.P. Isaksson & J.O. Janson. 1988. Pulsatile intravenous growth hormone (GH) infusion to hypophysectomized rats increases insulin-like growth factor I messenger ribonucleic acid in skeletal tissues more effectively than continuous infusion. *Endocrinology* **123**:2605–10.

Ishizaka, K., S. Kitahara, H. Oshima, P. Troen, B. Attardi & S.J. Winters. 1992. Effect of gonadotropin-releasing hormone pulse frequency on gonadotropin secretion and subunit messenger ribonucleic acids in perifused pituitary cells. *Endocrinology* **130**:1467–74.

Izumi, T., D.H. Walker & J.L. Maller. 1992. Periodic changes in phosphorylation of the *Xenopus* cdc25 phosphatase regulate its activity. *Mol. Biol. Cell* **3**:927–39.

Jacklet, J.W. 1977. Neuronal circadian rhythm: Phase shifting by a protein synthesis inhibitor. *Science* **198**:69–71.

Jacklet, J.W. (ed.) 1989a. *Neuronal and Cellular Oscillators.* Marcel Dekker, New York and Basel.

Jacklet, J.W. 1989b. Circadian neuronal oscillators. In: *Neuronal and Cellular Oscillators*, J.W. Jacklet, ed. Marcel Dekker, New York, pp. 483–527.

Jackson, F.R., T.A. Bargiello, S.H. Yun & M.W. Young. 1986. Product of *per* locus of *Drosophila* shares homology with proteoglycans. *Nature* **320**:185–8.

Jacob, F. & J. Monod. 1961. Genetic regulatory mechanisms in the synthesis of proteins. *J. Mol. Biol.* **3**:318–56.

Jacob, R. 1990a. Calcium oscillations in electrically non-excitable cells. *Biochim. Biophys. Acta* **1052**:427–38.

Jacob, R. 1990b. Imaging cytoplasmic free calcium in histamine stimulated endothelial cells and in fMet-Leu-Phe stimulated neutrophils. *Cell Calcium* **11**:241–9.

Jacob, R., J.E. Merritt, T.J. Hallam & T.J. Rink. 1988. Repetitive spikes in cytoplasmic calcium evoked by histamine in human endothelial cells. *Nature* **335**:40–5.

Jaffe, L. F. 1983. Sources of calcium in egg activation: A review and hypothesis. *Dev. Biol.* **99**:265–76.

Jaffe, L.F. 1991. The path of calcium in cytosolic calcium oscillations: A unifying hypothesis. *Proc. Natl. Acad. Sci. USA* **88**:9883–7.

Jaffe, L.F. 1993. Classes and mechanisms of calcium waves. *Cell Calcium* **14**:736–45.

Jafri, M.S., S. Vajda, P. Pasik & B. Gillo. 1992. A membrane model for cytosolic calcium oscillations. A study using *Xenopus* oocytes. *Biophys. J.* **63**:235–46.

554                         *References*

Jahnsen, H. & R. Llinas. 1984a. Electrophysiological properties of guinea-pig thalamic neurones: An *in vitro* study. *J. Physiol. (Lond.)* **349**:205–26.

Jahnsen, H. & R. Llinas. 1984b. Ionic basis for the electroreponsiveness and oscillatory properties of guinea-pig thalamic neurones *in vitro*. *J. Physiol. (Lond.)* **349**:227–47.

James, A.A., J. Ewer, P. Reddy, J.C. Hall & M. Rosbash. 1986. Embryonic expression of the *period* clock gene in the central nervous system of *Drosophila melanogaster*. *EMBO J.* **5**:2313–20.

Janssens, P.M.W. & P.J.M. Van Haastert. 1987. Molecular basis of transmembrane signal transduction in *Dictyostelium discoideum*. *Microbiol. Rev.* **51**:396–418.

Jansson, J.O., K. Albertsson-Wikland, S. Edén, K.G. Thorngren & O. Isaksson. 1982. Circumstantial evidence for a role of the secretory pattern of growth hormone in control of body growth. *Acta Endocrinol. (Copenh.)* **99**:24–30.

Janz, R.D., D.J. Vanacek & R.J. Field. 1980. Composite double oscillation in a modified version of the Oregonator model of the Belousov–Zhabotinsky reaction. *J. Chem. Phys.* **73**:3132–8.

Jerger, K. & S.S. Schiff. 1995. Periodic pacing an in vitro epileptic focus. *J. Neurophysiol.* **73**:876–9.

Jewett, M.E., R.E. Kronauer & C.A. Czeisler. 1991. Light-induced suppression of endogenous circadian amplitude in humans. *Nature* **350**:59–62.

Jinnah, H.A. & P.M. Conn. 1986. Gonadotropin-releasing hormone-mediated desensitization of cultured rat anterior pituitary cells can be uncoupled from luteinizing hormone release. *Endocrinology* **118**:2599–604.

Johnson, B.R., J.F. Griffith & S.K. Scott. 1991. Oscillations and chaos in $CO + O_2$ combustion. *Chaos* **1**:387–95.

Johnson, R.J. 1988. Diminution of pulsatile growth hormone secretion in the domestic fowl (*gallus domesticus*): evidence of sexual dimorphism. *J. Endocr.* **119**:101–9.

Johnston, D. & T.H. Brown. 1984. Mechanism of neuronal burst generation. In: *Electrophysiology of Epilepsy*. Academic Press, New York, pp. 277–301.

Juliani, M.H. & C. Klein. 1978. A biochemical study of the effects of cAMP pulses on aggregateless mutants of *Dictyostelium discoideum*. *Dev. Biol.* **62**:162–72.

Junge, D. & C.L. Stephens.1973. Cyclic variations of potassium conductance in a burst generating neuron in *Aplysia*. *J. Physiol. (Lond.)* **235**:155–81.

Kane, D.A., R.M. Warga & C.B. Kimmel. 1992. Mitotic domains in the early embryo of the zebrafish. *Nature* **360**:735–7.

Kaplan, D.T. & M. Talajic. 1991. Dynamics of heart rate. *Chaos* **1**:251–6.

Karakashian, M.W. & J.W. Hastings. 1963. The effects of inhibitors of macromolecular synthesis upon the persistent rhythm of luminescence in *Gonyaulax*. *J. Gen. Physiol.* **47**:1–12.

Karsch, F.J. 1987. Central actions of ovarian steroids in the feedback regulation of pulsatile secretion of luteinizing hormone. *Annu. Rev. Physiol.* **49**:365–82.

Karsenti, E., F. Verde & M.A. Félix. 1991. Role of type 1 and type 2A protein phosphatases in the cell cycle. *Adv. Protein Phosphat.* **6**:453–82.

Katchalsky, A. & P.F. Curran. 1965. *Nonequilibrium Thermodynamics in Biophysics*. Harvard Univ. Press, Cambridge, MA.

Katchalsky, A. & R. Spangler. 1968. Dynamics of membrane processes. *Quart. Rev. Biophys.* **1**:127–75.

Katz, B. & S. Thesleff. 1957. A study of 'desensitization' produced by acetylcholine at the motor end-plate. *J. Physiol. (Lond.)* **138**:63–80.

Kauffman, S. 1974. Measuring a mitotic oscillator: The arc discontinuity. *Bull. Math. Biol.* **36**:171–82.

Kauffman, S. & J.J. Wille. 1975. The mitotic oscillator in *Physarum polycephalum. J. Theor. Biol.* **55**:47–93.

Kaufman, M., J. Urbain & R. Thomas. 1985. Towards a logical analysis of the immune response. *J. Theor. Biol.* **114**:527–61.

Kawato, M. & R. Suzuki. 1980. Two coupled neural oscillators as a model of the circadian pacemaker. *J. Theor. Biol.* **86**:547–75.

Kay, S.A. & A.J. Millar. 1993. Circadian-regulated *cab* gene transcription in higher plants. In: *Molecular Genetics of Biological Rhythms*, M.W. Young, ed. Marcel Dekker, New York, pp. 73–89.

Keener, J.P. & J.J. Tyson. 1986. Spiral waves in the Belousov–Zhabotinsky reaction. *Physica* **21D**:307–24.

Keizer, J. & G.W. De Young. 1992. Two roles for $Ca^{2+}$ in agonist-stimulated $Ca^{2+}$ oscillations. *Biophys. J.* **61**:649–60.

Keizer, J. & G.W. De Young. 1994. Simplification of a realistic model of $IP_3$-induced $Ca^{2+}$ oscillations. *J. Theor. Biol.* **166**:431–42.

Keller, E.F. & L.A. Segel. 1970. Initiation of slime mold aggregation viewed as an instability. *J. Theor. Biol.* **26**:399–415.

Kessin, R.H. 1977. Mutations causing rapid development of *Dictyostelium discoideum. Cell* **10**:703–8.

Khalsa, S.B.S., D. Whitmore & G.D. Block. 1992. Stopping the circadian pacemaker with inhibitors of protein synthesis. *Proc. Natl. Acad. Sci. USA* **89**:10862–6.

Kimchi, A. 1992. Cytokine triggered molecular pathways that control cell cycle arrest. *J. Cell. Biochem.* **50**:1–9.

Kimmel, A.R. 1987. Different molecular mechanisms for cAMP regulation of gene expression during *Dictyostelium* development. *Dev. Biol.* **122**:163–71.

King, R.W., P.K. Jackson & M.W. Kirschner. 1994. Mitosis in transition. *Cell* **79**:563–71.

Kirschner, K. 1968. Allosteric regulation of enzyme activity. *Curr. Top. Microbiol.* **44**:123–46.

Kitajima, S., R. Sakakibara & K. Uyeda. 1983. Significance of phosphorylation of phosphofructokinase. *J. Biol. Chem.* **258**:13292–8.

Klein, C. 1976. Adenylate cyclase activity in *Dictyostelium discoideum* amoebae and its changes during differentiation. *FEBS Lett.* **68**:125–8.

Klein, C. 1979. A slowly dissociating form of the cell surface adenosine 3′:5′-monophosphate receptor of *Dictyostelium discoideum. J. Biol. Chem.* **254**:12573–8.

Klein, C. & M. Darmon. 1975. The relationship of phosphodiesterase to the developmental cycle of *Dictyostelium discoideum. Biochem. Biophys. Res. Commun.* **67**:440–7.

Klein, C. & M. Darmon. 1977. Effects of cyclic AMP pulses on adenylate cyclase and the phosphodiesterase inhibitor of *D. discoideum. Nature* **268**:76–8.

Klein, C., J. Lubs-Haukeness & S. Simons. 1985. cAMP induces a rapid and reversible modification of the chemotactic receptor in *Dictyostelium discoideum. J. Cell Biol.* **100**:715–20.

Klein, P., T.J. Sun, C.L. Saxe III, A.R. Kimmel, R.L. Johnson & P.N.

Devreotes. 1988. A chemo-attractant receptor controls development in *Dictyostelium discoideum*. *Science* **241**:1467–72.

Klein, P., A. Theibert, D. Fontana & P.N. Devreotes. 1985. Identification and cyclic-AMP induced modification of the cyclic AMP receptor in *Dictyostelium discoideum*. *J. Biol. Chem.* **260**:1757–64.

Kling, U. & G. Szekely. 1968. Simulation of rhythmic nervous activities. I. Function of networks with cyclic inhibitions. *Kybernetik* **5**:89–103.

Knobil, E. 1980. The neuroendocrine control of the menstrual cycle. *Rec. Progr. Horm. Res.* **36**:53–88.

Knobil, E. 1981. Patterns of hormone signals and hormone action. *New Engl. J. Med.* **305**:1582–3.

Knox, B.E., P.N. Devreotes, A. Goldbeter & L.A. Segel. 1986. A molecular mechanism for sensory adaptation based on ligand-induced receptor modification. *Proc. Natl. Acad. Sci. USA* **83**:2345–9.

Koch, C. & I. Segev. 1989. *Methods in Neuronal Modeling. From Synapses to Networks.* MIT Press, Cambridge, MA.

Kokoz, Y.M. & V.I. Krinskii. 1973. Analysis of equations of excitable membranes. II. Method of analyzing the electrophysiologic characteristics of the Hodgkin–Huxley membrane from the graphs of the zero-isoclines of a second-order system. *Biofizika* (Engl. transl.) **18**:937–44.

Komjati, M., P. Bratusch–Marrain & W. Waldhäusl. 1986. Superior efficacy of pulsatile *versus* continuous hormone exposure on hepatic glucose production *in vitro*. *Endocrinology* **118**:312–19.

Konijn, T.M. 1972. Cyclic AMP as first messenger. *Adv. Cyclic Nucleot. Res.* **1**:17–31.

Konijn, T.M. & Raper, K.B. 1961. Cell aggregation in *Dictyostelium discoideum*. *Dev. Biol.* **3**:725–56.

Konijn, T.M., J.G.C. Van de Meene, J.T. Bonner & D.S. Barkley. 1967. The acrasin activity of adenosine 3′,5′-cyclic phosphate. *Proc. Natl. Acad. Sci. USA* **58**:1152–4.

Konopka, R.J. 1979. Genetic dissection of the *Drosophila* circadian system. *Fed. Proc.* **38**:2602–5.

Konopka, R.J. & S. Benzer. 1971. Clock mutants of *Drosophila melanogaster*. *Proc. Natl. Acad. Sci. USA* **68**:2112–16.

Konopka, R.J., M.J. Hamblen-Coyle, C.F. Jamison & J.C. Hall. 1994. An ultrashort clock mutation at the *period* locus of *Drosophila melanogaster* that reveals some new features of the fly's circadian system. *J. Biol. Rhythms* **9**:189–216.

Konopka, R.J., C.S. Pittendrigh & D. Orr. 1989. Reciprocal behaviour associated with altered homeostasis and photosensitivity of *Drosophila* clock mutants. *J. Neurogenet.* **6**:1–10.

Kopell, N. & G.B. Ermentrout. 1986. Subcellular oscillations and bursting. *Math. Biosci.* **78**:265–91.

Koper, M.T.M. 1992. The theory of electrochemical instabilities. *Electrochem. Acta* **37**:1771–8.

Koper, M.T.M. & P. Gaspard. 1992. The modeling of mixed-mode and chaotic oscillations in electrochemical systems. *J. Chem. Phys.* **96**:7797–813.

Kornhauser, J.M., D.E. Nelson, K.E. Mayo & J.S. Takahashi. 1992. Regulation of *jun*-B messenger RNA and AP-1 activity by light and a circadian clock. *Science* **255**:1581–4.

Kort, A.A., M.C. Capogrossi & E.G. Lakatta. 1985. Frequency, amplitude, and propagation velocity of spontaneous $Ca^{2+}$-dependent contractile waves in intact adult rat cardiac muscle and isolated myocytes. *Circ. Res.* **57**:844–55.

Kosako, H., Y. Gotoh & E. Nishida. 1994. Mitogen-activated protein kinase kinase is required for the Mos-induced metaphase arrest. *J. Biol. Chem.* **269**:28354–8.

Koshland, D.E. Jr. 1977. A response regulator model in a simple sensory system. *Science* **196**:1055–63.

Koshland, D.E. Jr. 1979. A model regulatory system: bacterial chemotaxis. *Physiol. Rev.* **59**:811–62.

Koshland, D.E., Jr. 1980. Biochemistry of sensing and adaptation. *Trends Biochem. Sci.* **5**:297–302.

Koshland, D.E. Jr, A. Goldbeter & J.B. Stock. 1982. Amplification and adaptation in regulatory and sensory systems. *Science* **217**:220–5.

Koshland, D.E. Jr, G. Nemethy & D. Filmer. 1966. Comparison of experimental binding data and theoretical models in proteins containing subunits. *Biochemistry* **5**:365–85.

Krebs, E.G. & J.A. Beavo. 1979. Phosphorylation–dephosphorylation of enzymes. *Annu. Rev. Biochem.* **48**:923–59.

Krebs, H.A. 1972. The Pasteur effect and the relation between respiration and fermentation. In: *Essays in Biochemistry.* Vol. 8. P.N. Campbell & F. Dickens, eds. Academic Press, London. pp. 1–34.

Krek, W. & E.A. Nigg. 1991. Mutations of $p34^{cdc2}$ phosphorylation sites induce premature mitotic events in HeLa cells: evidence for a double block to $p34^{cdc2}$ kinase activation in vertebrates. *EMBO J.* **10**:3331–41.

Krinskii, V.I. & Y.M. Kokoz. 1973. Analysis of equations of excitable membranes. I. Reduction of the Hodgkin–Huxley equations to a second-order system. *Biofizika* (Engl. Transl.) **18**:533–9.

Kronauer, R.E. 1984. Modeling principles of human circadian rhythms. In: *Mathematical Models of the Circadian Sleep/Wake Cycle*, M.C. Moore-Ede & C.A. Czeisler, eds. Raven Press, New York, pp. 105–28.

Kronauer, R.E. 1990. A quantitative model for the effects of light on the amplitude and phase of the deep circadian pacemaker, based on human data. In: *Sleep '90* (Proceedings of the Tenth European Congress on Sleep Research, Strasbourg), J. Horne, ed. Pontenagel Press, Bochum, pp. 306–9.

Kronauer, R.E. & C.A. Czeisler. 1993. Understanding the use of light to control the circadian pacemaker in humans. In: *Light and Biological Rhythms in Man* (Wennergren International Series, Vol. 63), L. Wetterberg, ed. Pergamon Press, England, pp. 217–36.

Kronauer, R.E., C.A. Czeisler, S. Pilato, M.C. Moore-Ede & E.D. Weitzman. 1982. Mathematical model of the human circadian system with two interacting oscillators. *Am. J. Physiol.* **242**:R3–R17.

Krsmanovic, L.Z., S.S. Stojilkovic, F. Merelli, S.M. Dufour, M.A. Virmani & K.J. Catt. 1992. Calcium signaling and episodic secretion of gonadotropin-releasing hormone in hypothalamic neurons. *Proc. Natl. Acad. Sci. USA* **89**:8462–6.

Krsmanovic, L.Z., S.S. Stojilkovic, L.M. Mertz, M. Tomic & K.J. Catt. 1993. Expression of gonadotropin-releasing hormone receptors and autocrine regulation of neuropeptide release in immortalized hypothalamic neurons. *Proc. Natl. Acad. Sci. USA* **90**:3908–12.

Kuba, K. & S. Takeshita. 1981. Simulation of intracellular $Ca^{2+}$ oscillations in a sympathetic neurone. *J. Theor. Biol.* **93**:1009–31.

Kumagai, A. & W.G. Dunphy. 1991. The cdc25 protein controls tyrosine dephosphorylation of the cdc2 protein in a cell-free system. *Cell* **64**:903–14.

Kumagai, A. & W.G. Dunphy. 1992. Regulation of the cdc25 protein during the cell cycle in *Xenopus* extracts. *Cell* **70**:139–51.

Kyriacou, C.P., M.L. Greenacre, J.R. Thackeray & J.C. Hall. 1993. Genetic and molecular analysis of song rhythms in *Drosophila*. In: *Molecular Genetics of Biological Rhythms*, M.W. Young, ed. Marcel Dekker, New York, pp. 171–93.

Kyriacou, C.P. & J.C. Hall. 1980. Circadian rhythm mutations in *Drosophila melanogaster* affect the short-term fluctuations in the male's courtship song. *Proc. Natl. Acad. Sci. USA* **77**:6729–33.

Kyriacou, C.P. & J.C. Hall. 1986. Interspecific genetic control of courtship song production and reception in *Drosophila*. *Science* **232**:494–7.

Kyriacou, C.P., M.J. van den Berg & J.C. Hall. 1990. *Drosophila* courtship song cycles in normal and *period* mutant males revisited. *Behav. Genet.* **20**:617–44.

Labbé, J.C., A. Picard, G. Peaucellier, J.C. Cavadore, P. Nurse & M. Dorée. 1989. Purification of MPF from starfish: Identification as the H1 histone kinase p34$^{cdc2}$ and a possible mechanism for its periodic activation. *Cell* **57**:253–63.

Lacker, H.M. 1981. The regulation of ovulation number in mammals: An interaction law which controls follicle maturation. *Biophys. J.* **35**:433–54.

Lacker, H.M., M.E. Feuer & E. Akin. 1989. Cell to cell signalling through circulatory feedback: A mathematical model of the mechanism of follicle selection in the mammalian ovary. In: *Cell to Cell Signalling: From Experiments to Theoretical Models*. A. Goldbeter, ed. Academic Press, London, pp. 359–85.

Lakin-Thomas, P.L., S. Brody & G.G. Coté. 1991. Amplitude model for the effects of mutations and temperature on period and phase resetting of the *Neurospora* circadian oscillator. *J. Biol. Rhythms* **6**:281–97.

Lamba, P. & J.L. Hudson. 1985. Experimental evidence of multiple oscillatory states in a continuous reactor. *Chem. Eng. Commun.* **32**:369–75.

Landahl, H.D. 1969. Some conditions for sustained oscillations in biochemical chains with feedback inhibition. *Bull. Math. Biophys.* **31**:775–87.

Landahl, H.D. & V. Licko. 1973. On coupling between oscillators which model biological systems. *Int. J. Chronobiol.* **1**:245–52.

Lang, D.A., D.R. Matthews, M. Burnett, G.M. Ward & R.C. Turner. 1982. Pulsatile, synchronous basal insulin and glucagon secretion in man. *Diabetes* **31**:22–6.

Lang, D.A., D.R. Matthews, J. Peto & R.C. Turner. 1979. Cyclic oscillations of basal plasma glucose and insulin concentrations in human beings. *New Engl. J. Med.* **301**:1023–7.

LaPorte, D.C. & D.E. Koshland Jr. 1983. Phosphorylation of isocitrate dehydrogenase as a demonstration of enhanced sensitivity in covalent modification. *Nature* **305**:286–90.

Larter, R. 1990. Oscillations and spatial nonuniformities in membranes. *Chem. Rev.* **90**:355–81.

Larter, R., L.F. Olsen, C.G. Steinmetz & T. Geest. 1993. Chaos in biochemical systems: The peroxidase reaction as a case study. In: *Chaos in Chemistry and Biochemistry*. R.J. Field & L. Györgyi, eds. World Scientific, Singapore, pp. 175–224.

Laurent, G. & H. Davidowitz. 1994. Encoding of olfactory information with oscillating neural assemblies. *Science* **265**:1872–5.

Laurent, M., A.F. Chaffotte, J.P. Tenu, C. Roucous & F.J. Seydoux. 1978. Binding of nucleotides AMP and ATP to yeast phosphofructokinase: Evidence for distinct catalytic and regulatory sites. *Biochem. Biophys. Res. Commun.* **80**:646–52.

Laurent, M. & F. Seydoux. 1977. Influence of phosphate on the allosteric behavior of yeast phosphofructokinase. *Biochem. Biophys. Res. Commun.* **78**:1289–95.

Law, G.L., J.A. Pachter & P.S. Danies. 1989. Ability of repetitive $Ca^{2+}$ spikes to stimulate prolactin release is frequency dependent. *Biochem. Biophys. Res. Commun.* **158**:811–16.

Lebrun, P. & I. Atwater. 1985. Chaotic and irregular bursting electrical activity in mouse pancreatic β-cells. *Biophys. J.* **48**:529–31.

Lechleiter J. & D. Clapham D. 1992. Molecular mechanisms of intracellular calcium excitability in *X. laevis* oocytes. *Cell* **69**:283–94.

Lechleiter, J., S. Girard, E. Peralta & D. Clapham. 1991. Spiral calcium wave propagation and annihilation in *Xenopus laevis* oocytes. *Science* **252**:123–6.

Lee, H.C. 1993. Potentiation of calcium- and caffeine-induced calcium release by cyclic ADP-ribose. *J. Biol. Chem.* **268**:293–9.

Lee, H.C., R. Aarhus, R. Graeff, M.E. Gurnack & T. Walseth. 1994. Cyclic ADP ribose activation of the ryanodine receptor is mediated by calmodulin. *Nature* **370**:307–9.

Lee, K.J., W.D. McCormick, J.E. Pearson & H.L. Swinney. 1994. Experimental observation of self-replicating spots in a reaction-diffusion system. *Nature* **369**:215–18.

Lee, T.H., M.J. Solomon, M.C. Mumby & M.W. Kirschner. 1991. INH, a negative regulator of MPF, is a form of protein phosphatase 2A. *Cell* **64**:415–23.

Lees, J.A., K.J. Buchkovich, D.R. Marshak, C.W. Anderson & E. Harlow. 1991. The retinoblastoma protein is phosphorylated on multiple sites by human cdc2. *EMBO J.* **10**:4279–90.

Lefèbvre, P.J., G. Paolisso, A.J. Scheen & J.C. Henquin. 1987. Pulsatility of insulin and glucagon release: Physiological significance and pharmacological implications. *Diabetologia* **30**:443–52.

Lefever, R. 1981. Noise-induced transitions in biological systems. In: *Stochastic Nonlinear Systems*. R. Lefever & L. Arnold, eds. Springer, Berlin, pp. 127–36.

Lefever, R. & G. Nicolis. 1971. Chemical instabilities and sustained oscillations. *J. Theor. Biol.* **30**:267–84.

Lefever, R., G. Nicolis & P. Borckmans. 1988. The Brusselator: it does oscillate all the same. *J. Chem. Soc., Faraday Trans. 1* **84**:1013–23.

Lefever, R., G. Nicolis & I. Prigogine. 1967. On the occurrence of oscillations around the steady state in systems of chemical reactions far from equilibrium. *J. Chem. Phys.* **47**:1045–7.

Lefever, R. & J.W. Turner. 1986. Sensitivity of a Hopf bifurcation to multiplicative colored noise. *Phys. Rev. Lett.* **56**:1631–4.

Lefkowitz, R.J. & M.G. Caron. 1986. Regulation of adrenergic receptor function by phosphorylation. *J. Mol. Cell. Cardiol.* **18**:885–95.

Lemmer, B. (ed.) 1989. *Chronopharmacology: Cellular and Biochemical Interactions*. Marcel Dekker, New York and Basel.

Leng, G. (ed.) 1988. *Pulsatility in Neuroendocrine Systems.* CRC Press, Boca Raton, FL.

Lengyel, I. & I.R. Epstein. 1991. Diffusion-induced instability in chemically reacting systems: Steady-state multiplicity, oscillation, and chaos. *Chaos* 1:69–76.

Levi, F.A., R. Zidani, J.-M. Vannetzel, B. Perpoint, C. Focan, R. Faggiulo, P. Chollet, C. Garufi, M. Itzhaki, L. Dogliotti, S. Iacobelli, R. Adam, F. Kunstlinger, J. Gastiaburu, H. Bismuth, C. Jasmin & J.L. Misset. 1994. Chronomodulated versus fixed-infusion-rate delivery of ambulatory chemotherapy with oxaliplatin, fluorouracil, and folinic acid (leucovorin) in patients with colorectal cancer metastases: a randomized multi-institutional trial. *J. Natl. Cancer Inst.* 86:1608–17.

Levine, H. 1994. Modeling spatial patterns in *Dictyostelium. Chaos* 4:563–68.

Levine, H. & W. Reynolds. 1991. Streaming instability of aggregating slime mold amoebae. *Phys. Rev. Lett.* 66:2400–3.

Levitan, I.B. 1985. Phosphorylation of ion channels. *J. Membr. Biol.* 87:177–90.

Levitan, I.B. 1994. Modulation of ion channels by protein phosphorylation and dephosphorylation. *Annu. Rev. Physiol.* 56:193–212.

Levitan, I.B. & J.A. Benson. 1981. Neuronal oscillators in *Aplysia:* Modulation by serotonin and cyclic AMP. *Trends Neurosci.* 4:38–41.

Levitzki, A. & D.E. Koshland Jr. 1969. Negative cooperativity in regulatory enzymes. *Proc. Natl. Acad. Sci. USA* 62:1121–5.

Lewis, T.J. & M.R. Guevara. 1990. Chaotic dynamics in an ionic model of the propagated cardiac action potential. *J. Theor. Biol.* 146:407–32.

Lewy, A.J., R.L. Sack, L.S. Miller & T.M. Hoban. 1987. Antidepressant and circadian phase-shifting effects of light. *Science* 235:352–4.

Leyendecker, G.L., L. Wildt & M. Hansmann. 1980. Pregnancies following intermittent (pulsatile) administration of GnRH by means of a portable pump ('Zyklomat'): A new approach to the treatment of infertility in hypothalamic amenorrhea. *J. Clin. Endocr. Metab.* 51:1214–16.

Li, Y.X. 1992. Pulsatile hormonal signaling: A theoretical approach. Thèse de Doctorat en Sciences, Université Libre de Bruxelles.

Li, Y.X., D.F. Ding & J.H. Xu. 1984. Chaos and other temporal self-organization patterns in coupled enzyme-catalyzed systems. *Commun. Theor. Phys. (Beijing)* 3:629–38.

Li, Y.X. & A. Goldbeter. 1989a. Oscillatory isozymes as the simplest model for coupled biochemical oscillators. *J. Theor. Biol.* 138:149–74.

Li, Y.X. & A. Goldbeter. 1989b. Frequency specificity in intercellular communication: The influence of patterns of periodic signaling on target cell responsiveness. *Biophys. J.* 55:125–45.

Li, Y.X. & A. Goldbeter. 1990. Frequency encoding of pulsatile signals of cyclic AMP based on receptor desensitization in *Dictyostelium* cells. *J. Theor. Biol.* 146:355–67.

Li, Y.X. & A. Goldbeter. 1992. Pulsatile signaling in intercellular communication: Periodic stimuli are more efficient than random or chaotic signals in a model based on receptor desensitization. *Biophys. J.* 61:161–71.

Li, Y.X., J. Halloy, J.L. Martiel & A. Goldbeter. 1992a. Suppression of chaos and other dynamical transitions induced by intercellular coupling in a model for cAMP oscillations in *Dictyostelium* cells. *Chaos* 2:501–12.

Li, Y.X., J. Halloy, J.L. Martiel, B. Wurster & A. Goldbeter. 1992b. Suppression of chaos by periodic oscillations in a model for cAMP oscillations in *Dictyostelium* cells. *Experientia* 48:603–6.

Li, Y.-X. & J. Rinzel. 1994. Equations for Ins$P_3$ receptor-mediated [Ca]$_i$

oscillations derived from a detailed kinetic model: A Hodgkin–Huxley like formalism. *J. Theor. Biol.* **166**:461–73.

Li, Y.X., J. Rinzel, J. Keizer & S.S. Stojilkovic. 1994. Calcium oscillations in pituitary gonadotrophs: Comparison of experiment and theory. *Proc. Natl. Acad. Sci. USA* **91**:58–62.

Lieberman, D.N. & I. Mody. 1994. Regulation of NMDA channel function by endogenous $Ca^{2+}$-dependent phosphatase. *Nature* **369**:235–9.

Lin, B.T.-Y., S. Gruenwald, A.O. Morla, W.-H. Lee & J.Y.J. Wang. 1991. Retinoblastoma cancer suppressor gene product is a substrate of the cell cycle regulator cdc2 kinase. *EMBO J.* **10**:857–64.

Lincoln, D.W., H.M. Fraser, G.A. Lincoln, G.B. Martin & A.S. McNeilly. 1985. Hypothalamic pulse generators. *Rec. Progr. Horm. Res.* **41**:369–419.

Lipkin, E.W., D.C. Teller & C. de Haën. 1983. Dynamic aspects of insulin action: Synchronization of oscillatory glycolysis in isolated perifused rat fat cells by insulin and hydrogen peroxide. *Biochemistry* **22**:792–9.

Lipp, P. & E. Niggli. 1993. Microscopic spiral waves reveal positive feedback in subcellular calcium signaling. *Biophys. J.* **65**:2272–6.

Lipscombe, D., D.V. Madison, M. Poenie, H. Reuter, R.W. Tsien & R.Y. Tsien. 1988. Imaging of cytosolic $Ca^{2+}$ transients arising from $Ca^{2+}$ stores and $Ca^{2+}$ channels in sympathetic neurons. *Neuron* **1**:355–65.

Liu, G. & P.C. Newell. 1994. Regulation of myosin regulatory light chain phosphorylation via cyclic GMP during chemotaxis of *Dictyostelium. J. Cell Sci.* **107**:1737–43.

Liu, T.C. & Jackson, G.L. 1984. Long term superfusion of rat anterior pituitary cells: Effects of repeated pulses of gonadotropin-releasing hormone at different doses, durations and frequencies. *Endocrinology* **115**:605–13.

Liu, X., L.J. Zwiebel, D. Hinton, S. Benzer, J.C. Hall & M. Rosbash. 1992. The *period* gene encodes a predominantly nuclear protein in adult *Drosophila. J. Neurosci.* **12**:2735–44.

Llinas, R. 1984. Rebound excitation and the physiological basis for tremor: a biophysical study of the oscillatory properties of mammalian central neurones *in vitro.* In: *Movement Disorders: Tremor.* L.J. Findley & R. Capildeo, eds. Macmillan, London, pp. 165–82.

Llinas, R. 1988. The intrinsic electrophysiological properties of mammalian neurons: A new insight into CNS function. *Science* **242**:1654–64.

Llinas, R. & Y. Yarom. 1981a. Electrophysiology of mammalian inferior olivary neurones *in vitro.* Different types of voltage-dependent ionic conductances. *J. Physiol. (Lond.)* **315**:549–67.

Llinas, R. & Y. Yarom. 1981b. Properties and distribution of ionic conductances generating electroresponsiveness of mammalian inferior olivary neurones *in vitro. J. Physiol. (Lond.)* **315**:569–84.

Llinas, R. & Y. Yarom. 1986. Oscillatory properties of guinea-pig inferior olivary neurones and their pharmacological modulation: An *in vitro* study. *J. Physiol. (Lond.)* **376**:163–82.

Lloyd, D., A.L. Lloyd & L.F. Olsen. 1992. The cell division cycle: A physiologically plausible dynamic model can exhibit chaotic solutions. *Biosystems* **27**:17–24.

Lloyd, D. & M. Stupfel. 1991. The occurrence and functions of ultradian rhythms. *Biol. Rev.* **66**:275–99.

Lloyd, J.E. 1983. Bioluminescence and communication in insects. *Annu. Rev. Entomol.* **28**:131–60.

Longo, E.A., K. Tornheim, J.T. Deeney, B.A. Varnum, D. Tillotson, M. Prentki & B.E. Corkey. 1991. Oscillations in cytosolic free $Ca^{2+}$, oxygen consumption, and insulin secretion in glucose-stimulated rat pancreatic islets. *J. Biol. Chem.* **266**:9314–19.

Loomis, W.F. 1975. Dictyostelium discoideum: *A Developmental System.* Academic Press, New York.

Loomis, W.F. 1979. Biochemistry of aggregation in *Dictyostelium. Dev. Biol.* **70**:1–12.

Loomis, W.F. (ed.) 1982. *The Development of* Dictyostelium discoideum. Academic Press, New York.

Lorca, T., F.H. Cruzalegui, D. Fesquet, J.-C. Cavadore, J. Mery, A. Means & M. Dorée. 1993. Calmodulin-dependent protein kinase II mediates inactivation of MPF and CSF upon fertilization of *Xenopus* eggs. *Nature* **366**:270–3.

Lorca, T., D. Fesquet, F. Zindy, F. Le Bouffant, M. Cerruti, C. Brechot, G. Devauchelle & M. Dorée. 1991a. An okadaic acid-sensitive phosphatase negatively controls the cyclin degradation pathway in amphibian eggs. *Mol. Cell. Biol.* **11**:1171–5.

Lorca, T., S. Galas, D. Fesquet, A. Devault, J.C. Cavadore & M. Dorée. 1991b. Degradation of the protooncogene product $p39^{mos}$ is not necessary for cyclin proteolysis and exit from meiotic metaphase: requirement for a $Ca^{2+}$–calmodulin dependent event. *EMBO J.* **10**:2087–93.

Lorca, T., J.C. Labbé, A. Devault, D. Fesquet, J.P. Capony, J.C. Cavadore, F. Le Bouffant & M. Dorée. 1992. Dephosphorylation of cdc2 on threonine 161 is required for cdc2 kinase inactivation and normal anaphase. *EMBO J.* **11**:2381–90.

Lorenz, E.N. 1963. Deterministic nonperiodic flow. *J. Atmos. Sci.* **20**:130–41.

Lorenz, E.N. 1991. Dimension of weather and climate attractors. *Nature* **353**:241–4.

Loros, J.J., A. Lichens-Park, K.M. Lindgren & J.C. Dunlap. 1993. Molecular genetics of genes under circadian temporal control in *Neurospora.* In: *Molecular Genetics of Biological Rhythms*, M.W. Young, ed. Marcel Dekker, New York, pp. 55–72.

Lotka, A.J. 1920. Undamped oscillations derived from the law of mass action. *J. Am. Chem. Soc.* **42**:1595.

Lotka, A.J. 1925. *Elements of Physical Biology.* The Williams & Wilkins Co., Baltimore, MD.

Lubs-Haukeness, J. & C. Klein. 1982. Cyclic nucleotide-dependent phosphorylation in *Dictyostelium discoideum* amoebae. *J. Biol. Chem.* **257**:12204–8.

Lytton J. & S. Nigam. 1992. Intracellular calcium: molecules and pools. *Curr. Opinion Cell Biol.* **4**:220–6.

Mackay, S.A. 1978. Computer simulations of aggregation in *Dictyostelium discoideum. J. Cell Sci.* **33**:1–16.

Mackey, M.C. 1978. A unified hypothesis for the origin of aplastic anemia and periodic haematopoesis. *Blood* **51**:941–56.

Mackey, M.C. 1985. A deterministic cell cycle model with transition probability-like behaviour. In: *Temporal Order.* L. Rensing & N.I. Jaeger, eds. Springer, Berlin, pp. 315–20.

Mackey, M.C. & L. Glass. 1977. Oscillations and chaos in physiological control systems. *Science* **197**:287–9.

MacLeod, J.N. & B.H. Shapiro. 1989. Growth hormone regulation of hepatic drug-metabolizing enzymes in the mouse. *Biochem. Pharmacol.* **38**:1673–7.

Macnab, R.M. & D.E. Koshland Jr. 1972. The gradient sensory mechanism in bacterial chemotaxis. *Proc. Natl. Acad. Sci. USA* **69**:2509–12.

Maini, P.K. 1990. Superposition of modes in a caricature of a model for morphogenesis. *J. Math. Biol.* **28**:307–15.

Mäkela, T.P., J.-P. Tassan, E.A. Nigg, S. Frutiger, G.J. Hughes & R.A. Weinberg. 1994. A cyclin associated with the CDK-activating kinase MO15. *Nature* **371**:254–7.

Malgaroli, A., R. Fesce & J. Meldolesi. 1990. Spontaneous $[Ca^{2+}]_i$ fluctuations in rat chromaffin cells do not require inositol 1,4,5-trisphosphate elevations but are generated by a caffeine- and ryanodine-sensitive intracellular $Ca^{2+}$ store. *J. Biol. Chem.* **265**:3005–8.

Malgaroli, A. & J. Meldolesi. 1991. $[Ca^{2+}]_i$ oscillations from internal stores sustain exocytic secretion from the chromaffin cells of the rat. *FEBS Lett.* **283**:169–72.

Maller, J.L. 1990. *Xenopus* oocytes and the biochemistry of cell division. *Biochemistry* **29**:3157–66.

Mandelkow, E., E.-M. Mandelkow, H. Hotani, B. Hess & S. Müller. 1989. Spatial patterns from oscillating microtubules. *Science* **246**:1291–93.

Mandelkow, E.-M., G. Lange, A. Jagla, U. Spann & E. Mandelkow. 1988. Dynamics of the microtubule oscillator: Role of nucleotides and tubulin–MAP interactions. *EMBO J.* **7**:357–65.

Mann, S.K.O. & R.A. Firtel. 1987. Cyclic AMP regulation of early gene expression in *Dictyostelium discoideum*: Mediation via the cell surface cyclic AMP receptor. *Mol. Cell. Biol.* **7**:458–69.

Mann, S.K.O., C. Pinko & R.A. Firtel. 1988. cAMP regulation of early gene expression in signal transduction mutants of *Dictyostelium*. *Dev. Biol.* **130**:294–303.

Mano, Y. 1970. Cytoplasmic regulation and cyclic variation in protein synthesis in the early cleavage stages of the sea urchin embryo. *Dev. Biol.* **22**:433–60.

Mansour, T.E. 1972. Phosphofructokinase. *Curr. Top. Cell. Regul.* **5**:1–46.

Markus, M. & B. Hess. 1984. Transitions between oscillatory modes in a glycolytic model system. *Proc. Natl. Acad. Sci. USA* **81**:4394–8.

Markus, M. & B. Hess. 1990. Isotropic cellular automaton for modelling excitable media. *Nature* **347**:56–8.

Markus, M., D. Kuschmitz & B. Hess. 1984. Chaotic dynamics in yeast glycolysis under periodic substrate input flux. *FEBS Lett.* **172**:235–8.

Markus, M., D. Kuschmitz & B. Hess. 1985. Properties of strange attractors in yeast glycolysis. *Biophys. Chem.* **22**:95–105.

Marmillot, P., J.-F. Hervagault & G.R. Welch. 1992. Patterns of spatiotemporal organization in an 'ambiquitous' enzyme model. *Proc. Natl. Acad. Sci. USA* **89**:12103–7.

Martha, P.A. Jr, R.M. Blizzard, J.A. McDonald, M.O. Thorner & A.D. Rogol. 1988. A persistent pattern of varying pituitary responsivity to exogenous growth hormone (GH)-releasing hormone in GH-deficient children: Evidence supporting periodic somatostatin secretion. *J. Clin. Endocrinol. Metab.* **67**:449–54.

Martiel, J.L. 1988. Etude par un modèle de la génération périodique des signaux chimiotactiques chez *Dictyostelium discoideum*. Thèse de Doctorat d'Etat, Université de Paris-Sud (Orsay).

Martiel, J.L. & A. Goldbeter. 1981. Metabolic oscillations in biochemical systems controlled by covalent enzyme modification. *Biochimie* **63**:119–24.

Martiel, J.L. & A. Goldbeter. 1984. Oscillations et relais des signaux d'AMP cyclique chez *Dictyostelium discoideum*: Analyse d'un modèle fondé sur la désensibilisation du récepteur pour l'AMP cyclique. *C.R. Acad. Sci. (Paris) Sér. III* **298**:549–52.

Martiel, J.L. & A. Goldbeter. 1985. Autonomous chaotic behaviour of the slime mould *Dictyostelium discoideum* predicted by a model for cyclic AMP signalling. *Nature* **313**:590–2.

Martiel, J.L. & A. Goldbeter. 1987a. A model based on receptor desensitization for cyclic AMP signaling in *Dictyostelium* cells. *Biophys. J.* **52**:807–28.

Martiel, J.L. & A. Goldbeter. 1987b. Origin of bursting and birhythmicity in a model for cyclic AMP oscillations in *Dictyostelium* cells. *Lect. Notes Biomath.* **71**:244–55.

Marynissen, G., A. Sener & W.J. Malaisse. 1992. Oscillations in glycolysis: multifactorial quantitative analysis in muscle extract. *Mol. Cell. Biochem.* **113**:105–21.

Matsumoto, G., K. Aihara, M. Ichikawa & A. Tasaki. 1984. Periodic and nonperiodic responses of membrane potential in squid giant axons during sinusoidal current stimulation. *J. Theor. Neurobiol.* **3**:1–14.

Matthews, D.R., B.A. Naylor, R.G. Jones, G.M. Ward & R.C. Turner. 1983. Pulsatile insulin has greater hypoglycemic effect than continuous delivery. *Diabetes* **32**:617–21.

May, R.M. 1972. Limit cycles in predator–prey communities. *Science* **177**:900–2.

May, R.M. 1973. *Model Ecosystems*. Princeton Univ. Press, Princeton, NJ.

May, R.M. 1976. Simple mathematical models with very complicated dynamics. *Nature* **261**:459–67.

May, R.M. 1987. Chaos and the dynamics of biological populations. *Proc. R. Soc. Lond. A* **413**:27–44.

May, R.M. & G.F. Oster. 1976. Bifurcations and dynamic complexity in simple ecological models. *Am. Nat.* **110**:573–99.

McClung, C.R. 1993. The higher plant *Arabidopsis thaliana* as a model system for the molecular analysis of circadian rhythms. In: *Molecular Genetics of Biological Rhythms*, M.W. Young, ed. Marcel Dekker, New York, pp. 1–35.

McIntosh, J.E.A. & R.P. McIntosh. 1986. Varying the patterns and concentrations of gonadotropin-releasing hormone stimulation does not alter the ratio of LH and FSH release from perifused sheep pituitary cells. *J. Endocrinol.* **109**:155–61.

Meech, R.W. 1979. Membrane potential oscillations in molluscan 'burster' neurones. *J. Exp. Biol.* **81**:93–112.

Mees, A.I. & P.E. Rapp. 1978. Periodic metabolic systems: Oscillations in multiple-loop negative feedback biochemical control networks. *J. Math. Biol.* **5**:99–114.

Meier, K. & C. Klein, 1988. An unusual protein kinase phosphorylates the chemotactic receptor in *Dictyostelium discoideum*. *Proc. Natl. Acad. Sci. USA* **85**:2181–5.

Meijer, J.H. & W.J. Rietveld. 1989. Neurophysiology of the suprachiasmatic circadian pacemaker in rodents. *Physiol. Rev.* **69**:671–707.

Meinhardt, H. 1982. *Models for Biological Pattern Formation*. Academic Press, London.

Meinke, M.H., J.S. Bishop & R.D. Edstrom. 1986. Zero-order ultrasensitivity in the regulation of glycogen phosphorylase. *Proc. Natl. Acad. Sci. USA* **83**:2865–8.

Melki, R., M.-F. Carlier & D. Pantaloni. 1988. Oscillations in microtubule polymerization: The rate of GTP regeneration on tubulin controls the period. *EMBO J.* **7**:2653–9.

Menaker, M. (ed.) 1971. *Biochronometry*. National Academy of Sciences, Washington, DC.

Meyer, T. 1991. Cell signaling by second messenger waves. *Cell* **64**:675–8.

Meyer, T., P.I. Hanson, L. Stryer & H. Schulman. 1992. Calmodulin trapping by calcium-calmodulin-dependent protein kinase. *Science* **256**:1199–1202.

Meyer, T. & L. Stryer. 1988. Molecular model for receptor-stimulated calcium spiking. *Proc. Natl. Acad. Sci. USA* **85**:5051–5.

Meyer, T. & L. Stryer. 1990. Transient calcium release induced by successive increments of inositol 1,4,5-trisphosphate. *Proc. Natl. Acad. Sci. USA* **87**:3841–5.

Meyer, T. & L. Stryer. 1991. Calcium spiking. *Annu. Rev. Biophysics Biophys. Chem.* **20**:153–74.

Michel, S., M.E. Geusz, J.J. Zaritsky & G.D. Block. 1993. Circadian rhythm in membrane conductance expressed in isolated neurons. *Science* **259**:239–41.

Michelson, S. & H. Schulman. 1994. CaM kinase: A model for its activation and dynamics. *J. Theor. Biol.* **171**:281–90.

Mikhailov, A.S., V.A. Davydov & V.S. Zykov. 1994. Complex dynamics of spiral waves and motion of curves. *Physica* **70D**:1–39.

Millar, J.B.A. & P. Russell. 1992. The cdc25 M-phase inducer: An unconventional protein phosphatase. *Cell* **68**:407–10.

Millard, W.J., D.M. O'Sullivan, T.O. Fox & J.B. Martin. 1987. Sexually dimorphic patterns of growth hormone secretion in rats. In: *The Episodic Secretion of Hormones*. W.F. Crowley & J.G. Hofler, eds. Wiley, New York, pp. 287–304.

Mines, G.R. 1913. On dynamic equilibrium in the heart. *J. Physiol. (Lond.)* **46**:349–83.

Minorsky, N. 1962. *Nonlinear Oscillations*. Van Nostrand, Princeton, NJ.

Minshull, J., J. Pines, R. Golsteyn, N. Standart, S. Mackie, A. Coolman, J. Blow, J. Ruderman, M. Wu & T. Hunt. 1989. The role of cyclin synthesis, modification and destruction in the control of cell division. *J. Cell Sci. Suppl.* **12**:77–97.

Minshull, J., H. Sun, N.K. Tonks & A.W. Murray. 1994. A MAP kinase-dependent spindle assembly checkpoint in *Xenopus* egg extracts. *Cell* **79**:475–86.

Missiaen, L., H. De Smedt, G. Droogmans & R. Casteels. 1992. Luminal $Ca^{2+}$ controls the activation of the inositol 1,4,5-trisphosphate receptor by cytosolic $Ca^{2+}$. *J. Biol. Chem.* **267**:22961–6.

Miyazaki S., H. Shirakawa, K. Nakada, Y. Honda, M. Yuzaki, S. Nakade & K. Mikishiba. 1992a. Antibody to the inositol trisphosphate receptor blocks thimerosal-enhanced $Ca^{2+}$-induced $Ca^{2+}$ release and $Ca^{2+}$ oscillations in hamster egg. *FEBS Lett.* **309**:180–4.

Miyazaki, S., M. Yuzaki, K. Nakada, H. Shirakawa, S. Nakanishi, S. Nakade & K. Mikishiba. 1992b. Block of $Ca^{2+}$ wave and $Ca^{2+}$ oscillation by antibody to the inositol 1,4,5-trisphosphate receptor in fertilized hamster egg. *Science* **257**:251–3.

Miyazaki, S.-I. 1988. Inositol 1,4,5-trisphosphate-induced calcium release and guanine nucleotide-binding protein-mediated periodic calcium rise in golden hamster eggs. *J. Cell Biol.* **106**:345–53.

Miyazaki, S.-I., N. Hashimoto, Y. Yoshimoto, T. Kishimoto, Y. Igusa & Y. Hiramoto. 1986. Temporal and spatial dynamics of the periodic increase in intracellular free calcium at fertilization of golden hamster eggs. *Dev. Biol.* **118**:259–67.

Mizraji, E., L. Acerenza & J. Hernandez. 1988. Time delays in metabolic control systems. *BioSystems* **22**:11–17.

Mode, A., G. Norstedt, B. Simic, P. Eneroth & J.A. Gustafson. 1981. Continuous infusion of growth hormone feminises hepatic steroid metabolism in the rat. *Endocrinology* **108**:2103–8.

Moenter, S.M., R.M. Brand, A.R. Midgley & F.R. Karsch. 1992. Dynamics of gonadotropin-releasing hormone release during a pulse. *Endocrinology* **130**:503–10.

Monk, P.B. & H.G. Othmer. 1989. Cyclic AMP oscillations in suspensions of *Dictyostelium discoideum. Phil. Trans. R. Soc. Lond.* **323**:185–224.

Monk, P.B. & H.G. Othmer. 1990. Wave propagation in aggregation fields of the cellular slime mould *Dictyostelium discoideum. Proc. R. Soc. Lond. B* **240**:555–89.

Monod, J., J.P. Changeux & F. Jacob. 1963. Allosteric proteins and molecular control systems. *J. Mol. Biol.* **6**:306–29.

Monod, J., J. Wyman & J.P. Changeux. 1965. On the nature of allosteric transitions: A plausible model. *J. Mol. Biol.* **12**:88–118.

Moore, R.Y. 1983. Organization and function of a central nervous system circadian oscillator: The suprachiasmatic hypothalamic nucleus. *Fed. Proc.* **42**:2783–9.

Moore, R.Y. & V.B. Eichler. 1972. Loss of circadian corticosterone rhythm following suprachiasmatic lesions in the rat. *Brain Res.* **42**:201–6.

Moore-Ede, M.C. 1983. The circadian timing system in mammals: Two pacemakers preside over many secondary oscillators. *Fed. Proc.* **42**:2802–8.

Moore-Ede, M.C. & C.A. Czeisler (eds.) 1984. *Mathematical Models of the Circadian Sleep/Wake Cycle.* Raven Press, New York.

Moore-Ede, M.C., C.A. Czeisler & G.S. Richardson. 1983. Circadian timekeeping in health and disease. Part 2. Clinical implications of circadian rhythmicity. *New Engl. J. Med.* **309**:530–6.

Moore-Ede, M.C., F.M. Sulzman & C.A. Fuller. 1982. *The Clocks that Time Us. Physiology of the Circadian Timing System.* Harvard Univ. Press, Cambridge, MA.

Morales, M. & D. McKay. 1967. Biochemical oscillations in 'controlled' systems. *Biophys. J.* **7**:621–5.

Moran, F. & A. Goldbeter. 1984. Onset of birhythmicity in a regulated biochemical system. *Biophys. Chem.* **20**:149–56.

Moran, F. & A. Goldbeter. 1985. Excitability with multiple thresholds: A new mode of dynamic behavior analyzed in a regulated biochemical system. *Biophys. Chem.* **23**:71–7.

Moreno, S., J. Hayles & P. Nurse. 1989. Regulation of p34[cdc2] protein kinase during mitosis. *Cell* **58**:361–72.

Moreno, S. & P. Nurse. 1994. Regulation of progression through the G1 phase of the cell cycle by the *rum1*[+] gene. *Nature* **367**:236–42.

Moreno, S., P. Nurse & P. Russell.1990. Regulation of mitosis by cyclic accumulation of p80*cdc25* mitotic inducer in fission yeast. *Nature* **344**:549–52.

Morgan, D.O. 1995. Principles of CDK regulation. *Nature* **374**:131–4.

Morgan, E.T., C. MacGeoch & J.-A. Gustafsson. 1985. Hormonal and developmental regulation of expression of the hepatic microsomal steroid 16α-hydroxylase cytochrome P-450 apoprotein in the rat. *J. Biol. Chem.* **260**:11895–8.

Morla, A.O., G. Draetta, D. Beach & J.Y.J. Wang. 1989. Reversible tyrosine phosphorylation of cdc2: Dephosphorylation accompanies activation during entry into mitosis. *Cell* **58**:193–203.

Morris, C. & H. Lecar. 1981. Voltage oscillations in the barnacle giant muscle fiber. *Biophys. J.* **35**:193–213.

Morse, D.S., L. Fritz & J.W. Hastings. 1990. What is the clock? Translational regulation of circadian bioluminescence. *Trends Biochem. Sci.* **15**:262–5.

Motokura, T., T. Bloom, H.G. Kim, H. Jüppner, J.V. Ruderman, H.M. Kronenberg & A. Arnold. 1991. A novel cyclin encoded by a *bcl1*-linked candidate oncogene. *Nature* **350**:512–15.

Mullens, I. & P.C. Newell. 1978. cAMP binding to cell surface receptors of *Dictyostelium*. *Differentiation* **10**:171–6.

Mulliert, G., N. Kellershohn & J. Ricard. 1990. Dynamics of an open metabolic cycle at the surface of a charged membrane. II. Multiple steady states and oscillatory behavior generated by electric repulsion effects. *Physica* **46D**:380–91.

Murray, A.W. 1989a. The cell cycle as a *cdc2* cycle. *Nature* **342**:14–15.

Murray, A.W. 1989b. Cyclin synthesis and degradation and the embryonic cell cycle. *J. Cell Sci. Suppl.* **12**:65–76.

Murray, A.W. 1992. Creative blocks: Cell-cycle checkpoints and feedback controls. *Nature* **359**:599–604.

Murray, A.W. & T. Hunt. 1993. *The Cell Cycle: An Introduction.* Oxford Univ. Press, Oxford.

Murray, A.W. & M.W. Kirschner. 1989a. Dominoes and clocks: The union of two views of the cell cycle. *Science* **246**:614–21.

Murray, A.W. & M.W. Kirschner. 1989b. Cyclin synthesis drives the early embryonic cell cycle. *Nature* **339**:275–80.

Murray, A.W., M.J. Solomon & M.W. Kirschner. 1989. The role of cyclin synthesis and degradation in the control of maturation promoting factor activity. *Nature* **339**:280–6.

Murray, J.D. 1989. *Mathematical Biology.* Springer, Berlin.

Nakamura, S., K. Yokota & I. Yamazaki. 1969. Sustained oscillations in a lactoperoxidase, NADPH and $O_2$ system. *Nature* **222**:794.

Nanjundiah, V. 1973. Chemotaxis, signal relaying and aggregation morphology. *J. Theor. Biol.* **42**:63–105.

Nanjundiah, V. 1988. Periodic stimuli are more successful than randomly spaced ones for inducing development in *Dictyostelium discoideum*. *Biosci. Rep.* **8**:571–7.

Nanjundiah, V. & S. Saran. 1992. The determination of spatial pattern in *Dictyostelium discoideum*. *J. Biosci.* **17**:353–94.

Nanjundiah, V. & B. Wurster.1989. Is there a cyclic-AMP-independent oscillator in *Dictyostelium discoideum*? In: *Cell to Cell Signalling: From Experiments to Theoretical Models.* A. Goldbeter, ed. Academic Press, London, pp. 489–502.

Naor, Z., M. Katikineni, E. Loumaye, A.G. Vela, M.L. Dufau & K.J. Catt. 1982. Compartmentalization of luteinizing hormone pools: Dynamics of gonadotropin releasing hormone action in superfused pituitary cells. *Mol. Cell. Endocr.* **27**:213–20.

Naparstek, A., J.L. Romette, J.P. Kernevez & D. Thomas. 1974. Memory in enzyme membrane. *Nature* **249**:490–1.

Naparstek, A., D. Thomas & S.R. Caplan. 1973. An experimental enzyme–membrane oscillator. *Biochim. Biophys. Acta* **323**:643–6.

Nasmyth, K. 1993. Control of the yeast cell cycle by the Cdc28 protein kinase. *Curr. Opin. Cell Biol.* **5**:166–79.

Nathanson, M.H., M.B. Fallon, P.J. Padfield & A.R. Maranto. 1994. Localization of the type 3 inositol 1,4,5-trisphosphate receptor in the $Ca^{2+}$ wave trigger zone of pancreatic acinar cells. *J. Biol. Chem.* **269**:4693–6.

Nathanson, M.H., P.J. Padfield, A.J. O'Sullivan, A.D. Burgstahler & J.D. Jamieson. 1992. Mechanism of $Ca^{2+}$ wave propagation in pancreatic acinar cells. *J. Biol. Chem.* **267**:18118–21.

Negro-Vilar, A., M.D. Culler, M.M. Valença, T.B. Flack & G. Wisniewski. 1987. Pulsatile peptide secretion: Encoding of brain messages regulating endocrine and reproductive functions. *Environ. Health Perspect.* **75**:37–43.

Newell, P.C. 1977. How cells communicate: The system used by slime moulds. *Endeavour, New Ser.* **1**:63–8.

Newell, P.C. 1982. Cell surface binding of adenosine to *Dictyostelium* and inhibition of pulsatile signalling. *FEMS Microbiol. Lett.* **13**:417–21.

Newell, P.C. & F.M. Ross. 1982. Inhibition by adenosine of aggregation centre initiation and cyclic AMP binding in *Dictyostelium. J. Gen. Microbiol.* **128**:2715–24.

Newport, J.W. & M.W. Kirschner. 1984. Regulation of the cell cycle during early *Xenopus* development. *Cell* **37**:731–42.

Nicholas, J., K.R. Cameron & R.W. Honess. 1992. Herpesvirus saimiri encodes homologues of G protein-coupled receptors and cyclins. *Nature* **355**:362–5.

Nicolis, G. 1971. Dissipative structures in open systems far from equilibrium. *Adv. Chem. Phys.* **19**:209–324.

Nicolis, G., F. Baras & M. Malek-Mansour. 1984. Stochastic aspects of nonequilibrium transitions in chemical systems. In: *Nonequilibrium Dynamics of Chemical Systems.* A. Pacault & C. Vidal, eds. Springer, Berlin, pp. 184–99.

Nicolis, G. & J. Portnow. 1973. Chemical oscillations. *Chem. Rev.* **73**:365–84.

Nicolis, G. & I. Prigogine. 1971. Biological order, structure and instabilities. *Quart. Rev. Biophys.* **4**:107–48.

Nicolis, G. & I. Prigogine. 1977. *Self-Organization in Nonequilibrium Systems. From Dissipative Structures to Order through Fluctuations.* Wiley, New York.

Nissler, K., R. Kessler, W. Schellenberger & E. Hofmann. 1977. Binding of fructose-6-phosphate to phosphofructokinase from yeast. *Biochem. Biophys. Res. Commun.* **79**:973–8.

Noble, D. 1979. *The Initiation of the Heartbeat.* Oxford Univ. Press, Oxford.

Noble, D. 1984. The surprising heart: A review of recent progress in cardiac electrophysiology. *J. Physiol. (Lond.)* **353**:1–50.

Noble, D., D. DiFrancesco & J. Denyer. 1989. Ionic mechanisms in normal and abnormal cardiac pacemaker activity. In: *Neuronal and Cellular Oscillators.* J.W. Jacklet, ed. Marcel Dekker, New York and Basel, pp. 59–85.

Noble, D. & S.J. Noble. 1984. A model of sino-atrial node electrical activity based on a modification of the DiFrancesco–Noble (1984) equations. *Proc. R. Soc. Lond. B* **222**:295–304.

Noble, D. & T. Powell, eds. 1987. *Electrophysiology of Single Cardiac Cells.* Academic Press, London.

Noda, M., S. Shimizu, T. Tanabe, T. Takai, T. Kayano, T. Ikeda, H. Takahashi, H. Nakayama, Y. Kanaoka, N. Minamino, K. Kangawa, H. Matsuo, M. A. Raftery, T. Hirose, S. Inayama, H. Hayashida, T. Miyata & S. Numa. 1984. Primary structure of *Electrophorus electricus* sodium channel deduced from cDNA sequence. *Nature* **312**:121–7.

Noel, J., K. Fukami, A.M. Hill & T. Capiod. 1992. Oscillations of cytosolic free calcium concentration in the presence of intracellular antibodies to phosphatidylinositol 4,5-bisphosphate in voltage-clamped guinea-pig hepatocytes. *Biochem. J.* **288**:357–60.

Norbury, C., J. Blow & P. Nurse. 1991. Regulatory phosphorylation of the p34$^{cdc2}$ protein kinase in vertebrates. *EMBO J.* **10**:3321–9.

Norbury, C. & P. Nurse. 1992. Animal cell cycles and their control. *Annu. Rev. Biochem.* **61**:441–70.

Norel, R. & Z. Agur. 1991. A model for the adjustment of the mitotic clock by cyclin and MPF levels. *Science* **251**:1076–8.

Norstedt, G. & R. Palmiter. 1984. Secretory rhythm of growth hormone regulates sexual differentiation of mouse liver. *Cell* **36**:805–12.

Noshiro, M. & M. Negishi. 1986. Pretranslational regulation of sex-dependent testosterone hydroxylases by growth hormone in mouse liver. *J. Biol. Chem.* **261**:15923–7.

Nourse, J., E. Firpo, W.M. Flanagan, S. Coats, K. Polyak, M.-H. Lee, J. Massague, G.R. Crabtree & J.M. Roberts. 1994. Interleukin-2-mediated elimination of the p27$^{Kip1}$ cyclin-dependent kinase inhibitor prevented by rapamycin. *Nature* **372**:570–3.

Novak, B. & J.J. Tyson. 1993a. Modeling the cell division cycle: M-phase trigger, oscillations, and size control. *J. Theor. Biol.* **165**:101–34.

Novak, B. & J.J. Tyson. 1993b. Numerical analysis of a comprehensive model of M-phase control in *Xenopus* oocytes extracts and intact embryos. *J. Cell Science* **106**:1153–68.

Novak, B. & J.J. Tyson. 1995. Quantitative analysis of a molecular model of mitotic control in fission yeast. *J. Theor. Biol.* **173**:283–305.

Noyes, R.M. & R.J. Field. 1974. Oscillatory chemical reactions. *Annu. Rev. Phys. Chem.* **25**:95–119.

Noyes, R.M., R.J. Field & E. Köros. 1972. Oscillations in chemical systems. 1. Detailed mechanism in a system showing temporal oscillations. *J. Am. Chem. Soc.* **94**:1394.

Nurse, P. 1990. Universal control mechanism regulating onset of M-phase. *Nature* **344**:503–8.

Nurse, P. 1994. Ordering S phase and M phase in the cell cycle. *Cell* **79**:547–50.

O'Farrell, P.H., B.E. Edgar, D. Lakich & C.F. Lehner. 1989. Directing cell division during development. *Science* **246**:635–40.

O'Rahilly, S., R.C. Turner & D.R. Matthews. 1988. Impaired pulsatile secretion of insulin in relatives of patients with non-insulin-dependent diabetes. *New Engl. J. Med.* **318**:1225–30.

O'Rourke, B., B.M. Ramza & E. Marban. 1994. Oscillations of membrane current and excitability driven by metabolic oscillations in heart cells. *Science* **265**:962–6.

Obeyesekere, M.N., S.L. Tucker & S.O. Zimmerman. 1994. A model for regulation of the cell cycle incorporating cyclin A, cyclin ⱻ and their complexes. *Cell Prolif.* **27**:105–13.

Odell, G.M. 1980. Qualitative theory of systems of ordinary differential equations, including phase plane analysis and the use of the Hopf bifurcation theorem. In: *Mathematical Models in Molecular and Cellular Biology*, L.A. Segel, ed. Cambridge Univ. Press, Cambridge, pp. 649–727.

Ohmori, H. 1981. Development of intrinsic burst generation in identified neuron *R15* of *Aplysia*. *Brain Res.* **222**:383–7.

Olsen, L.F. 1978. The oscillating peroxidase–oxidase reaction in an open system: Analysis of the reaction mechanism. *Biochim. Biophys. Acta* **527**:212–20.

Olsen, L.F. 1979. Studies of the chaotic behaviour in the peroxidase-oxidase reaction. *Z. Naturforsch.* **34a**:1544–6.

Olsen, L.F. 1983. An enzyme reaction with a strange attractor. *Phys. Lett.* **94A**:454–7.

Olsen, L.F. & H. Degn. 1977. Chaos in an enzyme reaction. *Nature* **267**:177–8.

Olsen, L.F. & H. Degn. 1978. Oscillatory kinetics of the peroxidase–oxidase reaction in an open system. Experimental and theoretical studies. *Biochim. Biophys. Acta* **523**:321–34.

Olsen, L.F. & H. Degn. 1985. Chaos in biological systems. *Quart. Rev. Biophys.* **18**:165–225.

Olsen, L.F. & W.M. Schaffer. 1990. Chaos versus noisy periodicity: alternative hypotheses for childhood epidemics. *Science* **249**:499–504.

Osipchuk, Y.V., M. Wakui, D.I. Yule, D.V. Gallacher & O.H. Petersen. 1990. Cytoplasmic $Ca^{2+}$ oscillations evoked by receptor stimulation, G-protein activation, internal application of inositol trisphosphate or $Ca^{2+}$: Simultaneous microfluorimetry and $Ca^{2+}$ dependent $Cl^-$ current recording in single pancreatic acinar cells. *EMBO J.* **9**:697–704.

Othmer, H.G., P.B. Monk & P.E. Rapp. 1985. A model for signal-relay adaptation in *Dictyostelium discoideum*. II. Analytical and numerical results. *Math. Biosci.* **77**:79–139.

Ott, E., C. Grebogi & J.A. Yorke. 1990. Controlling chaos. *Phys. Rev. Lett.* **64**:1196–9.

Ozil, J.P. & K. Swann. 1995. Stimulation of repetitive calcium transients in mouse eggs. *J. Physiol.* **483.2**:331–46.

Pacault, A., P. Hanusse, P. De Kepper, C. Vidal & J. Boissonade. 1976. Phenomena in homogeneous chemical systems far from equilibrium. *Acc. Chem. Res.* **9**:438–45.

Page, T.L. 1982. Transplantation of the cockroach circadian pacemaker. *Science* **216**:73–5.

Painter, P.R. & J.J. Tyson. 1984. Periodic enzyme synthesis and the cell division cycle. In: *Cell Cycle Clocks*. L.N. Edmunds Jr, ed. Marcel Dekker, New York and Basel, pp. 173–92.

Palsson, B.O. & T.M. Groshans. 1988. Mathematical modelling of dynamics and control in metabolic networks. VI. Dynamic bifurcations in single biochemical control loops. *J. Theor. Biol.* **131**:43–53.

Paolisso, G., A.J. Scheen, A. Albert & P.J. Lefèbvre. 1989. Effects of pulsatile delivery of insulin and glucagon in humans. *Am. J. Physiol.* **257**:E686–E696.

Paolisso, G., A.J. Scheen, D. Giugliano, S. Sgambato, A. Albert, M. Varricchio, F. D'Onofrio & P.J. Lefèbvre. 1991. Pulsatile insulin delivery has greater metabolic effects than continuous hormone administration in man: Importance of pulse frequency. *J. Clin. Endocrinol. Metab.* **72**:607–15.

Pardee, A.B. 1989. G1 events and regulation of cell proliferation. *Science* **246**:603–8.

Parker, I. & I. Ivorra. 1990. Inhibition by $Ca^{2+}$ of inositol trisphosphate-mediated $Ca^{2+}$ liberation: A possible mechanism for oscillatory release of $Ca^{2+}$. *Proc. Natl. Acad. Sci. USA* **87**:260–4.

Parker, L.L., S. Atherton-Fessler & H. Piwnica-Worms. 1992. $p107^{wee1}$ is a dual-specificity kinase that phosphorylates $p34^{cdc2}$ on tyrosine 15. *Proc. Natl. Acad. Sci. USA* **89**:2917–21.

Parker, L.L. & H. Piwnica-Worms. 1992. Inactivation of the $p34^{cdc2}$-cyclin B complex by the human WEE1 tyrosine kinase. *Science* **257**:1955–7.

Parker, L.L., S.A. Walter, P.G. Young & H. Piwnica-Worms. 1993. Phosphorylation and inactivation of the mitotic inhibitor Wee1 by the *nim1/cdr1* kinase. *Nature* **363**:736–8.

Parnas, H. & L.A. Segel. 1978. A computer simulation of pulsatile aggregation in *Dictyostelium discoideum*. *J. Theor. Biol.* **71**:185–207.

Pavlidis, T. 1971. Populations of biochemical oscillators as circadian clocks. *J. Theor. Biol.* **33**:319–38.

Pavlidis, T. 1973. *Biological Oscillators: Their Mathematical Analysis*. Academic Press, New York.

Pavlidis, T. & W. Kauzmann. 1969. Toward a quantitative biochemical model for circadian oscillators. *Arch. Biochem. Biophys.* **132**:338–48.

Payne, R. & B.V.L. Potter. 1991. Injection of inositol trisphosphothioate into *Limulus* ventral photoreceptors causes oscillations of free cytosolic calcium. *J. Gen. Physiol.* **97**:1165–86.

Peng, B., V. Petrov & K. Showalter. 1991. Controlling chemical chaos. *J. Phys. Chem.* **95**:4957–9.

Perez-Armendariz, E., I. Atwater & E. Rojas. 1985. Glucose-induced oscillatory changes in extracellular ionized potassium concentration in mouse islets of Langerhans. *Biophys. J.* **48**:741–9.

Pérez-Iratxeta, C., J. Halloy, F. Moran, J.L. Martiel & A. Goldbeter. 1995. Multiple propagating wave fronts in a biochemical model: Link with birhythmicity and multiple excitability thresholds. Manuscript in preparation.

Perutz, M. 1990. *Mechanisms of Cooperativity and Allosteric Regulation in Proteins*. Cambridge Univ. Press, Cambridge.

Peter, M. & I. Herskowitz. 1994. Joining the complex: Cyclin-dependent kinase inhibitory proteins and the cell cycle. *Cell* **79**:181–4.

Petersen, O.H. & M. Wakui. 1990. Oscillating intracellular $Ca^{2+}$ signals evoked by activation of receptors linked to inositol lipid hydrolysis: Mechanism of generation. *J. Membr. Biol.* **118**:93–105.

Petrov, V., V. Gaspar, J. Masere & K. Showalter. 1993. Controlling chaos in the Belousov–Zhabotinsky reaction. *Nature* **361**:240–3.

Petrov, V., S.K. Scott & K. Showalter. 1992. Mixed-mode oscillations in chemical systems. *J. Chem. Phys.* **97**:6191–8.

Pietri, F., M. Hilly & J.P. Mauger. 1990. Calcium mediates the interconversion between two states of the liver inositol 1,4,5-trisphosphate receptor. *J. Biol. Chem.* **265**:17478–85.

Pikovsky, A.S. 1984. On the interaction of strange attractors. *Z. Phys. B* **55**:149–54.

Pittendrigh, C.S. 1954. On temperature independence in the clock system controlling emergence time in *Drosophila. Proc. Natl. Acad. Sci. USA* **40**:1018–29.

Pittendrigh, C. 1960. Circadian rhythms and the circadian organization of living systems. *Cold Spring Harbor Symp. Quant. Biol.* **25**:159–84.

Pittendrigh, C. 1961. On temporal organization in living systems. *Harvey Lect.* **56**:93–125.

Pittendrigh, C.S. 1965. On the mechanism of entrainment of a circadian rhythm by light cycles. In: *Circadian Clocks.* J. Aschoff, ed. North-Holland, Amsterdam, pp. 277–97.

Pittendrigh, C. 1967. Circadian systems. I. The driving oscillation and its assay in *Drosophila pseudoobscura. Proc. Natl. Acad. Sci. USA* **58**:1762–7.

Pittendrigh, C.S. 1993. Temporal organization: Reflections of a Darwinian clock-watcher. *Annu. Rev. Physiol.* **55**:17–54.

Plant, R.E. 1978. The effects of calcium$^{++}$ on bursting neurons. A modeling study. *Biophys. J.* **21**:217–37.

Plant, R.E. & M. Kim. 1976. Mathematical description of a bursting pacemaker neuron by modification of the Hodgkin–Huxley equations. *Biophys. J.* **16**:227–44.

Pohl, C.R., S.W. Richardson, J.S. Hutchison, J.A. Germak & E. Knobil. 1983. Hypophysiotropic signal frequency and the functioning of the pituitary-ovarian system in the rhesus monkey. *Endocrinology* **112**:2076–80.

Pool, R. 1989. Is it healthy to be chaotic? *Science* **243**:604–7.

Pralong, W.-F., A. Spät & C.B. Wollheim. 1994. Dynamic pacing of cell metabolism by intracellular $Ca^{2+}$ transients. *J. Biol. Chem.* **269**:27310–4.

Prentki, M., M.C. Glennon, A.P. Thomas, R.L. Morris, F.M. Matschinsky & B.E. Corkey, B.E. 1988. Cell-specific patterns of oscillating free $Ca^{2+}$ in carbamylcholine-stimulated insulinoma cells. *J. Biol. Chem.* **263**:11044–7.

Prigogine, I. 1967. *Introduction to Thermodynamics of Irreversible Processes.* Wiley, New York.

Prigogine, I. 1968. *Introduction à la Thermodynamique des Processus Irréversibles.* Dunod, Paris.

Prigogine, I. 1969. Structure, dissipation and life. In: *Theoretical Physics and Biology.* M. Marois, ed. North-Holland, Amsterdam, pp. 23–52.

Prigogine, I. & R. Balescu. 1956. Phénomènes cycliques dans la thermodynamique des processus irréversibles. *Bull. Cl. Sci. Acad. R. Belg.* **XLII**:256–65.

Prigogine, I., R. Lefever, A. Goldbeter & M. Herschkowitz-Kaufman. 1969. Symmetry-breaking instabilities in biological systems. *Nature* **223**:913–16.

Prigogine, I. & I. Stengers. 1979. *La Nouvelle Alliance.* Gallimard, Paris.

Putney, J.W. Jr. 1986. A model for receptor-regulated calcium entry. *Cell Calcium* **7**:1–12.

Putney, J.W. Jr. 1991. Capacitative calcium entry revisited. *Cell Calcium* **11**:611–24.

Pye, E.K. 1969. Biochemical mechanisms underlying the metabolic oscillations in yeast. *Can. J. Bot.* **47**:271–85.

Pye, E.K. 1971. Periodicities in intermediary metabolism. In: *Biochronometry.* M. Menaker, ed. National Academy of Sciences, Washington, DC, pp. 623–36.

Pye, E.K. & B. Chance. 1966. Sustained sinusoidal oscillations of reduced pyridine nucleotide in a cell-free extract of *S. carlsbergensis. Proc. Natl. Acad. Sci. USA* **55**:888–94.

Quinn, S.J., G.H. Williams & D.L. Tillotson. 1988. Calcium oscillations in single adrenal glomerulosa cells stimulated by angiotensin II. *Proc. Natl. Acad. Sci. USA* **85**:5754–8.

Raju, U., C. Koumenis, M. Nunez-Regueiro & A. Eskin. 1991. Alteration of the phase and period of a circadian oscillator by a reversible transcription inhibitor. *Science* **253**:673–75.

Ralph, M.R., R.G. Foster, F. Davis & M. Menaker. 1990. Transplanted suprachiasmatic nucleus determines circadian period. *Science* **247**:975–8.

Ralph, M.R. & M. Menaker. 1988. A mutation of the circadian system in golden hamsters. *Science* **241**:1225–7.

Raper, K.B. 1935. *Dictyostelium discoideum*, a new species of slime mold from decaying forest leaves. *J. Agricult. Res.* **50**:135–47.

Raper, K.B. 1940a. Pseudoplasmodium formation and organization in *Dictyostelium discoideum*. *J. Elisa Mitchell Sci. Soc.* **56**:241–82.

Raper, K.B. 1940b. The communal nature of the fruiting process in the acrasiae. *Am. J. Bot.* **27**:436–48.

Rapp, P.E. 1975. A theoretical investigation of a large class of biochemical oscillators. *Math. Biosci.* **25**:165–88.

Rapp, P.E. 1979. An atlas of cellular oscillators. *J. Exp. Biol.* **81**:281–306.

Rapp, P.E. 1980. Biological applications of control theory. In: *Mathematical Models in Molecular and Cellular Biology*, L.A. Segel, ed. Cambridge Univ. Press, Cambridge, pp. 146–247.

Rapp, P.E. 1987. Why are so many biological systems periodic? *Progr. Neurobiol.* **29**:261–73.

Rapp, P.E., T.R. Bashore, J.M. Martinerie, A.M. Albano & A.I. Mees. 1989. Dynamics of brain electrical activity. *Brain Topogr.* **2**:99–118.

Rapp, P.E. & M.J. Berridge. 1977. Oscillations in calcium–cyclic AMP control loops form the basis of pacemaker activity and other high frequency biological rhythms. *J. Theor. Biol.* **66**:497–525.

Rapp, P.E. & M.J. Berridge. 1981. The control of transepithelial potential oscillations in the salivary gland of *Calliphora erythrocephala*. *J. Exp. Biol.* **93**:119–32.

Rapp, P.E., A.I. Mees & C.T. Sparrow. 1981. Frequency encoded biochemical regulation is more accurate than amplitude dependent control. *J. Theor. Biol.* **90**:531–44.

Rapp, P.E., P.B. Monk & H.G. Othmer. 1985. A model for signal-relay adaptation in *Dictyostelium discoideum*. I. Biological processes and the model network. *Math. Biosci.* **77**:35–78.

Rashevsky, N. 1948. *Mathematical Biophysics*. Univ. of Chicago Press, Chicago, IL. (Reprinted as third revised edn., 1960, Dover, New York.)

Rasmussen, D.D. 1993. Episodic gonadotropin-releasing hormone release from the rat isolated median eminence *in vitro*. *Neuroendocrinology* **58**:511–18.

Reddy, P., A.C. Jacquier, N. Abovich, G. Petersen & M. Rosbash. 1986. The *period* clock locus of *D. melanogaster* codes for a proteoglycan. *Cell* **46**:53–61.

Reich, J.G. & E.E. Sel'kov. 1974. Mathematical analysis of metabolic networks. *FEBS Lett.* **40**:S119–S127.

Reich, J.G. & E.E. Sel'kov. 1981. *Energy Metabolism of the Cell: A Theoretical Treatise*. Academic Press, New York.

Reid, R.L., G.R. Leopold & S.S.C. Yen. 1981. Induction of ovulation and pregnancy with pulsatile luteinizing hormone-releasing factor: dosage and mode of delivery. *Fertil. Steril.* **36**:553–9.

Ricard, J. & J.M. Soulié. 1982. Self-organization and dynamics of an open futile cycle. *J. Theor. Biol.* **95**:105–21.

Richter, P.H. & J. Ross. 1980. Oscillations and efficiency in glycolysis. *Biophys. Chem.* **12**:285–97.

Rink, T.J. & R. Jacob.1989. Calcium oscillations in non-excitable cells. *Trends Neurosci.* **12**:43–6.

Rinzel, J. 1985. Excitation dynamics: Insights from simplified membrane models. *Fed. Proc.* **44**:2944–6.

Rinzel, J. 1987. A formal classification of bursting mechanisms in excitable systems. *Lect. Notes Biomath.* **71**:267–81.

Rinzel, J. & S.M. Baer. 1988. Threshold for repetitive activity for a slow stimulus ramp: A memory effect and its dependence on fluctuations. *Biophys. J.* **54**:551–5.

Rinzel, J. & B. Ermentrout. 1989. Analysis of neural excitability and oscillations. In: *Methods in Neuronal Modeling. From Synapses to Network.* C. Koch & I. Segev, eds. MIT Press, Cambridge, MA, pp. 135–69.

Rinzel, J. & Y.S. Lee. 1986. On different mechanisms for membrane potential bursting. In: *Nonlinear Oscillations in Biology and Chemistry.* H.G. Othmer, ed. Springer, Berlin, pp. 19–33.

Rinzel, J., A. Sherman & C.L. Stokes. 1992. Channels, coupling, and synchronized rhythmic bursting activity. In: *Analysis and Modeling of Neural Systems.* F.H. Eeckman, ed. Kluwer Academic Publishers, Boston, MA, pp. 29–46.

Rinzel, J. & W.C. Troy. 1982. Bursting phenomena in a simplified Oregonator flow system model. *J. Chem. Phys.* **76**:1775–89.

Rinzel, J. & W.C. Troy. 1983. A one-variable map analysis of bursting in the Belousov–Zhabotinsky reaction. *Contemp. Math.* **17**:411–27.

Robertson, A. & D.J. Drage. 1975. Stimulation of late interphase *Dictyostelium discoideum* amoebae with an external cyclic AMP signal. *Biophys. J.* **15**:767–75.

Robinson I. & R. Burgoyne. 1991. Characterization of distinct inositol 1,4,5-trisphosphate-sensitive and caffeine-sensitive calcium stores in digitonin-permeabilised adrenal chromaffin cells. *J. Neurochem.* **56**:1587–93.

Romond, P.C., J.-M. Guilmot & A. Goldbeter. 1994. The mitotic oscillator: Temporal self-organization in a phosphorylation–dephosphorylation enzymatic cascade. *Ber. Bunsenges. Phys. Chem.* **98**:1152–9.

Rooney, T.A., D.C. Renard, E.J. Sass & A.P. Thomas. 1991. Oscillatory cytosolic calcium waves independent of stimulated inositol 1,4,5-trisphosphate formation in hepatocytes. *J. Biol. Chem.* **266**:12272–82.

Rooney, T.A., E.J. Sass & A.P. Thomas. 1989. Characterization of cytosolic calcium oscillations induced by phenylephrine and vasopressin in single fura-2-loaded hepatocytes. *J. Biol. Chem.* **264**:17131–41.

Rooney, T.A., E.J. Sass & A.P. Thomas. 1990. Agonist-induced cytosolic calcium oscillations originate from a specific locus in single hepatocytes. *J. Biol. Chem.* **265**:10792–6.

Roos, W. & G. Gerisch. 1976. Receptor-mediated adenylate cyclase activation in *Dictyostelium discoideum. FEBS Lett.* **68**:170–2.

Roos, W., V. Nanjundiah, D. Malchow & G. Gerisch. 1975. Amplification of cyclic-AMP signals in aggregating cells of *Dictyostelium discoideum. FEBS Lett.* **53**:139–42.

Roos, W., C. Scheidegger & G. Gerisch. 1977. Adenylate cyclase oscillations as signals for cell aggregation in *Dictyostelium discoideum. Nature* **266**:259–61.

Rose, R.M. & P.R. Benjamin. 1981. Interneuronal control of feeding in the pond snail *Lymnea stagnalis*. II. The interneuronal mechanisms generating feeding cycles. *J. Exp. Biol.* **92**:202–28.

Rose, R.M. & J.L. Hindmarsh. 1985. A model for a thalamic neuron. *Proc. R. Soc. Lond. B* **225**:161–93.

Rose, R.M. & J.L. Hindmarsh. 1989a. The assembly of ionic currents in a thalamic neuron. I. The three-dimensional model. *Proc. R. Soc. Lond. B* **237**:267–88.

Rose, R.M. & J.L. Hindmarsh. 1989b. The assembly of ionic currents in a thalamic neuron. II. The stability and state diagram. *Proc. R. Soc. Lond. B* **237**:289–312.

Rose, R.M. & J.L. Hindmarsh. 1989c. The assembly of ionic currents in a thalamic neuron. III. The seven-dimensional model. *Proc. R. Soc. Lond. B* **237**:313–34.

Rössler, O.E. 1976. Chaotic behavior in simple reaction systems. *Z. Naturforsch.* **31a**:259–64.

Rossomando, E.F. & M. Sussman. 1973. A 5'-adenosine monophosphate-dependent adenylate cyclase and an adenosine 3':5'-cyclic monophosphate-dependent adenosine triphosphate pyrophosphohydrolase in *Dictyostelium discoideum*. *Proc. Natl. Acad. Sci. USA* **70**:1254–7.

Roth, B.J., S.V. Yagodin, L. Holtzclaw & J.T. Russell. 1995. A mathematical model of agonist-induced propagation of calcium waves in astrocytes. *Cell Calcium* **17**:53–64.

Roux, J.C. 1983. Experimental studies of bifurcations leading to chaos in the Belousov–Zhabotinsky reaction. *Physica* **7D**:57–68.

Roux, J.C. 1993. Dynamical systems theory illustrated: chaotic behavior in the Belousov–Zhabotinsky reaction. In: *Chaos in Chemistry and Biochemistry*. R.J. Field & L. Györgyi, eds. World Scientific, Singapore, pp. 21–46.

Roux, J.C., R.H. Simoyi & H.L. Swinney. 1983. Observation of a strange attractor. *Physica* **8D**:257–66.

Rubin, M.M. & J.P. Changeux. 1966. On the nature of allosteric transitions: Implications of non-exclusive ligand binding. *J. Mol. Biol.* **21**:265–74.

Ruelle, D. 1989. *Chaotic Evolution and Strange Attractors*. Cambridge Univ. Press, Cambridge.

Ruelle, D. 1994. Where can one hope to profitably apply the ideas of chaos? *Physics Today* **47** (July issue):24–30.

Rul'kov, N.F., A.R. Volkovskii, A. Rodriguez-Lozano, E. Del Rio & M.G. Velarde. 1992. Mutual synchronization of chaotic self-oscillators with dissipative coupling. *Int. J. Bif. Chaos* **2**:669–76.

Rusak, B. & I. Zucker. 1979. Neural regulation of circadian rhythms. *Physiol. Rev.* **59**:449–526.

Russell, P. & P. Nurse. 1986. $cdc25^+$ functions as a mitotic inducer in the mitotic control of fission yeast. *Cell* **45**:145–53.

Russell, P. & P. Nurse. 1987. Negative regulation of mitosis by $wee1^+$, a gene encoding a protein kinase homolog. *Cell* **49**:559–67.

Sachs, F., F. Qin & P. Palade. 1995. Models of $Ca^{2+}$ release channel adaptation. *Science* **267**:2010–11.

Saez, L., M.W. Young, M.K. Baylies, G. Gasic, T.A. Bargiello & D.C. Spray. 1992. *Per* – no link to gap junctions. *Nature* **360**:542.

Sagata, N., N. Watanabe, G.F. Vande Woude & Y. Ikawa. 1989. The c-*mos* proto-oncogene product is a cytostatic factor responsible for meiotic arrest in vertebrate eggs. *Nature* **342**: 512–17.

Sage, S.O. & T.J. Rink. 1987. The kinetics of changes in intracellular calcium concentration in fura-2-loaded human platelets. *J. Biol. Chem.* **262**:16364–9.

Sanchez-Bueno, A., C.J. Dixon, N.M. Woods, K.S.R. Cuthbertson & P.H. Cobbold. 1990. Inhibitors of protein kinase C prolong the falling phase of each free-calcium transient in a hormone-stimulated hepatocyte. *Biochem. J.* **268**:627–32.

Sanderson, M.J., A.C. Charles & E.R. Dirksen. 1990. Mechanical stimulation and intercellular communication increases intercellular $Ca^{2+}$ in epithelial cells. *Cell Regul.* **1**:585–96.

Santoro, N., M. Filicori & W.F. Crowley, Jr. 1986. Hypogonadotropic disorders in men and women: diagnosis and therapy with pulsatile gonadotropin-releasing hormone. *Endocr. Rev.* **7**:11–23.

Sassone-Corsi, P. 1994. Rhythmic transcription and autoregulatory loops: Winding up the biological clock. *Cell* **78**:361–4.

Schaap, P., T.M. Konijn & P.J.M. Van Haastert. 1984. cAMP pulses coordinate morphogenesis movement during fruiting body formation of *Dictyostelium minutum*. *Proc. Natl. Acad. Sci. USA* **81**:2122–6.

Schaap, P. & M. Wang. 1984. The possible involvement of oscillatory cAMP signaling in multi-cellular morphogenesis of the cellular slime molds. *Dev. Biol.* **105**:470–8.

Schaller, K.L., B.H. Leichtling, I.H. Majerfeld, C. Woffendin, E. Spitz, S. Kakinuma & H.V. Rickenberg. 1984. Differential cellular distribution of cAMP-dependent protein kinase during development of *Dictyostelium discoideum*. *Proc. Natl. Acad. Sci. USA* **81**:2127–31.

Schally, A.V., A. Arimura, A.J. Kastin, H. Matsuo, Y. Baba, T.W. Redding, R.M.G. Nair & L. Debeljuk. 1971. Gonadotropin releasing hormone: One polypeptide regulates secretion of luteinizing and follicle-stimulating hormones. *Science* **173**:1036–8.

Schellenberger, W., K. Eschrich & E. Hofmann. 1978. Dynamic properties of in vitro systems containing phosphofructokinase. *Acta Biol. Med. Germ.* **37**:1425–41.

Schiff, S.J., K. Jerger, D.H. Duong, T. Chang, M.L. Spano & W.L. Ditto. 1994. Controlling chaos in the brain. *Nature* **370**:615–20.

Schimz, A. & E. Hildebrand. 1992. Nonrandom structures in the locomotor behavior of *Halobacterium*: A bifurcation route to chaos? *Proc. Natl. Acad. Sci. USA* **89**:457–60.

Schirmer, T. & P.R. Evans. 1990. Structural basis of the allosteric behaviour of phosphofructokinase. *Nature* **343**:140–5.

Schmitz, R.A., K.R. Graziani & J.L. Hudson. 1977. Experimental evidence of chaotic states in the Belousov–Zhabotinskii reaction. *J. Chem. Phys.* **67**:3040–4.

Schulmeister, T. & E.E. Sel'kov. 1978. Folded limit cycles and quasi-stochastic self-oscillations in a third-order model of an open biochemical system. *Studia Biophys.* **72**:111–12 and Microfiche 1/24–37.

Schweiger, H.G., R. Hartwig & M. Schweiger. 1986. Cellular aspects of circadian rhythms. *J. Cell Sci. Suppl.* **4**:181–200.

Schweiger, H.G. & M. Schweiger. 1977. Circadian rhythms in unicellular organisms: An endeavor to explain the molecular mechanism. *Int. Rev. Cytol.* **51**:315–42.

Schwob, E., T. Böhm, M.D. Mendenhall & K. Nasmyth. 1994. The B-type cyclin kinase inhibitor p40$^{SIC1}$ controls the G1 to S transition in *S. cerevisiae. Cell* **79**:233–44.

Scott, S.K. 1991. *Chemical Chaos.* Oxford Univ. Press, Oxford.

Scott, S.K. & K. Showalter. 1992. Simple and complex propagating reaction-diffusion fronts. *J. Phys. Chem.* **96**:8702–11.

Segel, L.A. 1984. *Modeling Dynamic Phenomena in Molecular and Cellular Biology.* Cambridge Univ. Press, Cambridge.

Segel, L.A. 1988. On the validity of the steady state assumption of enzyme kinetics. *Bull. Math. Biol.* **50**:579–93.

Segel, L.A. & A. Goldbeter. 1994. Scaling in biochemical kinetics: Dissection of a relaxation oscillator. *J. Math. Biol.* **32**:147–60.

Segel, L.A., A. Goldbeter, P.N. Devreotes & B.E. Knox. 1985. A model for sensory response and exact adaptation mediated by receptor modification. In: *Sensing and Response in Microorganisms.* M. Eisenbach & M. Balaban, eds. Elsevier, Amsterdam, pp. 175–83.

Segel, L.A., A. Goldbeter, P.N. Devreotes & B.E. Knox. 1986. A mechanism for exact sensory adaptation based on receptor modification. *J. Theor. Biol.* **120**:151–79.

Sehgal, A., J.L. Price, B. Man & M.W. Young. 1994. Loss of circadian behavioral rhythms and *per* RNA oscillations in the *Drosophila* mutant *timeless. Science* **263**:1603–6.

Sel'kov, E.E. 1968a. Self-oscillations in glycolysis. 1. A simple kinetic model. *Eur. J. Biochem.* **4**:79–86.

Sel'kov, E.E. 1968b. Self-oscillations in glycolysis. Simple single-frequency model. *Mol. Biol. (Moscow)* **2**:208–21 (Engl. transl.).

Sel'kov, E.E. 1970. Two alternative autooscillating stationary states in thiol metabolism – two alternative types of cell reproduction: Normal and malignant one. *Biofyzika* **15**:1065–73.

Sel'kov, E.E. 1972a . Nonlinearity of multienzyme systems. In: *Analysis and Simulation of Biochemical Systems.* B. Hess & H.C. Hemker, eds. North-Holland, Amsterdam, pp. 145–61.

Sel'kov, E.E. 1972b. Nonlinear theory of regulation of the key step of glycolysis. *Studia Biophys.* **33**:167–76.

Selverston, A.I. & M. Moulins. 1985. Oscillatory neural networks. *Annu. Rev. Physiol.* **47**:29–48.

Sen, A.K. & W. Liu. 1990. Dynamic analysis of genetic control and regulation of amino acid synthesis: The tryptophan operon in *Escherichia coli. Biotechnol. Bioeng.* **35**:185–94.

Shaffer, B.M. 1956. Acrasin, the chemotactic agent in cellular slime moulds. *J. Exp. Biol.* **33**:645–57.

Shaffer, B.M. 1962. The acrasina. *Adv. Morphog.* **2**:109–82.

Shaffer, B.M. 1975. Secretion of cyclic AMP induced by cyclic AMP in the cellular slime mould *Dictyostelium discoideum. Nature* **255**:549–52.

Shangold, G.A., S.N. Murphy & R.J. Miller. 1988. Gonadotropin-releasing hormone-induced $Ca^{2+}$ transients in single identified gonadotropes requires both intracellular $Ca^{2+}$ mobilization and $Ca^{2+}$ influx. *Proc. Natl. Acad. Sci. USA* **85**:6566–70.

Shen, P. & R. Larter. 1995. Chaos in intracellular $Ca^{2+}$ oscillations in a new model for non-excitable cells. *Cell Calcium* **17**:225–32.

Sherman, A. & J. Rinzel. 1992. Rhythmogenic effects of weak electrotonic coupling in neuronal models. *Proc. Natl. Acad. Sci. USA* **89**:2471–74.

Sherman, A., J. Rinzel & J. Keizer. 1988. Emergence of organized bursting in clusters of pancreatic β-cells by channel sharing. *Biophys. J.* **54**:411–25.

Sherr, C.J. 1994. G1 phase progression: Cycling on cue. *Cell* **79**:551–5.

Shinbrot, T., C. Grebogi, E. Ott & J.A. Yorke. 1993. Using small pertubations to control chaos. *Nature* **363**:411–17.

Shoshan-Barnatz, V., G. Zhang, L. Garretson & N. Kraus-Friedman. 1990. Distinct ryanodine- and inositol 1,4,5-trisphosphate-binding sites in hepatic microsomes. *Biochem. J.* **268**:699–705.

Siegert, F. & C. Weijer. 1989. Digital image processing of optical density wave propagation in *Dictyostelium discoideum* and analysis of the effects of caffeine and ammonia. *J. Cell Sci.* **93**:325–35.

Siegert, F. & C. Weijer. 1991. Analysis of optical density wave propagation and cell movement in the cellular slime mould *Dictyostelium discoideum*. *Physica* **49D**:224–32.

Siegert, F. & C. Weijer. 1992. Three-dimensional scroll waves organize *Dictyostelium* slugs. *Proc. Natl. Acad. Sci. USA* **89**:6433–7.

Simon, M.-N., D. Driscoll, R. Mutzel, D. Part, J. Williams & M. Véron. 1989. Overproduction of the regulatory subunit of the cAMP-dependent protein kinase blocks the differentiation of *Dictyostelium discoideum*. *EMBO J.* **8**:2039–44.

Singer, W. 1993. Synchronization of cortical activity and its putative role in information processing and learning. *Annu. Rev. Physiol.* **55**:349–74.

Sinha, S. & R. Ramaswamy. 1988. Complex behaviour of the repressible operon. *J. Theor. Biol.* **132**:307–18.

Siwicki, K.K., C. Eastman, G. Petersen, M. Rosbash & J.C. Hall. 1988. Antibodies to the *period* gene product of *Drosophila* reveal diverse tissue distribution and rhythmic changes in the visual system. *Neuron* **1**:141–50.

Skarda, C.A. & W.J. Freeman. 1987. How brain makes chaos in order to make sense of the world. *Behav. Brain Sci.* **10**:161–95.

Smith, M.A., M.H. Perrin & W.W. Vale. 1983. Desensitization of cultured pituitary cells to gonadotropin-releasing hormone: Evidence for a post-receptor mechanism. *Mol. Cell. Endocr.* **30**:85–96.

Smith, M.A. & W.W. Vale. 1981. Desensitization to gonadotropin-releasing hormone observed in superfused pituitary cells on cytodex beads. *Endocrinology* **108**:752–9.

Smith, R.F. & R.J. Konopka. 1982. Effects of dosage alterations at the *per* locus on the period of the circadian clock of *Drosophila*. *Mol. Gen. Genet.* **185**:30–6.

Smith, R.G., K. Cheng, W.R. Schoen, S.S. Pong, G. Hickey, T. Jacks, B. Butler, W.W.S. Chan, L.Y.P. Chaung, F. Judith, J. Taylor, M.J. Wyvratt & M.H. Fisher. 1993. A nonpeptidyl growth hormone secretagogue. *Science* **260**:1640–3.

Smolen, P. 1995. A model for glycolytic oscillations based on skeletal muscle phosphofructokinase kinetics. *J. Theor. Biol.* **174**:137–48.

Smolen, P. & J. Keizer. 1992. Slow voltage inactivation of $Ca^{2+}$ currents and bursting mechanisms for the mouse pancreatic beta-cell. *J. Membrane Biol.* **127**:9–19.

Smythe, C. & J.W. Newport. 1992. Coupling of mitosis to the completion of the S phase in Xenopus occurs via modulation of the tyrosine kinase that phosphorylates $p34^{cdc2}$. *Cell* **68**:787–97.

Snaar-Jagalska, B.E. & P.J.M. Van Haastert. 1990. Pertussis toxin inhibits cAMP-induced desensitization of adenylate cyclase in *Dictyostelium discoideum*. *Mol. Cell. Biochem.* **92**:177–89.

Sneyd, J., A.C. Charles & M.J. Sanderson. 1994. A model for the propagation of intercellular calcium waves. *Am. J. Physiol.* **266**:C293–C302.

Sneyd, J., S. Girard & D. Clapham. 1993. Calcium wave propagation by calcium-induced release: An unusual excitable system. *Bull. Math. Biol.* **55**:315–44.

Solomon, M.J., T. Lee & M.W. Kirschner. 1992. Role of phosphorylation in p34$^{cdc2}$ activation: identification of an activating kinase. *Mol. Biol. Cell* **3**:13–27.

Somogyi, R. & J.W. Stucki. 1991. Hormone-induced calcium oscillations in liver cells can be explained by a simple model. *J. Biol. Chem.* **266**:11068–77.

Spangler, R.A. & F.M. Snell. 1961. Sustained oscillations in a catalytic chemical system. *Nature* **191**:457–61.

Sparrow, C. 1982. *The Lorenz Equations: Bifurcation, Chaos, and Strange Attractors*. Springer, New York.

Sporn, M.B. & G.T. Todaro. 1980. Autocrine secretion and malignant transformation of cells. *New Engl. J. Med.* **303**:878–80.

Springer, M.S., M.F. Goy & J. Adler. 1979. Protein methylation in behavioral control mechanisms and in signal transduction. *Nature* **280**:279–84.

Stehle, J.H., N.S. Foulkes, C.A. Molina, V. Simonneaux, P. Pévet & P. Sassone-Corsi. 1993. Adrenergic signals direct rhythmic expression of transcriptional repressor CREM in the pineal gland. *Nature* **365**:314–20.

Steinbock, O., H. Hashimoto & S.C. Müller. 1991. Quantitative analysis of periodic chemotaxis in aggregation patterns of *Dictyostelium discoideum*. *Physica* **49D**:233–9.

Steinmetz, C.G., T. Geest & R. Larter. 1993. Universality in the peroxidase–oxidase reaction: Period doublings, chaos, period three and unstable limit cycles. *J. Phys. Chem.* **97**:5649–53.

Stephan, F.K. & I. Zucker. 1972. Circadian rhythms in drinking behavior and locomotor activity are eliminated by suprachiasmatic lesions. *Proc. Natl. Acad. Sci. USA* **54**:1521–7.

Steriade, M., E.G. Jones & R.R. Llinas. 1990. *Thalamic Oscillations and Signaling*. Wiley Interscience, New York.

Stern, M.D., M.C. Capogrossi & E.G. Lakatta.1988. Spontaneous calcium release from the sarcoplasmic reticulum in myocardial cells: Mechanisms and consequences. *Cell Calcium* **9**:247–58.

Stojilkovic, S.S. & K.J. Catt. 1992. Calcium oscillations in anterior pituitary cells. *Endocrine Rev.* **13**:256–80.

Stojilkovic, S.S., M. Kukuljan, M. Tomic, E. Rojas & K.J. Catt. 1993. Mechanism of agonist-induced [Ca$^{2+}$]$_i$ oscillations in pituitary gonadotrophs. *J. Biol. Chem.* **268**:7713–20.

Stojilkovic, S.S., E. Rojas, A. Stutzin, S.-I. Izumi & K.J. Catt. 1989. Desensitization of pituitary gonadotropin secretion by agonist-induced inactivation of voltage-sensitive calcium channels. *J. Biol. Chem.* **264**:10939–42.

Stokes, C.L. & J. Rinzel. 1993. Diffusion of extracellular K$^+$ can synchronize bursting oscillations in a model islet of Langerhans. *Biophys. J.* **65**:597–607.

Strausfeld, U., A. Fernandez, J.P. Capony, F. Girard, N. Lautredou, J. Derancourt, J.-C. Labbé & N.J.C. Lamb. 1994. Activation of p34$^{cdc2}$ protein kinase by microinjection of human cdc25C into mammalian cells.

Requirement for prior phosphorylation of cdc25C by p34$^{cdc2}$ on sites phosphorylated at mitosis. *J. Biol. Chem.* **269**:5989–6000.

Strausfeld, U., J.C. Labbé, D. Fesquet, J.C. Cavadore, A. Picard, K. Sadhu, P. Russell & M. Dorée. 1991. Dephosphorylation and activation of a p34$^{cdc2}$/cyclin B complex *in vitro* by human CDC25 protein. *Nature* **351**:242–5.

Strogatz, S.H. 1986. The mathematical structure of the human sleep–wake cycle. *Lect. Notes Biomath.* **69**.

Strogatz, S.H. 1987. A comparative analysis of models of the human sleep-wake cycle. In: *Some Mathematical Questions in Biology. Circadian Rhythms* (Lectures on Mathematics in the Life Sciences. Vol. 19). The American Mathematical Society, Providence, Rhode Island, pp. 1–37.

Stucki, J.W. & R. Somogyi. 1994. A dialogue on Ca$^{2+}$ oscillations: An attempt to understand the essentials of mechanisms leading to hormone-induced intracellular Ca$^{2+}$ oscillations in various kinds of cell on a theoretical level. *BBA Bioenerg.* **1183**:453–72.

Sturis, J., K.S. Polonsky, E. Mosekilde & E. Van Cauter. 1991. Computer model for mechanisms underlying ultradian oscillations of insulin and glucose. *Am. J. Physiol.* **260**:E801–E809.

Swann, K. & J.P. Ozil. 1994. Dynamics of the calcium signal that triggers mammalian egg activation. *Intern. Rev. Cytol.* **152**:183–222.

Sweeney, B. 1969. *Rhythmic Phenomena in Plants.* Academic Press, New York.

Sweeney, B.M. & J.W. Hastings. 1960. Effect of temperature upon diurnal rhythms. *Cold Spring Harbor Symp. Quant. Biol.* **25**:87–104.

Swillens, S., L. Combettes & P. Champeil. 1994. Transient inositol 1,4,5-trisphosphate-induced Ca$^{2+}$ release: A model based on regulatory Ca$^{2+}$-binding sites along the permeation pathway. *Proc. Natl. Acad. Sci. USA* **91**:10074–8.

Swillens, S. & J.E. Dumont. 1977. The mobile receptor hypothesis in hormone action: A general model accounting for desensitization. *J. Cyclic Nucleot. Res.* **3**:1–10.

Swillens, S. & D. Mercan. 1990. Computer simulations of a cytosolic calcium oscillator. *Biochem. J.* **271**:835–8.

Swinney, H.L. & V.I. Krinsky (eds.) 1991. Waves and patterns in chemical and biological media. *Physica* **49D** (1&2).

Takahashi, J.S. 1992. Circadian clock genes are ticking. *Science* **258**:238–40.

Takahashi, J.S. 1993. Circadian-clock regulation of gene expression. *Curr. Opin. Genet. Dev.* **3**:301–9.

Takahashi, J.S., H. Hamm & M. Menaker. 1980. Circadian rhythms of melatonin release from individual superfused chicken pineal glands *in vitro. Proc. Natl. Acad. Sci. USA* **77**:2319–22.

Takahashi, J.S., J.M. Kornhauser, C. Koumenis & A. Eskin. 1993. Molecular approaches to understanding circadian oscillations. *Annu. Rev. Physiol.* **55**:729–53.

Takahashi, J.S., N. Murakami, S.S. Nikaido, B.L. Pratt & L.M. Robertson. 1989. The avian pineal, a vertebrate model system of the circadian oscillator: Cellular regulation of circadian rhythms by light, second messengers, and macromolecular synthesis. *Recent Progr. Horm. Res.* **45**:279–352.

Takamatsu, T. & W.G. Wier. 1990. Calcium waves in mammalian heart: Quantification of origin, magnitude, waveform, and velocity. *FASEB J.* **4**:1519–25.

Takesue, S. & K. Kaneko. 1984. Fractal basin structure. *Progr. Theor. Phys.* **71**:35–49.

Takeuchi, I., T. Noce & M. Tasaka. 1986. Prestalk and prespore differentiation during development of *Dictyostelium discoideum*. *Curr. Top. Dev. Biol.* **20**:243–56.

Tang, Y. & H.G. Othmer. 1994a. A G-protein based model of adaptation in *Dictyostelium discoideum*. *Math. Biosci.* **120**:25–76.

Tang, Y. & H.G. Othmer. 1994b. A model of calcium dynamics in cardiac myocytes based on the kinetics of ryanodine-sensitive calcium channels. *Biophys. J.* **67**:2223–35.

Tang, Z., T.R. Coleman & W.G. Dunphy. 1993. Two distinct mechanisms for negative regulation of the Wee1 protein kinase. *EMBO J.* **12**:3427–36.

Tannenbaum, G.S. & N. Ling. 1984. The interrelationship of growth hormone (GH)-releasing factor and somatostatin in generation of the ultradian rhythm of GH secretion. *Endocrinology* **115**:1952–7.

Tannenbaum, G.S. & J.B. Martin. 1976. Evidence for an endogenous ultradian rhythm governing growth hormone secretion in the rat. *Endocrinology* **98**:562–70.

Taylor, C.T. 1994. Calcium signals and human oocyte activation: Implications for assisted conception. *Hum. Reprod.* **9**:980–3.

Taylor, W.R., J.C. Dunlap & J.W. Hastings. 1982a. Inhibitors of protein synthesis on 80s ribosomes phase-shift the *Gonyaulax* clock. *J. Exp. Biol.* **97**:121–36.

Taylor, W., R. Krasnow, J.C. Dunlap, H. Broda & J.W. Hastings. 1982b. Critical pulses of anisomycin drive the circadian oscillator in *Gonyaulax* towards its singularity. *J. Comp. Physiol.* **148**:11–25.

Termonia, Y. & J. Ross. 1981. Oscillations and control features in glycolysis: Numerical analysis of a comprehensive model. *Proc. Natl. Acad. Sci. USA* **78**:2952–56.

Tesarik, J. 1994. How the spermatozoon awakens the oocyte: Lessons from intracytoplasmic sperm injection. *Hum. Reprod.* **9**:977–8.

Tesarik, J., M. Sousa & J. Testart. 1994. Human oocyte activation after intracytoplasmic sperm injection. *Hum. Reprod.* **9**:511–18.

Thalabard, J.C., D. Rotten, M.C. Colin & J.P. Gautray. 1984. Rythmes et reproduction. *Pour la Science* **77**:92–102.

Theeuwes, F. 1989. Triggered, pulsed and programmed drug delivery. In: *Novel Drug Delivery and its Therapeutic Application*. L.F. Prescott & W.S. Nimmo, eds. John Wiley & Sons, New York, pp. 323–40.

Theeuwes, F., S.I. Yum, R. Haak & P. Wong. 1991. Systems for triggered, pulsed, and programmed drug delivery. In *Temporal Control of Drug Delivery*. W.J.M. Hrushesky, R. Langer & F. Theeuwes, eds. *Ann. N.Y. Acad. Sci.* **618**:428–40.

Theibert, A. & P.N. Devreotes. 1983. Cyclic 3′,5′-AMP relay in *Dictyostelium discoideum*: Adaptation is independent of activation of adenylate cyclase. *J. Cell Biol.* **97**:173–7.

Thomas, A.P., D.C. Renard & T.A. Rooney. 1991. Spatial and temporal organization of calcium signalling in hepatocytes. *Cell Calcium* **12**:111–26.

Thomas, R. 1978. Logical analysis of systems comprising feedback loops. *J. Theor. Biol.* **73**:631–56.

Thomas, R. (ed.) 1979. Kinetic logic. A Boolean approach to the analysis of complex regulatory systems. *Lect. Notes Biomath.* **29**.

Thomas, R. 1991. Regulatory networks seen as asynchronous automata. *J. Theor. Biol.* **153**:1–23.

Thomas, R. & R. d'Ari. 1990. *Biological Feedback.* CRC Press, Boca Raton, FL.

Thron, C.D. 1991. The secant condition for instability in biochemical feedback control. I. The role of cooperativity and saturability. *Bull. Math. Biol.* **53**:383–401.

Thron, C.D. 1994. Theoretical dynamics of the cyclin B-MPF system: A possible role for p13$^{suc1}$. *Biosystems* **32**:97–109.

Tomchik, K.J. & P.N. Devreotes. 1981. Adenosine 3′,5′-monophosphate waves in *Dictyostelium discoideum.* A demonstration by isotope dilution fluorography. *Science* **212**:443–6.

Tomita, K. & H. Daido. 1980. Possibility of chaotic behaviour and multi-basins in forced glycolytic oscillations. *Phys. Lett.* **79A**:133–7.

Tomita, K. & I. Tsuda. 1979. Chaos in the Belousov–Zhabotinsky reaction in a flow system. *Phys. Lett.* **71A**:489–92.

Tornheim, K. 1979. Oscillations of the glycolytic pathway and the purine nucleotide cycle. *J. Theor. Biol.* **79**:491–541.

Tornheim, K. 1988. Fructose 2,6-bisphosphate and glycolytic oscillations in skeletal muscle extracts. *J. Biol. Chem.* **263**:2619–24.

Tornheim, K., V. Andrés & V. Schulz. 1991. Modulation by citrate of glycolytic oscillations in skeletal muscle extracts. *J. Biol. Chem.* **266**:15675–8.

Tornheim, K. & J.M. Lowenstein. 1974. The purine nucleotide cycle. IV. Interactions with oscillations of the glycolytic pathway in muscle extracts. *J. Biol. Chem.* **249**:3241–7.

Tornheim, K. & J.M. Lowenstein. 1975. The purine nucleotide cycle. V. Control of phosphofructokinase and glycolytic oscillations in muscle extracts. *J. Biol. Chem.* **250**:6304–14.

Törnquist K. 1992. Evidence for receptor-mediated calcium entry and refilling of intracellular calcium stores in FRTL-5 rat thyroid cells. *J. Cell. Physiol.* **150**:90–8.

Touitou, Y. & E. Haus (eds.) 1992. *Biologic Rhythms in Clinical and Laboratory Medicine.* Springer, Berlin.

Tracqui, P. 1993. Homoclinic tangencies in an autocatalytic model of interfacial processes at the bone surface. *Physica* **62D**:275–89.

Tracqui, P., A.M. Perault-Staub, G. Milhaud & J.F. Staub. 1987. Theoretical study of a two-dimensional autocatalytic model for calcium dynamics at the extracellular fluid–bone interface. *Bull. Math. Biol.* **49**:597–13.

Tregear, R.T., A.P. Dawson & R.F. Irvine. 1991. Quantal release of $Ca^{2+}$ from intracellular stores by Ins$P_3$: Tests of the concept of control of $Ca^{2+}$ release by intraluminal $Ca^{2+}$. *Proc. R. Soc. Lond. B* **243**:263–8.

Tribe, R.M., M.L. Borin & M.P. Blaustein. 1994. Functionally and spatially distinct $Ca^{2+}$ stores are revealed in cultured vascular smooth muscle cells. *Proc. Natl. Acad. Sci. USA* **91**:5908–12.

Troy, W.C. 1978. The bifurcation of periodic solutions in the Hodgkin–Huxley equations. *Quart. Appl. Math.* **36**:73–83.

Tse, A., F.W. Tse, W. Almers & B. Hille. 1993. Rhythmic exocytosis stimulated by GnRH-induced calcium oscillations in rat gonadotropes. *Science* **260**:82–4.

Tsien, R.W., R.S. Kass & R. Weingart. 1979. Cellular and subcellular mechanisms of cardiac pacemaker oscillations. *J. Exp. Biol.* **81**:205–15.

Tsien, R.W. & R.Y. Tsien. 1990. Calcium channels, stores, and oscillations. *Annu. Rev. Cell Biol.* **6**:715–60.

Tsunoda, Y. 1991. Oscillatory $Ca^{2+}$ signalling and its cellular function. *New Biologist* **3**:3–17.

Turek, F.W. 1985. Circadian neural rhythms in mammals. *Annu. Rev. Physiol.* **47**:49–64.

Turek, F.W. 1987. Pharmacological probes of the mammalian circadian clock: Use of the phase response curve approach. *Trends Pharmacol. Sci.* **8**:212–17.

Turek, F.W., J. Swann & D.J. Earnest. 1984. Role of the circadian system in reproductive phenomena. *Rec. Progr. Horm. Res.* **40**:143–83.

Turing, A.M. 1952. The chemical basis of morphogenesis. *Phil. Trans. R. Soc. Lond. B* **237**:37–72.

Tyson, J.J. 1973. Some further studies of nonlinear oscillations in chemical systems. *J. Chem. Phys.* **58**:3919–30.

Tyson, J.J. 1976. The Belousov–Zhabotinskii reaction. *Lect. Notes Biomath.* **10**.

Tyson, J.J. 1977. Analytic representation of oscillations, excitability, and traveling waves in a realistic model of the Belousov–Zhabotinskii reaction. *J. Chem. Phys.* **66**:905–15.

Tyson, J.J. 1983. Periodic enzyme synthesis and oscillatory repression: Why is the period of oscillation close to the cell cycle time? *J. Theor. Biol.* **103**:313–28.

Tyson, J.J. 1989. Cyclic-AMP waves in *Dictyostelium*: Specific models and general theories. In: *Cell to Cell Signalling: From Experiments to Theoretical Models*, A. Goldbeter, ed. Academic Press, London, pp. 521–37.

Tyson, J.J. 1991. Modeling the cell division cycle: cdc2 and cyclin interactions. *Proc. Natl. Acad. Sci. USA* **88**:7328–32.

Tyson, J.J., K.A. Alexander, V.S. Manoranjan & J.D. Murray. 1989. Spiral waves of cyclic AMP in a model of slime mold aggregation. *Physica* **34D**:193–207.

Tyson, J.J. & P.C. Fife. 1980. Target patterns in a realistic model of the Belousov–Zhabotinsky reaction. *J. Chem. Phys.* **73**:2224–37.

Tyson, J. & S. Kauffman. 1975. Control of mitosis by a continuous biochemical oscillator: synchronization of spatially inhomogeneous oscillations. *J. Math. Biol.* **1**:289–310.

Tyson, J.J. & J.P. Keener. 1988. Singular perturbation theory of traveling waves in excitable media (A review). *Physica* **32D**:327–61.

Tyson, J.J. & J.D. Murray. 1989. Cyclic AMP waves during aggregation of *Dictyostelium* amoebae. *Development* **106**:421–6.

Tyson, J. & W. Sachsenmaier. 1978. Is nuclear division in *Physarum* controlled by a continuous limit cycle oscillator? *J. Theor. Biol.* **73**:723–38.

Tyson, J. & W. Sachsenmaier. 1984. The control of nuclear division in *Physarum polycephalum*. In: *Cell Cycle Clocks*. L.N. Edmunds Jr, ed. Marcel Dekker, New York and Basel, pp. 253–70.

Tyson, J.J. & H.G. Othmer. 1978. The dynamics of feedback control circuits in biochemical pathways. In: *Progress in Theoretical Biology*. Vol. 5. F. Snell & R. Rosen, eds. Academic Press, New York, pp. 2–62.

Valdeolmillos, M., R.M. Santos, D. Contreras, B. Soria & L.M. Rosario. 1989. Glucose-induced oscillations of intracellular $Ca^{2+}$ concentration resembling bursting electrical activity in single mouse islets of Langerhans. *FEBS Lett.* **259**:19–23.

Van Cauter, E. 1987. Pulsatile ACTH secretion. In: *Episodic Hormone Secretion: From Basic Science to Clinical Application.* T.O.F. Wagner & M. Filicori, eds. TM-Verlag, Hameln, pp. 65–75.

Van Cauter, E. & J. Aschoff. 1989. Endocrine and other biological rhythms. In: *Endocrinology*, Vol. 3, L.J. DeGroot, ed. W.B. Saunders, Philadelphia, PA, pp. 2658–705.

van der Pol, B. & J. van der Markt. 1928. The heart beat considered as a relaxation oscillation, and an electrical model of the heart. *Phil. Mag.* 6:763–75.

Van Haastert, P.J.M. 1984. Guanine nucleotides modulate cell surface cAMP binding-sites in membranes from *Dictyostelium discoideum. Biochem. Biophys. Res. Commun.* 124:597–604.

Vanden Driessche, T. 1980. Circadian rhythmicity: General properties – as exemplified by *Acetabularia* – and hypotheses on its cellular mechanism. *Arch. Biol.* 91:49–76.

Varghese, A. & R.L. Winslow. 1994. Dynamics of abnormal pacemaking activity in cardiac Purkinje fibers. *J. Theor. Biol.* 168:407–20

Vaughan, R. & P.N. Devreotes. 1988. Ligand-induced phosphorylation of the cAMP receptor from *Dictyostelium discoideum. J. Biol. Chem.* 263:14538–43.

Veneriatos, D. & A. Goldbeter. 1979. Allosteric oscillatory enzymes: Influence of the number of protomers on metabolic periodicities. *Biochimie* 61:1247–56.

Vitaterna, M.H., D.P. King, A.-M. Chang, J.M. Kornhauser, P.L. Lowrey, J.D. McDonald, W.F. Dove, L.H. Pinto, F.W. Turek & J.S. Takahashi. 1994. Mutagenesis and mapping of a mouse gene, *clock*, essential for circadian behavior. *Science* 264:719–25.

Volkov, E.I. & M.N. Stolyarov. 1991. Birhythmicity in a system of two coupled identical oscillators. *Phys. Lett. A* 159:61–66.

Volterra, V. 1926. Fluctuations in the abundance of a species considered mathematically. *Nature* 118:558–60.

Von Klitzing, L. & A. Betz. 1970. Metabolic control in flow systems. I. Sustained glycolytic oscillations in yeast suspensions under continuous substrate infusion. *Arch. Mikrobiol.* 71:220–5.

von Schilcher, F. 1976. The function of sine song and pulse song in the courtship of *Drosophila melanogaster. Anim. Behav.* 24:622–5.

Vosshall, L.B., J.L. Price, A. Sehgal, L. Saez & M.W. Young. 1994. Block in nuclear localization of *period* protein by a second clock mutation, *timeless. Science* 263:1606–9.

Wagner, J. & J. Keizer. 1994. Effects of rapid buffers on $Ca^{2+}$ diffusion and $Ca^{2+}$ oscillations. *Biophys. J.* 67:447–56.

Wagner, T.O.F. (ed.) 1985. *Pulsatile LHRH Therapy of the Male.* TM-Verlag, Hameln.

Wagner, T.O.F. & M. Filicori (eds.) 1987. *Episodic Hormone Secretion: From Basic Science to Clinical Application.* TM-Verlag, Hameln.

Wagner, T.O.F., G. Wenzel, C. Dette, G. Daehne, A. Goehring, O. Vosmann, I. Messerschmidt, J. Bruns, R. Wünsch & A. von zur Muhlen. 1989. Episodic hormone secretion in the regulation and pathophysiology of the reproductive system in man. In: *Cell to Cell Signalling: From Experiments to Theoretical Models*, A. Goldbeter, ed. Academic Press, London, pp. 407–14.

Wakui, M., Y.V. Osipchuk & O.H. Petersen. 1990. Receptor-activated cytoplasmic $Ca^{2+}$ spiking mediated by inositol trisphosphate is due to $Ca^{2+}$-induced $Ca^{2+}$ release. *Cell* **63**:1025–32.

Wakui, M., B.V.L. Potter & O.H. Petersen. 1989. Pulsatile intracellular calcium release does not depend on fluctuations in inositol trisphosphate concentration. *Nature* **339**:317–20.

Walter, C. 1970. The occurrence and the significance of limit cycle behavior in controlled biochemical systems. *J. Theor. Biol.* **27**:259–72.

Walton P., J. Airey, J. Sutko, C. Beck, G. Mignery, T. Südhof, T. Deerinck & M. Ellisman. 1991. Ryanodine and inositol trisphosphate receptors coexist in avian cerebellar Purkinje neurons. *J. Cell Biol.* **113**:1145–57.

Waltz, D. & S.R. Caplan. 1987. Consequence of detailed balance in a model for sensory adaptation based on ligand-induced receptor modification. *Proc. Natl. Acad. Sci. USA* **84**:6152–6.

Wang, J.Y.J. 1992. Oncoprotein phosphorylation and cell cycle control. *Biochim. Biophys. Acta* **1114**:179–92.

Watanabe, N., G.F. Vande Woude, Y. Ikawa & N. Sagata. 1989. Specific proteolysis of the c-*mos* proto-oncogene product by calpain on fertilization of *Xenopus* eggs. *Nature* **342**:505–11.

Waxman, D.J., N.A. Pampori, P.A. Ram, A.K. Agrawal & B.H. Shapiro. 1991. Interpulse interval in circulating growth hormone patterns regulates sexually dimorphic expression of hepatic cytochrome P450. *Proc. Natl. Acad. Sci. USA* **88**:6868–72.

Waxman, D.J., P.A. Ram, S.H. Park & H.K. Choi. 1995. Intermittent plasma growth hormone triggers tyrosine phosphorylation and nuclear translocation of a liver-expressed, Stat5-related DNA binding protein. Proposed role as an intracellular regulator of male-specific gene transcription. *J. Biol. Chem.* **270**:13262–70

Webb, G.F. 1990. Resonance phenomena in cell population chemotherapy models. *Rocky Mountain J. Math.* **20**:1195–1216.

Weigle, D.S. 1987. Pulsatile secretion of fuel-regulatory hormones. *Diabetes* **36**:764–75.

Weigle, D.S. & C.J. Goodner. 1986. Evidence that the physiological pulse frequency of glucagon secretion optimizes glucose production by perifused rat hepatocytes. *Endocrinology* **118**:1606–13.

Weigle, D.S., D.J. Koerker & C.J. Goodner. 1984. Pulsatile glucagon delivery enhances glucose production by perfused rat hepatocytes. *Am. J. Physiol.* **247**:E564–E568.

Welch, P.J. & J.Y.J. Wang. 1992. Coordinated synthesis and degradation of cdc2 in the mammalian cell cycle. *Proc. Natl. Acad. Sci. USA* **89**:3093–7.

Welsh, J.P., E.J. Lang, I. Sugihara & R. Llinas. 1995. Dynamic organization of motor control within the olivocerebellar system. *Nature* **374**:453–7.

Wetsel, W.C., M.M. Valença, I. Merchenthaler, Z. Liposits, F.J. Lopez, R.I. Weiner, P.L. Mellon & A. Negro-Vilar. 1992. Intrinsic pulsatile secretory activity of immortalized luteinizing hormone-releasing hormone-secreting neurons. *Proc. Natl. Acad. Sci. USA* **89**:4149–53.

Wever, R.A. 1965. A mathematical model for circadian rhythms. In: *Circadian Clocks*, J. Aschoff, ed. North Holland, Amsterdam, pp. 47–63.

Wever, R.A. 1966. The duration of re-entrainment of circadian rhythms after phase shifts of the zeitgeber. A theoretical consideration. *J. Theor. Biol.* **13**:187–201.

Wever, R.A. 1972. Virtual synchronization towards the limits of the range of entrainment. *J. Theor. Biol.* **36**:119–32.

Wever, R.A. 1987. Mathematical models of circadian one- and multi-oscillator systems. In: *Some Mathematical Questions in Biology. Circadian Rhythms* (Lectures on Mathematics in the Life Sciences, vol. 19). The American Mathematical Society, Providence, Rhode Island, pp. 205–65.

Whim, M.D. & P.E. Lloyd. 1989. Frequency-dependent release of peptide cotransmitters from identified cholinergic motor neurons in *Aplysia*. *Proc. Natl. Acad. Sci. USA* **86**:9034–8.

Wilders, R., H.J. Jongsma & A.C.G. van Ginneken. 1991. Pacemaker activity of the rabbit sinoatrial node. A comparison of mathematical models. *Biophys. J.* **60**:1202–16.

Wildt, L., A. Haüsler, G. Marshall, J.S. Hutchison, T.M. Plant, P.E. Belchetz & E. Knobil. 1981. Frequency and amplitude of gonadotropin releasing hormone stimulation and gonadotropin secretion in the rhesus monkey. *Endocrinology* **109**:376–85.

Wildt, L., G. Marshall & E. Knobil. 1980. Experimental induction of puberty in the infertile female rhesus monkey. *Science* **207**:1373–5.

Wille, J.J., C. Scheffey & S.A. Kauffman. 1977. Novel behaviour of the mitotic clock in *Physarum*. *J. Cell Sci.* **27**:91–104.

Williams, K.L. & P.C. Newell. 1976. A genetic study of aggregation in the cellular slime mould *Dictyostelium discoideum* using complementary analysis. *Genetics* **82**:287–307.

Wilson, R.C., J.S. Kesner, J.M. Kaufman, T. Uemura, T. Akema & E. Knobil. 1984. Central electrophysiologic correlates of pulsatile luteinizing hormone secretion in the rhesus monkey. *Neuroendocrinology* **39**:256–60.

Winer, L.M., M.A. Shaw & G. Baumann. 1990. Basal plasma growth hormone levels in man: New evidence for rhythmicity of growth hormone secretion. *J. Clin. Endocrinol. Metab.* **70**:1678–86.

Winfree, A.T. 1970. Integrated view of resetting a circadian clock. *J. Theor. Biol.* **28**:327–74.

Winfree, A.T. 1972a. Spiral waves of chemical activity. *Science* **175**:634–6.

Winfree, A.T. 1972b. Oscillatory glycolysis in yeast: The pattern of phase resetting by oxygen. *Arch. Biochem. Biophys.* **149**:388–401.

Winfree, A.T. 1974. Patterns of phase compromise in biological cycles. *J. Math. Biol.* **1**:73–95.

Winfree, A.T. 1980. *The Geometry of Biological Time*. Springer, New York (Reprinted as Springer Study Edition, 1990, Springer, Berlin).

Winfree, A.T. 1982. Human body clocks and the timing of sleep. *Nature* **297**:23–7.

Winfree, A.T. 1984. Discontinuities and singularities in the timing of nuclear division. In: *Cell Cycle Clocks*. L.N. Edmunds Jr, ed. Marcel Dekker, New York and Basel, pp. 63–80.

Winfree, A.T. 1987. *When Time Breaks Down. The Three-Dimensional Dynamics of Electrochemical Waves and Cardiac Arrhythmias*. Princeton Univ. Press, Princeton, NJ.

Winfree, A.T. 1991a. Alternative stable rotors in an excitable medium. *Physica* **49D**:125–40.

Winfree, A.T. 1991b. Varieties of spiral wave behavior: An experimentalist's approach to the theory of excitable media. *Chaos* **1**:303–34 and (*Erratum*) **2**:273 (1992).

Wojtowicz, J. 1973. Oscillatory behavior in electrochemical systems. In: *Modern Aspects of Electrochemistry*, Vol. 8, J.O'M. Bockris & B.E. Conway, eds. Butterworth & Co., London, pp. 47–120.

Woods, N.M., K.S.R. Cuthbertson & P.H. Cobbold. 1986. Repetitive transient rises in cytoplasmic free calcium in hormone-stimulated hepatocytes. *Nature* **319**:600–2.

Woods, N.M., K.S.R. Cuthbertson & P.H. Cobbold. 1987. Agonist-induced oscillations in cytoplasmic free calcium concentration in single rat hepatocytes. *Cell Calcium* **8**:79–100.

Wu, L. & P. Russell. 1993. Nim1 kinase promotes mitosis by inactivating Wee1 tyrosine kinase. *Nature* **363**:738–41.

Wurster, B. 1982. On induction of cell differentiation by cyclic AMP pulses in *Dictyostelium discoideum*. *Biophys. Struct. Mechan.* **9**:137–43.

Wurster, B. 1988. Periodic cell communication in *Dictyostelium discoideum*. In: *From Chemical to Biological Organization*. M. Markus, S.C. Müller & G. Nicolis, eds. Springer, Berlin, pp. 255–60.

Wurster, B. & R. Mohn. 1987. Spike-shaped oscillations in the absence of measurable changes in cyclic AMP concentration in a mutant of *Dictyostelium discoideum*. *J. Cell Science* **87**:723–30.

Xiong, Y., G.J. Hannon, H. Zhang, D. Casso, R. Kobayashi & D. Beach. 1993. p21 is a universal inhibitor of cyclin kinases. *Nature* **366**:701–4.

Yagodin, S., L. Holtzclaw, C. Sheppard & T. Russell. 1994. Nonlinear propagation of agonist-induced cytoplasmic calcium waves in single astrocytes. *J. Neurobiol.* **25**:265–80.

Yamazaki, I., K. Yokota & R. Nakajima. 1965. Oscillatory oxidations of reduced pyridine nucleotide by peroxidase. *Biochem. Biophys. Res. Commun.* **21**:582–6.

Yew, N., M.L. Mellini & G.F. Vande Woude. 1992. Meiotic initiation by the *mos* protein in *Xenopus*. *Nature* **355**:649–52.

Young, M.W. (ed.) 1993. *Molecular Genetics of Circadian Rhythms*. Marcel Dekker, New York and Basel.

Yu, Q., H.V. Colot, C.P. Kyriacou, J.C. Hall & M. Rosbash. 1987a. Behaviour modification by *in vitro* mutagenesis of a variable region within the *period* gene of *Drosophila*. *Nature* **326**:765–9.

Yu, Q., A.C. Jacquier, Y. Citri, M. Hamblen, J.C. Hall & M. Rosbash. 1987b. Molecular mapping of point mutations in the *period* gene that stop or speed up biological clocks in *Drosophila melanogaster*. *Proc. Natl. Acad. Sci. USA* **84**:784–8.

Zacchetti D., E. Clementi, C. Fasolato, P. Lorenzon, M. Zottini, F. Grohovaz, G. Fumagalli, T. Pozzan & J. Meldolesi. 1991. Intracellular $Ca^{2+}$ pools in PC12 cells. *J. Biol. Chem.* **266**:20152–8.

Zaikin, A.N. & A.M. Zhabotinsky. 1970. Concentration wave propagation in two-dimensional liquid-phase self-oscillating system. *Nature* **225**:535–7.

Zehring, W.A., D.A. Wheeler, P. Reddy, R.J. Konopka, C.P. Kyriacou, M. Rosbash & J.C. Hall. 1984. P-element transformation with *period* locus DNA restores rhythmicity to mutant arrhythmic *Drosophila melanogaster*. *Cell* **39**:369–76.

Zeitler, P., G.S. Tannenbaum, D.K.. Clifton & R.A. Steiner. 1991. Ultradian oscillations in somatostatin and growth hormone-releasing hormone mRNAs in the brains of adult male rats. *Proc. Natl. Acad. Sci. USA* **88**:8920–4 and (*Erratum*) **89**:1997 (1992).

Zeng, H., P.E. Hardin & M. Rosbash. 1994. Constitutive overexpression of the *Drosophila period* protein inhibits *period* mRNA cycling. *EMBO J.* **13**:3590–8.

Zerr, D.M., M. Rosbash, J.C. Hall & K.K. Siwicki. 1990. Circadian fluctuations of *period* protein immunoreactivity in the CNS and the visual system of *Drosophila. J. Neurosci.* **10**:2749–62.

Zhabotinsky, A.M. 1964. Periodic process of the oxidation of malonic acid in solution. Study of the kinetics of Belousov's reaction. *Biofizika* **9**:1306.

Zhang, D., L. Györgyi & W.R. Peltier. 1993. Deterministic chaos in the Belousov–Zhabotinsky reaction: Experiments and simulations. *Chaos* **3**:723–45.

Zhang, R. & R. Diasio. 1994. Pharmacologic basis for circadian pharmacodynamics. In *Circadian Cancer Therapy*, W.J.M. Hrushesky, ed. CRC Press, Boca Raton, FL, pp. 61–103.

Zhang, R., Z. Lu, T. Liu, S.-J. Soong & R.B. Diasio. 1993. Relationship between circadian-dependent toxicity of 5-fluorodeoxyuridine and circadian rhythms of pyrimidine enzymes: Possible relevance to fluoropyrimidine chemotherapy. *Cancer Res.* **53**:2816–22.

Zheng, X.-F. & J.V. Ruderman. 1993. Functional analysis of the P box, a domain in cyclin B required for the activation of cdc25. *Cell* **75**:155–64.

Zholos, A.V., S. Komori, H. Ohashi & T.B. Bolton. 1994. $Ca^{2+}$ inhibition of inositol trisphosphate-induced $Ca^{2+}$ release in single smooth muscle cells of guinea-pig small intestine. *J. Physiol.* **481.1**:97–109.

Zilberstein, M., H. Zakut & Z. Naor. 1983. Coincidence of down-regulation and desensitization in pituitary gonadotrophs stimulated by gonadotropin-releasing hormone. *Life Sci.* **32**:663–9.

Zwiebel, L.J., P.E. Hardin, X. Liu, J.C. Hall & M. Rosbash. 1991. A post-transcriptional mechanism contributes to circadian cycling of a per–β-galactosidase fusion protein. *Proc. Natl. Acad. Sci. USA* **88**:3882–6.

# Index

589

Printed in the United States
By Bookmasters